R00219 52474

D1783470

REF
QH
601
.B53
v.4A
cop. 1

FORM 125M

NATURAL SCIENCES
and USEFUL ARTS

e Chicago Public Library

Received _____ JUL 9 1975 _____

BIOMEMBRANES
Volume 4A

BIOMEMBRANES

A series edited by
Lionel A. Manson
The Wistar Institute
Philadelphia, Pennsylvania

1971 · Biomembranes · Volume 1
Articles by M. C. Glick, Paul M. Kraemer, Anthony Martonosi, Milton R. J. Salton, and Leonard Warren

1971 · Biomembranes · Volume 2
Proceedings of the Symposium on Membranes and the Coordination of Cellular Activities
Edited by Lionel A. Manson

1972 · Biomembranes · Volume 3
Passive Permeability of Cell Membranes
Edited by F. Kreuzer and J. F. G. Slegers

1974 · Biomembranes · Volume 4A
Intestinal Absorption
Edited by D. H. Smyth

1974 · Biomembranes · Volume 4B
Intestinal Absorption
Edited by D. H. Smyth

1974 · Biomembranes · Volume 5
Articles by Richard W. Hendler, Stuart A. Kauffman, Dale L. Oxender, Henry C. Pitot, David L. Rosenstreich, Alan S. Rosenthal, Thomas K. Shires and Donald F. Hoelzl Wallach

In preparation   Biomembranes   Volume 6
Bacterial Membranes in the Respiratory Cycle
Edited by N. S. Gel'man, M. A. Lukoyanova, and D. N. Ostrovskii

A Continuation Order Plan is available for this series. A continuation order will bring delivery of each new volume immediately upon publication. Volumes are billed only upon actual shipment. For further information please contact the publisher.

BIOMEMBRANES, Volume 4A

INTESTINAL ABSORPTION

Edited by
D. H. SMYTH
Department of Physiology
The University
Sheffield S10 2TN
England

PLENUM PRESS · LONDON - NEW YORK

Library of Congress Catalog Card Number: 72-77043
ISBN 0-306-39891-5

Copyright ©1974 by Plenum Publishing Company Ltd
Plenum Publishing Company Ltd
4a Lower John Street
London W1R 3PD
Telephone 01-437 1408

U.S. Edition published by
Plenum Publishing Corporation
227 West 17th Street
New York, New York 10011

All Rights Reserved

No part of this book may be reproduced, stored in a retrieval system, or transmitted, in any form or by any means, electronic, mechanical, photocopying, microfilming, recording or otherwise, without written permission from the Publisher

Printed in Great Britain by R. & R. Clark Ltd

Contributors to Volume 4A

D. Chapman	Reckitt & Colman Ltd., Dansom Lane, Hull, Yorkshire, England.
R. K. Crane	Department of Physiology, Rutgers Medical School, Rutgers State University, New Brunswick, New Jersey, USA.
B. Creamer	Department of Medicine, St. Thomas's Hospital Medical School, London SE1, England.
Z. Lojda	Laboratory of Histochemistry, Hlava Institute of Pathology, Charles IV University, 2039 Albertov, Praha 2, Czechoslovakia.
I. G. Morris	Department of Zoology, University College of North Wales, Bangor, Caernarvonshire, Wales.
S. G. Schultz	Department of Physiology, University of Pittsburgh, Pittsburgh, Pennsylvania, 15212, USA.
D. H. Smyth	Department of Physiology, The University, Sheffield, S10 2TN, England.
A. M. Ugolev	Pavlov Institute of Physiology, Academy of Sciences of the USSR, Nab. Makarova 6, Leningrad V-164, USSR.
G. Wiseman	Department of Physiology, The University, Sheffield, S10 2TN, England.
E. M. Wright	Department of Physiology, University of California Medical School, Los Angeles, California, USA.

Preface

It might be asked if there is a need for yet another large review on Intestinal Absorption, and the answer is that this is still a rapidly expanding field of interest both from the medical and scientific points of view. There is ample evidence for this in the number of papers which continue to be published, and the bulletin on Intestinal Absorption issued by the Biomedical Information Project of the University of Sheffield lists about 150 titles per month, and there is still no sign of any diminution in this rate. There are in fact so many papers that those interested in intestinal absorption have to be specialists in one particular field, but must at the same time be aware of the general developments in the subject as a whole. The last major review was the excellent volume in the American Handbook published in 1968, already six years ago, and indeed a number of the contributors to that volume have taken part in the present work.

Some observations made in the introduction to a volume of the British Medical Bulletin on Intestinal Absorption some years ago are still pertinent. Progress in the experimental sciences is not continuous, but proceeds in phases of rapid expansion alternating with periods of slower growth. This is partly because of a fundamental law governing the progress in experimental science which states that if you think of anything easy to do which has not been done before, further investigation shows either that it is not easy, or that it has been done before. One way of escaping from the grip of this law is to avoid finding out or to ignore what has been done before. This book is not intended for those seeking this solution. But this rigorous law is periodically relaxed, and this happens when a new technique is discovered. There is then a sudden surge of publications to exploit the new technique. It is easy to date the present tide of advance in intestinal absorption to the introduction of an effective *in vitro* technique by R. B. Fisher and D. S. Parsons in 1949, followed by the development of the everted sac by T. H. Wilson and G. Wiseman in 1954. While these workers popularized *in vitro* techniques, they did not in fact introduce them, and this was done more than fifty years ago by Weymouth Reid, whose remarkable work seems to have escaped serious notice by the physiologists of the day.

But perhaps the real credit for *in vitro* intestinal studies and indeed *in vitro* studies in everything should go to Sidney Ringer, who first introduced the idea of replacing the life-giving blood with a salt solution and hence led the way for the highly unphysiological *in vitro* experiments. It was indeed the introduction of salt solutions for keeping isolated tissues alive that made modern physiology and biochemistry possible, and it is well that Ringer should be remembered chiefly by Ringer solution rather than by the experiments he did with it, important though they were. Ringer's most famous lineal scientific descendant is Hans Krebs, whose name, although associated with at least two major discoveries in biochemistry, is still probably most widely used in referring to Krebs' solution, and indeed it is Krebs' bicarbonate saline which has mainly been used for the *in vitro* intestine. In the early days of the *in vitro* intestine a great many things were said about the unphysiological nature of the preparation and particularly when it was exposed to the insult of being turned inside out in the everted sac technique. But unphysiological approaches are paradoxically the way to advances in physiological knowledge, and most major advances in our knowledge of how living tissues work have come from using living tissues in conditions very different from their normal ones. Ringer was the great apostle of unphysiological experiments, and his disciples do not need to make apologies for continuing his tradition.

Early studies of the intestine emphasize the important role of the cells lining the gut, and Hiedenheim spoke of the 'Triebkraft' or driving force of these cells. Hiedenheim was involved in the old controversy on vitalism, and his unfashionable vitalistic term perhaps prevented full recognition of the importance of his ideas on the intestinal cell. A later generation was explaining the movement of fluid in terms of classical osmosis, and did not require the Triebkraft of the epithelial cell. Modern work has fully substantiated Heidenheim's idea and we now know that movement of water depends on forces generated by the activity of the living cell. The undesirable connotation of vital forces of Latin derivation (vita = life) has been neatly avoided by substituting biophysical forces of Greek derivation (βίος = life) to everyone's complete satisfaction.

The study of intestinal absorption offers opportunities to people of very widely different skills, varying from those who try to formulate the problems in terms of irreversible thermodynamics to those who think in terms of the clinical problems of the person unable to absorb enough of the nutrient substances he requires. Between these are the large number who think of one aspect of the absorptive process, and try to formulate the problems in such terms as is possible by their limited knowledge of fundamental science and their awareness of the dangers in making too many approximations and assumptions to make biological observations fit mathematical expressions. These volumes contain therefore many different approaches to the problems of the intestine. It purposely does not include detailed discussion of clinical problems, as these have been the subject of many symposia and many discussions in recent years. If it encourages its readers to broaden their interests and make an effort to come to grips with new and unfamiliar expertise, it will have served its purpose.

D. H. Smyth

Contents

Contributors to Volume 4A		v
Preface		vii
Contents of Volume 4B		xii
Chapter 1	Intestinal Structure in Relation to Absorption B. Creamer	1
Chapter 2	Cytochemistry of Enterocytes and of Other Cells in the Mucous Membrane of the Small Intestine Z. Lojda	43
Chapter 3	Biological Membranes D. Chapman	123
Chapter 4	The Passive Permeability of the Small Intestine E. M. Wright	159
Chapter 5	Irreversible Thermodynamics S. G. Schultz	199
Chapter 6	Methods of Studying Intestinal Absorption D. H. Smyth	241
Chapter 7	Membrane (Contact) Digestion A. M. Ugolev	285
Chapter 8	Absorption of Protein Digestion Products G. Wiseman	363
Chapter 9	Immunological Proteins I. G. Morris	483
Chapter 10	Intestinal Absorption of Glucose R. K. Crane	541
Subject Index to Volume 4A		i

Contents of Volume 4B

Chapter 11	Fat Digestion and Absorption B. Borgström	555
Chapter 12	The Intracellular Phase of Fat Absorption D. N. Brindley	621
Chapter 13	Transport of Short Chain Fatty Acids M. J. Jackson	673
Chapter 14	Salts and Water C. J. Edmonds	711
Chapter 15	Iron Absorption S. T. Callender	761
Chapter 16	Calcium H. E. Harrison and H. C. Harrison	793
Chapter 17	Absorption of Water-Soluble Vitamins D. M. Matthews	847
Chapter 18	Electrical Activity of the Intestine R. J. C. Barry and J. Eggenton	917
Chapter 19	Hereditary Disorders of Intestinal Transport H. D. Milne	961

Subject Index to Volume 4B   *i*

CHAPTER 1

Intestinal Structure in Relation to Absorption

BRIAN CREAMER

Physician, St. Thomas' Hospital,
Senior Lecturer in Medicine, St. Thomas's Hospital Medical School

		Page
1.1	GENERAL STRUCTURAL FEATURES	2
	1.1.1 *Introduction*	2
	1.1.2 *Surface area*	2
	1.1.3 *The mucus layer*	3
	1.1.4 *Villous shape*	3
	1.1.5 *Cell turnover*	6
	1.1.6 *Cell turnover and villous shape*	8
	1.1.7 *Compensatory changes*	9
1.2	HISTOLOGICAL APPEARANCE	11
	1.2.1 *Epithelial layer*	11
	1.2.2 *Vascular structure of the mucosa*	13
1.3	ULTRASTRUCTURE	14
	1.3.1 *Scanning electron microscopy*	14
	1.3.2 *The cell membrane*	14
	1.3.3 *Junctional complex*	15
	1.3.4 *Brush border*	17
	1.3.5 *The apical region*	23
	1.3.6 *Endoplasmic reticulum*	23
	1.3.7 *Golgi apparatus*	25
	1.3.8 *Mitochondria*	25
	1.3.9 *Lysosomes*	27
	1.3.10 *F-Bodies*	27
	1.3.11 *Nucleus*	27
1.4	OTHER CELLS	29
	1.4.1 *Goblet cells*	29
	1.4.2 *Crypt cells*	29
	1.4.3 *Paneth cells*	31
	1.4.4 *Argentaffin cells*	35
1.5	OTHER COMPONENTS OF THE MUCOSA	35
	1.5.1 *The basement membrane*	35
	1.5.2 *Capillaries*	35
	1.5.3 *Lymphatics*	36
	REFERENCES	36

1.1 GENERAL STRUCTURAL FEATURES

1.1.1 Introduction

This section attempts to assemble those elements of the structure of the small intestine that are most pertinent to absorption. At the moment of absorption almost every substance is in a form smaller than the resolution of the electron-microscope, most cell 'pores' are also invisible and their sizes hypothetical and most substances are absorbed without any change being seen at light or ultrastructural level. Perhaps the most meaningful factor that structure could contribute to an understanding of absorption is a measurement of surface area, but as Wilson [1] rightly says, 'Despite wide interest in the small intestine brought about by modern biopsy techniques and absorption studies, quantitative knowledge of the basic structure of the small intestine remains slight' [1]. The relation of epithelial cell to capillary and lymphatic are probably crucial though little is known about blood flow and its control in this area. The ultrastructure of the epithelial cell is documented in great detail though this does not necessarily help in understanding absorptive processes. Of more value is the localization of enzymes, and this well shown by histochemical techniques (see chapter 2). Recent work using histochemical and immune labelling methods with electron microscopy has helped to place certain enzymes with greater accuracy, so that the gap between the membrane structure of the brush border as conceived by the physiologist and visualized by the microscopist is narrowing.

1.1.2 Surface area

This highly important factor has received scant attention, particularly in human disease. Surface area is increased by the mucosal folds of Kerkring (valvulae conniventes), by villi and by the microvilli of the brush border. Mucosal folds are prominent in man but absent in many smaller laboratory animals. Villi are strikingly longer in the jejunum than the ileum. It has been estimated that all these structures increase the surface area of the small intestine by a factor of 600 [2]. By careful measurements from histological sections of the length of the surface outline and the mucosal-serosal ratios enough data has been obtained to enable a calculation of surface area to be made. The results in the dog, cat and rat show a diminution of surface area from the jejunum to the ileum; the fall in surface area is linear with distance along the intestine [3, 4, 5]. Wilson has made a detailed study of surface area in man from post-mortem specimens [1]. There is a fourfold difference between jejunum and ileum but this is not linear; the large surface area of the upper jejunum falls away sharply, while in the ileum the decline is much less marked.

The microvilli of the brush border make a big increment in surface area; estimates have been made of 14 to 39 times [6, 7]. Whether all the surface area so calculated is available for absorption is unknown. Jejunal villi are closely packed and the lower parts may not be exposed to luminal contents. Furthermore the glycocalyx of the brush border creates a microenvironment and similarly it cannot be assumed that all the surface area of this part of the cell is available. More work is needed to elucidate this problem.

1.1.3 The mucus layer
Mucoprotein secreted by goblet cells coats the epithelium and envelops the luminal contents. The function of this is poorly understood. Freeze dried secretions of the whole gut with contents intact shows the relation of mucus to epithelium and lumen. In the duodenum and jejunum mucus is scanty and wispy and the contents are fluid and well dispersed. By contrast, in the ileum there is a thick layer of mucus between the villi and the luminal content, which is usually more solid. This layer is situated above the villi so that little content is ever seen between the villi. Bacteria are frequently seen in the mucus layer and it has been suggested that this is the true habitat of the flora rather than the lumen [8].

1.1.4 Villous shape
The text-book picture of a finger-shaped villus is not invariably found in animals or man. The new-born of almost all mammals do indeed have these villi, but at the time of weaning most small animals suffer a change in villous shape to triangular or leaf-like villi [9] (Fig. 1.1). In the rat these may become joined so that ridges with triangular projections run across the long axis of the bowel. This change is more marked in the jejunum than the ileum. It is due to environmental factors, as can be demonstrated by a reversion to finger villi in closed segments of

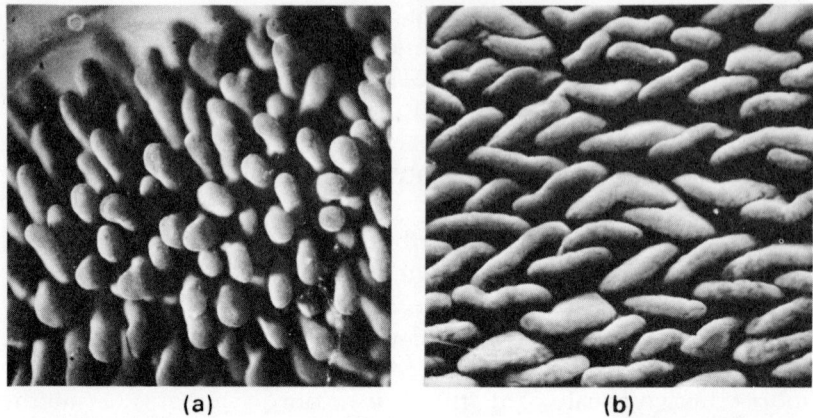

Fig. 1.1. Dissecting microscope photographs of rat jejunal mucosa. A, at ten days old and B, in the adult. × 41.

intestine that have been surgically separated from the continuity of the bowel [10].

In adult man, in countries with a Western culture, the small intestinal structure is little altered from the foetal structure (Fig. 1.2). Leaf-like villi are frequently encountered in the duodenum and upper jejunum mixed with finger forms. Occasionally convolutions are found in the normal population and there may be some small local variation even within the British Isles [11]. In other parts of the world the 'normal' mucosal structure may be a convoluted or a mixture of leaf and convoluted mucosal shapes. This has been reported in Thailand [12], East [13] and West Pakistan [14], India [15], Uganda [16] and parts of the Carribean [18]. Thus the majority of the human population probably has a small intestinal structure with a comparatively diminished surface area. Again there is good reason to believe that this is environmental in origin. Diminished absorption of xylose and other evidence of marginal malabsorption has been demonstrated in as much as a third of the apparently healthy population in some areas [13, 14].

In some animals that might be used for absorption studies striking changes of villous shape have been reported due to infection and infestation. Pout documents extreme and extensive mucosal change in lambs producing at times a flat mucosa in association with coccidia and helminths [18].

INTESTINAL STRUCTURE IN RELATION TO ABSORPTION 5

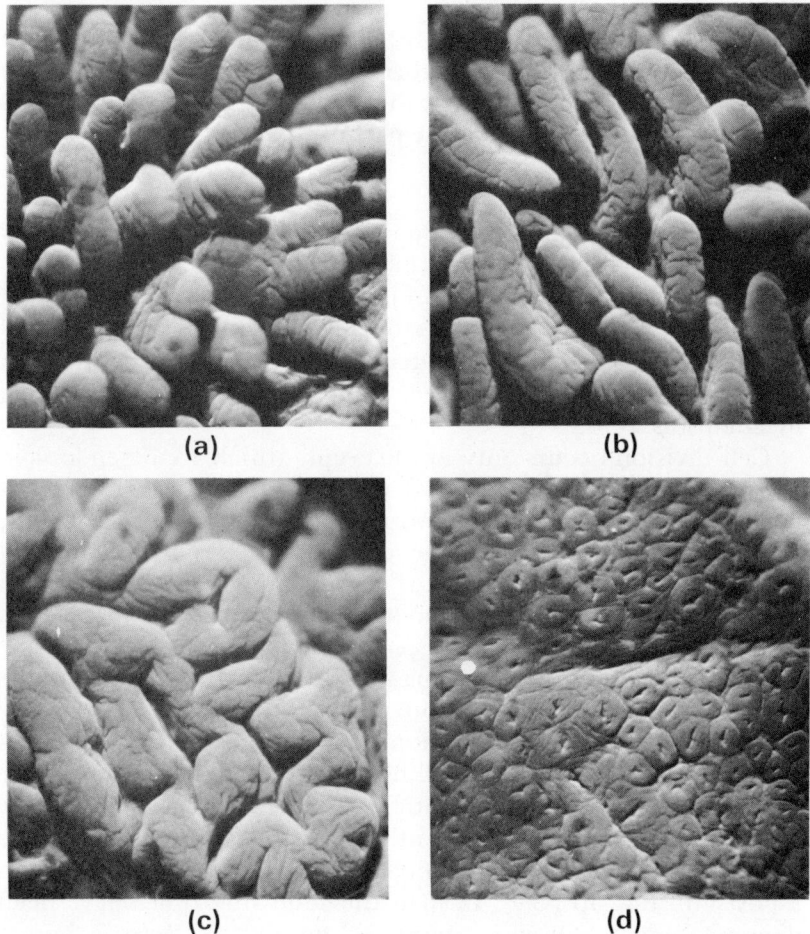

Fig. 1.2. Dissecting microscope photographs of human jejunal mucosa. A, finger-shaped villi; B, leaf-shaped villi; C, convolutions and D, a flat mucosa. A and B are within normal limits for Western man. × 33.

In human disease states mucosal changes can be found. The most important is the coeliac syndrome (adult coeliac disease, idiopathic steatorrhoea) where the upper small intestinal mucosa becomes diffusely flattened (Fig. 1.2). The lesion is usually marked and definitive, total villous atrophy, though occasionally milder changes, partial villous atrophy, are encountered. The lesion gradually regresses in the distal bowel so that in the terminal ileum normal appearances are found. The extent of the lesion usually correlates with the severity of the malabsorption so that it is the integrity of the ileum that determines much of the disease pattern.

1.1.5 Cell turnover

The small intestinal epithelium, in common with the rest of the gut, is highly dynamic being in a continuous state of division, migration and loss [16, 20, 21]. It is, however, much more active in this respect than the stomach or colon. The patterns of cell turnover in the small intestine are well documented but there is almost complete ignorance about controlling mechanisms.

Cell division occurs only in the crypts which are often looked upon as areas totally devoted to replication, though there is considerable evidence of secretion in this zone. Mitotic activity is readily visible and may be counted to arrive at a mitotic index. If the total population of cells is known and the duration of mitosis is measured or assumed then the turnover time can be calculated. Blocking of mitosis at the metaphase with colchicine is frequently used for increasing the efficiency of mitotic counts and estimating mitotic duration.

By the use of a combination of mitotic counting and autoradiography after pulse labelling with tritiated thymidine the individual events in the mitotic cycle can be timed [22]. The sequence is composed of a synthetic (S) phase in which DNA is doubled, a post-synthetic gap (G2), the act of mitosis and then a post-mitotic gap (G1). In man the following times have been documented: S phase is about 14 hours, G2 2 to 7 hours, mitosis 1 hour, while G1 is very variable as some cells will cease dividing while others may enter a prolonged interphase (G0) [23]. The production of new cells can be expressed as a birthrate; in the human ileum 0.8 cells/100 crypt cell/hour and in the jejunum 1 cell/100 crypt cells/hour [23, 24, 25].

The cells undergo several divisions in the crypts as they migrate up the gland and at the crypt-villous junction undergo a rapid maturation to become an adult epithelial cell. This is marked by a more columnar appearance of the cell and a prominent brush border. Goblet cells appear to arise from the same stem cell as epithelial cells but mature earlier and are seen fully distended with mucus in the crypts. They migrate in the same way as epithelial cells.

Complete maturation of cells is not accomplished until they are between a third and half-way up the villous as judged by histochemical findings. The migration of cells is readily visualized by autoradiography with tritiated thymidine which enables accurate migration times to be measured. The cells appear to flow continuously up the villi and in most small animals the turnover time is of the order of 2 days. In man a turnover time of 4 to 5 days has been recorded but there have been few measurements [23, 24, 25]. The rate of movement of cells up the villus can be expressed as a flow rate. In man this is of the order of 1 cell position per hour in the jejunum and 0.8 cell positions per hour in the ileum [23, 25]. There is now good evidence that not only do the epithelial cells divide and migrate but the fibroblasts which cover the basement membrane of the crypts divide at the same rate [26, 27]. These labelled cells migrate synchronously with the epithelial cells. It is not known whether the basement membrane also migrates but the simultaneous movement of epithelial cell and fibrocyte would give support to this idea. If so, then the whole absorbing unit, epithelial cells, basement membrane and fibrocyte, would maintain a constant relationship.

At the villous tips the epithelial cells are shed from distinct sites called extrusion zones. These can usually be seen on light microscopy as either a cleft or a small streamer of cells entering the lumen. The cells appear somewhat shrivelled or pyknotic in this area but are otherwise shed intact into the lumen.

Another approach to measuring cell turnover is to quantitate the cell loss into the lumen. If all the cells can be recovered and if the nuclei are intact then a chemical measurement of DNA should act as a cell count [28]. This approach has been perfected by Croft and gives a fairly reproducible method that can be used in animal or man [29]. The main difficulty is to be sure that all the cells have been washed from the segment of the lumen being studied, as it is easy to imagine that cells may be caught in mucus. Using this technique in man, the cell loss from a 5 cm segment of jejunum is about half a million cells per minute and therefore for the whole small intestine may be calculated to be between 20 and 50 million per minute [30]. A different approach to measuring cell loss has been developed by

Clarke [31]. In the rat he has been able to trap the cells shed into the mucus overlying villi during a timed period. With suitable staining the cells can be directly counted over each villus and this gives a highly sensitive method of calculating turnover.

The mass of cells shed is impressive. In man, Croft's data gives a figure of 250 gms per day which is in close agreement with Leblond's estimate of half a pound [21]. When faecal excretion of a substance is being used in balance experiments to measure absorption this endogenous component may be a significant factor.

Cell turnover is a highly sensitive process and may be altered by a large number of physical, nutritional and chemical changes. Irradiation, antimitotic and antimetabolic drugs will inhibit mitosis so that division will be stopped or slowed [32, 33]. Deficiencies of vitamin B_{12} [34] and folic acid also reduce mitosis, but iron deficiency apparently has no effect [35]. A dimunition of mitosis is not accompanied by an immediate dimunition of cell loss and villi shrink, sometimes to a striking degree. Physical or chemical trauma to villi increases cell loss and cell turnover undergoes a compensatory increase [36]. However, division may be unable to match loss and the adult cell population will fall, again producing smaller villi though these are now of a different shape as will be outlined below. The flora of the small intestine influences cell turnover; in the germ-free animal cell division is halved but gell life is doubled and the turnover time is therefore twice as long [37, 38]. Abrupt alterations of flora, as at weaning, may also change cell turnover [39]. Nutritional factors are primarily seen in pregnancy and lactation where hyperphagia is the determining factor; villi become larger and cell turnover is increased [40]. By contrast, in protein malnutrition and starvation cell division decreases [41, 42]. Even brief periods of starvation may slow cell turnover and there is a demonstrable increase after normal feeding [43].

1.1.6 *Cell turnover and villous shape*
In states of diminished cell turnover villi become smaller but retain their shape; by contrast, with increased cell turnover villi metamorphose into the range of shapes already described. These changes are usually brought about by physical or chemical changes in the lumen or by infestation which cause an increased

cell loss with a compensatory increase in cell division. Cell turnover and villous shape are closely linked so that turnover can be predicted from the three-dimensional appearance [40].

An understanding of this process comes from the relationship of crypts to villi; an area which cannot normally be seen under the dissecting microscope and cannot be visualized from a single section. The relationship can be recognized by examining and reconstructing serial sections [44] but much more easily and directly by using the autolysed mucosa under the dissecting microscope. Autolysis simply removes the epithelial layer and leaves the basement membrane intact revealing the basic architecture of villi and uncovering the crypt orifices. In man there are three times as many crypts as villi and some crypts are not in direct contact with a villous (Fig. 1.3). In between the crypts run fine ridges, intervillous ridges, which carry the migrating cells to the villous base. These ridges respond swiftly to an increase in cell turnover by hypertrophying and some become thick structures. On the hypertrophied ridges the villi become broader and leaf-like, with a further increase in cell turnover the bigger ridges themselves are all that is left, convolutions. A flat mucosa is merely a network of these low ridges. In experimental models the turnover time of the various villous shapes can be measured. Finger-shaped villi, as in man, have a turnover time of four days; leaves, as in the rat, take two days, while a convoluted mucosa takes twelve hours and a flat mucosa about six hours [45]. As cells normally take at least twelve hours to mature on their passage up the villi these changes may have considerable implications in absorption studies.

1.1.7 *Compensatory changes*

If part of the small intestine is resected, compensatory change may occur in the remaining segment. Gross enlargement and microscopic hypertrophy have been repeatedly documented [46]. This takes place only after a considerable resection and the alteration is most strikingly seen in the ileum after jejunal resection. Almost all measurements are increased; the diameter and weight of the gut and the villous height which in the residual ileum nearly approaches that found in the jejunum. The mechanism for this hypertrophy appears to be mainly nutritional as transposition of jejunum and ileum brings about the same change in the ileum [46]. This emphasizes the

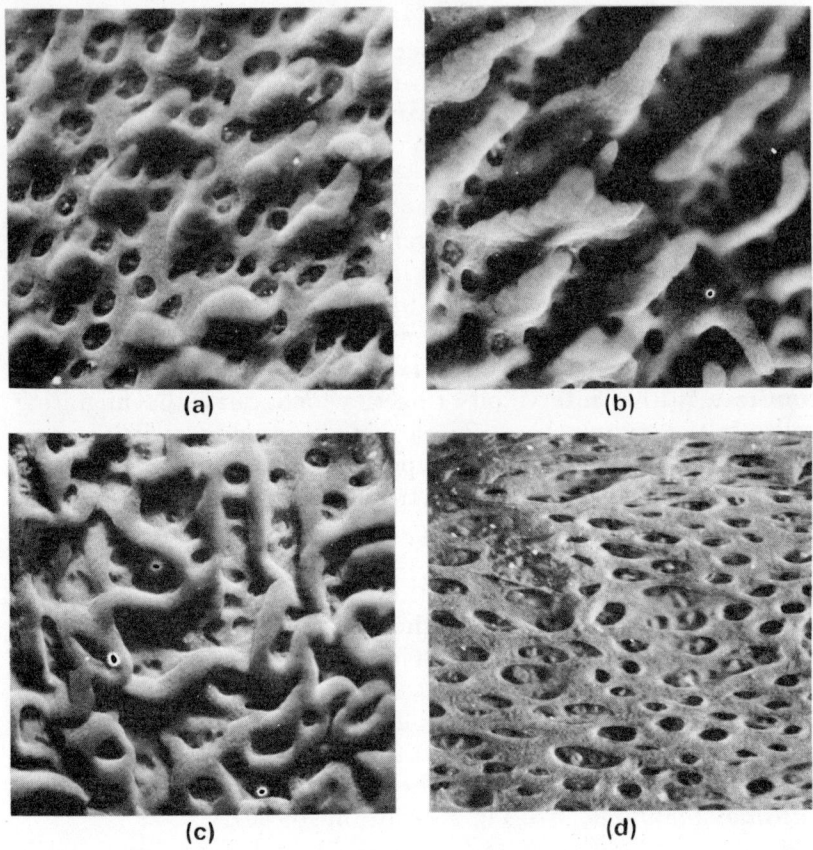

Fig. 1.3. Dissecting microscope photographs of human small intestinal mucosa after autolysis to show the basic structure. A, finger villi; B, triangular, leaf-shaped villi; C, convolutions and D, a flat mucosa. The crypt orifices and intervillous ridges can be seen. × 26.

concept that the small intestinal mucosa receives much of its nutrition from the lumen [47].

This structural hypertrophy has been shown to be accompanied by an increased capacity for absorption, particularly of glucose [46]. Similar findings have been reported in patients following massive resection [48].

1.2 HISTOLOGICAL APPEARANCE

1.2.1 Epithelial layer

Under conventional light microscopy the epithelial layer is visible as a continuous sheet covering the villi and dipping down into the crypts (Fig. 1.4). The immature crypt cells are basophilic due to the presence of RNA and have large nuclei. At the crypt villous junction a change to the mature epithelial cell is seen within a few cell positions. Paneth cells are present at the base of the crypts and are clearly visible with large eosinophilic granules above the nucleus which are frequently seen to be extruding into the crypt lumen. Argentaffin cells (enterochromaffin cells) are also present in the crypts lying between the immature cells with their secretory granules below the nucleus and against the basement membrane. Mature goblet cells are found in the crypts.

The villi are covered with tall columnar epithelial cells with a basally placed nucleus. The brush border is a prominent and regular structure and the cells are always closely joined at this level. Goblet cells are interspersed in this layer but diminish in number towards the villous tip. Here the extrusion zone is usually visible and is the only site where a gap may be regularly found between cells.

The whole epithelium rests on the basement membrane and this in turn is covered by fibroblasts and fibrocytes so that the whole absorbing unit is composed of three layers. The crypts are completely invested by fibroblasts, but within the villi the fibrocytes are spread out and their cytoplasm extended into foot processes so that only a small part of the basement membrane is covered. PAS stains show the mucopolysaccharide content of brush border, goblet cells and, to a lesser extent, basement membrane (Fig. 1.5).

Beneath the basement membrane the capillary network is clearly visible; this will be described in more detail later. The lacteal is seen in well-cut villi; it lies in the centre of the villus and does not come into close apposition to the epithelial basement membrane as do the capillaries.

The lamina propia contains many wandering cells, mostly immunocytes. These are lymphocytes and plasma cells and can be shown by immunofluorescent techniques to be dominantly Immunoglobulin A (IgA) secreting. The number of these cells is

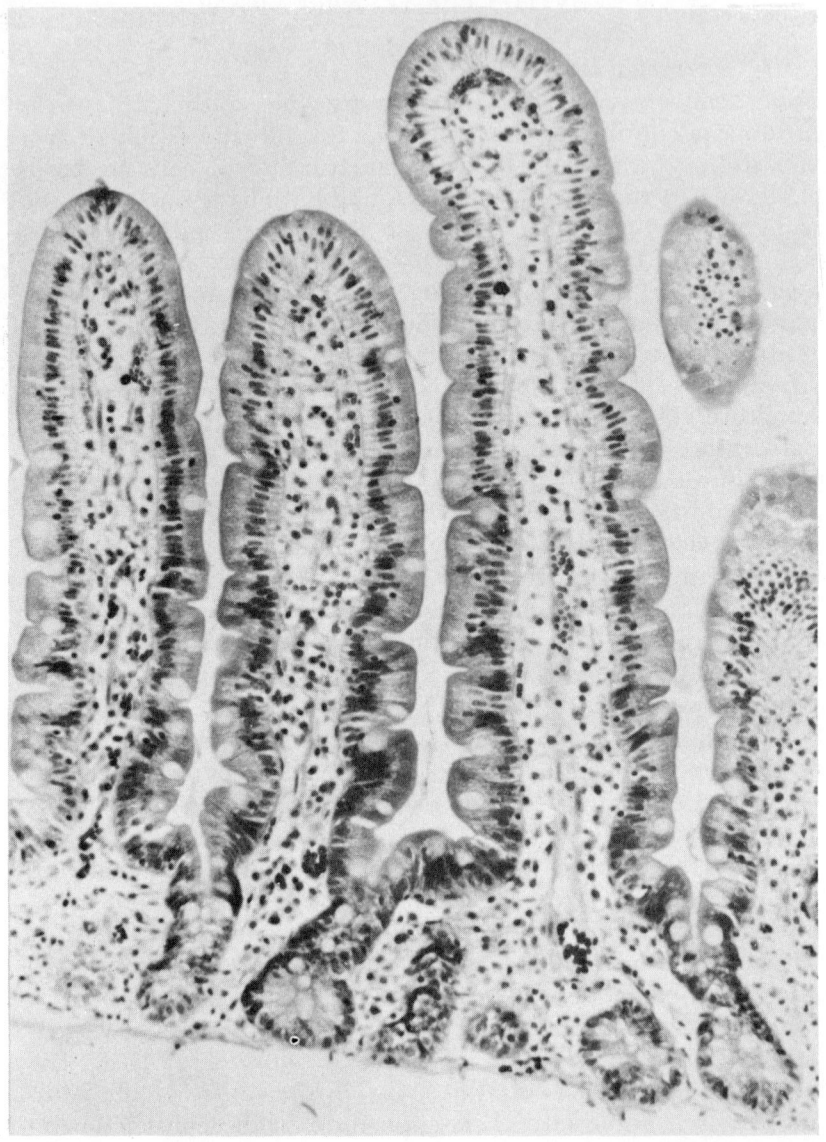

Fig. 1.4. A section of human jejunal mucosa stained with haematoxylin and eosin showing normal villous structure. X 196.

INTESTINAL STRUCTURE IN RELATION TO ABSORPTION 13

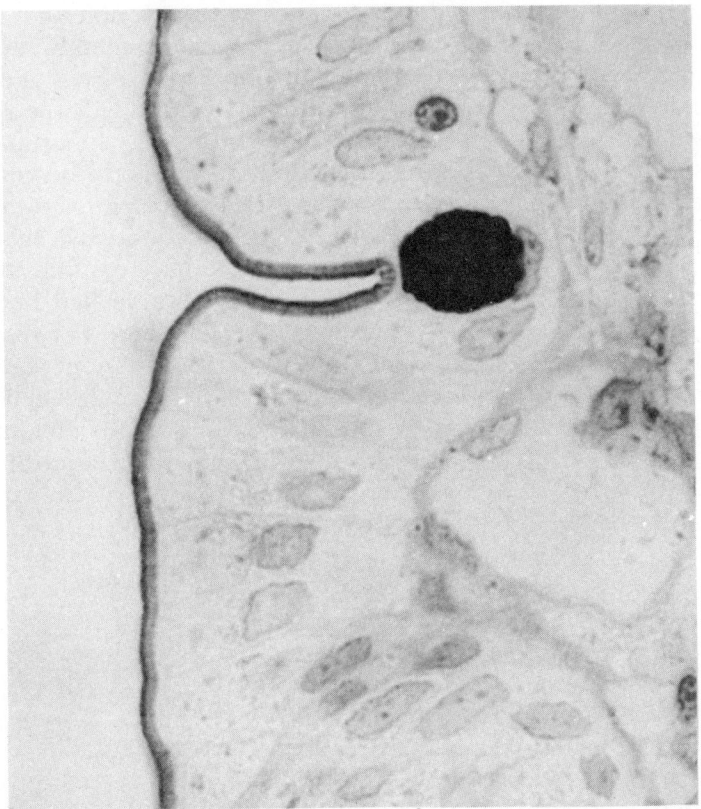

Fig. 1.5. A light micrograph of epithelial cells stained with P.A.S. The tissue has been fixed in gluteraldehyde, dehydrated in acetone, post-fixed in OsO_4 in carbon tetrachloride and embedded in Epon. The brush border is visible as a dark band with an increased intensity at the luminal surface. One goblet cell is visible. × 1,900.

mostly determined by the flora of the intestine and in disease by the permeability of the epithelial layer to antigens. Brunner's glands are only found in the duodenum. They lie directly beneath the muscularis mucosa and are characterized by acini of cuboidal mucus secreting cells that connect by ducts to the crypts of Lieberkühn. They secrete an alkaline watery mucoid fluid.

1.2.2 Vascular structure of the mucosa

All villi have a rich blood supply. This has been studied in detail by Miller and his coworkers [49] using silicone rubber injection techniques. Finger-shaped villi have a central arteriole that runs right to the villous tip beneath the basement membrane. A network of capillaries connects the marginal vessels and lie beneath the basement membrane. In leaf-shaped villi the same arrangement is present with the marginal vessels lying at the edges of the leaf. At the base the capillaries drain into venules. As leaf villi change into ridges the basic vascular anatomy is preserved with at first a series of leaves connected by the capillary network at the base. A true ridge may contain many arterioles running straight to the top as if there had been a number of villi in a row, but now the capillary network is totally interconnected. Not all the epithelial cells are in contact with a capillary; there are groups of five or more cells in between the capillary connections. The difficulties of measuring mucosal blood flow have been reviewed by Hamilton *et al.* [50].

1.3 ULTRASTRUCTURE

1.3.1 *Scanning electron microscopy*

A new development which produces the most startling vivid pictures is scanning electron microscopy [52]. The principle is easily stated: a tissue is coated with an electron dense layer and is then scanned by an electron beam and the reflection recorded; quite different from the conventional transmission models. It produces pictures of fine resolution (about 25 Å) and a good depth of focus, with magnification up to × 100,000. Mucosa can be surveyed from the surface as under the dissecting microscope or the cut surface can be examined.

The surface structure is revealed in great detail and in particular the depth of focus allows the crypt orifices to be seen. The brush border is visible as a closely packed array of the tips of microvilli. Goblet cells can be seen as empty craters or discharging masses. Apart from the clarity of illustration no new structures have been revealed. On the cut surface the granules of Paneth cells and the red cells within capillaries can be seen in three-dimensional appearance.

1.3.2 *The cell membrane*

The overall appearance of the intestinal epithelial cell is

characterized by its tall columnar shape and prominent brush border of microvilli. The lateral surfaces are closely opposed to their neighbours over the luminal half of the cell with frequent interlocking folds, while the basal half is often separated. The nucleus is basally placed and the cell stands on a well defined basement membrane. Most of the organelles lie between the brush border and the nucleus.

The basic structure of the plasma membrane is common to most cells, but shows striking modification in the brush border. The typical trilaminar plasma membrane is seen in every site; it is recognized by two electron dense layers with a clear layer in the middle. At the lateral sides of the cell the plasma membrane width varies from 70 to 90 Å (mean 80 Å), made up of the two dense strata each about 25 Å wide and the central light stratum about 30 Å wide. The membrane is thought to be a single bimolecular leaflet of lipid covered by a fully spread monolayer of protein on both sides.

1.3.3 Junctional complex
As mentioned above it is only in the upper, luminal, part of the cell that neighbouring cells are opposed and connected. For most of the opposing area there is a true intercellular space about 200 Å wide. In the region immediately below the brush border are several structures connecting the adjacent cells together that are known as the junctional complex [53]. They consist of a tight junction, and intermediate zone and a desmosome (Fig. 1.6). The tight junction (zona occludens) lies immediately beneath the brush border and is the only site where the plasma membranes actually show contact. Over a distance of 0.2 to 0.5 μ the outer leaflets fuse, though this may not be completely homogeneous. Below this is the intermediate junction (zona adherens), a distance of 0.3 to 0.5 μ with an intercellular space of about 200 Å. About this area the cytoplasm appears denser and probably represents the terminal bar as seen with the light microscope. This area ends in a desmosome (macula adherens), though frequently more desmosomes are seen on the upper lateral cell border. A desmosome is about 0.2 to 0.3 μ in length and stands out by the increased density of the structure. The plasma membranes are slightly further apart than usual, about 240 Å, and the interval is filled with relatively dense material. The cytoplasm about a

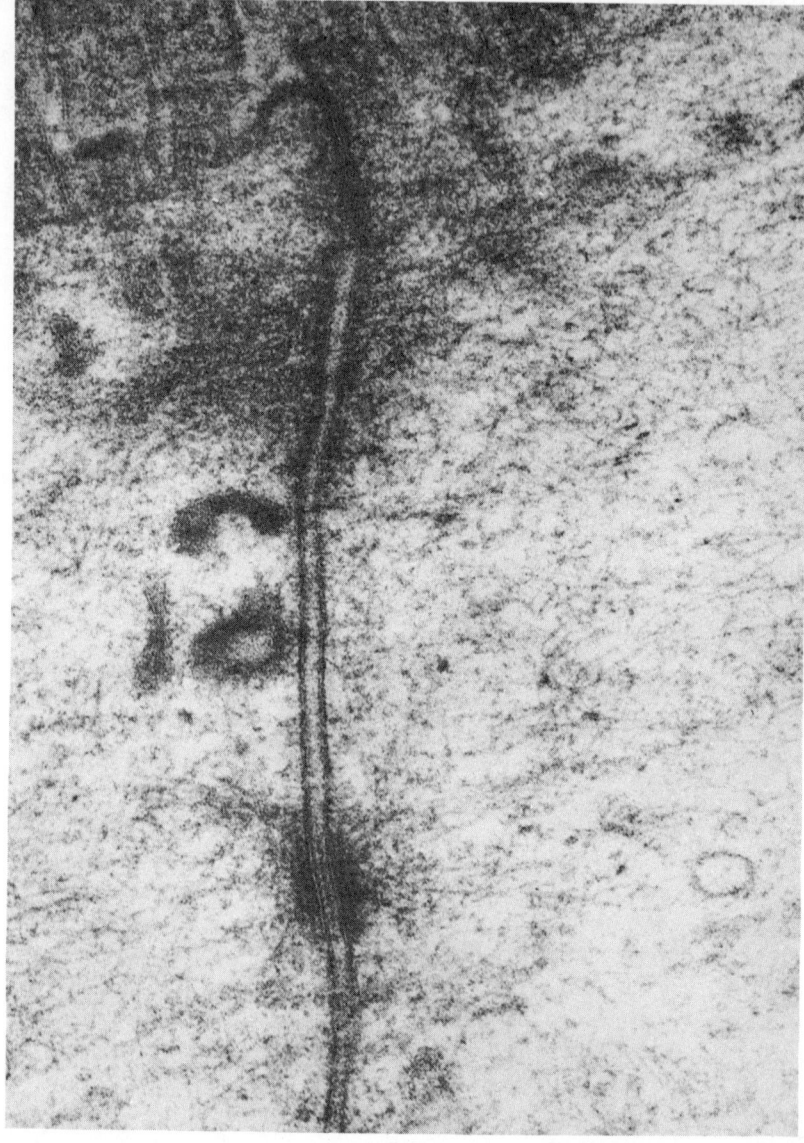

Fig. 1.6. Electron micrograph of junctional complex between two epithelial cells. At the top of the picture are the bases of microvilli and in the middle a tight junction is visible. Below this is the intermediate junction with the dense area about it known as the terminal bar. At the bottom of the picture is a desmosome. × 74,000.

INTESTINAL STRUCTURE IN RELATION TO ABSORPTION 17

desmosome appears condensed and fibrils are visible radiating from it.

These junctional complexes undoubtedly function as connecting areas between cells and may have a function in acting as a diffusion barrier. The tight junction can be seen to act as a barrier to large molecules but its permeability to small molecules is controversial.

1.3.4 Brush border

The structure and function of this area has excited great interest and over the last few years, and interplay between electron-microscopists and physiologists has produced a visual interpretation of the membrane function. It basically consists of an array of closely packed microvilli, about 1000 to each absorptive cell, with a specialized adaptation of the outer leaflet of the plasma membrane by a mucopolysaccharide-protein surface coat [55, 54]. There is nothing unique in this as microvilli and a mucopolysaccharide coat are widely found in nature in a variety of organs and species.

In health the microvilli are regular in shape and size with a length of from 0.75 to 1.5 μ (mean 1.0 μ) and a breadth of 0.1 to 0.2 μ (Fig. 1.7). The intermicrovillous space is comparatively narrow, about 0.01 to 0.05 μ and on cross-section they are packed in a hexagonal pattern. Each microvillous has a rounded luminal end and at the base may be joined to its fellow or show an apical pit. Immature cells in the crypts show a relatively sparse and diminutive brush border.

The interior of the microvilli contains a bundle of straight fibrils that run vertically to the top and downwards to the terminal web [55]. There are from ten to fifty to each microvillus arranged in the central area leaving a clear peripheral zone. On careful examination each fibril is tubular and the dimensions have been measured by microdensitometry. They are usually 60 Å but may be as much as 110 to 150 Å in width.

The plasma membrane runs continuously over the microvilli and the outer leaflet is covered by the surface coat, glycocalyx or fuzzy layer [56]; the whole structure being called 'the greater membrane' (Fig. 1.8). Over the microvilli the plasma membrane is somewhat thicker by about 20 Å than elsewhere (range 85 to 110 Å compared with 73 to 95 Å). There has been some disagreement about the symmetry of the trilaminar

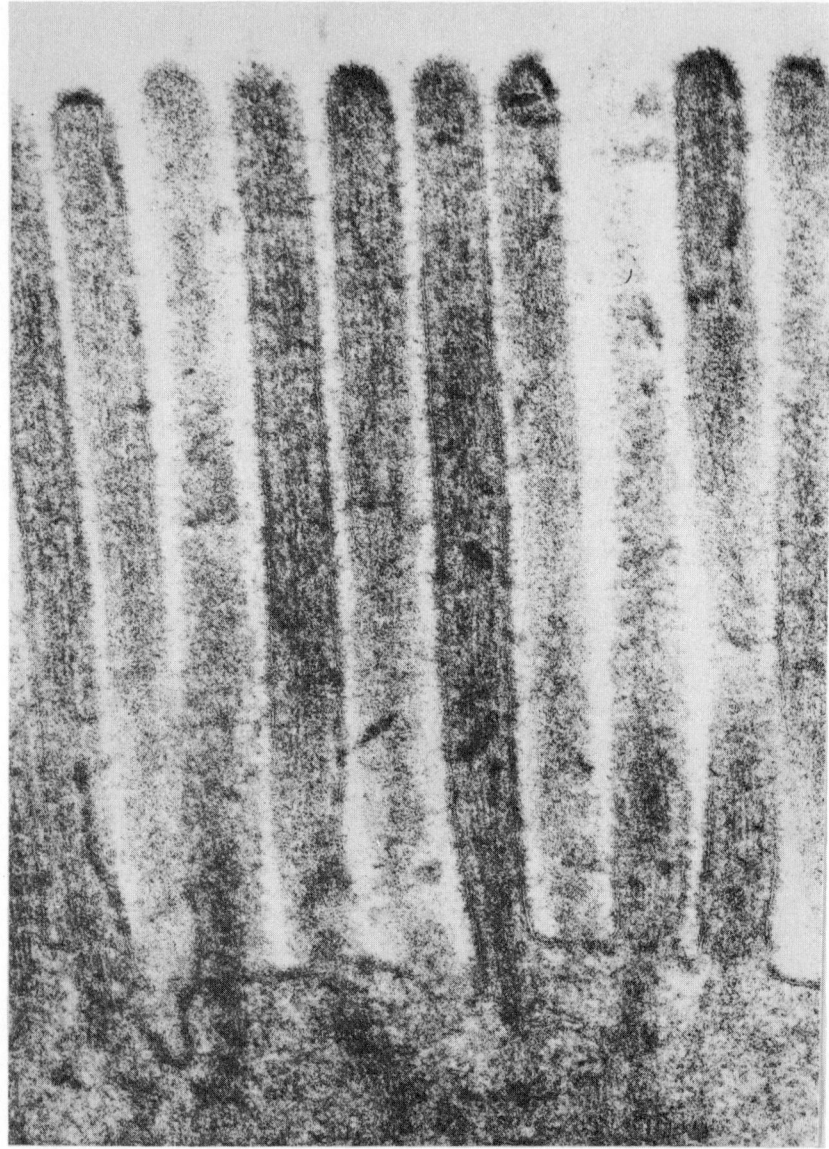

Fig. 1.7. Electron micrograph of the intestinal brush border. The microvilli show the trilaminar membrane, the fibrillary core and rootlets extending into the cell. × 65,000.

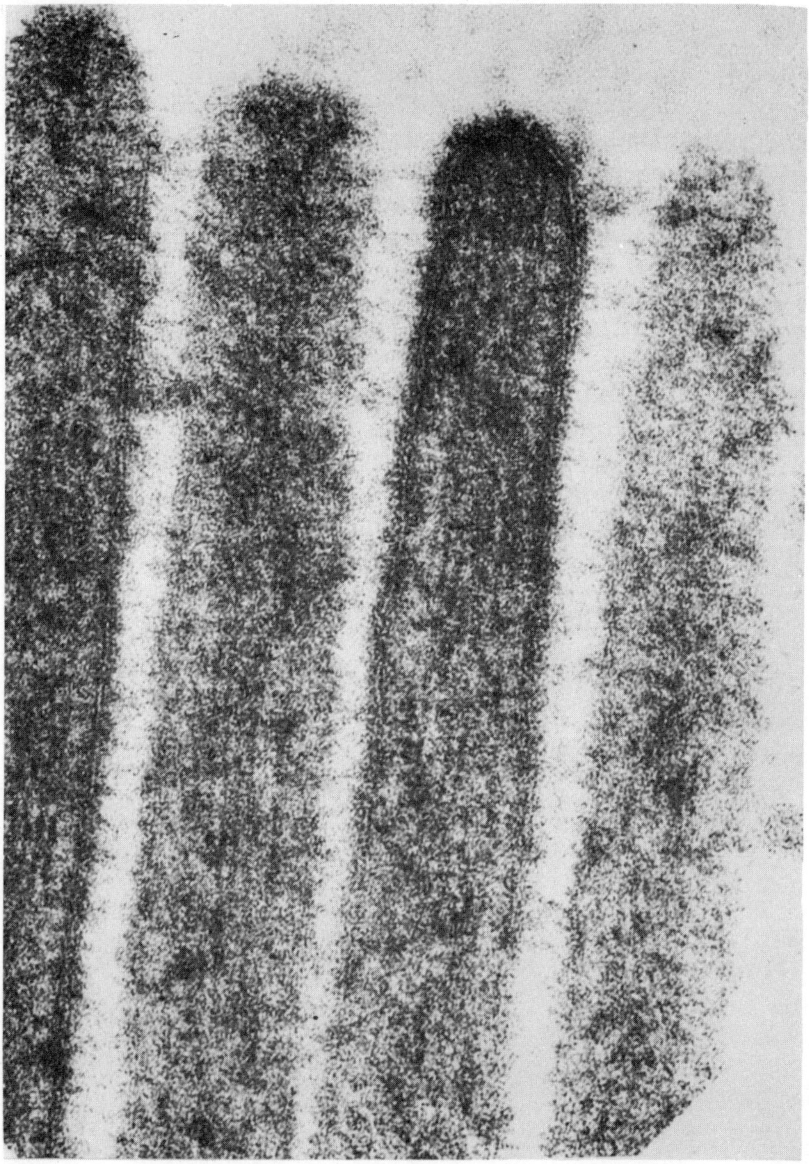

Fig. 1.8. Electron micrograph of the luminal part of a few microvilli. The membrane and surface coat are clearly shown. × 144,000.

membrane over the brush border; it has been reported that the outer layer is thicker in the mouse. However, not all species show this and furthermore the fixation, buffering and embedding may affect it and the appearance can alter during absorption so that no definite conclusions may be drawn [57, 55].

The surface coat is firmly attached to the outer leaflet and is resistant to washing, mucolytic and most proteolytic digests; only papain is known to dislodge it [58]. This layer is the PAS positive area first noted by Leblond [59] and first seen under the electron microscope by Ito [56]. Its appearance varies with different fixation and staining; it is best seen with gluteraldehyde fixation followed by post fixation with osmium tetroxide. As will be mentioned later the surface coat binds heavy metals and the osmium is well taken up from phosphate-buffered OsO_4 but not from carbon tetrachloride. The resultant picture shows the coat as radiating fibrils being marked at the microvillous tips and sparse on the microvillous slides. These form a dense band across the area of the luminal surface of the brush border (Fig. 1.9). Stains for the light microscope show it as a homogeneous

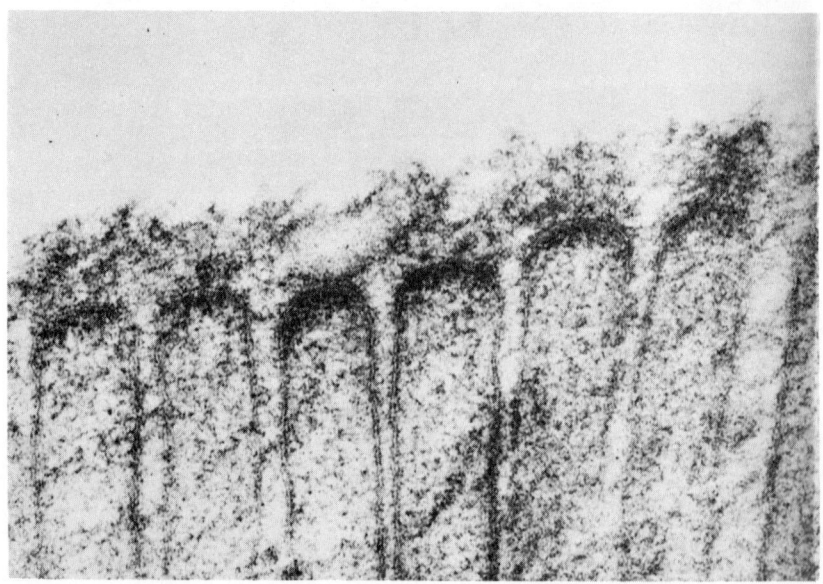

Fig. 1.9. Electron micrograph of the apex of microvilli. The increased amount of surface coat on the luminal aspect is shown. This corresponds to the outer dense line in the P.A.S. stain in Fig. 1.5. × 80,000.

band. It is strongly PAS positive, indicating a high content of acid mucopolysaccharides, as does the staining with Alcian blue and Hale's colloidal iron. Under the electron microscope the strong affinity for heavy metals can be further shown; iron, thorium and ruthenium red are all selectively localized to the surface coat [55].

There is no doubt that this surface coat of protein and mucopolysaccharide is an integral part of the greater membrane and not a layer of mucus from nearby goblet cells. Extensive studies on the turnover and origin of the components have been made by the use of isotopically labelled compounds combined with autoradiography or cell fractionation. Sodium acetate, sulphate, glucose, glucosamine, serine, mannose and galactose can be shown to be incorporated into the surface coat [60]. Furthermore, labelled mannose and galactose were well taken up into the surface coat but poorly into goblet cell mucus. The pattern of labelling was also distinct from the basement membrane and intercellular material. Both cell fractionation studies and autoradiography indicate that the precursors are initially concentrated in the Golgi apparatus and then move to the brush border region, probably in vesicles. They appear in the surface coat within 30 to 60 minutes, suggesting that this is the time required for the synthesis and transport of polysaccharides. The label remains in the surface coat from 4 to 10 hours. It has been shown that leucine is readily incorporated into the brush border region, certainly within six hours, so it is reasonable to suppose both plasma membrane and surface coat are highly dynamic [61].

The whole question of the localization of enzymes in the greater membrane has been given an impetus by the recognition of its granular nature. By the use of negative staining the fibrillary structure appears to consist of knobs 60 Å in diameter. The surface coat can be detached by a papain digest and a preparation obtained that appears to be pure 60 Å particles, sometimes called elementary particles [62]. Almost all the maltase, sucrase and leucine amino-peptidase can be demonstrated in this fraction [63]. Furthermore, after purification on a sephadex column the leucine amino peptidase could be precipitated and stained and was revealed as protein particles of about 60 Å. By contrast, magnesium ion activated adenosine triphosphatase can be shown to be localized to the inner leaflet of the microvillous plasma membrane. Furthermore, if a preparation is made of the core of brush borders it can be shown to contain about 30% of the protein but no specific enzyme. It therefore seems almost certain that disaccharidases and some dipeptidases are localized in the surface coat [64, 65, 66]. This fits well with Crane's hypothesis of a spatial relationship between disaccharidase and monoaccharide transport mechanism that probably lies in the plasma membrane [67]. The views of Ugolev [68] (see chapter 7) on contact digestion are also of great interest in this connection.

In addition to the physical presence of enzyme the surface coat has other properties that may be of great importance in absorption; it has been shown to be able to bind certain substances. There is evidence that glucose is bound to brush borders in isolated preparations [69]. D-glucose is specifically bound and this binding can be inhibited by other actively transported monosaccharides, calcium, lithium and D-glucosamine [70]. Non-actively absorbed sugars, amino-acids, sodium, ammonium and potassium ions were without this inhibitory effect. However, other studies have revealed that this binding takes place at different pH values from those required for active transport and the Km values obtained may also be different. At the moment the role of the brush border in binding glucose must be considered non-proven under physiological conditions.

By contrast, the binding of vitamin B_{12} is clearly established [71]. This has been shown to occur specifically in the ileum and to be facilitated by the presence of intrinsic factor. Furthermore, antibodies can be prepared to brush border of the distal small intestine in the hamster which inhibit the binding of intrinsic factor–vitamin B_{12} complexes. Antibodies prepared against brush borders from the proximal intestine are without this effect. Therefore a specific receptor or binding site for the intrinsic factor–vitamin B_{12} exists in the surface coat.

Iron is probably also bound by brush borders. Isolated preparations show that this occurs dominantly in the proximal small intestine [72]. The degree of binding can be influenced by the iron status of the animals, it is decreased in iron overload and increased in iron deficiency.

The surface coat is also a barrier that protects the epithelial cell and may act as a filtration system or like an ion exchange column. Evidence for the protective nature of the surface coat has come from experiments on hypovolaemic shock on the small intestine of the dog. Following a period of volume depletion the mucosa becomes susceptible to tryptic digestion with an increase in permeability to curare. This correlates with a loss of the PAS positive coat on the brush border [73].

The current view of the brush border is therefore of a structure of extreme complexity with numerous functional correlates. The surface coat of protein and mucopolysaccharide is highly organized spatially and is envisaged as a mosaic of enzymes and binding sites. It almost certainly exerts a barrier-like action against the diffusion of certain molecules. It seems likely that the next few years will bring confirmation and amplification of these concepts.

1.3.5 The apical region
Beneath the brush border is a narrow zone free of organelles that was recognized by the light microscopists and called the terminal web. It was envisaged as a felt work of fibres that stiffened the apex of the cell and condensed at the lateral margin to form the terminal bar. Many electron microscopic descriptions have perpetuated this impression, but a recent reappraisal of this area suggests a different structure [74].

The fibrils of the microvilli continue some 0.5 to 0.7μ into the apical cytoplasm forming rootlets. These filaments then abruptly stop and do not branch or splay sideways into a web. In between the rootlets is an amorphous substance without any fibrillary structure though a very occasional filament may be seen. The desmosomes are lateral condensations forming part of the junctional complex and these have radiating fibrils that may run some distance, usually just below the area of the terminal web. However, there is little doubt that this area has a skeletal function. In spite of osmotic or vibrational trauma the brush border and apex remains intact even though separated from the rest of the cell. The tight junctions are also resistant to separation by stretching.

Between this area and the nucleus lie the endoplasmic reticulum, Golgi apparatus, mitochondria and lysosomes.

1.3.6 Endoplasmic reticulum
This structure is easily recognizable but is not so highly developed as in cells specialized for protein secretion, e.g.

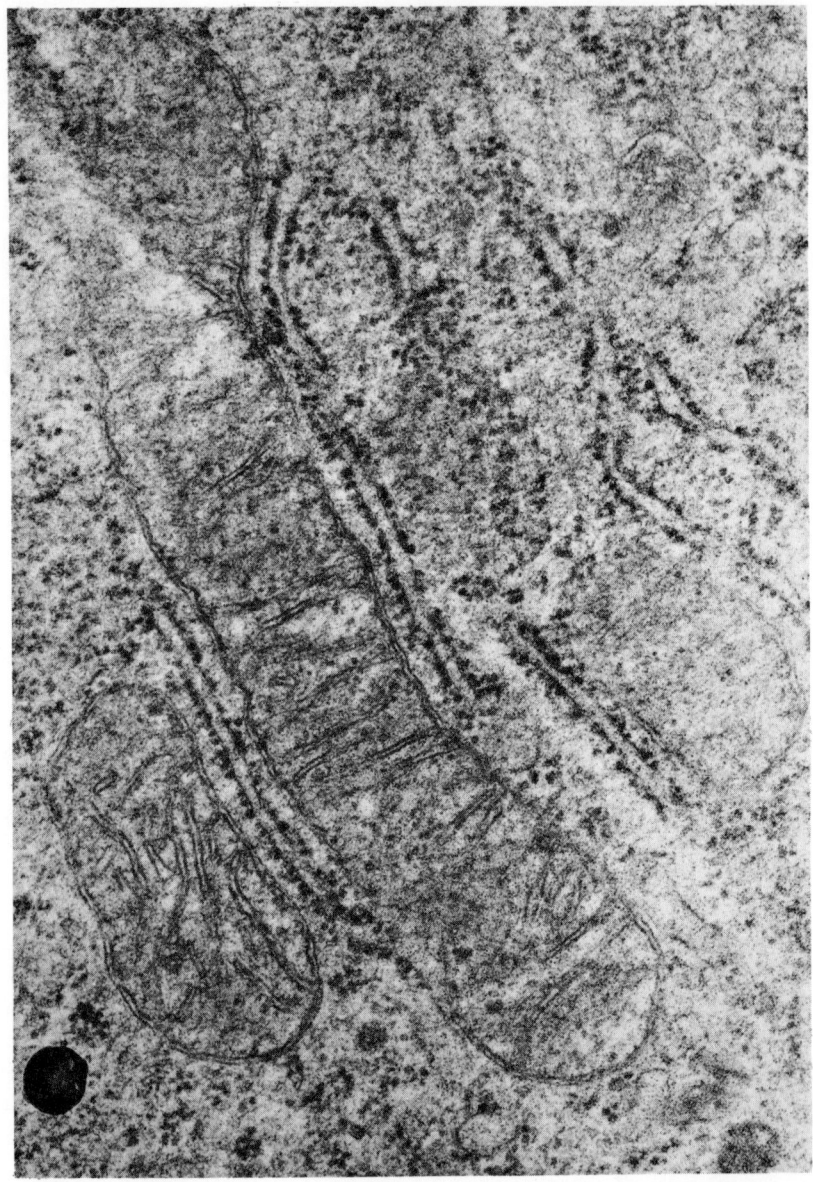

Fig. 1.10. Electron micrograph showing mitochondria and endoplasmic reticulum. The ribosomes are clearly visible studded along the channels of endoplasmic reticulum. × 46,000.

Paneth cells. It consists of a system of canals or cisterns seen in cross-section as short tubes or vesicles. Part of this has ribosomes studded along the outer side of the membrane, the rough or granular reticulum, while part is without ribosomes, the smooth or agranular reticulum. As in other cells the endoplasmic reticulum is the site of protein synthesis which may be used for enzymes or coating chylomicrons. It also provides a canal system for transport within the cell (Fig. 1.10).

1.3.7 Golgi apparatus

The Golgi complex is situated in the supra-nuclear region. It appears as a stack of tubes lying horizontally above the nucleus with outriding vesicles. Three-dimensionally it is a series of flat cisterns that buds off vesicles and almost certainly communicates with the endoplasmic reticulum. The membranes covering the complex are smooth without ribosomes (Fig. 1.11). The cisterns and vesicles often contain electron dense material. Functionally the Golgi apparatus is connected with the synthesis of polysaccharide and also the storage and intracellular transport of substances.

1.3.8 Mitochondria

Numerous mitochondria are present throughout the cytoplasm. They are usually rod-shaped, but they may appear spherical or even branched. In the apical part of the cell they usually lie in the long axis, but elsewhere their orientation is random. They are readily identified by the double membrane and prominent cristae, reduplications of the inner membrane that projects into the interior like a stack of shelves (Fig. 1.10). The interior contains a homogeneous lightly electron dense material and also a few really dense bodies about 200 Å in diameter that are situated in the outer layer or within cristae. They have been found in the duodenum and jejunum. In the rat it has been noted that these increase in size and density about 1 to 2 hours after the instillation of iron compounds in the lumen. Although they become somewhat granular they do not show the specific appearance of ferritin. Mitochondria are metabolically highly active and can carry out the whole of the Kreb's tricarboxylic acid cycle.

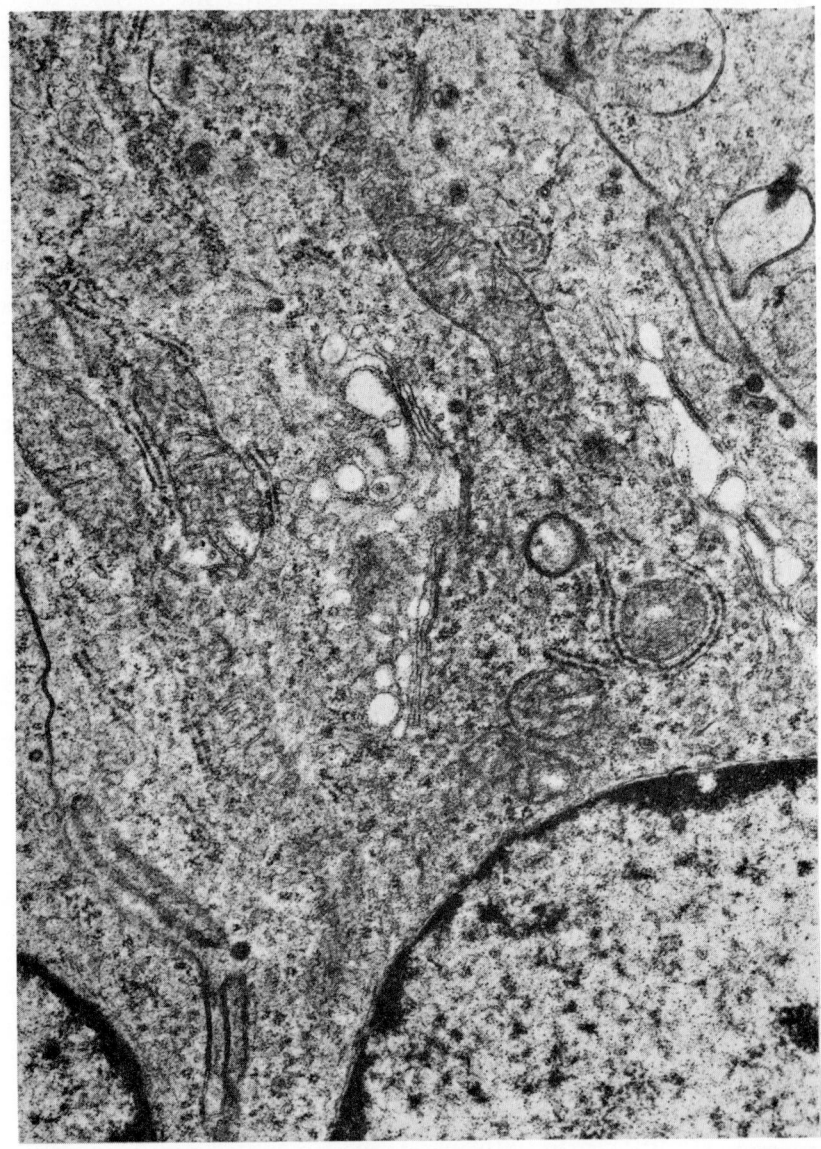

Fig. 1.11. Electron micrograph of the supra-nuclear portion of an epithelial cell showing the vesicles of the Golgi complex. × 18,500.

1.3.9 Lysosomes

Lysosomes are membrane-bounded structures containing enzymes, particularly acid phosphatase. Morphologically they show some variation but all have a distinct double membrane of lipoprotein that restricts access of substances into and enzymes out of the lysosome (Fig. 1.12). The interior is relatively electron dense but they may contain vesicles or degenerating endoplasmic reticulum or other cell organelle remnants. The whole concept of lysosomes as segregated concentrations of enzymes was formulated by de Duve, who described them as a digestive system within the cell [75]. They are visualized as structures potent in lysing unwanted or degenerate material but may also serve as storage areas. Their disruption as a disease mechanism (suicide bags) is well documented. There is some evidence that they may arise from the Golgi complex.

1.3.10 F-Bodies

These have some similarities to lysosomes but contain ferritin [76]. They are usually spherical or oval and are 0.5 to 1.0 μ in diameter with a dense structure and a single or occasionally double membrane. Ferritin is a protein structure containing six units of iron that are usually visible as four small closely placed dots because of the three-dimensional arrangement of the hexad. This makes it uniquely recognizable under the electron microscope. F-bodies contain a variable amount of ferritin usually arranged about the periphery though the whole structure may be densely impregnated with them. Occasionally free ferritin is seen in the cytoplasm. The number of F-bodies and the amount of ferritin they contain is related to the iron status; in iron deficiency they are few and relatively empty, while in iron overload or after iron supplementation in man they are plentiful and full. Preliminary observations in man suggest that they are absent in haemachromatosis where iron absorption is known to be excessive [77]. They may therefore be linked in some way with iron absorption.

1.3.11 Nucleus

The nucleus lies in the basal half and appears as an oval homogeneous structure. The nucleus is invested by a double membrane which is clearly fenestrated, though these pores are bridged by a thin diaphragm. The outer membrane is studded

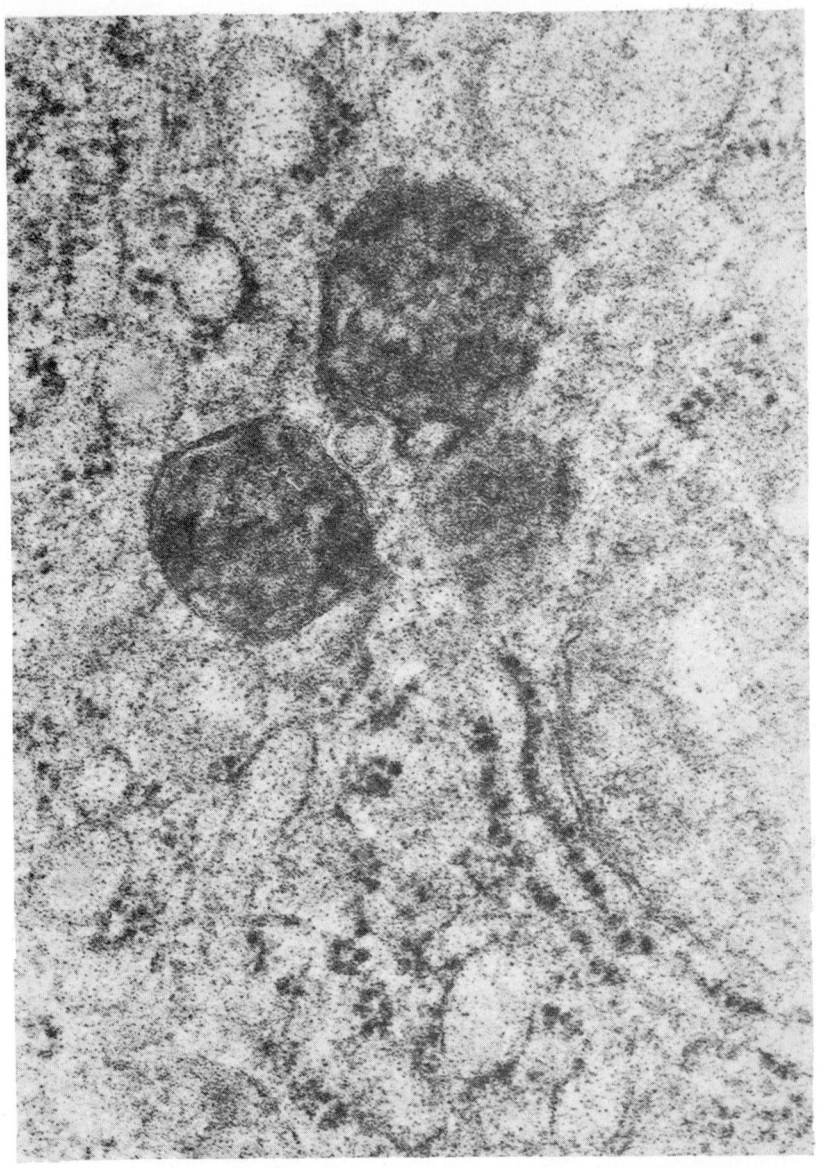

Fig. 1.12. Electron micrograph of two lysosomes from the apical region of an epithelial cell. × 75,000.

INTESTINAL STRUCTURE IN RELATION TO ABSORPTION

with ribosomes. The nuclear material is usually finely granular and without structure, though some dense aggregates of chromatin are occasionally seen. One or two nucleoli are usually seen in any section and appear as a dense, finely granular sphere.

1.4 OTHER CELLS

1.4.1 Goblet cells

Goblet cells are epithelial cells specialized for mucus production and secretion and when distended by mucus have a typical appearance (Fig. 1.13). The upper two-thirds of the cell is usually swollen with mucigen granules so that the structures described in the simple epithelial cell are sparse and displaced basally. A brush border of microvilli is present, but this is either thinner than normal or ruptured by the mucus discharge. Granular endoplasmic reticulum is present and is usually situated on either side of the nucleus. A prominent Golgi complex lies above the nucleus. There is good evidence that mucigen is formed here and packed into granules. These granules form the mucus mass, usually apparent as spheres 1-3 μ in diameter and often of varying electron density. Each granule is encased in a thin membrane envelope though, in places, this may not be intact and granules appear to merge. Secretion occurs by rupture of the luminal surface and discharge of the granules. There is some controversy as to whether goblet cells secrete mucus continuously or have one or more massive discharges. Autoradiography with radioactive sulphur shows a continuous migration of the label from the Golgi to the apex in 12 hours, whereas some studies of morphology show the appearance of an abrupt evacuation.

1.4.2 Crypt cells

The crypts of Lieberkühn contain undifferentiated epithelial cells and many mature goblet cells, these latter have the same appearance as described above. The undifferentiated cells have an immature brush border with fewer and shorter microvilli. The cell margins are straight with only occasional interlocking folds. The area of the terminal web is not yet developed and in its place are ribosomes, vesicles and secretory granules. A Golgi

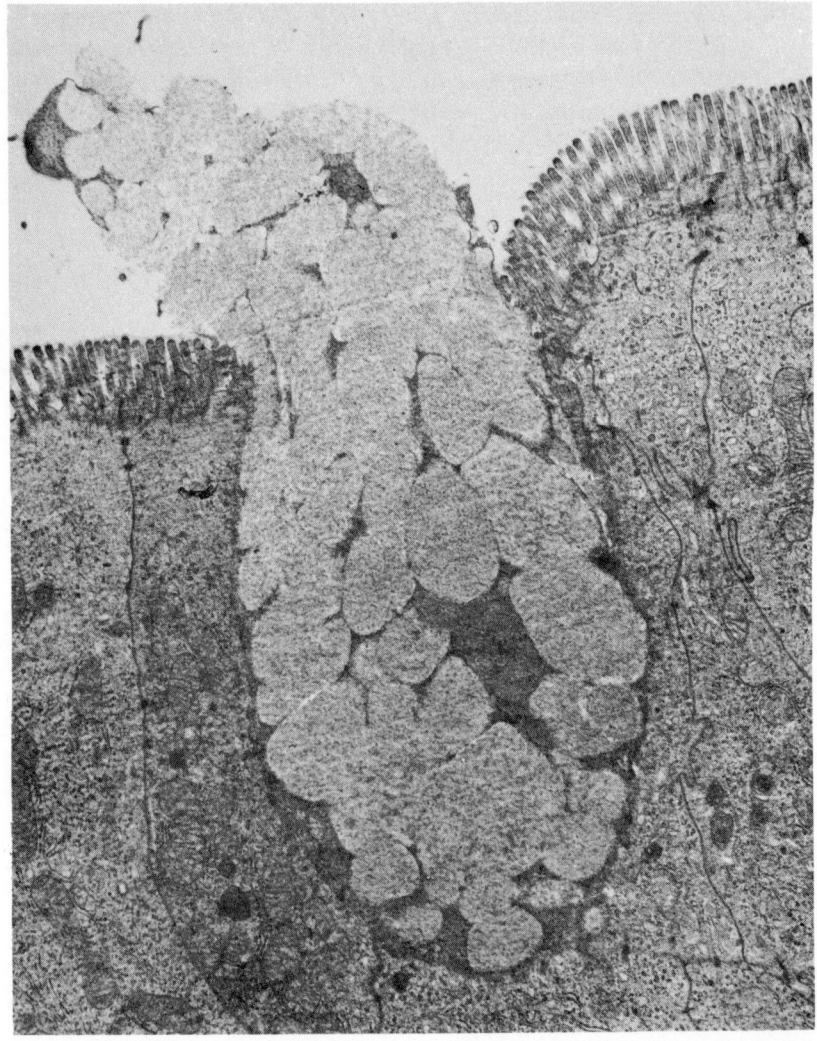

Fig. 1.13. Electron micrograph of a goblet cell discharging mucus. × 5,700.

complex and endoplasmic reticulum are present. Lysosomes occur and may arise from the Golgi complex. The nucleus is larger than in the adult absorption cell and usually appears lobulated. Particularly in man the apical part of the cell contains numerous secretory granules, electron dense bodies bounded by membrane, about 0.1 to 1.5 μ in diameter [78].

These granules are strongly PAS positive when viewed under the light microscope. They are undoubtedly secretory in nature and can be seen entering the crypt lumen. Trier (1964) has shown that under the effect of pilocarpine in man intense secretory activity takes place in the crypts, both apocrine and merocrine [79]. Large pseudopodia bud off into the lumen carrying cell membrane, cytoplasm, endoplasmic reticulum and the secretory granules. Within the crypt lumen these disintegrate, releasing their contents, although to what purpose is unknown. In rodents this apocrine secretion is not seen and the granules are discharged through a more prominent terminal web area by merocrine secretion.

1.4.3 Paneth cells

These are situated at the bases of crypts in the same layer as the immature crypt cells. They are readily identified by light microscopy on account of the eosinophilic granules which usually distend the supra-nuclear part of the cell. They are only found in mammals and have a curious distribution; by and large they are absent in carnivores. Paneth cells are present in man, monkeys and small laboratory animals whereas they are absent in cats, dogs and pigs. There seems no rational explanation for this variation.

They are clearly differentiated for active secretion and the granules can be seen entering the crypt lumen. Under the electron microscope the granules appear as spherical electron dense bodies bounded by a membrane [80] (Fig. 1.14). In the mouse the granules have a halo of acid mucopolysaccharide, while in man and the rat the granules are composed of a polysaccharide-protein complex which is PAS positive. Towards the apex of the cell the granules are large, 2 to 4 μ in man. They can be seen entering the lumen by merocrine secretion, while in man there is also evidence of apocrine secretion as in the immature crypt cells [81]. Mitochondria and lysosomes are prominent, but most of the basal part of the cell is packed with a tightly layered granular endoplasmic reticulum as in other protein-producing cells. There is a well marked Golgi complex above the nucleus from which vesicles can be seen budding off (Figs. 1.15 and 1.16). In these vesicles is material that is clearly the beginning of a secretory granule. Autoradiographic studies have shown the origination of protein in the endoplasmic reticulum and its subsequent migration to the Golgi complex and

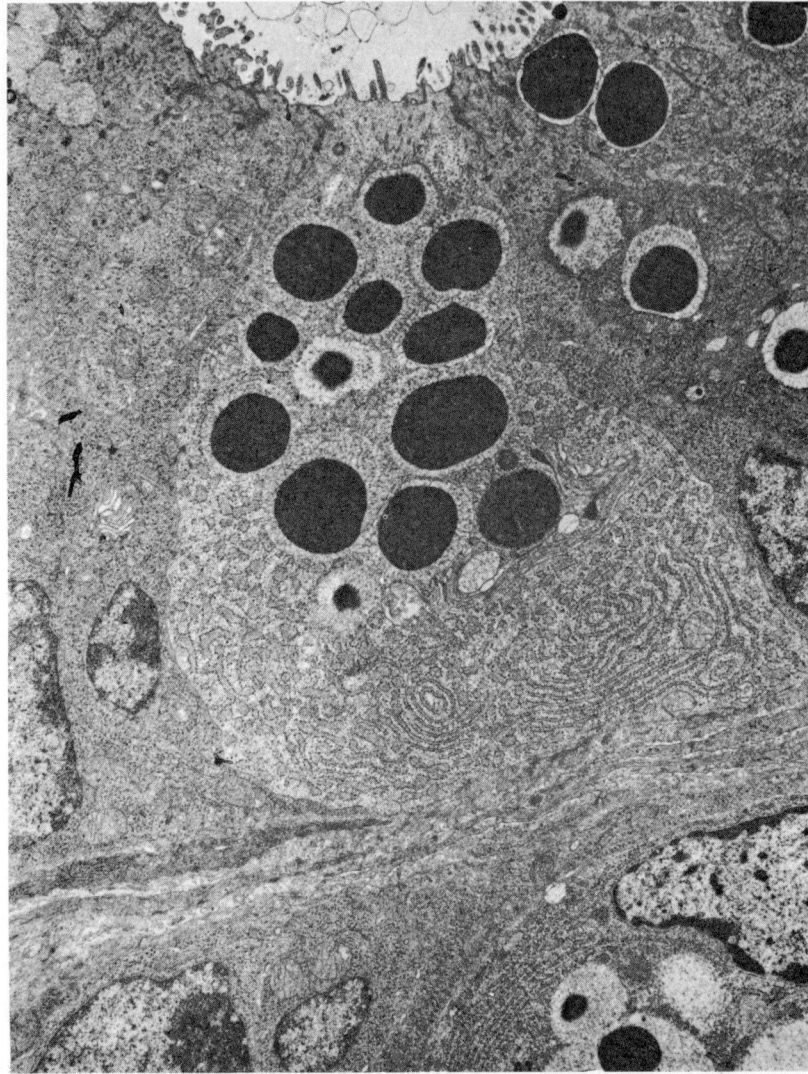

Fig. 1.14. Electron micrograph of a mouse Paneth cell. The granules have a clear halo in this species. At the base of the cell the closely packed rough endoplasmic reticulum is apparent. X 4,400.

then the secretory granule, the whole process taking about five hours, while by six hours label was seen in the crypt lumen [82]. This process can be speeded by pilocarpine so that label is in the lumen by three to four hours.

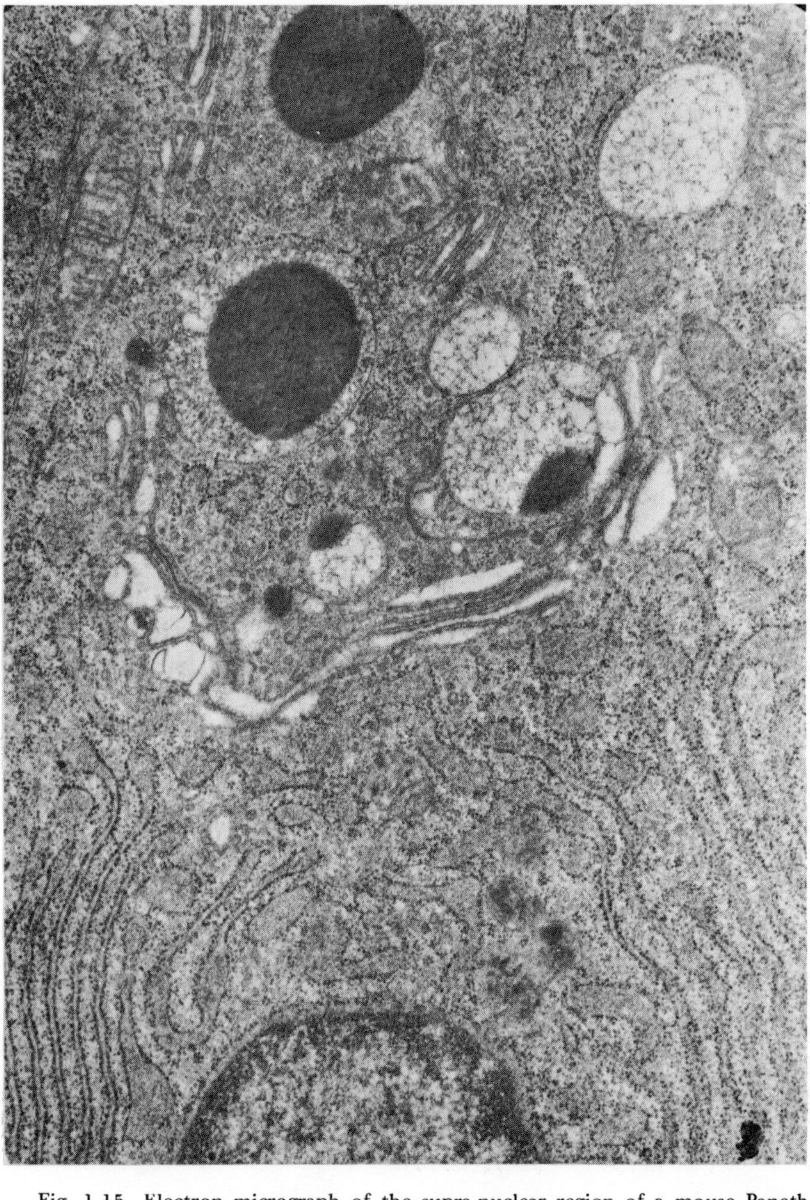

Fig. 1.15. Electron micrograph of the supra-nuclear region of a mouse Paneth cell. Above the endoplasmic reticulum the Golgi complex is seen with granules arising in the vesicles. × 19,600.

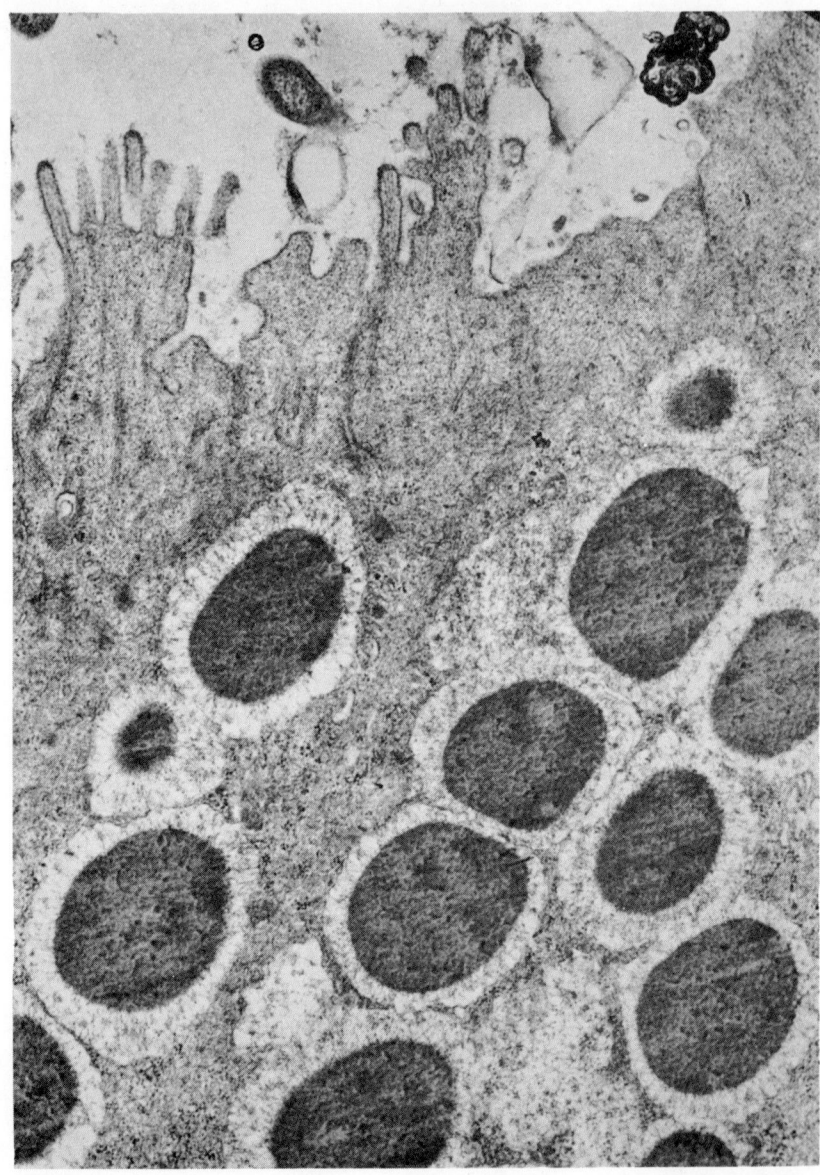

Fig. 1.16. Electron micrograph of the apical region of a mouse Paneth cell. The granules are closely packed in this area. The rudimentary microvilli are typical of crypt cells. × 19,600.

1.4.4 Argentaffin cells

These cells, also called enterochromaffin or Kultschitzky cells, are usually found in the basal region of the crypts. They are widely present throughout the gastrointestinal tract and are neurosecretory cells that are probably migrants from the neural crest rather than specialized elements of the intestinal epithelium. They are characterized by granules situated mostly in the infra-nuclear region against the basement membrane. Electron-microscopy reveals them as triangular-shaped cells with the base resting on the basement membrane and the apex stretching towards the lumen, occasionally actually reaching the crypt lumen. The granules are membrane-bounded and characterized by a very variable electron density so that some appear pale and some almost black (Fig. 1.17). Because of the predominantly basal position of the granules most of the other organelles lie about or above the nucleus. There is an endoplasmic reticulum and a Golgi apparatus but neither of these is well developed. The cells are known to contain considerable amounts of 5-hydroxytryptamine, but there is no evidence as to how this is secreted or in which direction.

Somewhat similar cells are found in the duodenum and there is considerable speculation current as to whether these cells may secrete secretin or pancreozymin or some of the other gastrointestinal hormones. It is probable that all these cells can secrete an amine and a polypeptide hormone.

1.5 OTHER COMPONENTS OF THE MUCOSA

1.5.1 The basement membrane

Crypt and villus are invested with a basement membrane on which the epithelial cell rest. This is a homogenous sheet about 300 Å thick and is a protein-polysaccharide structure. It is not impenetrable and lymphocytes and chylomicrons appear to migrate across it with ease, often leaving gaps. It may be thought of as analogous with the surface coat of the brush border and there is some evidence that it is secreted by the basal part of each cell.

1.5.2 Capillaries

These are seen in close association with the epithelial cell basement membrane, usually about 0.5 μ from it but some-

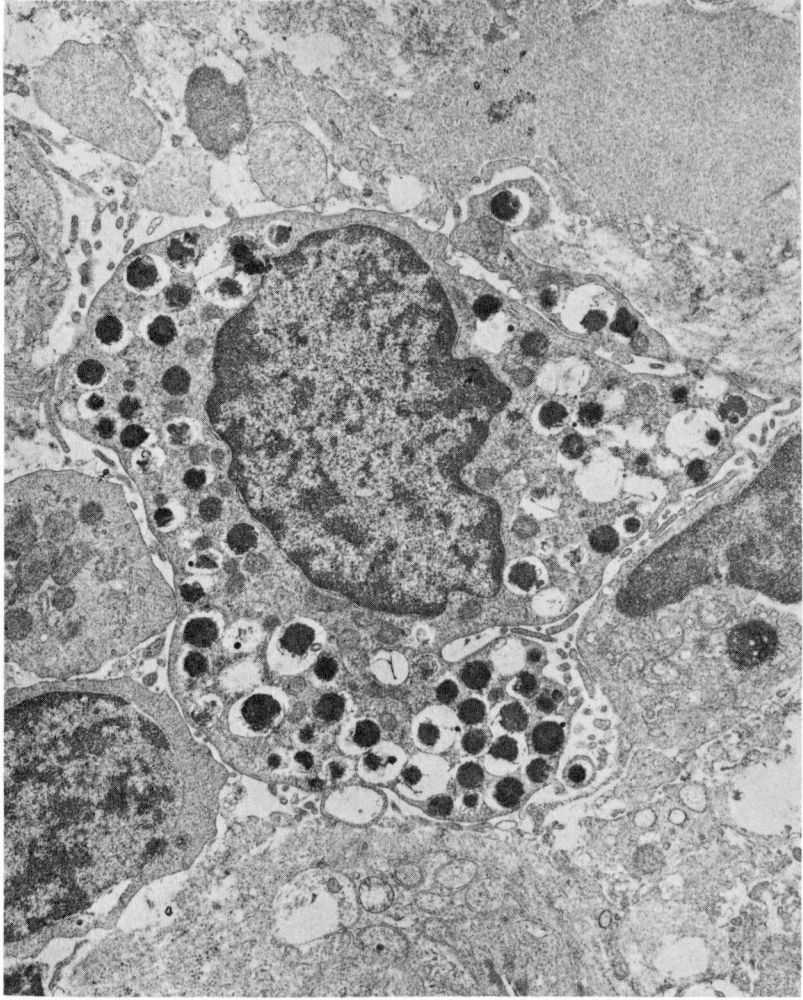

Fig. 1.17. Electron micrograph of an argentaffin cell. The granules are of varying electron density. × 9,600.

times closer. They are composed of a single layer of endothelial cells and a thin basement membrane. The cytoplasm of the endothelial cells is thin, apart from the area containing the nucleus, and fenestrated [83] (Fig. 1.18). The fenestrae are arranged in clusters so that over this area about a third of the cross-section is composed of the fenestrations; about one tenth

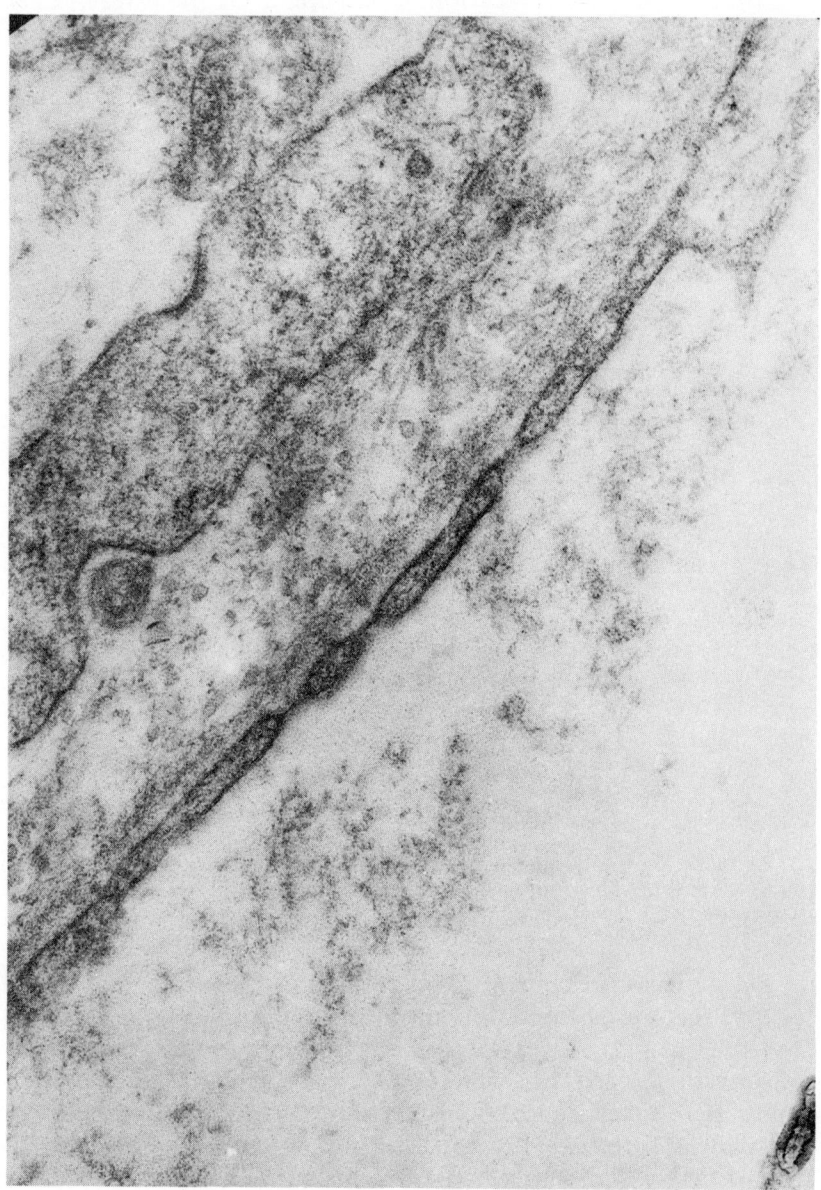

Fig. 1.18. Electron micrograph of capillary wall from a human villus. The fenestrations are clearly shown and each is spanned by a diaphragm. X 53,000.

of the whole cell in section is occupied by fenestrae. Each fenestra is about 350-450 Å in diameter and closed by a fine diaphragm. The plasma membrane of the endothelial cell can be followed round the edge of a fenestration from the luminal to the outer side; the diaphragm takes off at the mid point. The structure of the diaphragm is uncertain, it is a single layer 20 to 40 Å in width, but sometimes appears uneven. The endothelial cells contain vesicles similar to those in lacteals (Fig. 1.19).

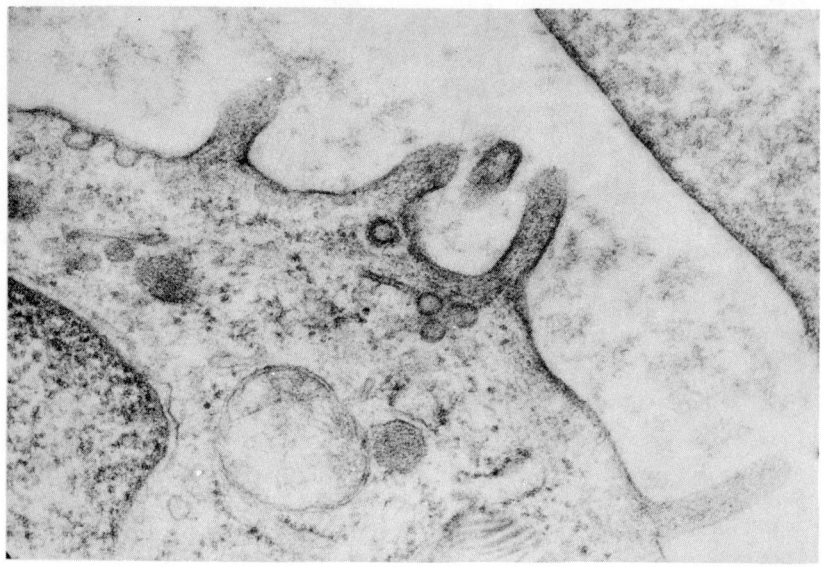

Fig. 1.19. Electron micrograph of part of capillary wall near the endothelial nucleus. The projections into the lumen are well shown and vesicles are present at the surface within the cytoplasm. × 37,000.

By a series of ingenious experiments Clementi and Palade (1969) have used the visible macromolecules, peroxidase (50 Å) and ferritin (110 Å), as probes to characterize the permeability of intestinal capillaries [83]. Both pass through the fenestrations; peroxidase in one minute and ferritin in three minutes. There was little evidence of transport by vesicles. If the endothelial cells are disrupted, the basement membrane will still act as a diffusion barrier so that molecules over 200 Å cannot escape [84]. It therefore appears that the intestinal capillaries are freely permeable to most molecules that may be absorbed, but anything over 200 Å, such as chylomicrons, will be unable to enter.

1.5.3 Lymphatics

Each villus has a centrally placed lymphatic, the lacteal. Under the electron microscope it has a well defined structure and can usually be differentiated from capillaries. Basically it is composed of endothelial cells and a thin and sometimes incomplete basement membrane, but there are no investing pericytes. The endothelial cells are thicker than those of capillaries and do not have fenestrations. Folds or projections of cytoplasm enroach into the lumen and vesicles are frequent [85]. There are also projections from the outer surface of the endothelial cells. Collagen and muscle cells are found in association with the lacteal.

The endothelial cells do not have fully developed junctional complexes and small gaps are visible at some junctions. There is no certainty as yet as to the route of entry of chylomicrons into the lacteal. Occasionally chylomicrons are observed in the intercellular junctional gap, suggesting that this is the route of entry; but they may also be seen within vesicles, raising the possibility of transport by this means.

REFERENCES

1. J. P. WILSON, *Gut*, **8**, 618 (1967).
2. T. HASTINGS WILSON, Intestinal Absorption, p. 2. W. B. Saunders & Co. (1962).
3. R. WARREN, *Anat. Rec.*, **75**, 427 (1939).
4. H. O. WOOD, *J. Anat. (Lond.)*, **78**, 103 (1944).
5. R. B. FISHER and D. S. PARSONS, *J. Anat.*, **84**, 272 (1950).
6. S. L. PALAY and L. J. KARLIN, *J. Biophys. and Biochem. Cytol.*, **5**, 363 (1959).
7. A. L. BROWN, Jr., *J. Cell Biol.*, **12**, 623 (1962).
8. R. DUBOS, R. W. SCHAEDLER, R. W. COSTELLO and R. P. HOET, *J. Exp. Med.*, **122**, 67 (1965).
9. S. J. BAKER, V. I. MATHAN and V. CHERIAN, *Lancet*, **1**, 860 (1963).
10. C. LOEHRY and B. CREAMER. Unpublished observations (1970).
11. J. MARKS and S. SHUSTER, *Gut*, **11**, 281 (1970).
12. H. SPRINZ, R. SRIBHIBHADA, E. J. GANGAROSA, C. BENYAJATI, D. KUNDEL and S. HALSTEAD, *Am. J. Clin. Path.*, **38**, 43 (1962).

13. J. LINDENBAUM, A. K. M. J. ALAM and T. H. KENT, *Brit. Med. J.*, ii, 1616 (1966).
14. P. K. RUSSELL, M. A. AZIZ, N. AHMAD, T. H. KENT and E. J. GANGAROSA, *Am. J. Dig. Dis.*, 11, 296 (1966).
15. S. J. BAKER, M. IGNATIUS, V. I. MATHAN, S. K. VAISH and C. C. CHACKO, Intestinal Biopsy. (Ciba, London, 1962.)
16. J. G. BANWELL, M. S. R. HUTT and R. TUNNICLIFFE, *East African Medical Journal*, 41, 46 (1964).
17. V. L. SWANSON and D. V. M. THOMASSEN, *Am. J. Path.*, 46, 511 (1965).
18. D. D. POUT, *Br. Vet. J.*, 126, 357 (1970).
19. C. P. LEBLOND and C. E. STEVENS, *Anat. Rec.*, 100, 357 (1948).
20. C. P. LEBLOND and B. MEISSIER, *Anat. Rec.*, 132, 247 (1958).
21. C. P. LEBLOND and B. E. WALKER, *Phys. rev.*, 36, 255 (1956).
22. M. LIPKIN and H. QUASTLER, *J. Clin. Invest.*, 41, 141 (1962).
23. M. LIPKIN, P. SHERLOCK and B. BELL, *Gastroenterology*, 45, 721 (1963).
24. W. C. MacDONALD, J. S. TRIER and N. B. EVERETT, *Gastroenterology*, 46, 405 (1964).
25. R. G. SHORTER, C. G. MOERTEL, J. L. TITUS and R. J. REITEMEIER, *Am. J. Dig. Dis.*, 9, 760 (1964).
26. R. R. PASCAL, G. I. KAYE and N. LANE, *Gastroenterology*, 54, 835 (1968).
27. G. I. KAY, N. LANE and R. R. PASCAL, *Gastroenterology*, 54, 852 (1968).
28. D. N. CROFT and M. LUBRAN, *Biochem. J.*, 95, 612 (1965).
29. D. N. CROFT, D. J. POLLACK and N. F. COGHILL, *Gut*, 7, 333 (1966).
30. D. N. CROFT, C. A. LOEHRY, J. F. N. TAYLOR and J. COLE, *Lancet*, 2, 70 (1968).
31. R. M. CLARKE, *Gut*, 11, 1051 (1970).
32. J. S. TRIER and T. H. BROWNING, *J. Clin. Invest.*, 45, 194 (1966).
33. J. S. TRIER, *Gastroenterology*, 42, 295 (1962).
34. P. FOROOZAN and J. S. TRIER, *New Eng. J. Med.*, 277, 553 (1967).
35. R. M. BANNERMAN, B. CREAMER and K. B. TAYLOR. Unpublished observations.
36. B. CREAMER, *Brit. Med. J.*, 2, 1373 (1964).
37. G. D. ABRAMS, H. BAUER and H. SPRINZ, *Lab. Invest.*, 12, 355 (1963).
38. S. LESHER, H. E. WALBURG and G. A. SACHER, *Nature*, 202, 884 (1964).
39. R. KENWORTHY and W. D. ALLEN, *J. Comp. Path.*, 76, 291 (1966).
40. C. A. LOEHRY and B. CREAMER, *Gut*, 10, 116 (1969).
41. M. G. DEO and V. RAMALINGASWAMI, *Gastroenterology*, 49, 150 (1965).
42. C. STEVENS HOOPER and M. BLAIR, *Exp. Cell Res.*, 14, 175 (1958).
43. J. P. A. McMANUS and K. J. ISSELBACHER, *Gastroenterology*, 59, 214 (1970).

44. A. E. COCCO, M. J. DOHRMANN and T. R. HENDRIX, *Gastroenterology*, **51**, 24 (1966).
45. C. A. LOEHRY, R. GRACE and B. CREAMER. Unpublished observations (1970).
46. R. H. DOWLING and C. C. BOOTH, *Clin. Sci.*, **32**, 139 (1967).
47. D. H. SMYTH, *Gastroenterology*, **42**, 76 (1962).
48. R. H. DOWLING and C. C. BOOTH, *Lancet*, **2**, 146 (1966).
49. D. S. MILLER, M. A. RAHMAN, R. TANNER, V. I. MATHAN and S. J. BAKER, *Scan. J. Gastroenterology*, **4**, 477 (1969).
50. J. D. HAMILTON, A. M. DAWSON and J. WEBB, *Gut*, **8**, 509 (1967).
51. M. N. MARSH, J. A. SWIFT and E. D. WILLIAMS, *Brit. Med. J.*, **4**, 95 (1968).
52. P. G. TONER, K. E. CARR, A. FERGUSON and C. MACKAY, *Gut*, **11**, 471 (1970).
53. M. G. FARQUHAR and G. E. PALADE, *J. Cell Biol.*, **17**, 375 (1963).
54. B. GRANGER and R. F. BAKER, *Anat. Rec.*, **107**, 423 (1950).
55. W. D. DOBBINS, *Am. J. Med. Sci.*, **258**, 150 (1969).
56. S. ITO, *J. Cell Biol.*, **27**, 475 (1965).
57. G. B. DERMER, *J. Ultrastruct. Res.*, **20**, 331 (1967).
58. C. F. JOHNSON, *Science*, **155**, 1670 (1967).
59. C. P. LEBLOND, *Am. J. Anat.*, **86**, 1 (1950).
60. G. G. FORSTNER, *Am. J. Med. Sci.*, **258**, 172 (1969).
61. R. HOLMES and R. K. CRANE, *Gut*, **9**, 365 (1968).
62. C. F. JOHNSON, *Fed. Proc.*, **28**, 26 (1969).
63. T. ODA, S. SEKI and S. WATANABE, *Acta Med. Okayame*, **23**, 357 (1969).
64. A. EICHHOLZ and R. K. CRANE, *J. Cell Biol.*, **26**, 687 (1965).
65. J. OVERTON, A. EICHHOLZ and R. K. CRANE, *J. Cell Biol.*, **26**, 693 (1965).
66. G. C. FORSTNER, S. M. SABESIN and K. J. ISSELBACHER, *Biochem. J.*, **106**, 381 (1968).
67. R. K. CRANE, *Gastroenterology*, **50**, 254 (1966).
68. A. A. UGOLEV, Physiology and Pathology of Membrane Digestion. Leningrad, 1967.
69. A. EICHHOLZ, *Fed. Proc.*, **28**, 30 (1969).
70. N. J. GREENBERGER, *Am. J. Med. Sci.*, **258**, 144 (1969).
71. I. L. MACKENZIE, R. M. DONALDSON, W. L. KOPP and J. S. TRIER, *J. Exp. Med.*, **128**, 375 (1968).
72. N. J. GREENBERGER, S. P. BALCERZAK and G. A. ACKERMAN, *J. Lab. Clin. Med.*, **73**, 711 (1969).
73. G. BOUNOUS, A. H. McARDLE, D. M. HODGES, L. G. HAMPSON and F. N. GURD, *Am. Surg.*, **164**, 13 (1966).
74. O. BRUNSER and J. H. LUFT, *J. Ultrastruct. Res.*, **31**, 291 (1970).
75. C. DE DUVE, 'The Lysosome Concept'. In Lysosomes. Ciba Foundation Symposium, London (1963).
76. R. S. HARTMAN and W. CROSBY, *Blood*, **22**, 397 (1963).
77. I. J. PINK. Unpublished observations (1970).
78. J. S. TRIER, *J. Cell Biol.*, **18**, 599 (1963).
79. J. S. TRIER, *Gastroenterology*, **47**, 480 (1964).

80. A. D. HALLY, *J. Anat.*, 92, 268 (1958).
81. O. BEHNKE and H. MOE, *J. Cell Biol.*, 22, 633 (1964).
82. J. S. TRIER, V. LORENZSONN and K. GROEHLER, *Gastroenterology*, 53, 240 (1967).
83. F. CLEMENTI and G. E. PALADE, *J. Cell Biol.*, 41, 33 (1969).
84. F. CLEMENTI and G. E. PALADE, *J. Cell Biol.*, 42, 706 (1969).
85. G. E. PALADE and R. R. BURNS, *J. Cell Biol.*, 37, 633 (1968).

CHAPTER 2

Cytochemistry of Enterocytes and of Other Cells in the Mucous Membrane of the Small Intestine

ZDENĚK LOJDA

*Research Laboratory of Angiology and
1st Department of Pathology,
Faculty of General Medicine,
Charles University,
Prague, CSSR*

		Page
2.1	INTRODUCTION	44
2.2	THE MUCOUS MEMBRANE OF THE NORMAL SMALL INTESTINE	48
	2.2.1 *Enterocytes*	48
	2.2.1.1 *Brush Border*	48
	2.2.1.2 *Lateral Plasma Membrane*	65
	2.2.1.3 *Terminal Web*	66
	2.2.1.4 *Lysosomes*	66
	2.2.1.5 *Mitochondria*	69
	2.2.1.6 *Golgi Complex*	72
	2.2.1.7 *Endoplasmic Reticulum*	73
	2.2.1.8 *Nucleus*	75
	2.2.2 *Undifferentiated cells*	76
	2.2.3 *Goblet cells*	76
	2.2.4 *Paneth cells*	78
	2.2.5 *Endocrine cells (E cells)*	79
	2.2.5.1 *EC Cells*	81
	2.2.5.2 *L Cells*	83
	2.2.5.3 *S Cells*	84
	2.2.6 *Lamina propria*	84
2.3	CYTOCHEMISTRY OF ENTEROCYTES DURING DEVELOPMENT	86
	2.3.1 *Microvillous zone*	86
	2.3.2 *Lysosomes and related particles*	88
	2.3.3 *Endoplasmic reticulum*	89
	2.3.4 *Mitochondria*	90

2.4	ABSORPTIVE CELLS OF THE INTESTINAL MUCOSA IN EXPERIMENTAL CONDITIONS	90
	2.4.1 Changes during absorption of Nutrients and water	91
	2.4.2 The effect of diet	92
	2.4.3 The action of hormones	93
	2.4.4 The injury by X-rays	94
	2.4.5 The effect of antibiotics, colchicine and triparanol	96
2.5	CYTOCHEMISTRY OF THE HUMAN JEJUNAL MUCOSA IN MALABSORPTION SYNDROME	96
	2.5.1 Celiac sprue	97
	2.5.2 Tropical sprue	106
	2.5.3 Malabsorption of disaccharides	107
	2.5.4 Glucose-galactose malabsorption	107
	2.5.5 Gongenital α-β-lipoproteinemia	108
	2.5.6 Whipple's disease	108
	2.5.7 Other cases of secondary malabsorption syndrome	109
2.6	CONCLUSIONS	110
	REFERENCES	111

2.1 INTRODUCTION

The importance of intestinal function has stimulated a growing interest in clinical and experimental studies of the intestine. A better understanding of absorption and other functions of the intestine requires a more detailed knowledge of the morphology of individual components of the intestine (including ultrastructure), of its chemical composition as well as of biochemical processes in the intestine under normal and experimentally changed conditions and in pathological disturbances.

With the introduction of peroral enterobiopsy [1, 2, 3, 4, 5, 6] fresh human material has been made readily available for investigation so that work does not need to be restricted to animal material and to rare human specimens occasionally obtained by surgical intervention.

It is necessary to emphasize that findings obtained by different approaches must be correlated. Hence the ultrastructural details must be interpreted in biochemical terms, and the biochemical processes must be localized in cells. This is the goal of biochemical cytology. Cytochemistry has an important position in this field, even if discrepancies exist about the meaning of cytochemistry. In the language of a morphologist cytochemistry means the distribution of substances in cells visualized by special methods in tissue sections and observed by a light or electron microscope. For a biochemist cytochemistry means the determination of various substances in cells and their fractions previously separated and studied *extra situm*.

There are thus two main approaches which can be used to reach the goal of biochemical cytology. Both have their advantages and shortcomings. These may be largely overcome by mutual comparison of results of both approaches applied to the same material. For localization purposes the biochemists use the technique of analysis of fractions separated by differential ultracentrifugation of homogenates. This technique was also adapted to the small intestine. The brush border fraction is of greatest importance for understanding the absorptive processes and was isolated first. This consists of the microvillous zone, the terminal web and some subjacent part of the cytoplasm of the enterocyte. The procedure was originated by Miller and Crane [7] and was applied in the original or in a modified form by other authors [8, 9, 10, 11, 12, 13]. The brush border fraction can be further osmotically disrupted and various subfractions, particularly the brush border membrane fraction and the fibrillar fraction, separated by centrifugation in glycerol density gradients [9] or sucrose gradients [13]. The limitation of this procedure is that the brush border fraction (particularly that prepared according to the procedure of Hübscher *et al.* [10], is not pure [14, 15] and the isolation procedure, particularly that of Eichholz and Crane [9], is deleterious for some enzymes and for the glycocalyx [16, 15]. In the course of the isolation of brush borders according to the technique of Eichholz and Crane [9], other subcellular particles are disrupted so that they cannot be isolated from the same homogenate. By the technique of Hübscher *et al.* [10] it is possible to obtain also the mitochondrial and lysosomal fractions, the microsomal fraction and the soluble fraction. As was stated for the brush border fraction the purity of these fractions is always questionable. Moreover the particles can be damaged and substances released from them. Even if the fractions would be prepared in an ideal state of purity one has to realize that all cells from the homogenate contribute to them, i.e. not only the epithelial cells but also the cells of the propria which, particularly in the case of lysosomes [10, 17, 18, 19], must always be taken into consideration, and that one obtains average values which can be distorted by the calculation basis. Nevertheless much information was obtained by this approach.

These studies on isolated fractions cannot provide any information on the distribution of substances in epithelial and connective tissue cells nor in cells of villi and crypts.

There have been several attempts to isolate epithelial cells for biochemical analysis. The isolation can be achieved by the treatment of intestinal segments with trypsin-pancreatin [20] or with hyaluronidase [21]. A simple method was recently devised by Sjöstrand [22]. To obtain a collection of cells the position of which in the intestine is known the vibration technique was applied [23, 24]. During these procedures cells can be damaged.

The cells from various heights of villi and crypts may be obtained by sectioning the opened intestine parallel with the luminal surface. A special 'intestinal planing apparatus' was devised for this purpose by Imondi et al. [25] which makes it possible to cut the mounted blocks by sliding the micrometer-mounted razor blade horizontally across the tissue. The cut material is collected from the blade and is homogenized. Successive cutting of layers of about 0.1 mm are made and the villi completely removed. The crypt region is then gently scraped with a glass slide.

A very useful adaptation of the technique of Linderstrøm-Lang and Mogensen [26] has been used [27, 28, 29, 30], in which the intestine samples are quickly frozen so that the autolytic changes are excluded, and cut in a cryostat in sections parallel to the surface. The major portion of sections is collected for biochemical analysis. Morphological control of single sections before and after each collection of sections used for biochemical analysis reveals the position of the analysed sample in the intestine.

These procedures make it possible to localize substances in villi and crypts. In the samples individual epithelial cells and connective tissue elements are analysed together and this is always to be taken into consideration in the evaluation of results. The value of the information obtained could be improved by using histochemical staining reactions instead of pure histological staining. The procedures already referred to for the *extra situm* analysis (excluding the procedure of Dahlqvist and Nordström [27] require relatively large quantities of material so that they can hardly be applied in the tissue samples obtained by peroral enterobiopsy and on embryological material. None of them can show the differences between individual cells.

The distribution pattern of many substances in human and animal intestine under normal, experimentally changed and pathological conditions can be studied very well by histochemical 'staining' methods applied to sections of properly treated samples of the intestine. The advantage of these methods is that they can be performed in relatively minute samples of tissue such as peroral enterobiopsies and that they can demonstrate very easily the localization of substances in cells along the villus, and the heterogeneity in the distribution of substances in individual epithelial and connective tissue cells. It is the only approach which can be used for the analysis of some epithelial cells, e.g. endocrine cells. Even an intracellular localization can be achieved by many of these methods. It is to be emphasized that due attention must be paid to the preparation of material, which differs according to the substance to be demonstrated, and to the methods used for localization. In comparison with biochemistry, limitations of this approach lie in the smaller number of substances which can be demonstrated, in the relatively low specificity of some methods, and in the difficulty and sometimes even the impossibility of quantitative evaluations. Nevertheless much useful information was obtained by this approach in the intestine. The contribution of the *extra situm* analysis particularly to the knowledge of the brush border has been recently reviewed [31, 32, 33, 34, 35].

In this paper data on the biochemical cytology of the intestine are presented. These are based mainly on our own studies of human enterobiopsies and also of some animal material by histochemical 'staining' methods [36, 37, 38, 39, 40, 41, 42, 43, 44, 45, 46, 47, 48, 49]. The findings are correlated and supplemented with data from other laboratories obtained by the analysis *in situ* and with most important data obtained by the analysis *extra situm*. The methodological problems which are very important will not be dealt with and the reader has to consult our papers or textbooks of histochemistry [50, 51, 52]. No effort has been made to compile a complete bibliography but the references provide the reader with enough information for further study.

2.2 THE MUCOUS MEMBRANE OF THE NORMAL SMALL INTESTINE

The absorptive surface of the mucous membrane of the normal small intestine is increased by mucosal folds—valves of Kerkring, and finger-like outgrowths—villi. Both increase very substantially the absorption surface [53, 54, 55]. The height of villi is dependent on the animal species, on the degree of contraction and relaxation of smooth muscle cells within the villus core and on the degree of distension of the bowel. Between the bases of the villi, openings of tubular gland-like crypts of Lieberkühn are found.

The epithelium which covers the surface of the mucous membrane is simple columnar. On the villi are found two kinds of cells: enterocytes and goblet cells. In addition undifferentiated cells, Paneth cells and endocrine cells can be recognized in crypts. The electron microscopy of these epithelial cells has been reviewed [56, 54, 55, 57, 58] and in Chapter 1 of this book.

2.2.1 *Enterocytes*

Absorptive cells, brush border cells, chief cells (Fig. 2.1a) are the predominant type of cells and the most important from the point of view of absorption. They are at the apex of the villi, cylindrical, and in the human 20—30 high and 6—9 wide. The free surface of these cells is provided with a *brush (striated) border*. This part has been studied extensively. Electron microscopy showed that the striated border consists of very regular finger-like projections—microvilli. Their axis runs parallel with the long axis of the cells. It is evident that this structure greatly increases the surface area. Microvilli are enclosed by a somewhat modified plasma membrane which displays some species differences in its thickness [32, 55, 57]. To the outer leaflet of the plasma membrane a polysaccharide-rich coat is attached. This coat has been termed the glycocalyx [59], or fuzzy layer [60, 61]. It is prominent in human, rat and feline enterocytes whereas in other species it is much thinner [16]. Its

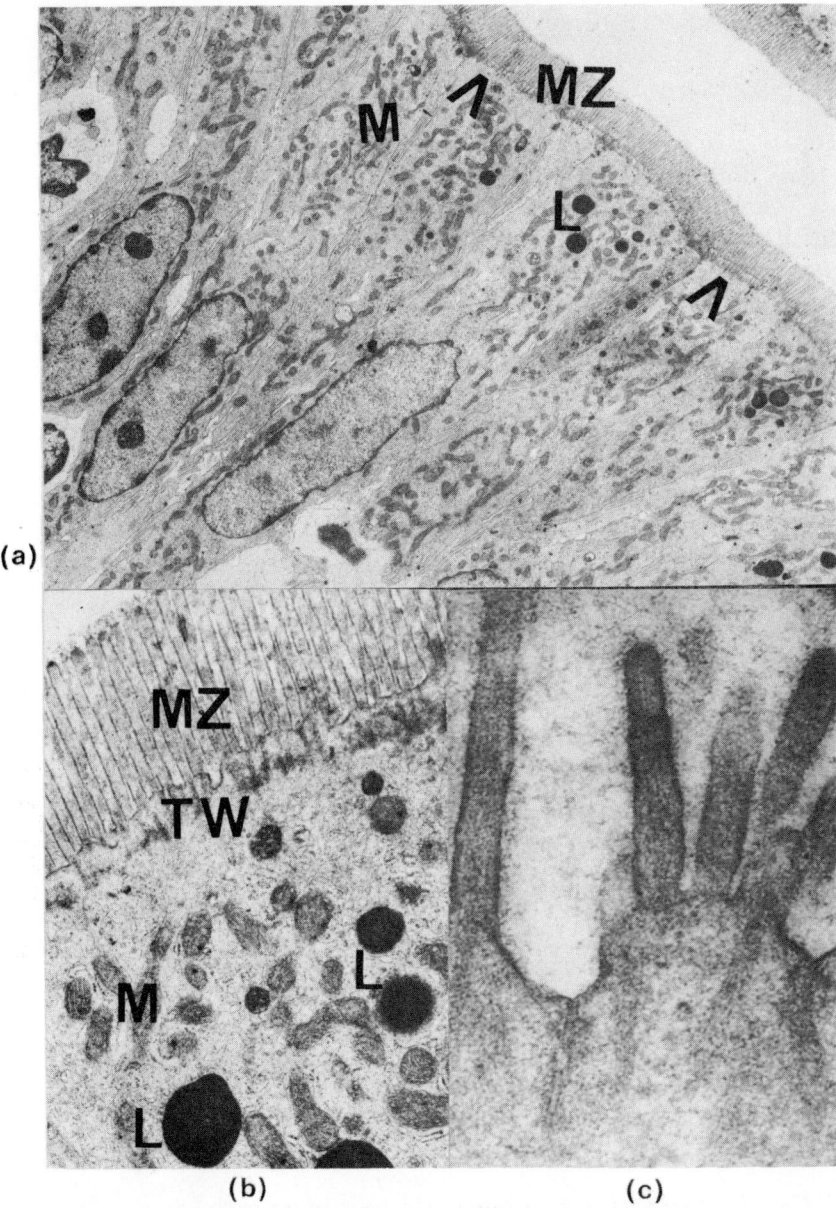

Fig. 2.1a. Electron micrograph of enterocytes covering the upper part of a villus side in normal human jejunum. L—lysosomes, M—mitochondria, MZ—microvillous zone. Arrows point to the terminal web. ×3,500.

Fig. 2.1b. Apical part of a human enterocyte under a higher magnification. L—lysosomes, M—mitochondria, MZ—microvillous zone, TW—terminal web. ×17,500.

Fig. 2.1c. Electron micrograph of the apical part of enterocytes of a patient with celiac sprue who had very low activities of all disaccharidases. The glycocalyx covering the microvilli is well apparent. ×50,000.

appearance varies according to the preparation procedure [62]. After double fixation in glutaraldehyde and osmium tetroxide and staining with lead citrate it appears to be composed of fine filamentous material extending from the outer leaflet of the plasma membrane. It is not known whether this appearance exists also in the fresh unfixed state.

In the central part of the microvillous core 10—50 closely packed, parallel microfilaments, or microtubules [32] are found. These microtubules extend from a region close to the tip of the microvilli into the zone of the terminal web (Fig. 2.1b). The terminal web is composed of microfilaments or microtubules which run parallel with the surface but are not straight.

Neutral and acidic mucosubstances can be demonstrated in the brush border by histochemical reactions, e.g. a positive PAS-reaction (Fig. 2.2a) which is blocked by acetylation and

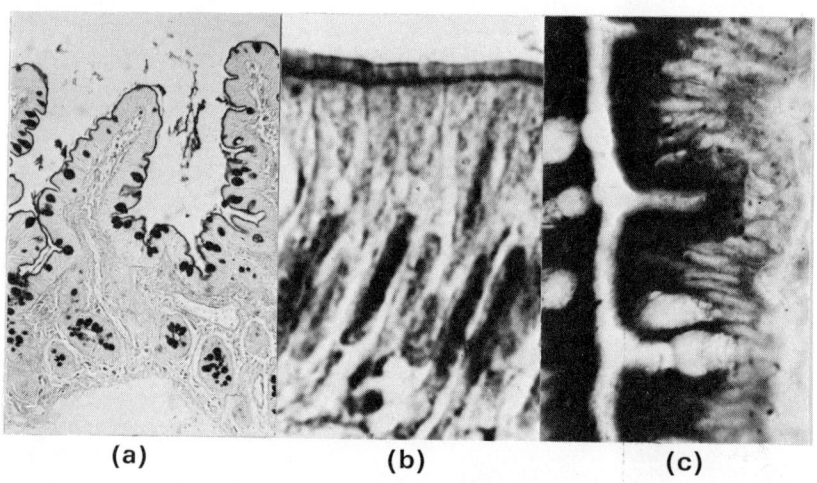

(a)   (b)   (c)

Fig. 2.2a. PAS-reaction in the mucous membrane of the normal jejunum. Mucopolysaccharides of the brush border of enterocytes covering the villi and of goblet cells react very strongly. X60.

Fig. 2.2b. Reaction for SH-groups in enterocytes of normal human jejunum. Strong positivity in the terminal web and nuclei, weaker positivity in mitochondria. X1800.

Fig. 2.2c. OTAN (= osmium tetroxide-α-naphthylamine) reaction in villi enterocytes of normal human jejunum. Strongest coloration is apparent in the supranuclear part of enterocytes (due to mitochondria phospholipids). Brush border displays a weaker staining. X600.

becomes positive after saponification, and positive staining with Hale's reaction or alcian blue at low pH. The zone stained by alcian blue does not entirely correspond to the zone giving the PAS reaction. Combined staining with the PAS reaction and alcian blue in the human intestine shows that the outer layer stains blue while the inner layer stains more red [47]. The staining with alcian blue and Hale's reaction is blocked by methylation, and becomes positive again after demethylation, the intensity of coloration being only very slightly reduced. The positivity is therefore chiefly caused by carboxylic groups probably of sialic acid, and free sulphate groups are not present in substantial numbers. The mucus of goblet cells which covers the inner surface of the intestine differs from the mucosubstances of the brush border in that it contains more sulphate groups. Our failure to demonstrate a significant amount of sulphate groups in mucosubstances of the brush border of the human jejunum is at variance with the observations of Ito and Revel [63], who demonstrated the incorporation of radioactive sulphate into the surface coat of the cat ileum. Further studies are required to explain this discrepancy.

The brush border takes acid stains, gives a positive coupled tetrazonium reaction, a positive reaction for tyrosine, tryptophan, SH- (Fig. 2.2b), and SS-groups, showing that proteins containing these amino acids are present. Because a substantially stronger reaction, particularly in the case of coupled tetrazonium reaction and reaction for SH-groups, can be found in the terminal web (Fig. 2.2b) it is probable that microtubules of the microvillous cores are mainly responsible for this reaction.

The brush border region is weakly stained by Sudan Black B, particularly after glutaraldehyde fixation. However, it is difficult to decide how much the lipid component and how much proteins are responsible for it. The presence of choline-containing lipids is proved well by the OTAN (osmium tetroxide-α-naphthylamine) reaction in which the brush border is stained orange brown, although much less than mitochondria (Fig. 2.2c). The staining remains unchanged after acetone pre-extraction and is almost abolished after chloroform-methanol pre-extraction. Due to the low resolving power of the light microscope it is not possible to localize these reactions more precisely.

However, there are methods for localizing mucosubstances

with electron microscopy. It was demonstrated that with colloidal iron [64, 65, 66], colloidal thorium dioxide [67, 61, 68, 69] and ruthenium red [69a, 32], the surface coat can be selectively stained. These reactions demonstrate acid groups in the mucosubstances. The coat also reacts positively with methods analogous to the PAS method in light microscopy [68, 66]. Comparison of these reactions points to a somewhat stronger binding of colloidal iron at the apices of microvilli, whereas the periodate positive material is evenly distributed along the microvillous membrane (cf. Figs. 2.3 and 2.4 in Bradbury and Stoward [66]). Electron microscopy shows

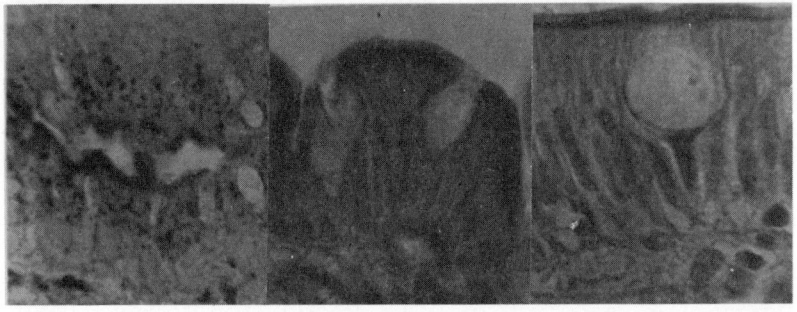

Fig. 2.3a. Azo-coupling reaction for α-D-glucosidase (maltase) according to Lojda [36] in enterocytes of normal human jejunum. Nuclei counterstained with haematoxylin. Only the brush border reacts positively. X 600.

Fig. 2.3b. Indigogenic reaction with 4-Cl-5-Br-3-indolyl-β-D-glucoside according to Lojda [38] in enterocytes of normal human jejunum. Nuclei counterstained with nuclear fast red. Note a strong staining of the brush border. X 600. (From Lojda [38], courtesy of Springer-Verlag.)

Fig. 2.3c. Combined reaction for alkaline phosphatase (blue) and thermostable esterase (brown) in differentiating crypt enterocytes in the upper part of a crypt. X 600.

that the mucosubstances, as demonstrated in the brush border in the light microscope, are in fact part of the glycocalyx.

Many comparative chemical analyses of the surface coat have not been carried out due to difficulties in its isolation in pure form free from goblet cell mucus. Forstner [70] considers the presence of significant amounts of acid mucopolysaccharides, such as chondroitin sulphate, in the glycocalyx unlikely. He demonstrated a substantial amount of glucosamine-^{14}C label in

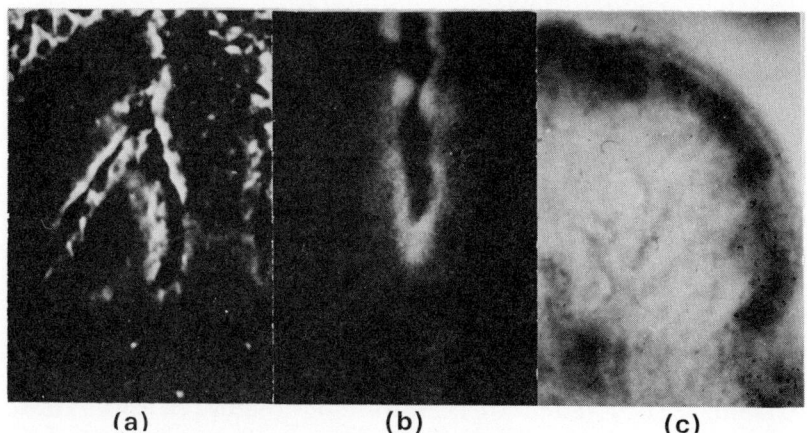

Fig. 2.4a. Immunohistochemical demonstration of lactase in rat jejunum. Specific fluorescence can be seen in the brush border of differentiated enterocytes covering the villi. × 130.

Fig. 2.4b. Immunohistochemical demonstration of sucrase I in rat jejunum. Specific fluorescence can be seen in the brush border. × 600.

Fig. 2.4c. Indigogenic reaction with 4-Cl-5-Br-3-indolyl-β-D-galactoside according to Lojda [39] at pH 5.5 in enterocytes of normal human jejunum. Positive reaction in the brush border (lactase) and in lysosomes (acid β-D-galactosidase) of enterocytes. In the left-hand corner macrophages react positively (acid β-D-galactosidase). × 700.

sialic acid. There is thus no evidence in favour of a greater amount of sulphate groups in the surface coat.

Analysing fractions of isolated brush borders Eichholz [33] found a large quantity of SH-groups in the fraction derived from the microvillous core. This is in agreement with findings *in situ*.

Phospholipids and cholesterol were found in isolated brush borders [71] and a detailed analysis of lipids of the rat intestinal microvillous membrane is available [72, 73].

Ito and Revel [63] succeeded in autoradiographic labeling of the surface coat in the cat with glucose-^{3}H, glucosamine-^{14}C, galactose-^{3}H, mannose-^{3}H, serine-^{3}H, sulphate-^{35}S and acetate-^{3}H and demonstrated that the coat is a dynamic cell component which is continuously synthesized in the enterocyte. The labelling is first apparent in the Golgi region, but it is not clear how the surface coat material is transported from there to the cell surface [16]. A transport in vesicles was suggested by Forstner [70].

The presence of *enzyme activities* in the brush border is of major interest, and the list of these is continually growing. So far the following enzymes were found in the isolated brush border fraction, mainly in its membrane subfraction: sucrase, maltase, isomaltase, lactase [7, 9, 13], trehalase [74], γ-amylase and dextranase, the latter being identical with isomaltase [75], β-D-glucosidase (phlorizin hydrolase) [76], alkaline phosphatase [7, 9], ATP-ase [77], leucyl-β-naphthylamidase, tri- and to a lower degree diipeptide hydrolases [78, 79, 10, 80, 15], enterokinase [80a, b] cholesterol and retinol ester hydrolases [81, 82], palmitate thiokinase, mono- and diglyceride acylases [83], sphyngomyelinase and ceramidase [84]. Some of these findings require corroboration.

Analysis of cryostat sections cut parallel to the surface showed that alkaline phosphatase, disaccharidases and dipeptidases of the rat small intestine have much higher activities on villi than in crypts [27, 29, 30]. It is worth mentioning that the brush border enzymes are relatively strongly associated with the structure so that they can be easily detected even in unfixed cryostat sections. This is in accordance with the experiences of biochemists [33, 85, 86] that homogenates must be specially treated to solubilize the enzymes (either with papain digestion or treatment with Triton X-100). This has the important implication in practice that it is necessary to determine the activity of these enzymes in whole homogenates and not in supernatants unless a special solubilization procedure is applied [87].

(a) Glycosidases. Of these *disaccharidases* have been of most interest because they are responsible for splitting the disaccharides in food. Defects of these enzymes in intestinal mucosa, either as an inborn error or as an acquired disturbance, is a cause of malabsorption of disaccharides in man [88]. (See also Chapter 18.) A classification of these enzymes is given by Semenza [85].

Two basic approaches have been used for the *in situ* localization of these enzymes: localization of the enzymatic protein and localization of the enzymatic activity. The fluorescence antibody technique was used by Doell *et al.* [89], for localization of sucrase and lactase in the suckling rat, and the specific fluorescence was localized in the brush border of enterocytes covering the villi. The enzymes used for immunization in these studies were not highly purified, and the possibility could not be excluded that they localized a nonenzymatic component of the brush border [90]. In our own unpublished studies with Kraml and Kolínská highly purified lactase from the intestine of suckling rats and sucrase 1 and 2 from adult rats were used as antigens for immunization of guinea pigs and subsequent preparation of antisera. In all these cases a strong specific fluorescence was observed in the brush border of enterocytes covering the villi of the intestine of adult rats and 11-day-old rats (Fig. 2.4a, b). A weak fluorescence was confined to the brush border of differentiating enterocytes in the upper part of crypts. No specific brush border fluorescence was found when antisera against rat intestinal acid β-D-galactosidase were used in these experiments. The findings of Doell et al. [89] were thus confirmed. It is interesting that a weak fluorescence can be observed also in the apical cytoplasm subjacent to the brush border in which also a positive staining occurs when methods with 5-Br-4-Cl-3-indolyl-β-D-glucoside and β-D-fucoside are used [38, 48]. Compare Fig. 2.4b and 2.3b).

The antibody technique can be used with electron microscopy. Gitzelmann et al. [91] studied the localization of rabbit intestinal sucrase-isomaltase with ferritin-antibody conjugates. They concluded that the enzyme is localized in the outer layer of the microvillous membrane and that it is not a part of the glycocalyx of the enterocytes as was claimed by Johnson [92, 93]. Johnson's conclusion was made because by digestion with activated papain followed by differential ultracentrifugation the glycocalyx of the enterocytes was removed as were the disaccharidases and could be recovered from Sephadex columns in the same fraction as disaccharidases. Also, from the resemblance of globular particles revealed by negative staining in the glycocalyx with the doughnut structure of particles of purified sucrase [94], it was concluded that disaccharidases are localized in the glycocalyx. These particles have been considered as a fixed component of the microvilli membrane attached to the outer leaflet of its trilaminar structure [95]. Even if the assumption of Johnson is very suggestive the reported findings are insufficient to prove the localization of the disaccharidases in the glycocalyx unequivocally. Furthermore the glycocalyx can be found in enterocytes of patients with celiac sprue (Fig. 2.1c) in which the activity of sucrase and other disaccharidases is greatly reduced. There is also no direct proportionality between the activity of sucrase and other disaccharidases and the degree of evolution of the surface coat in the enterocytes of different animal species. Further studies are needed to settle this problem.

Even if all methodological prerequisites for the immunohistochemical (immunocytochemical) localization are met and the problem of antigenic specificity or similarity of disaccharidases of different animal species is solved, the question, whether the localized enzymatic protein is active, cannot be answered by such studies. Therefore methods localizing the activities of these enzymes must be employed. These methods are also convenient for routine practice.

They can be divided into two groups: (i) Methods with artificial substrates. (ii) Methods using disaccharides as substrates.

Glycosides of 6-Br-2-naphthol were used as substrates [96, 97, 98, 99] because it was assumed that for the specificity of disaccharidases, like in other glycosidases, the type of the glycosidic linkage and the glycon only are decisive [100, 101, 102]. In the case of disaccharidases the influence of the aglycon on the specificity is more marked [103, 85]. Nevertheless some artificial substrates can be used in studies of disaccharidases.

6-Br-2-naphthol-α-D-glucoside is split by two intestinal maltases [103]. Using this substrate in our simultaneous azo-coupling method with hexazonium para-rosanilin [36] an unequivocal localization of enzyme activity splitting this substrate in the brush border of differentiated enterocytes was achieved for the first time (Fig. 2.3a). Simultaneous coupling is a prerequisite for this localization because all methods using postincubation coupling with this and other glycosides of 6-Br-2-naphthol are useless for the demonstration of the intracellular localization [36]. The brush border localization obtained with 6-Br-2-napthol-α-D-glucoside in paraffin sections can be inhibited by maltase [18] 10^{-2} M but not by other disaccharides used in the same concentration [42, 47]. In recent experiments (Lojda, 1972), it was shown that in unfixed cryostat sections the reaction is inhibited by both maltose and sucrose. The discrepancy between results in paraffin and cold microtome sections can be explained by the heat inactivation of sucrase during the embedding in paraffin. In experiments carried out in cooperation with Slabý, Kraml and Kolínská it was shown that 6-Br-2-naphthol-α-glucoside is split by glucoamylase and also by sucrase isolated from the human intestine.

Substantial progress in the localization of lactase was achieved by indigogenic methods using 5-Br-4-Cl-3-indolyl-β-D-fucoside, β-D-galactoside and β-D-glucoside [38, 39, 48]. In these studies it was demonstrated that 'neutral' β-D-galactosidase (lactase) released from homogenates of the intestine of suckling rats and separated by chromatography on Sephadex G-200 cleaved all three glycosides. This was shown in the test tube and also in the immunoprecipitation lines obtained by Ouchterlony's technique with lactase and its appropriate rabbit antiserum (Fig. 2.5). Similar results were obtained with isolated

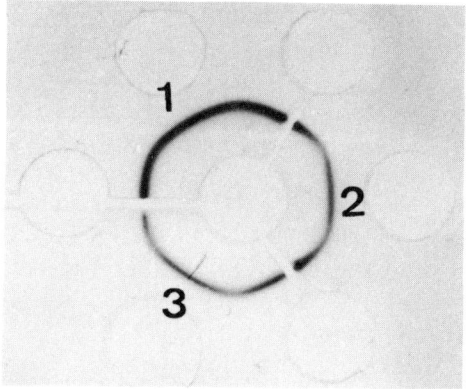

Fig. 2.5. Ouchterlony test with lactase isolated from the small intestine of suckling rats (peripheral wells) and precipitated with rabbit antiserum against lactase (central well). The plate was divided into three parts which were processed separately with the indigogenic media according to Lojda [38, 39] and Lojda and Kraml [78] at pH 6. The most intense staining was obtained with β-D-fucoside (1, incubation 1 hour), followed by β-D-glucoside (2, incubation 3 hours) and β-D-galactoside (3, incubation 18 hours).

human lactase.* From this it is evident that the fucoside is the best artificial substrate for lactase and the galactoside the worst.

* Lojda, Slabý and Kraml 1972, to be published.

In this connection it should be mentioned that Kraml *et al.* [104] found that lactase of suckling rats is the major fraction of intestine hetero-β-D-glucosidase. Hetero-β-D-glucosides are also split by a 'minor' fraction of β-glucosidase [=phorizin hydrolase].

In a study by Lojda and Kraml [48], it was demonstrated that lactase is localized in the brush border of enterocytes in the upper part of crypts. It reaches its maximum in differentiated enterocytes covering the sides and tops of villi in the intestine of suckling (the highest activity) and adult rats, and of human (Fig. 2.12), monkey and hamster jejunum. In the cock jejunum traces of brush border staining were seen only after 24 hours of incubation. These histochemical findings are not only in good accordance with the biochemical data [85, 105, 30, 106] but they amplify them in localization.

With 5-Br-4-Cl-3-indolyl-β-D-glucoside a similar distribution pattern was obtained [38, 48]. However, this substrate is split not only by lactase but also by other enzymes. In experiments with Slabý and Kraml (to be published) it was shown that the glucoside is split also by isolated human hetero-β-D-galactosidase. The splitting by phlorizin hydrolase must be taken into consideration as well. It is not known to what extent these enzymes participate in cleaving the substrate in various animals and at different ages of the same animal.

The demonstration of two β-galactosidases *in situ* was achieved for the first time with one and the same substrate using 5-Br-4-Cl-3-indoly-β-D-galactoside [39]. 'Neutral' β-D-galactosidase (lactase) was found in the brush border of differentiated enterocytes of animal species mentioned above. Similarly as with β-D-fucoside, which is the best substrate for its visualization, a gradient in the intestine was found: the highest activity resides in the brush border of differentiated enterocytes of the jejunum decreasing in the aboral direction, has a pH optimum of 5.5-6, and was not demonstrated at pH 3.5.

The acid β-D-galactosidase, which has a pH optimum about 3.5 (at pH 5.5 its activity is weaker but still demonstrable, Fig. 3.4c), was demonstrated in lysosomes of enterocytes and in macrophages of the propria. The mutual ratios of activity in enterocytes and macrophages are different in different animal species and are not the same along the gut. This is to be borne in mind in explaining the biochemical data on this enzyme.

Both activities are considerably inhibited by galactonolactone (5×10^{-3} M) and lactose (8×10^{-2} M). The enzyme in the brush border is inhibited also by gluconolactone (4×10^{-4} M) and cellobiose (8×10^{-2} M).

Biochemical findings of different authors reached in different animal species (for references see Lojda [39, 48]) can be now shown *in situ*. The demonstration of a third β-galactosidase, with the same pH activity curve as lactase, which is soluble and which splits only synthetic β-D-galactosides but not lactose [109, 107, 108, 108a] was not attempted.

Natural disaccharides were used as substrates in the method with coupled oxidation of glucose [110]. In this multistep method glucose released by the action of endogenous disaccharidase is oxidized by means of exogenous glucose oxidase which reduces phenazine methosulphate and this in turn a tetrazolium salt. Dahlqvist and Brun [110] reported on the basis of results obtained by this method that the digestive disaccharidases are associated with small cytoplasmic granules rather than with the brush border. Previously it was shown that these disaccharidase granules are artifacts due to the diffusion of the liberated glucose [36]. To reduce this diffusion agar or gelatine was added to the modified medium, the concentration of the tetrazolium salt increased and to prevent the oxygen from competing with phenazine methosulphate the incubation was carried out under anaerobic conditions [36]. Even under these conditions the diffusion artifacts could not be entirely prevented. Nevertheless the method is able to show well the cellular localization. It proved useful in studies of the distribution of sucrase, trehalase, lactase and maltase (provided that the sample of glucose oxidase did not contain any concomitant activity of maltase) in the animal intestine and in human enterobiopsies. Even if we take into consideration that reactions with substrates from which arise two molecules of glucose (maltose and trehalose) must be more intense than with substrates from which only one molecule of glucose is obtained, the activity of maltase and trehalase is higher than that of sucrase in the rat intestine. The activity of lactase is the weakest and can be demonstrated under anaerobic conditions only [36].

The activity of all disaccharidases is confined to differentiated enterocytes covering the villi and the highest activity resides in the jejunum. In the duodenum and ileum the activities are lower. No activity of sucrase was shown in suckling rats up to 11 days of age. Trehalase was definite and lactase was very active. In human enterobiopsies this method demonstrated well a decrease of activities of all disaccharidases in patients with celiac sprue. These disaccharidase activities were restored after a gluten-free diet [42, 46, 47]. Also a primary deficiency of lactase could be shown by this method [47].

Jos et al. [111, 112], developed a more cumbersome modification of this method. With this modification a more distinct staining of the brush border of differentiated enterocytes was observed in the reactions for maltase, sucrase and trehalase, but not for lactase. Diffusion artifacts could not be avoided, however. Even with this method we were able to achieve only a cellular localization. We therefore adhered to our simpler method which enabled well also parallel studies of zymograms. All disaccharidases of the human intestine displayed two bands: one cathodal and one (less intense) anodal band [113]; see also Fig. 2.12 of this paper. Because the cathodal bands were weaker in patients with celiac sprue in which the activities of brush border disaccharidases were very much reduced, we assume that the cathodal band originates in the brush border. The origin of the remaining bands is not certain and further studies are required to explain them.

Recently, we developed a different method for disaccharidases in which peroxidatic oxidation of 3,3' diaminobenzidine was coupled to the method with glucose oxidase [114]. Intracellular localization is not achieved by this method, but it may well serve as a very good screening test for disaccharidase deficiencies, because it is simple and requires only one cryostat section $10-15\mu$ thick for the visualization of one disaccharidase, which is much less than is necessary for biochemical methods *extra situm* (Fig. 2.13). The biological significance of disaccharidases in the digestion of disaccharides is evident but it may not be their only function.

(b) Alkaline phosphatase and ATP-ase. Nonspecific alkaline phosphomonoesterase is the enzyme which was first demonstrated in the brush border of differentiated enterocytes by Gomori [115] and later by many others. In the pH range of 6.0–9.6 it splits various substrates: glycerophosphate (Fig. 2.6b, 2.14), phenylphosphate, naphthylphosphates, phosphates

of the naphthol AS series (Fig. 2.6a), nucleoside tri-, di-, and mono-phosphates, etc. The reaction with most substrates appears in enterocytes at the opening of crypts (Fig. 2.14). With nucleoside triphosphates a weak reaction in the microvillous zone of crypt enterocytes is visible as well (Fig. 2.6c). It

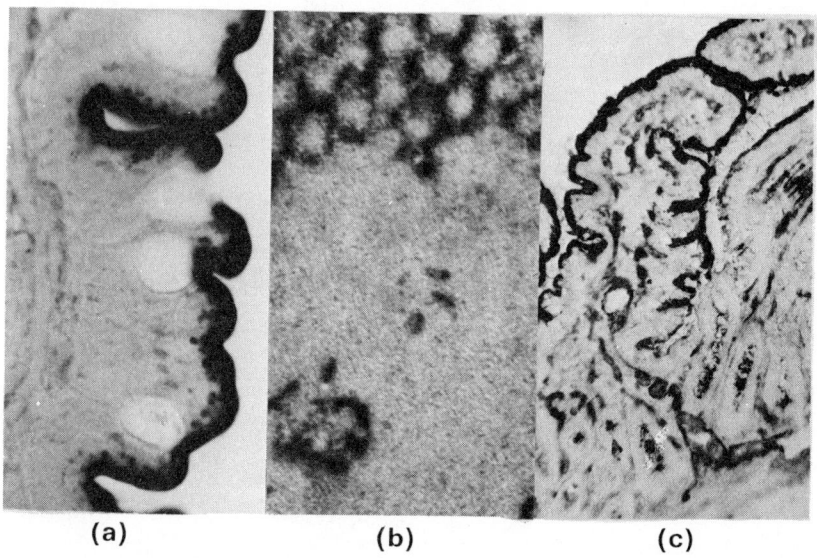

(a) (b) (c)

Fig. 2.6a. Alkaline phosphatase demonstrated with azo-coupling reaction in normal human jejunum. Strongly positive reaction in the brush border and lysosomes, weaker reaction in the Golgi apparatus of enterocytes and in the capillary endothelium. X 900.

Fig. 2.6b. Alkaline phosphatase in a part of an enterocyte of the normal human jejunum on the electron microscopical level. A positive reaction is on the outer side of cross-sectioned microvilli, in small invagination of plasma membrane, in pinocytic vacuoles and in a dense body (lysosome). X 3,800.

Fig. 2.6c. ATP splitting enzymes at pH 7.2 in normal human jejunum. A strongly positive reaction in the brush border of villi enterocytes and a weaker reaction in the brush border of crypt enterocytes can be seen. Capillary endothelium in the propria reacts also strongly. X 120.

remains to be answered to what degree nonspecific alkaline phosphatase and to what degree specific nucleoside triphosphatase (ATP-ase) is responsible for the reaction with nucleoside triphosphates. The varying intensity of reactions using equimolar concentration of ATP and glycerophosphate is suggestive of the presence of an ATP-ase. On the other hand the reaction with ATP is highly inhibited by L-phenylalanine similarly as the reaction with glycerophosphate (see also Pearse and Riecken [116]), a matter also referred to later. It is well known that the isolated brush border of the intestine possesses in addition to alkaline phosphatase at least one Na^+-K^+ dependent ATP-ase activity [117, 118] which is generally regarded as the *in vitro* expression of an *in vivo* functioning Na^+-pump (see Crane [89]). This is very important in relation to the Na^+-dependent active transport. It is possible that there are two ATP-ases differing in their localization within the brush border [74]. The possible relation of these ATP-ases to the 'ATP-ase' demonstrated *in situ* remains to be established.

The activity of alkaline phosphatase in the microvillous zone displays a stereospecific inhibition by 0.05 M L-phenylalanine, contrary to the activity in vascular endothelium which is more resistant to it. The results of the inhibition are dependent on the substrates and on the types of methods used for the demonstration. Our results in man confirm those obtained in rat enterocytes [119], L-histidine, L-tryptophan and L-arginine are also inhibitory. The activity of alkaline phosphatase is inhibited also by some fractions of gluten prepared by Dr. Hekkens in Leiden. It is not known whether the inhibition with L-amino acids and with gliadin has any physiological meaning. L-forms of amino acids originate by proteolysis in the lumen of the intestine from proteins of food and gliadin is cleaved by alkaline phosphatase.

The inhibition with L-phenylalanine may be used in the evaluation of bands in zymograms. In our experiments with human jejunum two bands of activity of alkaline phosphatase were demonstrated after electrophoretic separation in agar gel [120, 113] (see also Fig. 2.14). The cathodic fraction, which is more intense, is inhibited more by L-phenylalanine and the weaker anodic band is more resistant. This means that the cathodic fraction apparently comes from the brush border. The anodic fraction may be derived from the capillary endothelium. This conclusion is supported by the facts that (a) in biopsies from the stomach, where alkaline phosphatase is localized in the capillary endothelium, the cathodic fraction in zymograms is absent [120], and (b) in enterobiopsies of celiac sprue patients in which the activity of the brush border is diminished the cathodic fraction is reduced (Fig. 2.14d). There are species differences in the electrophoretic mobility of bands of alkaline phosphatase [121]. In the rat intestine two bands were demonstrated [122]. Two classes of alkaline phosphatase isoenzymes of the mouse duodenum which differ in immunochemical, electrophoretic, chromatographic and kinetic properties were described by Moog and co-workers in several papers which were then summarized [123]. The first class was found in suckling as well as adult mice whereas the second class is present in the duodenum of adult mice and accounts for the majority of alkaline phosphatase activity. Both of them are localized in the brush border of enterocytes covering the villi as was demonstrated by the immunofluorescence technique [124]. Four isoenzymes were described in the cock intestine [125] by chromatographic separation. Because the activity of alkaline phosphatase in the lysosomes, endoplasmic reticulum and Golgi apparatus of enterocytes is much lower than that of the brush border the cathodic fraction in zymograms of the human jejunal mucosa reflects the activity of the brush border and can be considered a measure of the digestive absorptive surface of the studied sample.

In the light microscope an exact localization of alkaline phosphatase in the microvillous zone is not evident because both in Gomori and azodye methods the zone appears to be stained diffusely (Figs. 2.6a, 2.14a). In the electron microscope the reaction product with both glycerophosphate and ATP as substrates is localized on the outer side of the cellular membrane (Fig. 12.6b). The reaction product can be detected also in invaginations between the microvilli, in pinocytic vesicles and in lysosomes. Our results are similar to those reported in the intestine of different animal species [126, 127, 128, 129, 130, 131, 132, 133]. On the other hand Oda and Seki [95] described the localization of magnesium-activated 'ATP-ase' in the inner leaflet of the microvillous membrane. The influence of various preparation procedures on the localization of the activity of phosphatases on the electron microscopical level in the intestine was not suffciently investigated. In the kidney it was shown that the localization of the reaction product on the inner or outer side of the membrane depends on the fixative and on the histochemical method used [134].

The biological function of alkaline phosphatase in enterocytes is obscure. Verzár and McDougal [53] suggested it might play a role in sugar absorption but this assumption was denied [31, 135], and Miller and Crane [7] suggested a digestive function for it because it catalyses hydrolysis of a wide variety of phosphorylated compounds which do not readily penetrate cell membranes, whereas their non-mineral moiety does [90]. A positive correlation exists between the activity of alkaline phosphatase in the cell membrane and the active transport across it [136]. This does not prove, however, any causal relationship. Some authors [137, 138, 139] suggest a relationship of alkaline phosphatase to the absorption of lipids because the intestinal alkaline phosphatase can be demonstrated in lymph and blood after fatty meals.

In the microvillous zone *aminopeptidease(s)* splitting leucyl-(methoxy)-β-naphthylamide, alanyl-β-naphthylamide, γ-glutamyl-β-naphthylamide but not benzoyl-arginyl-β-naphthylamide can be demonstrated. The activity splitting leucyl- and alanyl-β-naphthylamide was first demonstrated by Burstone and Folk [140] and then by many others. For a good localization in the microvillous zone some prerequisites are necessary [141], for otherwise a staining of the apical part of the enterocytes is inevitable. It cannot be decided whether this staining is entirely due to diffusion artifacts and whether enzymes present primarily in other localizations participate in this picture (see also Pearse and Riecken [116]). A similar distribution pattern in pig intestine was described by use of an immunofluorescence method [142]. Hopsu-Havu and Ekfors [143] described in the brush border of rat enterocytes a dipeptide naphthylamidase different from leucyl-β-naphthylamidase. In man this activity has not been described so far.

In zymograms of jejunal mucosa two fractions of aminopeptidase can be demonstrated [113], one anodic and one stronger cathodic. Owing to the fact that the cathodic band is rather weak in celiac sprue patients, which goes parallel with the diminishing of the brush border activity as demonstrated *in situ*, we suppose that the cathodic band originates in the brush border.

These localization studies are in agreement with biochemical findings [89, 144], localizing leucyl-β-naphthylamidase to the brush border membrane. Also γ-glutamyl transpeptidase was found in the brush border fraction [145].

The analysis *extra situm* gave much information on other peptidases which, however, cannot be demonstrated by the analysis *in situ*, e.g. the localization of significant amounts of tripeptide hydrolase activities and smaller amounts of dipeptide hydrolase activities in brush border [15].

The biological significance of the demonstrated leucyl-β-naphthylamidase activity lies probably in the hydrolysis of smaller peptides. Its relation to other peptide hydrolysing enzymes requires further study. For the γ-glutamyl transpeptidase a function in amino acid transport was suggested [146]. No reliable methods exist for the *in situ* demonstration of other enzymes and of various transport components demonstrated biochemically in the brush border fraction (see Crane [89]).

The microvillous zone thus represents a highly specialized cell surface with proteins, lipids and polysaccharides in which are bound various enzymes taking part in terminal digestion and in active transport. Crane [147, 89] suggested that these enzymes are arranged in a mosaic. In recent studies it was shown that through the action of papain a multienzyme particle containing maltase, sucrase and invertase is released. Further treatment leads to the liberation of leucyl-β-naphthylamidase whereas trehalase is bound more firmly [33]. Whether all enzymes are bound to the plasma membrane itself or whether some of them are a part of the glycocalyx remains to be proved unequivocally. To the surface coat pancreatic enzymes can be adsorbed so that digestion takes place in contact with the brush border [148]. The coat also contains specific receptors, e.g. for vitamin B_{12} in the distal ileum (see [149, 35]). And last but not least the coat represents a protective barrier keeping various noxious bacteria and particulate substances out of contact with the plasma membrane proper.

2.2.1.2 Lateral plasma membrane

The cellular membrane on the sides and on the base of enterocytes takes acid stains and gives a positive tetrazonium reaction and the reaction for SH-groups. It stains weakly by the PAS-reaction which is caused by polysaccharides. In electron microscopic cytochemistry the polysaccharide coat was demonstrated very clearly [68]. Of enzymatic activities only a weak activity using ATP as substrate can be detected on the lateral membrane. However, it was not assessed to what degree this positivity is caused by nonenzymic hydrolysis of ATP by lead ions [150]. After prolonged incubation the membrane stains also in the reaction for alkaline phosphatase. It is likewise difficult to decide to what degree it is a result of genuine activity and to what degree it is a diffusion artifact. Nevertheless it was suggested that this activity is important for the adhesivity of cells and for the transport of ions and water [58].

2.2.1.3 Terminal web
In the region of the *terminal web* which gives quite strong reactions for proteins no enzyme activity was demonstrated with the exception of small invaginations of plasma membrane and pinocytic vesicles in which an activity cleaving ATP and glycerophosphate can be demonstrated.

2.2.1.4 Lysosomes
Beneath the terminal web in a narrow zone (in humans, Fig. 2.7, and guinea pigs) or in a broader supranuclear part (rat, pig) *lysosomes*, particles $0.1-1.5\mu$ in size are located. A polysaccharide material can be demonstrated in them with the PAS reaction. Although ferritin can be found in these particles, particularly after massive administration of ferrous iron [151], the histochemical reactions for iron in the light microscope are usually negative. In appropriately prepared material the following enzymes were demonstrated in them. acid phosphatase (Fig. 2.7c, d), E600 resistant, relatively thermostable esterase (Fig. 2.7a), acetyl-β-D-glucosaminidase (Fig. 2.7e) and β-D-glucuronidase. This was demonstrated in the human intestine in the light microscope [49, 42, 45, 46, 47, 116, 152, 153, 154]. The activity of acid phosphatase can be well demonstrated in the electron microscope. It was first shown in animal material [155, 156, 157, 158, 130, 132], and later also in human material [159, 131, 43, 47]. It was not possible to demonstrate the activity of sulphatase in lysosomes with naphthol AS-BI sulphate, although the activity of this enzyme was clearly shown in lysosomes of animal and human enterocytes [158, 159] using p-nitro-catecholsulphate as substrate even on the electron microscopical level. Recently, the presence of acid β-D-galactosidase was shown in lysosomes of animal and

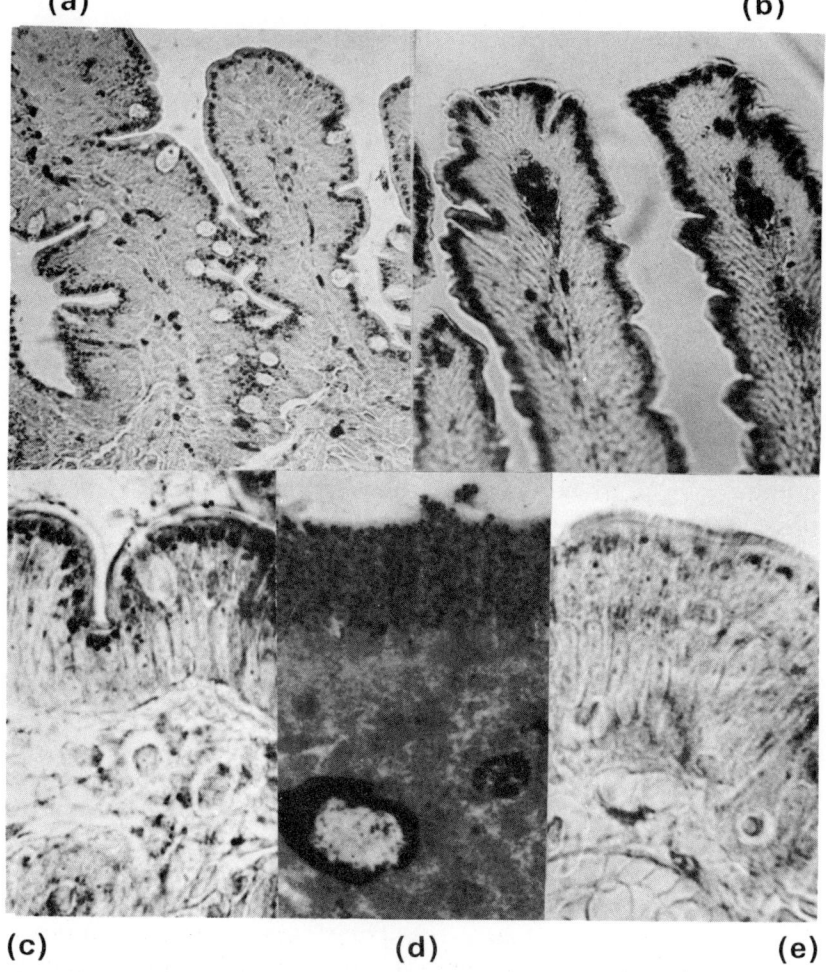

Fig. 2.7. Localization of some lysosomal enzymes.

(a) Thermostable esterase (azo-coupling reaction) in the normal human jejunum. Lysosomes of differentiated enterocytes and macrophages react positively. × 300.

(b) Acid β-D-galactosidase (indigogenic reaction) in guinea pig jejunum. Positive reaction in lysosomes of enterocytes and in macrophages, particularly at the tips of villi. × 300. (From Lojda [39], courtesy of Springer-Verlag.)

(c) Acid phosphatase (Gomori reaction) in normal human jejunum. Strong reaction in lysosomes of enterocytes, a weaker reaction in lysosomes of stromal cells. × 500.

(d) Electronmicrograph of (c). The reaction in lysosomes is clearly visible. The precipitate in the brush border is not due to the acid phosphatase. × 15,000.

(e) Acetyl-β-D-glucosaminidase in normal human jejunum. Reaction in lysosomes of enterocytes is clearly discernible. × 800. (From Lojda et al. [47], courtesy of Springer-Verlag.)

human enterocytes with the indigogenic method using 4-Cl-5-Br-3-indolyl-β-D-galactoside as substrate [39] (Figs. 2.4a, 2.7b). The enzyme of some species can also cleave β-D-fucoside. It has an activity even at pH 5.5—6, i.e. in the range of pH optimum of neutral β-D-galactosidase (lactase) so that at this pH both enzymes are visualized (see above). With glycosides of 6-Br-2-naphthol used in postincubation coupling a positive staining is obtained in the supranuclear zone of enterocytes with α-D-glucoside, α-and β-D-galactoside and a very weak reaction with β-D-glucoside. Although it is known that glycosidases splitting these substrates are also found in the lysosomal fraction [160], the demonstrated staining does not prove this localization, because it is not possible to obtain a correct localization by this type of method [36]. In lysosomes of human as well as animal enterocytes the presence of alkaline phosphatase can be also revealed both with azodye [45, 46, 47] and Gomori methods [128, 158, 159, 130, 131, 45, 46, 47] (Fig. 6a, b). In the human there may be some relationship of this activity to blood groups [138].

In this connection it should be mentioned that there are differences in the staining intensity of lysosomes in enterocytes and macrophages in the propria. The mutual ratios of activities differ for individual enzymes and for the same enzyme in various animals.

The arrangement of lysosomes in enterocytes and degree of enzyme activity in them change during the process of differentiation. This can be well shown in the human intestine: in crypt enterocytes lysosomes are smaller and dispersed in the supranuclear part of cells (Fig. 2.3c). Their enzyme activity appears weaker. The final arrangement is reached usually in enterocytes at the bases of villi. Lysosomes acquire enzyme activities probably from the Golgi apparatus or endoplasmic reticulum with the exception of alkaline phosphatase which is derived most probably from the cell membrane of the microvillous zone (see also Hugon and Borgers [130]).

At least two fractions of acid phosphatase and β-glucuronidase can be revealed in zymograms of human jejunal mucosa [113]. Contrary to our results in the vascular wall [44] those in human jejunum do not suggest a different contribution of individual cells to the pattern of lysosomal enzymes in zymograms.

In the lysosomal fraction of the animal intestine several acid hydrolases were found [17, 18, 19]. Pteroyl polyglutamate hydrolase, formerly known as folate conjugase, was discovered in this fraction recently [161, 162].

It must be borne in mind that lysosomal fraction is derived from different epithelial and also connective tissue cells so that the contribution of individual cells cannot be reflected by such studies.

The significance of lysosomes in enterocytes is not entirely clear. They were claimed to be connected with the absorption of fats [155], but the majority of fat droplets in enterocytes has no acid phosphatase activity [157]. The number of lysosomes containing ferritin increases after administration of iron [151] and this finding may be connected with a regulatory function of lysosomes in iron absorption. The position of lysosomes beneath the terminal web and their content of lytic enzymes suggest that they may take part in the storage and digestion of various substances which entered the cell by pinocytosis and phagocytosis. This hypothesis is supported by the fact that the number of lysosomes increases during infectious diseases of the intestine [56]. In our material we found an increased activity of acid phosphatase in lysosomes of enterocytes in patients with Whipple's disease [46, 47]. The significance of lysosomal acid hydrolases in connection with regressive changes of enterocytes in celiac sprue patients was suggested [152]. Their possible role in folate absorption [162] must be studied.

2.2.1.5 Mitochondria

Mitochondria of enterocytes are usually elongated with their axis running parallel to that of the cell and occupy the supranuclear, paranuclear and infranuclear zone. Their position in enterocytes is not unchangeable and may be connected with processes of absorption. With staining methods phospholipids are clearly revealed, particularly by the OTAN method [47], the intensity of which in mitochondria surpasses that of the brush border (Fig. 2.2c). Because the staining intensity is somewhat reduced after acetone extraction some quantity of hydrophobic lipids (cholesterolesters, triglycerides) is responsible also for the staining. From the reactions for protein amino acids coupled tetrazonium reaction and the reaction for SH-groups are remarkable. Concerning the lipid and protein composition the analysis *in situ* cannot furnish such information as the analysis *extra situm* (for references see Korn [163]).

The following enzymes can be localized in mitochondria of enterocytes [45, 46, 47]: succinate dehydrogenase (SDH), menadione activated α-glycerophosphate dehydrogenase (αGPDH), both tetrazolium reductases (high activities), β-hydroxybutyrate dehydrogenase and cytochrome oxidase (medium activities). Usual histochemical techniques also localize in the mitochondria the other dehydrogenase activities which are very high: lactate dehydrogenase (LDH), malate dehydrogenase, isocitrate dehydrogenase, glucose-6-phosphate dehydrogenase (GPDH), and 6-phosphogluconate dehydrogenase. The exclusive mitochondrial localization of these enzymes, however, is artificial, because these enzymes are soluble and the usual histochemical techniques always localize the last step of the system, i.e. the respective tetrazolium reductase. This, of course, in enterocytes has a mitochondrial localization [164, 47]. The activity of dehydrogenases in enterocytes in the human intestine is not evenly distributed: more intensive reactions are found in the apical part of the cells rather than in the infranuclear region (Fig. 2.8a, b). These differences are not so marked in all species. The gradient in the human enterocyte is apparent not only when using the usual detection technique (via tetrazolium reductase) but also when gel media with phenazine methosulphate are used [164], so that tetrazolium reductase is not a rate limiting factor in the detection. The distribution of formazan in enterocytes is more diffuse, when using gel media. Higher activity in the apical part of the cell is apparently connected with higher requirements for energy. Differences in the distribution of dehydrogenases are also found on the tissue level and in the case of LDH are demonstrable by both techniques: activities in differentiated enterocytes on the sides and tips of villi are higher than the activities at the bases of villi and in crypts. G6 PDH is distributed more evenly, showing a high activity also in crypts (Fig. 2.8c). The activity of enterocytes is higher than the activity in other kinds of epithelial cells or cells of connective tissue of the propria, and so enterocytes contribute to the total activity of dehydrogenases in homogenates of normal intestinal mucosa in the greatest degree.

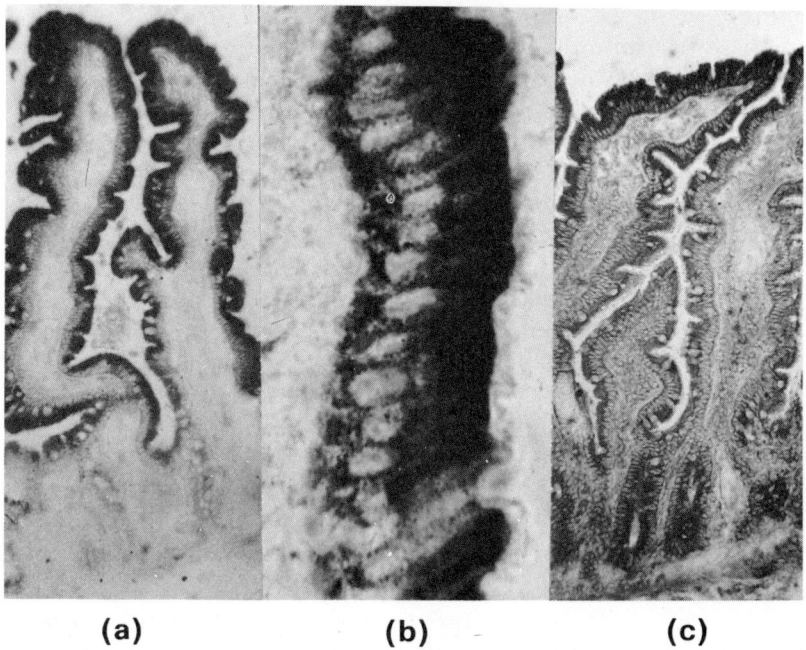

(a) **(b)** **(c)**

Fig. 2.8a. Reaction for $NADPH_2$ tetrazolium reductase in normal human jejunum. The reaction in enterocytes covering the villi (particularly in the supranuclear part) is more intense than in crypt enterocytes. × 100.

Fig. 2.8b. Reaction for menadione-activated (mitochondrial) α-glycerophosphate dehydrogenase in enterocytes of a villus side of normal human jejunum. The reaction in the supranuclear part is much stronger than at the cell basis. × 600. (Compare with Fig. 2.2c.)

Fig. 2.8c. Reaction for glucose-6-phosphate dehydrogenase (gel medium Lojda [164]) in normal human jejunum. The reaction in crypt enterocytes and in Paneth cells is also high. × 100.

In the detection of LDH isoenzymes using gel media [164] with iodo-nitrotetrazolium, after separation of homogenates of human jejunal biopsies by electrophoresis in agar-gel and densitometric evaluation [165, 113], LDH_5 comprises 48%, LDH_4 22%, LDH_3 16%, LDH_2 10%, and LDH_1 4% of the total activity. An example of a zymogram of LDH in normal intestinal mucosa is given in Fig. 2.16c. By calculating according to the tetramer hypothesis [113], the 'M' subunits account for about 75%, and the 'H' subunits for 25% of the total activity. By using gel medium for LDH with inhibitors (pyruvate for inhibition of 'H' subunits, and 50% acetone or urea for inhibition of 'M' subunits), substrate analogues (α-hydroxybutyrate and α-hydroxyvalerate instead of lactate) or a coenzyme analogue (acetyl pyridine NAD instead of NAD) it is possible to detect these basic types of LDH *in situ* on the cellular level [165a]. It was demonstrated that both in enterocytes in crypts and on villi both basic types of LDH are present. Differentiated enterocytes at the sides and tips of villi are somewhat richer in 'M' subunits than the crypt cells.

From this it is evident that in enterocytes there is a high activity of the glycolytic cycle, the Krebs cycle, and the pentose cycle. The energetic yield is needed for active absorption and proliferation. Judging by the relatively low activity of β-hydroxybutyrate dehydrogenase the oxidation of fatty acids in enterocytes does not seem to be as high as, e.g., in the liver or myocardium. The oxidation of amino acids does not seem to be high either. These data are in agreement with biochemical studies which brought more information on the economy of the enterocytes.

In enterocytes on the apex and sides of villi there is a high activity of monoamine oxidase. The activity in the human enterocytes is higher than in liver cells and probably serves to detoxicate monoamines which are formed in the lumen of the intestine. In the rat, in which the activity of all enzymes is very high, the activity of monoamine oxidase is lower [166].

Only quantitative differences in dehydrogenases and oxidases among mammals were reported [166, 167]. A decreasing gradient of activity of NADH tetrazolium reductase from the duodenum to the ileum in the rat, guinea pig, and rabbit was found [166]. Strong alcohol dehydrogenase activity was demonstrated in enterocytes [168] and the distribution of 17-β-hydroxysteroid dehydrogenase was described in detail [169]. Additional information on dehydrogenases is available elsewhere [171, 172, 173].

2.2.1.6 Golgi complex

A well developed *Golgi complex* is located immediately above the nucleus and consists of flattened cisternae and vesicles. PAS-positive substances other than glycogen are found in the Golgi region. Also hydrophobic lipids were described here, especially after administration of fat [174, 116]. In enterocytes of rats and mice in the Golgi apparatus a high activity of alkaline phosphatase can be detected [175, 176, 177]. Hugon and Borgers [130] demonstrated the activity of alkaline phosphatase electron microscopically in Golgi cisternae and vesicles of enterocytes in the rodent duodenum. The activity of alkaline phosphatase in man is much lower [47]. The same holds true for thiamine pyrophosphatase and nucleoside diphosphatase. The activity of thiamine pyrophosphatase was demonstrated in Golgi cisternae of mouse [180] and rat [180a] jejunal epithelium electron microscopically. In mice, acid phosphatase was shown in the Golgi saccules electron microscopically [157]. We were not able to detect it electron microscopically in mature enterocytes in man, although Hugon and Borgers [131] describe the activity in Golgi vesicles of enterocytes in the human duodenum.

From experiments with labelled glucose and other substances which appear in a very short time after administration in the Golgi region it has been inferred that the Golgi region is connected with the synthesis of glycoproteins [178, 63]. Also lysosomal enzymes are packed into primary lysosomes here [152, 153]. The accumulation of fat particles in this region points to its participation in fat absorption. Its exact nature is not known, however. Przelecka *et al.* [179] reported a response of Golgi-associated alkaline phosphatase to fat feeding in enterocytes of the honeycomb moth (*Galleria mellonella*).

2.2.1.7 Endoplasmic reticulum
The endoplasmic reticulum of differentiated enterocytes is both agranular (without ribosomes) and granular (with ribosomes). The ribosomes which also occur free are responsible for the staining with basic dyes (azure and pyronine are most popular) observable in the light microscope which can be removed by ribonuclease. The quantity of RNA in differentiated enterocytes is not high. The following enzymes can be demonstrated in the endoplasmic reticulum of enterocytes:

(a) *Glucose-6-phosphatase,* a biochemical marker of microsomes which are derived mainly from endoplasmic reticulum, is in enterocytes quite strong although weaker than in liver cells. It is localized mainly in the supranuclear part of mature enterocytes and the difference between enterocytes in crypts compared to those on the sides and apices of villi is very distinct. Also the electron microscopical localization was shown [180, 180a]. The activity of this enzyme was found biochemically in the microsomal fraction of the animal intestine [10, 181]. Because it catalyses also phosphotransferase reactions a role in a specialized mechanism for glucose absorption was suggested for it [181].

(b) Also *alkaline phosphatase* was demonstrated in numerous smooth reticulum profiles, especially in the rat [130]. The function of alkaline phosphatase in endoplasmic reticulum is unknown.

(c) Concerning the activity of nucleoside diphosphatase we confirmed earlier observations [155, 127] on its presence in the supranuclear part of the rat enterocyte. Its activity in the human enterocyte is much weaker than in the rat. Its function is unknown. The relationship of this enzyme to glucose-6-phosphatase is to be studied.

(d) The endoplasmic reticulum of human enterocytes displays a high activity of *nonspecific esterase* splitting thiolacetic acid [37]. The reaction is positive even in endoplasmic reticulum at the base of the cells. In the light microscope it can clearly be seen that the activity is higher in the upper part of the enterocytes than at their bases. The distribution of nonspecific esterase when using thiolacetic acid has in the light microscope a similar diffuse character as when other substrates are used, i.e. α-naphthyl acetate, -propionate, -butyrate, 5-Br-indoxyl acetate, -propionate, -butyrate, naphthyl AS acetate. It is very probable that enzymes splitting these substrates are also localized in the endoplasmic reticulum. Since the used substrates are split by a number of enzymes (nonspecific esterase, cholinesterases and even some peptidases, see Pearse [50]) it is important to differentiate these reactions with the aid of inhibition tests. Following the application of 10^{-4} M eserine the reaction is weakened, but the diffuse character of the reaction in enterocytes is unaffected. The weakening is due to the presence of cholinesterase. Its contribution to the staining is not high, however. Following inhibition with diethyl-p-nitrophenyl phosphate (E 600) the diffuse character of the reaction is suppressed and only the reaction in lysosomes is preserved which is otherwise more apparent only when substrates derived from naphthols AS are used. Esterases localized outside of lysosomes are E 600-sensitive in nature. We observed that with the increasing length of the fatty acid the staining was weaker [47]. This is at variance with the findings in the rat [182]. Using naphthol AS-nonanoate as substrate and taurocholate as an activator, Abe et al. [183] found that 'lipase' is distributed similarly as nonspecific esterase. The older literature on nonspecific esterase is carefully reviewed by Arvy [184].

Even the reaction for nonspecific esterases is more intensive in mature enterocytes on the apex and sides of villi than in crypts. The reaction in enterocytes is more intense than in other cells and contributes the most to the activity of nonspecific esterase in homogenates.

In zymograms of homogenates of human jejunal mucosa after separation in agar-gel a series of 3—4 cathodic bands and 1 anodic band appears. The position and intensity of these bands depend on the substrate used [120, 113].

Eserin inhibits to some degree the anodic and the first two cathodic bands. After E 600 one to two cathodic (the slowest) and one anodic band persist. Thus there are at least two E 600-resistant esterases in the human jejunum. Further studies will be required to produce more data on the characteristics of these enzymic activities and show their localization *in situ*. Five bands of 'nonspecific esterase' in the rat intestine were revealed with starch gel electrophoresis [182]. The exact physiological role of the demonstrated activities is not known. It is interesting that an increase in esterase activity, largely confined to two molecular fractions, occurs in the intestinal lymph following a meal of corn oil [185]. This points to a role of this enzyme in fat absorption.

Biochemical studies on isolated fractions demonstrated that the endoplasmic reticulum is the most important subcellular structure for glyceride biosynthesis [186]. The endoplasmic reticulum is also significant for the transport of electrolytes, lipids, glucose and amino acids [56]. Ribosomes are sites of proteosynthesis.

2.2.1.8 *Nucleus*

The nucleus of the enterocytes is oval and is situated usually in the lower half of the cell. Deoxyribonucleic acid is demonstrable in the nucleus along the nuclear membrane and is dispersed in fine grains. In the nucleus are found one or two nucleoli containing ribonucleic acid. By histochemical methods no hydrolase or dehydrogenase activities can be explicitly demonstrated in the nucleus. The staining seen after Gomori type methods is an artifact.

2.2.2 Undifferentiated cells

Enterocytes arise in crypts by division of *undifferentiated cells* which differ in many parameters from enterocytes [54, 55]. They are then displaced towards the apex of villi and during this time they differentiate. Reaching the extrusion zone at the tip of the villi they desquamate into the gut lumen. This was demonstrated in animals with the aid of isotope labels [187]. The time required for complete renewal of epithelium is 2 to 5 days [188] and is dependent on the age and species of the animal and on many other factors [188, 189]. During maturation, ultrastructural and cytochemical changes take place in enterocytes. Of the ultrastructural changes it is the formation of regular microvilli and of the terminal web, a decrease of free ribosomes and an increase in the agranular endoplasmic reticulum. These ultrastructural changes are reflected cytochemically [190, 116, 46, 47]. The microvillous zone can be appreciated according to the thickness of the PAS positive band and according to the reactions for alkaline phosphatase, leucyl-β-naphthylamidase, and glycosidases. Lactase is the most important enzyme for practical purposes in the human intestine. Evaluation of the terminal web is made possible by an increased staining for some protein amino acids (SH-groups). The decrease of ribosomes is reflected by the decrease of pyroninophilia or basophilia removable by ribonuclease. Lysosomes, which are small and widely scattered in the supranuclear part of the crypt enterocytes in man (Fig. 2.3c), become arranged in a narrow band beneath the terminal web (Fig. 2.7). This can be well shown by the reactions for lysosomal enzymes. The intensity of reactions for dehydrogenases and oxidases increases and the differences of activities as compared with crypts are conspicuous. On the other hand the activity of enzymes of the Golgi complex (at least in man) decreases. There are also changes in enzymes connected with nucleoprotein metabolism [25, 191]; these cannot be studied *in situ,* however.

2.2.3 Goblet cells

Their electron microscopy [54, 57] and cytochemistry [192,

116, 193, 47] have been well reviewed. The cell filled with secretory granules is goblet-shaped. In the cytoplasm at the base of the cell, paranuclearly and also supranuclearly are mitochondria (granular and filamentous), giving reactions for proteins and phospholipids and similarly as in enterocytes reactions for cytochrome oxidase, and dehydrogenases, which are weaker. The pyroninophilia and basophilia digestible with ribonuclease is located infranuclearly and above the nucleus beneath the granules. Of the hydrolytic enzymes E 600-sensitive esterase has a weak diffuse activity. In the supranuclear zone is a well developed Golgi complex. In this area incorporation of ^{35}S-sulphate begins [194, 193]. The activity of thiamine pyrophosphatase and nucleoside diphosphatases is low. Above the Golgi complex there are electron dense bodies giving a reaction for acid phosphatase [195], E 600-resistant esterase, and acetyl-β-D-glucosaminidase [47], β-D-galactosidase [39] and so having the character of lysosomes. Most of the supranuclear zone consists of secretory granules of 'mucinogen'. They give a PAS reaction which is resistant to digestion by amylase, is blocked by acetylation and is restituted upon saponification. It is therefore caused by vicinal glycol groups of mucosubstances. Most of the granules also stain intensively with alcian blue and Hale's reaction. This can be prevented by methylation. After demethylation the colouring is variably restored. These reactions, together with the not entirely alcohol-resistant metachromasia with toluidine blue or with azure, suggest the presence of carboxylic groups and sulphate groups in mucosubstances. In the elaboration of secretory material neutral mucosubstances precede the appearance of acidic mucosubstances. The character of mucosubstances shows some species differences and changes also along the small intestine (increase of acidic character in aboral direction). Reactions for protein amino acids (tyrosine, histidine, SH-groups) are weakly positive in plasmatic septa between the granules. Proteins can be demonstrated indirectly by a slight increase in basophilia following brief proteolysis [196].

Secretion is autonomous, but can be influenced by various substances and the nervous system [197]. At least some goblet cells in the intestinal crypts are capable of proliferation. They may be derived from undifferentiated and transitional cells [198].

Mucus, secreted by goblet cells and mucus-secreting duodenal glands, the character of which displays great species differences [192], covers the mucosa with a thin layer, protects it against mechanical insults and lubricates its surface. No findings, with the exception of that of Geyer and Nietzold [199], indicate that secretion of the goblet cells actively takes part in the digestive processes.

2.2.4 Paneth cells

Paneth cells are pyramid-shaped cells found on the bottom of crypts where they form groups. Their electron microscopy has been described [54, 55, 57]. A complete review of the information up to 1936 is given by Patzelt [200] and their cytochemistry is described for the rat [201], for the mouse [65] and for the human [47]. Reactions for proteins in the cell membrane are very weak and from membrane enzymes only a positive reaction is obtained when ATP is used as substrate. The round or oval nucleus containing usually one nucleolus is found at the base of the cell. In the infranuclear part of the cell is a very well developed granular endoplasmic reticulum (ergastoplasm) which corresponds to a distinct pyroninophilia which can be removed by ribonuclease. Mitochondria are present chiefly in the basal part of the cell where a high activity of all dehydrogenases demonstrated in enterocytes can be detected. In the supranuclear zone is a well developed Golgi complex which has activities splitting nucleoside diphosphates and thiamine pyrophosphate. These are much lower in man than in the rat. Above the Golgi complex are lysosomes giving a reaction for acid phosphatase and E 600-resistant esterase. The supranuclear zone of the cell is filled with characteristic strongly birefringent granules which have a high affinity for acid stains (basic nature of the proteins which is confirmed also by a positive reaction for arginine) and are also stained by the Gram method. The granules give a strongly positive coupled tetrazonium reaction and a reaction for tyrosine and tryptophan. No distinct positivity of the reaction for SH-groups was obtained in man, although in mouse and in rat this reaction is clearly positive. The granules are only weakly stained with PAAS reaction (peracetic acid-Schiff), suggesting that disulphide groups are not present in large amounts. In the mouse a layer containing sulpho- and carboxy mucins can be demonstrated on the surface of granules [65]. Such a layer cannot be found in man. Granules are weakly PAS-positive and this positivity is blocked by acetylation which shows its glycide nature [192]. In granules in the apical part of Paneth cells we were not able to demonstrate any enzymatic activities. The granules contain a protein-polysaccharide complex [201, 65, 116]. In the granules of rat and some other animals large amounts of a heavy metal (zinc) demonstrable both with dithizone and sulphide silver methods were demonstrated [202, 201]. Novikoff [202a] considers the granules to be derivatives of lysosomes. Because we demonstrated the activity of acid phosphatase, E 600-resistant esterase, and acetyl-β-D-glucosaminidase only in granules located in a narrow zone above the Golgi apparatus and sometimes amidst secretory granules but not in the secretory granules themselves we do not consider the granules to be lysosomal in nature (see also Behnke and Moe, 203).

Although the Paneth cell is a typical example of a secretory cell there exist only few exact data about the physiological role of its secretory product. The older hypothesis such as a role in the digestion of cellulose, fats, in the elaboration of secretin, erepsin, enterokinase, in the regulation of pH are summarized in Patzelt [200]. Secretion changes depending on food intake were also described [204, 205] and Selzman and Liebelt [206] claimed that the secretion contains a dipeptidase. The assumption that Paneth cells may participate in the secretion of lysosomal enzymes into the lumen of the small intestine [201] does not seem to be justified. Creamer [207] suggested a trophic function for the differentiating crypt cells.

The first proved enzymic content of Paneth cell granules was published by Speece [208] who found cytochemically a lysosyme activity in them. This finding was corroborated biochemically in an isolated fraction of mouse intestine containing Paneth cell granules [209]. This finding can have a functional significance in relation to the bacterial population in the intestine and this aspect must be further studied.

2.2.5 *Endocrine cells (E cells)*

These cells are dispersed singly in Lieberkühn crypts amidst Paneth and undifferentiated cells. Sometimes they can be found even between enterocytes on the sides and tips of villi. They are known under several names which are not always synonymous:

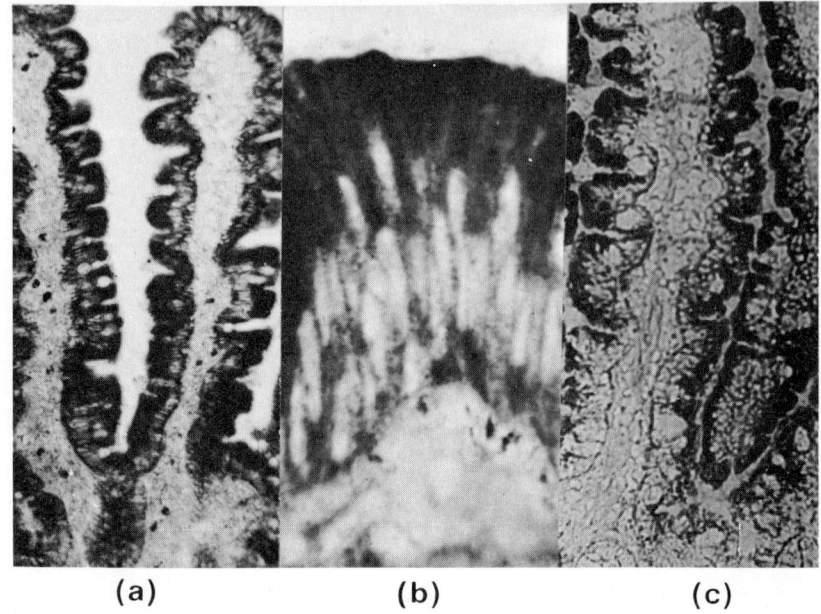

(a) (b) (c)

Fig. 2.9a. Nonspecific esterase (azo-coupling reaction) in normal human jejunum. Strongly positive reaction in the cytoplasm of enterocytes covering the villi (particularly in their supranuclear part), somewhat weaker reaction in crypt cells. In the propria some macrophages react positively. X 120.

Fig. 2.9b. A detail of the preceding picture. X 850.

Fig. 2.9c. The reaction for glucose-6-phosphatase in the normal human jejunum. Note the strong positivity of the supranuclear part of villi enterocytes. X 140. (From Lojda et al. [47], courtesy of Springer-Verlag.)

enterochromaffin, Kultschitzki, argentaffin, argyrophil, basigranular, APUD-cells. Their cytochemical properties were reviewed recently [210]. Progress in their understanding was reached in recent years mainly due to the contributions of three research groups, i.e.: (a) the Pavia group [211, 212, 213, 214, 215, 216, 221, 222] (b) the London group [217, 218] and (c) the Geneva group [219, 220, 220a]. The terminologic differences between these groups were largely overcome at the Merck conference, 'Origin, Chemistry, Physiology and Pathophysiology of the Gastrointestinal Hormones', held in Wiesbaden in 1969 [220b]. At least three types of endocrine cells can now be identified electron microscopically in the small intestine of studied mammals, including man.

2.2.5.1 EC cells

Enterochromaffin cells (EC cells, argentaffin cells) in the electron microscope (Fig. 2.10a, b) have granules which are highly variable in shape (round, oval, discoid, crescent-shaped,

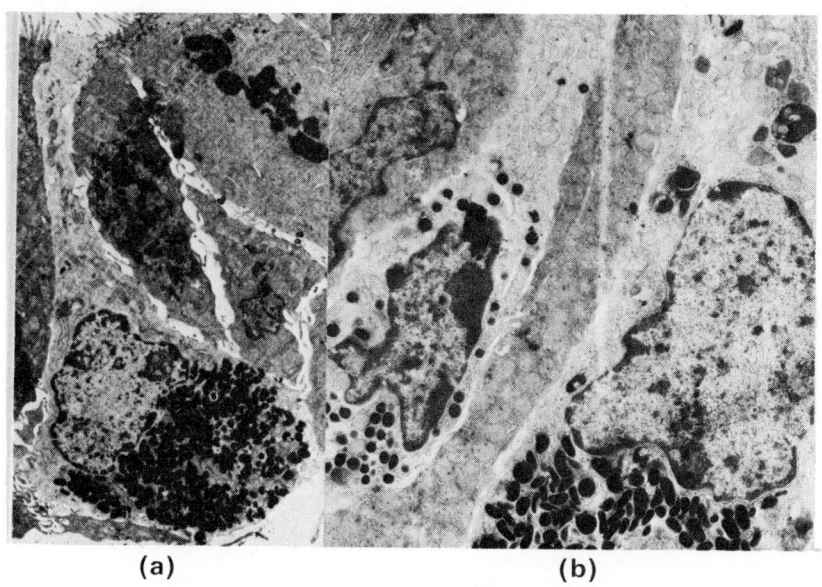

Fig. 2.10a. Enterochromaffin cell of the normal human jejunum. Note microvilli, lysosomes of the supranuclear part and specific granules of various shape in the infranuclear part. × 3500.

Fig. 2.10b. Two endocrine cells in the same section of a crypt of the normal human jejunum. The cell on the left is an L cell; the cell on the right is an EC cell. × 12,000.

biconcave, elongated), about 400 mμ in diameter. They are usually highly osmiophilic, more than those of other E cells, but sometimes less osmiophilic material can be found in them. The supranuclearly situated Golgi complex and rough endoplasmic reticulum are not specially remarkable. Free ribosomes and polysomes and small mitochondria are scattered in the cell body. Lysosomes are found mostly in the apical and supranuclear part of them. Although Carvalheira et al. [217] stated that these cells do not reach the lumen, we found in the human jejunum that it can be so (Fig. 2.10a). The cells then have well developed microvilli which are smaller than those of enterocytes. Their coat is not so developed as in enterocytic microvilli.

The specific granules which are concentrated infranuclearly give a positive diazo reaction after aldehyde fixation, and Gibb's indophenol reaction. They are argentaffin and of course argyrophilic, give a positive xanthydrol reaction, are stained dark blue by lead haematoxylin and display masked purple metachromasia. They also give strong yellow fluorescence after condensation of freeze-dried material with formaldehyde vapours. There is no doubt that the granules contain serotonin. This may be not the only product, however. Because these cells share some reactions of polypeptide secreting cells, Pearse et al. [218] suggested that they might also produce a polypeptide incretin. The enzymatic equipment of EC cells in the human jejunum was studied by Lojda and Frič [41] and Lojda et al. [47]. They found lysosomal enzyme activities (acid phosphatase, E 600-resistant esterase, acetyl-β-D-glucosaminidase, β-D-galactosidase) in the apical, supranuclear and sometimes even the infranuclear part of cells (Fig. 2.11b, c), diffuse

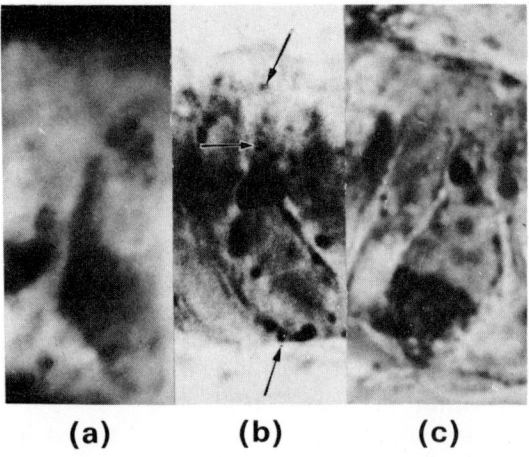

Fig. 2.11a. Reaction for some enzymes in EC cells of the human jejunum. (a) Thiamine pyrophosphatase in the Golgi apparatus. (b) Acid phosphatase in lysosomes (arrows). (c) E 600-resistant esterase (one lysosome in the supranuclear part is visible). × 1,500. (From Lojda and Frič [41], courtesy of Springer-Verlag.)

activity of nonspecific esterase (a part of this activity was due to nonspecific cholinesterase) and activities of dehydrogenases similar to those in enterocytes but of lower activity.

According to the pattern of lysosomal enzymes it was suggested that EC cells receive some substances from the lumen which could perhaps influence the formation or release of the secretory product [41].

2.2.5.2 L cells
The L cells (Lg cells, argyrophilič cells) have specific granules also located infranuclearly. These granules are more regular than those of EC cells and their profile is mostly round. In the human jejunum they are smaller and somewhat less electron-dense than the granules in EC cells (Fig. 2.10b). The Golgi complex is also located supranuclearly. Arrays of well organized rough endoplasmic reticulum are scarce. Lysosomes are usually less numerous than in EC cells. Free ribosomes and polyribosomes are rather numerous and are responsible for pyroninophilia removable with ribonuclease. Mitochondria are more numerous than in EC cells. These cells were reported to reach the lumen very often. In such cases they are then also provided with microvilli.

The granules are stained with lead haematoxylin, are argyrophilic and display masked metachromasia. They do not give a positive diazo reaction. After the performance of the reaction for menadione-activated α-glycerophosphate dehydrogenase in the human jejunum a very high activity appears in about one-half of E cells giving a strong nonspecific esterase staining [47]. We assume that the majority of these cells giving the reaction for α-glycerophosphate dehydrogenase belong to the L cells.

Carvalheira *et al.* [217] stated that these cells have a very high diffuse reaction for nonspecific cholinesterase and nonspecific esterase which should be negative in EC cells. In the human jejunum we found also a diffuse reaction in EC cells so that the differences are only quantitative.

The cytochemical characteristics suggest that the L cells produce polypeptides. Orci *et al.* [220] suggested that in the rat they produce enteroglucagon. Pearse *et al.* [218] suggest on the basis of a close structural relationship between glucagon and secretin that a case can be made out for connecting these cells with secretin production. Vassallo *et al.* [222] assume that L cells in the cat produce cholecystokinin-pancreozymin.

2.2.5.3 S cells

The S cells (Sg cell) can be differentiated clearly from the L cells only electron microscopically. They differ from the L cells in that their granules are very small. They are argyrophilic and display usually a blue staining with toluidine blue after treatment with HCl. They are stained very probably also by lead haematoxylin. The diazo reaction is negative. Also these cells belong to polypeptide-producing cells. Vassallo *et al.* [222] suggested a working hypothesis that these cells might produce secretin. On the other hand Pearse *et al.* [218] assume that S cells produce cholecystokinin-pancreozymin.

The distribution of E cells along the small bowel shows regional and probably also species differences. For man the ratio of the three endocrine cell types in the duodenum was EC : L : S = 1 : 4 : 1, in the jejunum 4 : 1 : 1 [218]. Vassallo *et al.* [222] did not find any S cells in the cat jejunum. In 20 human jejunal biopsies investigated so far for E cells we found several S cells only. Whereas in the cat duodenum the L cells were four times less numerous than EC cells, in the ileum they were the predominant type of E cells [222]. Although great progress has been made in this field much work is needed to settle the function of E cells unequivocally.

2.2.6 Lamina propria

Cells of the *lamina propria,* in addition to blood and lymph vessels and intact innervation, are also very important for the normal course of absorptive processes. The following cells are found in the lamina propria: reticulum cells, macrophages, fibroblasts, plasma cells, lymphocytes, eosinophils and smooth muscle cells. *Macrophages* are cytochemically most interesting. They have relatively high activities of acid phosphatase, β-glucuronidase, acetyl-β-D-glucosaminidase and acid β-D-galactosidase. The activity of nonspecific esterase can also be detected. The activities mentioned are localized in lysosomes but extralysosomal staining also occurs. We wish to draw attention to the differences in activities of lysosomal enzymes in enterocytes and macrophages which are not the same in individual enzymes and differ also in individual species [42, 47]. This is a good example of the importance of the *in situ* detection of enzymatic activities for the correct interpretation of data.

Macrophages phagocytose particular substances which reach the propria, and therefore they are more numerous at the tips of villi where enterocytes perish and where therefore the chance for the penentration of heterologous substances is the greatest (Fig. 7b). In these macrophages the activities of lysosomal enzymes are the highest.

Debray *et al.* [223] judge that macrophages are important in the metabolism of iron which can be released from them depending on the needs of the organism. In our material no iron could be detected in macrophages of the normal human mucosa [47].

Plasmocytes are found even in human jejunal mucosa in close contact with macrophages. Deane [224] suggested that heterologous material is processed in the macrophages and subsequently passes on to plasmocytes where it gives information about the need for antibody formation. By the immunofluorescence technique IgA immunoglobulin in large amounts, IgM immunoglobulin in smaller amounts and IgG immunoglobulin in the smallest amounts were demonstrated in plasmocytes of the normal human jejunal mucosa [225, 226, 227]. As immunocytes are absent from the fetal intestine and in germ-free animals they were not detected, it was concluded that IgA immunocytes owe their existence to local antigenic stimulation [227].

Just beneath the epithelium are blood capillaries, the endothelium of which gives reactions for alkaline phosphatase which is more resistant to the inhibition by L-phenylalanine and a strongly positive staining when nucleoside tri- and diphosphates are used as substrates. In the walls of lymphatic (chylous) vessels no characteristic enzyme activities were detected.

The structures described lie in the ground substance, the amorphous part of which can hardly be demonstrated in usual paraffin sections. In thicker cryostat sections metachromasia with azure A is visible which is stronger along the basement membranes. It is not alcohol resistant. This suggests the presence of hyaluronic acid; sulphated mucosubstances are probably not present in greater amounts.

2.3 CYTOCHEMISTRY OF ENTEROCYTES DURING DEVELOPMENT

In the development of the intestinal functions in mammals three main periods can be distinguished: (a) the prenatal period during which the intestine is preparing to take over the role of absorbing nutrients, (b) the early postnatal period (suckling) during which nutrition depends on mother's milk, (c) later postnatal period (weaning) during which the transition from breast milk to other foods occurs.

It is obvious that a combined morphological, biochemical and histo-(cyto-)chemical investigation is a prerequisite for a better understanding of intestinal functions even in these periods. Most work has been carried out in experimental animals. As far as the human intestine is concerned more data from the period of postnatal development require to be collected. In the present article it is not possible to review the work done in this field completely. Only an outline restricted to enterocytes can be given. The reader has to consult also the monographs of Koldovský [228] and Vollrath [229] or the quoted papers for more detailed information on this subject.

Although there are substantial species differences concerning the length of individual periods, the degree of development at the time of birth (some mammals are born with a more mature intestine than others, e.g. guinea pig as against rat) and differences which are the basis for those in adult animals, some common traits can be found. These traits are: (a) different behaviour of individual parts of the intestine, (b) the influence which is exerted upon development by hormones of the suprarenal cortex in the perinatal period, (c) the influence of food composition and bacterial flora on the pattern of the intestine in the postnatal period.

2.3.1 Microvillous zone

The morphological differentiation of the *microvillous zone* of enterocytes covering the villi is attained just in the prenatal period. Although the glycocalyx is somewhat less distinct than in enterocytes of adult animals it is well visible (see Vollrath, 229). The enzymatic equipment of this zone in the perinatal period is somewhat different from that in adult differentiated enterocytes.

(a) From *glycosidases* a high activity of lactase (higher than in adult animals) is present in all mammals except two species of sea-lions [230]. This is related to the content of lactose in mother's milk. On the other hand the remaining brush border glycosidases are either absent (e.g. rat, mouse), or their activity is lower than in adult animals (e.g. guinea pig, man). Almost all data were obtained by biochemical analysis. On superficial observation of data presented by Koldovský [228] one could gain the impression that the activity of lactase is higher in the ileum than in the jejunum. These higher values are not due to brush border lactase but to the lysosomal β-D-galactosidase which also splits lactose at pH 5.5. When this activity is inhibited by p-chloro-mercuribenzoate [231] the activity of lactase is lower in the ileum than in the jejunum. This can be well shown *in situ* by improved indigogenic methods [38, 39, 48]. The differences between the activity in ileum and jejunum are not as great as in adult animals. With these methods it was demonstrated that the activity is localized in the brush border of enterocytes covering the villi. Near the top of villi a broader zone of a weaker staining is sometimes observed than in adult animals. The greater extent of this zone, which corresponds to the extrusion zone, is most probably due to a lower migration rate of cells from crypts towards villi which was demonstrated by Koldovský *et al.* [232] in suckling rats. The findings *in situ* amplify the observations of Nordström *et al.* [30] obtained by biochemical analysis of cryostat sections cut parallel to the surface. The activity of lactase was demonstrated in the brush border also with the immunofluorescent method [88].

With the beginning of weaning a sharp decrease of lactase activity is observed (see Koldovský, 228). Continued feeding of lactose past the weaning period fails to prevent the postnatal decline of lactase activity unless high doses of lactose are given [233]. While the activities of lactase decrease those of other disaccharidases increase. This biochemical observation was corroborated by the demonstrations *in situ* of maltase [229]-sucrase and trehalase-Lojda, unpublished results. These changes in disaccharidase activities during the weaning period can be experimentally induced by cortisone or influenced negatively by adrenalectomy.

(b) *Alkaline phosphatase* displays a different development in individual parts of the small intestine [234]. The proximal—

distal gradient of alkaline phosphatase apparent in adult animals was reported to be inverted in certain developmental stages [235], so that the maximal activity was found in the ileum. It remains to be ascertained to what extent the activity in lysosomes and vacuoles of enterocytes and to what extent that of the brush border participate in this pattern. The fine localization of alkaline phosphatase in the microvillous zone on the electron microscopical level is similar to that in the adult animal [235a]. The activity of alkaline phosphatase during the development of the mouse intestine was studied very extensively by Moog and her group (see Moog et al. [123]). They found in infant mouse duodenum a low activity distributed uniformly over the surface of the villi, which hydrolyses β-glycerophosphate more rapidly than phenylphosphate. Between the 11th and 14th day the activity increases, particularly when assayed with phenylphosphate. With the beginning of weaning there was a sharp rise of activity demonstrated with phenylphosphate, especially at the tips of the villi, and a gradient of activity rising from the base to the tip became established. The activity of alkaline phosphatase can be influenced by cortisone the effect of which was reported to be different in different parts of the intestine: in the duodenum it causes an increase, in jejunum a slight increase and in ileum a decrease of activity [236]. Also the administration of actinomycin, puromycin and cycloheximide increases the activity of alkaline phosphatase in duodenum in the second and third week of life [237, 238].

(c) The activity of *leucyl-β-naphthylamidase* was demonstrated in the prenatal period of rats later than alkaline phosphatase [229]. After birth a decrease of activity was observed followed by a sharp increase in the weaning period. A similar increase could be induced with hydrocortisone earlier.

(d) Activities of *dipeptidases* as studied biochemically were correlated with the degree of differentiation of microvilli [239]. This correlation is insufficient for the localization.

2.3.2 Lysosomes and related particles

The development of lysosomes and related particles has been described in the rat [229, 235a, 240] and in man [241]. In these particles a high activity of acid phosphatase and E 600-resistant esterase [242, 243, 244, 240, 235a, 245, 229,

246] was observed. A high activity of acid β-galactosidase and β-glucuronidase was demonstrated in them by indigogenic methods [39, 40]. Also Williams and Beck [246] reported a high activity of β-glucuronidase in these particles. The presence of these hydrolases and also of arylsulphatase and acid deoxyribonuclease in the lysosomal fraction of suckling rat small intestine has been demonstrated [247]. Much higher activity of all these enzymes resides in enterocytes of the ileum than in the jejunum. This is in agreement with biochemical data [228]. Although there are some biochemical data on the influence of cortisone on the postnatal development of activities of lysosomal enzymes, a detailed histochemical study is lacking.

In lysosomes and in vesicular and vacuolar apparatus which develops in enterocytes during suckling alkaline phosphatase is also localized [242, 243, 240, 235a, 246]. The activity of this enzyme is also found in invaginations of the brush border membrane.

2.3.3 *Endoplasmic reticulum*

Endoplasmic reticulum of enterocytes, both agranular and granular, increases during the prenatal and early postnatal periods. On the other hand the number of free ribosomes and polysomes, responsible for the pyroninophilia removable with ribonuclease, decreases.

(a) *Glucose-6-phosphatase* was not studied so extensively in mammals as in the chick embryo [248]. It was found in enterocytes covering the villi particularly in their supranuclear part. The distinction was similar to glycogen which occurs in enterocytes of fetal intestine of all animals in rather large amounts. The activity of glucose-6-phosphatase increased before hatching. It is interesting that it could be influenced by insulin.

(b) *Nonspecific esterase* was studied more extensively also in mammals. It was demonstrated histochemically in fetal intestine [249, 243, 229] and its activity increases towards birth. A further marked increase was found in the weaning period. The distribution pattern is roughly similar to that in the adult animal, the proximal–distal gradient in the weaning period is similar to that of adult animals [228]. A part of the activity is due to cholinesterase. The influence of cortisone on the activity of this enzyme has been studied [250].

2.3.4 Mitochondria

The changes of *mitochondria* are reflected in changes of enzymes localized in them. Also other dehydrogenases which are not exclusively mitochondrial or are predominantly extra-mitochondrial display changes in their activity. These were studied histochemically by Kubát et al. [251], and Vollrath [229]. The activities of all *mitochondrial oxidases and dehydrogenases* increase during fetal life. At birth the overall activity is lower than in adult animals and the difference on the histological level (higher activity in villi enterocytes than in crypt enterocytes) and those on the cellular level are less conspicuous than in adult animals. They become more expressed in early postnatal life, however. During this period the activity of most dehydrogenases increases with the exception of malate dehydrogenase, lactate dehydrogenase, α-glycerophosphate dehydrogenase and β-hydroxybutyrate dehydrogenase which were reported not to increase [251] or even to decrease temporarily after birth [229]. No gradient in activities was found. All dehydrogenases were demonstrated with aqueous media in the system of the respective tetrazolium reductases and no attempts were made to eliminate the diffusion of enzymes and limiting influence of tetrazolium reductases. This has to be done. It is interesting that cortisone induces a precocious development in the pattern of dehydrogenases as well. Hence cortisone induces not only changes of digestive enzymes but also of those connected with the metabolism of the enterocyte.

2.4 ABSORPTIVE CELLS OF THE INTESTINAL MUCOSA IN EXPERIMENTAL CONDITIONS

Much experimental work has been carried out to elucidate (a) the mechanism of absorption of individual nutrients, water and electrolytes, (b) to determine the influence of food composition and its amount and of the damage to the intestine by radiation, antimitotic drugs, antibiotics and other substances on, (c) the functional capacity of enterocytes, and (d) to understand the mechanism of changes in absorption in endocrine disorders.

It is beyond the scope and possibilities of this chapter to give even an outline of all known changes which were reported in various animals in various experimental conditions. Only some examples are presented.

2.4.1 Changes during absorption of nutrients and water

It must be emphasized that there are great differences in the pattern between infant and adult animals. In adult animals no distinct morphological and cytochemical changes during *carbohydrate and amino acid absorption* were reported (see Trier [55] and Merker [58] for references). During water and electrolyte absorption dilation of intercellular spaces, lymphatics and capillaries was described and the possible role of ATP-ase in the lateral membrane in absorption was suggested [252].

Absorption of lipids attracted much more attention due to the ease with which the lipid material can be demonstrated both in light and electron microscopy. The interpretation of these findings is difficult, however, owing to the difficulties in the more subtle identification of lipid material *in situ*. Lipid absorption was studied by light microscopy in the human intestine [174] and droplets of hydrophobic lipids were observed in the apical cytoplasm of enterocytes within 3 minutes after administration of fat. Several minutes later fat droplets were found in the Golgi region and after 30 minutes between enterocytes and in the lamina propria. The entrance of lipids into the cell was a matter of some discussion. The role of pinocytosis emphasized by Palay and Karlin [252a, b] and others has now been denied by Shiner *et al.* [252c], Phelps *et al.* [252d] and Vodovar and Planzy [252e] (see Toner [5] for details). Nowadays it is accepted that molecules of the micellar solution, arising by the action of pancreatic lipase on fats in the presence of bile acids in the lumen of the intestine, enter the enterocytes. In the endoplasmic reticulum triglycerides are resynthetized, and with the participation of Golgi apparatus transported in the endoplasmic reticulum through the cell. The various enzymes shown biochemically to participate in the lipid absorption and resynthesis cannot be demonstrated *in situ*. The possible role of alkaline phosphatase and nonspecific esterase was mentioned in a previous section as was also acid phosphatase.

In contrast to adult animals pinocytosis plays an important role in lipid absorption in the perinatal period [228, 229]. After the first intake of the mother's milk the enterocytes at the tips of villi, particularly in the jejunum, are filled with lipid droplets which have the tendency to coalesce, the largest ones being in the Golgi region. The staining reactions show that nonacidic lipids are involved [253]. Chromatographic analysis [254, 255] showed that a high amount of triglycerides is present, the fatty acids of which have shorter chains as in the mother's milk. Changes in enzymes activities after the fat load were reported in rats [229]. Alkaline phosphatase, nonspecific esterase, β-hydroxybutyrate dehydrogenase, glucose-6-phosphate dehydrogenase, succinate dehydrogenase increased, whereas leucyl-β-naphthylamidase, α-glycerophosphate dehydrogenase and glutamate dehydrogenase decreased.

In *protein absorption* there are also pronounced differences between suckling and adult animals in that the former absorb intact proteins (see Koldovský [228], Kraehenbuhl and Campiche [245] and Chapter 9 for references). There are marked differences between the absorption rate of individual proteins and the time during which this phenomenon can be observed is different in different species (it was correlated with changes in alkaline phosphatase activity [256]). Species differences are also apparent in the cytochemical pattern of enterocytes. Proteins enter the enterocytes by pinocytosis [257, 245, 229, 240]. They are then transported across the epithelium in vacuoles or trigger a lysosomal response, and are trapped and digested in lysosomes. This is accompanied by an increase of activities of acid phosphatase and of other lysosomal enzymes.

2.4.2 The effect of diet

Because enzymes are decisive in the determination of functional capacities of enterocytes some of the recent work has centred on *the effect of diet* on them. The 'demand-and-supply' relationship was not always found.

The influence of *diets with high and extremely low carbohydrate content* on disaccharidase activities was studied biochemically in animals and man (see Rosenweig and Herman [258], for a review). It was found that sucrase and maltase are readily adaptable to an increased content of carbohydrates in the diet whereas lactase is not. No parallel histochemical and electron microscopical studies were carried out.

It is noteworthy that Prosper *et al.* [259] failed to demonstrate ultramicroscopical changes in absorptive cells and a decrease of disaccharidase activities in young rats deprived of dietary proteins and express some doubts whether the observations on the morphological and biochemical changes in kwashiorkor are due to protein malnutrition only. Similar results were reported [260] concerning the brush border and its enzymes. On the other hand the activity of acid phosphatase and succinate dehydrogenase was reported to be significantly reduced.

The changes are very marked when the caloric deprivation is of a greater degree. Knudsen et al. [261] observed a decrease of activities of sucrase, maltase, lactase and alkaline phosphatase during periods of total caloric deprivation of patients whereas the mucosa appeared normal in the light microscope. Berkel et al. [262] reported significant reduction of activities of lactase, maltase and invertase in the jejunal mucosa in infantile malnutrition.

Of the vitamins the effect of *vitamin D* was studied [263, 264]. It was found that it caused an increase of activity of brush border alkaline phosphatase and Ca^{2+} and Mg^{2+} stimulated ATP-ase, whereas other brush border enzymes remained unchanged.

The mucosal pattern is influenced also by the quantity of ingested food. Lojda and Fábry [265] described an increase of alkaline phosphatase determined quantitatively *in situ* with Lojda's method [266] in the brush border and Golgi apparatus of enterocytes apparent in the whole small intestine (the greatest increase was recorded in the jejunum) of rats in intermittent starvation. Also the activity of nonspecific esterase was increased. Basically similar changes were reported in animals fed high-bulk diet [267].

2.4.3 *The action of hormones*

An important field of work is the elucidation of *the action of hormones* on the digestive and absorptive functions and metabolism of the small intestine. This topic was recently reviewed by Levin [268]. The influence of thyroid gland, islets of Langerhans, adrenal gland and hypophysis was studied.

The 'enteroctrophic activity' [269] of thyroxine was studied by Wall et al. [270]. Hypertrophy was apparent in the intestinal mucosa, particularly in the jejunum. The enterocytes displayed an increased height of microvilli and activities splitting ATP and thiamine pyrophosphate in the alkaline range and glycerophosphate and naphthol AS-BI phosphate in the acid range. The activity with the latter substrates was reported unchanged in the alkaline pH range. This unusual behaviour, if corroborated, would indicate some changed properties of alkaline phosphatase caused by thyroxine. The activity of lysosomal and mitochondrial enzymes was unchanged with the exception of monoamine oxidase which was decreased.

The effect of alloxan diabetes, insulin and glucagon on various functions of the intestine was reported but parallel histochemical studies were not carried out.

The influence of the adrenal cortex on the intestine have been studied very extensively. Adrenalectomy causes the decrease of activity of alkaline phosphatase [271, 234, 272] and 'ATP-ase' [273]. The activities of these enzymes are restored by cortisone. Chiquoine [274] did not find any histochemical evidence for changes in glucose-6-phosphatase. Levin et al. [275] reported a reduction of maltase and peptidase and Rotgers et al. [276] a reduction of sucrase. The profound influence of cortisone on the absorptive cells of suckling animals was mentioned in section 3.

The effect of hypophysectomy which causes an atrophy of the intestine in many species (see Levin [268]) is complicated. It is certainly caused through the lack of other hormones, but a direct action of some hypophyseal hormones, e.g. somatotrophin, must be also considered. No pertinent histochemical studies were carried out.

2.4.4 The injury by X-rays

Walsh [277] draws attention to the injury of the small intestine by X-rays and since this time radiation damage to the intestine has been subject of many studies. There are some species differences and also differences dependent on the age of the animal. The picture of the intestine is dependent on the dose of the X-rays.

The histological changes in mammals have been well documented both with the light microscope [278, 279, 280, 281] and the electron microscope [282, 283, 129, 284, 285]. The sensitivity of the intestinal epithelium is understandable because radiation interferes with the physiological regenerative processes in the intestine. A high dose of X-rays leads to an arrest of mitoses of crypt cells followed by regressive changes so that the cells perish. They desquamate and denudation of the propria occurs which is connected with loss of electrolytes, water and proteins. When the dose is not lethal in crypts so-called 'omega' defective cells appear, and later they are replaced by healthier cells and the epithelial lining is restored.

The ultrastructural changes are represented by changes in mitochondria, alterations of microvilli, changes in nuclei, formation of cytolysomes, and loss of ribosomes. The histochemical studies were carried out in rats by Jonek *et al.* [286], who reported a great increase of acid phosphatase and E 600-resistant esterase in enterocytes attaining the maximum in animals with intestinal symptoms. Ansari *et al.* [287] described in the guinea pig also an increased activity of β-glucuronidase.

Spiro and Pearse [288] investigated histochemically the intestine of irradiated mouse and studied the activity of many enzymes. They emphasized the resistance of enzyme activities to direct damage by radiation and recorded the dynamic changes in enzymes displaying increased activity, i.e. nonspecific esterase and its E 600-resistant part soon after irradiation followed by an increase of aminopeptidase and monoamine oxidase. Afterwards an irregular distribution pattern of enzyme activities and later their decrease was observed. Otoupal and Kunstýř [289] studied histochemically the intestine of irradiated pigs and emphasized a considerable decrease of alkaline phosphatase activity in lysosomal localization which may point to a decreased pinocytosis. A decrease of nonspecific esterase was also observed.

Fine structural localization of phosphatases in X-irradiated mice [129] showed a transitional increase of acid phosphatase activity localized in small vesicles 'emitted' from the Golgi zone which then fused in lysosome-like bodies and clustered into cytolysomes. Activities of 'ATP-ase' and thiamine pyrophosphatase were decreased. Despite the differences in the mechanism of action of X-rays and actinomycins on the cell their effect on the small intestine is strikingly similar [290]. More combined studies are needed to establish the definite development of enzymatic changes in absorptive and other cells and their relationship to changed functions of the intestine.

2.4.5 The effect of antibiotics, colchicine and triparanol

Investigations of the effect of antibiotics (neomycin, tetracyclin, terramycin, penicillin) on the intestine are scanty. The disruption of microvilli has been described [291, 292], while biochemical studies have shown a slight decrease of disaccharidase activities in the mouse intestine [293]. A similar decrease was reported in the rat using both biochemical and histochemical methods [294]. The interaction of antibiotics with resynthesis of triglycerides in enterocytes was also pointed out [295]. Hence antibiotics appear to cause a transient disaccharidase deficiency and to interfere with lipid absorption. Both are restored after the cessation of treatment.

The effects of colchicine have been examined because of its ability to arrest mitoses in metaphase. Histochemical investigations showed a decrease of activity of many dehydrogenases [172]. Alkaline phosphatase was also reported to be reduced [296]. Activities of disaccharidases are significantly decreased [297].

The changes in the intestine after administration of triparanol which causes the atrophy of villi [298, 299] were studied by Riecken et al. [300] in the rat. These authors found only minor histological changes which contrasted with a striking generalized alteration of the cytochemical pattern of enterocytes: decrease of activities of brush border enzymes, decrease of activities of lysosomal enzymes and the appearance of the enzymatic reactions in the Golgi zone, a slight decrease of activities of dehydrogenases and a marked decrease of thiamine pyrophosphatase in the Golgi apparatus. These changes correlated well with the decreased active transport of glucose.

2.5 CYTOCHEMISTRY OF THE HUMAN JEJUNAL MUCOSA IN MALABSORPTION SYNDROME

The analysis of enterobiopsies of patients with malabsorption syndrome has furnished data which help not only in a better understanding of the malabsorption but also in the diagnosis. The electron microscopical findings were reviewed recently [56, 54, 55], and also the histochemical and cytochemical findings [301, 46, 47]. In these articles more detailed information on the topic and references can be found.

2.5.1 Celiac sprue

Celiac sprue, (celiac disease of children, nontropical sprue of adults, gluten enteropathy, primary malabsorption syndrome).

Changes in the intestinal mucosa depend on the stage of the disease and on treatment. Even when the small intestine is diffusely affected the severity of changes decreases aborally. The characteristic morphological feature is flattening of the mucosal surface caused by the changed villi which are very short and broad and sometimes not visible at all (so-called subtotal and total atrophy). With the light microscope, deepening of crypts, thickening of the basement membrane and in the lamina propria edema and enhanced cellularization consisting of macrophages, plasmocytes, lymphocytes and eosinophils are found beside the described changes of villi. In later stages of the disease *fibrotization* appears and capillaries are removed from the epithelium.

Attention was focused particularly on the absorptive cells the morphology and cytochemistry of which depend on their position in crypts and on the surface [302, 47].

The most conspicuous changes in the enterocytes are as follows:

Their microvillous zone is usually reduced, microvilli are irregular (Fig. 2.1c) and shorter. The terminal web is imperfectly developed. An increased number of free ribosomes and imperfectly developed smooth endoplasmic reticulum can be observed. Changes of mitochondria which are swollen and shorter were also described. Very severe changes are found in enterocytes which border directly with the lumen where they are lower. Their often pycnotic nuclei are found in varying heights giving the impression of a pseudostratified epithelium. In the cytoplasm are numerous vacuoles which in frozen sections, at least in part, are filled with hydrophobic lipids. Mitochondria are swollen and cisternae of the endoplasmic reticulum are greatly dilated.

The described changes have their cytochemical correlation. The band of polysaccharides in the brush border region is narrowed. A weaker staining of protein amino acids points to the defect of the terminal web. Most prominent are enzymatic defects which are sometimes more marked than morphology alone can suggest.

The activity of all disaccharidases is decreased. Lactase and trehalase are the most affected enzymes (Figs. 2.12, 2.13). In our comparative biochemical, histochemical and morphological study [87] we found very good agreement between histochemistry and biochemistry provided that in the biochemical

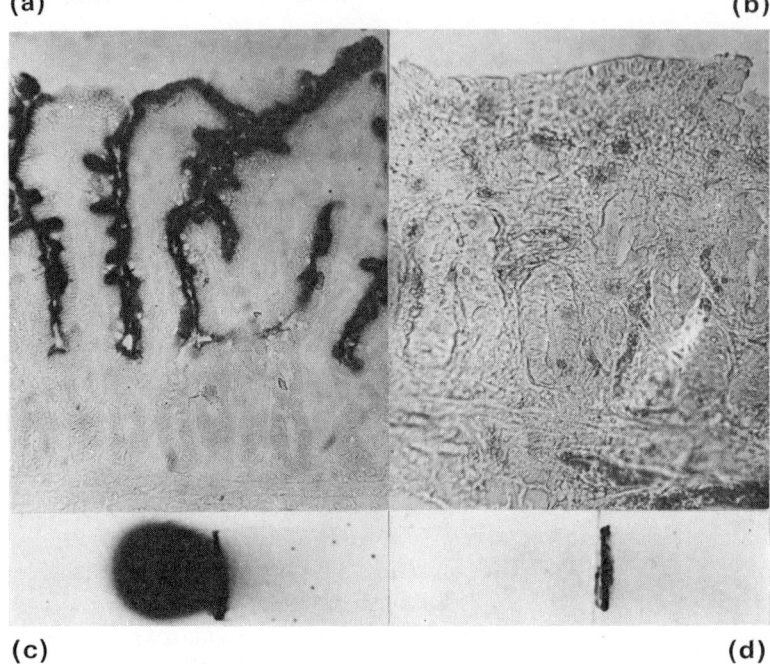

Fig. 2.12. Indigogenic reaction for lactase (according to Lojda and Kraml [48]) in the normal human jejunum (a) and in jejunum of a patient with celiac sprue (b). × 80. (c) and (d): Zymograms of lactase (method of the coupled oxidation of glucose according to Lojda [36]) in the same samples. Cathode is to the left. Note the negative reaction in the enterocytes of celiac sprue patient (b) and corresponding weakening of the cathodic band (d). (Fig. (a) and (b) from Lojda and Kraml [48], courtesy of Springer-Verlag.)

determination the activities of disaccharidases were assayed in complete homogenates and not in supernatants as has been mostly done. The reason for it is described on page 000.

In all our patients with active disease we observed a decrease in the activity of *alkaline phosphatase* (Fig. 2.14) which has been sometimes described as mostly unchanged [170, 171, 303, 301]. In zymograms there is a defect of cathodic fractions of the mentioned enzymes (Figs. 2.12d, 2.14d) [120].

CYTOCHEMISTRY OF ENTEROCYTES 99

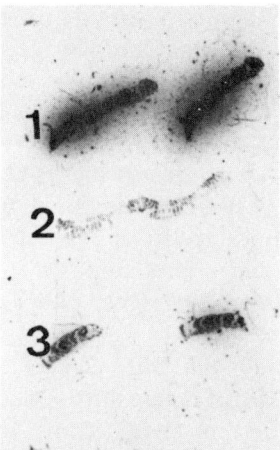

Fig. 2.13. Macrophotograph of a slide with three pairs of cold microtome sections of normal human jejunum (1), florid celiac sprue (2) and celiac sprue after a gluten-free diet (3) in which the reaction for trehalase according to Lojda [114] was performed. The degree of the black colour corresponds to the activity of trehalase.

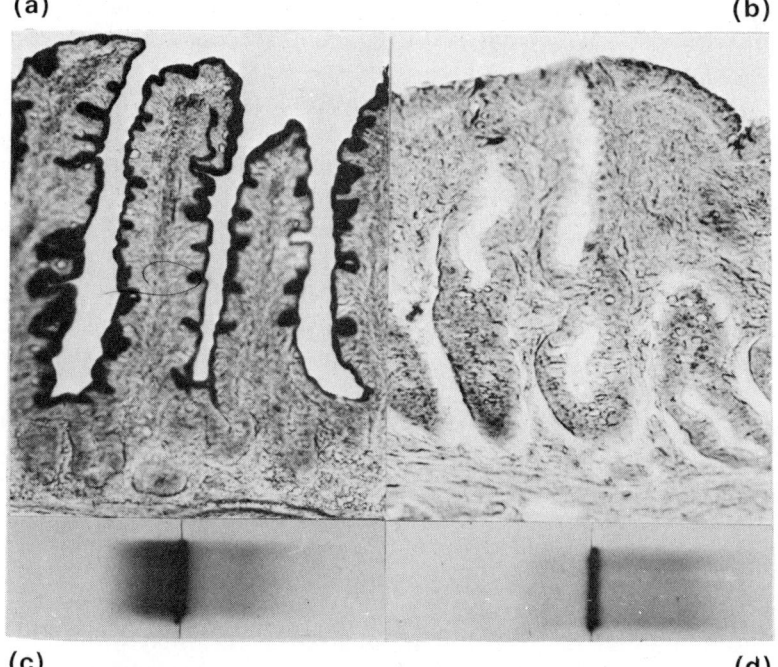

Fig. 2.14. Reaction for alkaline phosphatase (Gomori) in normal human jejunum (a) and in the jejunum of a patient with florid celiac sprue (×100) (c) and (d): Corresponding zymograms (cathode to the left). Note the strongly decreased staining of the brush border (b) which is reflected by a great diminution of the cathodic band (d).

A weaker staining than in control subjects was obtained also when ATP was used as substrate. Similarly, as in the normal jejunum, it is very difficult to decide to what extent the positive results are due to nonspecific phosphomonoesterase and to what extent to an ATP-ase.

Schenck *et al.* [304] point out a constant defect (decrease or disappearance) of 'ATP-ase' activity in untreated celiac sprue patients. Although the activity of this enzyme in enterocytes of patients with celiac sprue in comparison with that of normal enterocytes was reduced, we never observed a complete defect of this activity.

The activity of *leucyl-β-naphthylamidase* was also lowered in all our patients with florid celiac sprue. Likewise a defect of this enzyme was confirmed in zymograms in which the cathodic fraction is reduced in patients with celiac sprue [113]. In celiac sprue patients a lowered activity of *peptidases* studied biochemically with many substrates was reported [305, 306, 307].

Lysosomes of enterocytes and their enzymes (acid phosphatase, E 600-resistant esterase, acetyl-β-D-glucosaminidase, β-glucuronidase and acid β-D-galactosidase) were also affected. Activities of all these enzymes in the surface epithelium were decreased and in some patients with florid disease were virtually absent (Fig. 2.15b). In less severely affected patients the activities were present and lysosomes appeared coarser (Fig. 2.15a). This could be shown best in frozen sections of fixed specimens whereas in cryostat sections of unfixed biopsies the activity was always more diffuse in character and the decrease of activity was not always so expressed [46, 47]. As was emphasized in our paper [47] it cannot be decided whether this pattern reflects the actual state *in vivo* (damaged lysosomes and release of enzymes) or whether it is an artifact produced by the procedure used. In the case of fixed sections the extralysosomal activity might be suppressed by fixation and only lysosomal activity preserved.

In the upper part of the crypts and in some cases also in the surface zone the distribution of lysosomal enzymes was changed. The granules either occupied the supranuclear part of the cell similarly as in normal crypt enterocytes (Fig. 2.3c) or showed bizonal distribution (beneath the terminal web and in the Golgi zone, Fig. 2.15a). The activity of the Golgi zone was seen particularly in the case of acid phosphatase. This pattern

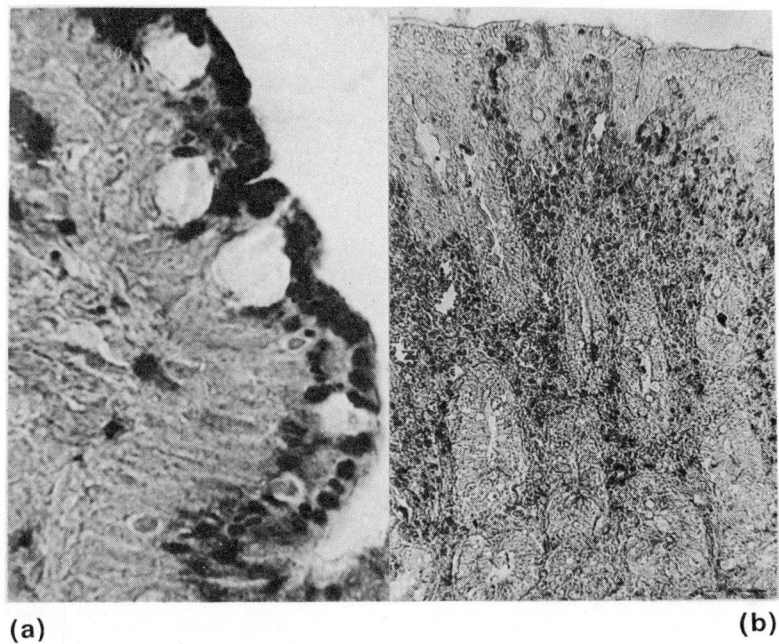

(a) (b)

Fig. 2.15a. Reaction for acid phosphatase in the enterocytes of a reduced villus of a patient with celiac sprue. Note bizonal distribution of enzyme activity (Golgi zone and coarser lysosomes at the cell apex). X 800.

Fig. 2.15b. Reaction for β-glucuronidase in a patient with untreated celiac sprue. The reaction in enterocytes is very weak, whereas macrophages in the propria react very strongly. X 140. (From Lojda et al. [47], courtesy of Springer-Verlag.)

points to an enhanced production of these enzymes [308]. The changes in lysosomal enzymes were considered in connection with damage of these organelles which release highly destructive enzymes leading to cellular autodigestion in the superficial layer [152]. It remains to be decided whether this damage is primary or secondary in nature. No constant changes were recorded in zymograms of lysosomal enzymes [113].

The activities of thiamine pyrophosphatase and nucleoside diphosphatases in the *Golgi apparatus* of celiac sprue enterocytes in crypts and sometimes also in the surface epithelium were somewhat increased. This fact points to a higher activity of this organelle [45].

Smooth *endoplasmic reticulum* is not as developed and its cisternae may be dilated and contain droplets of hydrophobic lipids. The activities of glucose-6-phosphatase and nonspecific esterase are lower than in normal enterocytes. The decreased enzyme activities contrast with an increased amount of ribonucleic acid mainly in free ribosomes (Fig. 2.17b).

The changes in *mitochondria* have their counterpart in changed activities of oxidases and dehydrogenases. Although we did not find any qualitative defect in the activities of the oxidoreductases studied [46, 47] there were quantitative changes (decrease of activities) in all cases. These were remarkable, particularly in the case of monoamine oxidase. It is striking that there are not such great differences in the activities of cells in crypts and on the surface which are so characteristic of the normal intestine.

Even the activities of *dehydrogenases* which are *not firmly bound to structures* so that their correct intracellular localization cannot be assessed (LDH–Fig. 2.16a, b—malate dehydrogenase, glucose-6-phosphate dehydrogenase and 6-phosphogluconate dehydrogenase) are reduced.

Constant changes were found in zymograms of LDH (Fig. 2.16c, d) [165, 113]. In comparison with zymograms of normal mucosa a significant shift in favour of H subunits was found. Of the total activity of LDH only 25% goes to LDH_5 (normally 48%), 30% to LDH_1 (normally 22%), 22% to LDH_3 (normally 16%), 17% to LDH_2 (normally 10%) and 6% to LDH_1 (normally 4.8%). Analysing the basic forms of LDH *in situ* [165a] we found that enterocytes are responsible for this shift.

These findings show that the enterocyte in celiac sprue is defective in its digestive capacities towards disaccharides and peptides, in detoxication abilities, in its energetic economy, and in its transport capacity.

The nature of mucosubstances produced by *goblet cells* the number of which was sometimes increased [47] was either unchanged or a somewhat greater amount of more neutral mucosubstances was found than normally. No cytochemical changes were found in *Paneth cells* [47].

Great variation is observed in the quantity of E cells. In about half of our patients with celiac sprue an increased number of enterochromaffin cells was found [41, 47] often with increased activities of lysosomal enzymes. This was considered as one of the reasons for an increased serotoninemia found in

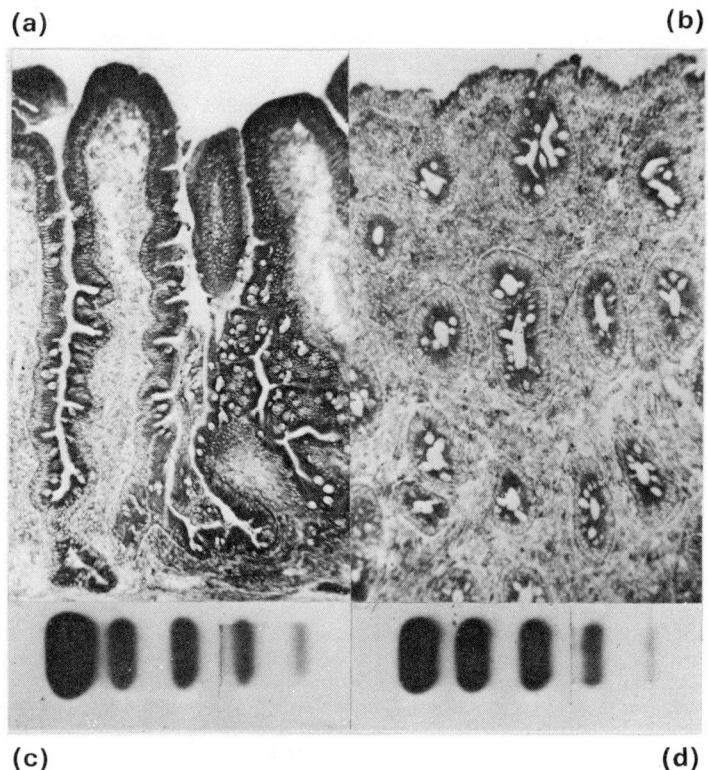

Fig. 2.16. Reaction for lactase dehydrogenase (gel medium according to Lodja [164]) in normal human jejunum (a) and in jejunum of a patient with untreated celiac sprue (b). X 120. (c) and (d): Corresponding zymograms (cathode to the left). The reaction in enterocytes of the celiac sprue patient is weaker and differences in the intensity between enterocytes of crypts and villi are not so marked (b). The weakening of the 5th band accompanied by an increase of staining in the 4th and 3rd bands is well apparent (d).

these patients. A more detailed analysis of all E cells in celiac sprue patients is badly needed.

Macrophages of the propria, especially in the upper part of it, in the neighbourhood of the lumen, have a very high activity of lysosomal enzymes (acid phosphatase, β-glucuronidase—Fig. 2.15b—acetyl-β-D-glucosaminidase, acid β-D-galactosidase) and E 600-resistant esterase. Of the dehydrogenases the activity of α-glycerophosphate dehydrogenase is quite high [47].

Plasma cells are usually larger than normally with a conspicuous amount of ribonucleic acid. In some of them a rather high activity of acid phosphatase can be shown.

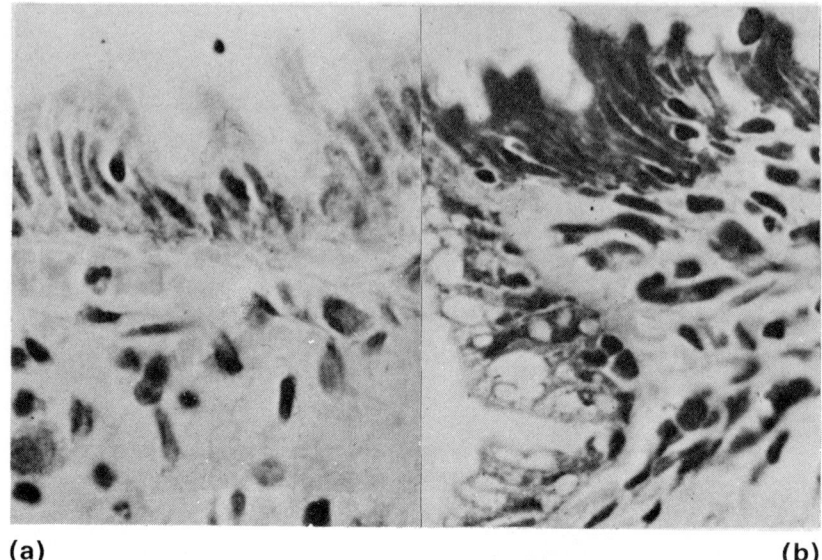

(a) (b)

Fig. 2.17. Reaction with methyl-green pyronine in enterocytes of a villus side of the normal human jejunum (a) and in enterocytes near the surface and on the surface of the mucous membrane of a patient with celiac sprue (b). In celiac sprue the cytoplasmic staining of enterocytes due to RNA is much stronger. Regressive changes are apparent, in enterocytes (lipids can be detected in the majority of vacuoles). X 700. (From Lojda et al. [47], courtesy of Springer-Verlag.)

There are several reports on observations that there is a histological, ultrastructural and cytochemical recovery of the jejunal mucosa after a *gluten-free diet* [47] and after corticosteroid therapy [47, 309].

Activities of some enzymes (membrane enzymes, enzymes of the endoplasmic reticulum, mitochondrial enzymes) increased rather quickly (two weeks of treatment) and this amelioration was correlated with a rapid clinical response. It should be noted, however, that the restitution of all cytochemical parameters was not reached and that enzymes in some locations improved and in others they did not. Of the cytochemical parameters an increased quantity of cytoplasmic ribonucleic acid in enterocytes, a double-layered appearance of lysosomes and decreased activity of lactase persisted most often [47]. We did not observe any absolute normalization in adult patients. Only in 3 children this recovery was absolute [47]. It is doubtful whether the diet which our patients received was absolutely free of gluten and on the other hand whether some abnormalities could not be found in normalized cases reported in the literature when the spectrum of methods applied should have been broader. Because the cytochemical parameters, changed in celiac sprue, improve after a gluten-free diet and after corticosteroid therapy the recorded changes are most probably secondary in nature.

Detailed knowledge of the pathogenesis of celiac sprue is obscure. Although the role of gluten is well known the mechanism of its effect is still a matter of hypothesis [47].

It is possible to assume a genetical transmission of a certain metabolic or immunological defect which is made worse by gluten peptides. Enterocytes are damaged and lose their normal enzymatic equipment. Gluten peptides are not degraded and step by step they evoke an allergic reaction. If gluten is removed the pattern is restored but a new load of gluten leads to a prompt allergic reaction.

McDonald *et al.* [310] described changes in biopsies of some parents of children with celiac sprue and suggested that the disease might be inherited through a dominant gene of low penetrance. In our biopsies of asymptomatic parents of children with celiac sprue we could show some abnormalities [310a]: irregular configuration of villi, increased cellulization of the propria, increased quantity of cytoplasmic RNA in enterocytes, decreased activities of membrane enzymes, shift of LDH subunits in favour of the 'H' form and a decreased activity of lactase and cellobiase. These changes were observed more often in mothers than in fathers. The demonstrated abnormalities differentiated these subjects from controls in our geographic conditions. These changes are nonspecific, however, and in no case the diagnosis of celiac sprue could be done on their basis. On the other hand changes of similar character were seen in some of our patients with celiac sprue after a gluten-free diet and corticosteroid therapy. These changes may be a mild expression of the above-mentioned metabolic defect which is inherited. However. Hoffman *et al.* [311] reported that in monozygotic twins only one sibling suffered from gluten enteropathy.

An important question is whether the described morphological and cytochemical pattern of gluten enteropathy is characteristic for this disease only. Celiac-like lesions have been produced in experimental animals by acid and physical trauma and drugs [312, 189, 313, 300] and occur also in pigs with transmissible gastroenteritis [314]. A similar type of lesions with a varying degree of similarlity has been described in tropical sprue [315, 304, 316, 301, 317], lymphoma [318], giardiasis (lambliasis) [319], agammaglobulinemia [320], after gastrectomy [315], in endocrine disorders [321], after administration of triparanol [298], in radiation enteritis [322], in infant malnutrition [323].

Two important points have to be considered in this connection [47]. (1) A latent form of celiac sprue can manifest itself in the course of the above-mentioned disorders [324]. (2) In the interpretation of mucosal abnormalities the evolutionary approach has to be applied [325] and hyperregeneratory and hyporegeneratory types distinguished. A combined type may also occur. In central Europe the described hyperregeneratory type of lesions is usually found only in gluten enteropathy [326].

Schenk *et al.* [304] attempted to differentiate tropical and nontropical sprue by histochemical methods. They maintain that in tropical sprue the defect in 'ATP-ase' is of a lesser degree than in gluten enteropathy and that lipids in tropical sprue are found in a thickened basement membrane and not in enterocytes as in nontropical sprue. As was pointed out in our previous paper [47] in our cases of gluten enteropathy we found that the defect in 'ATP-ase' does not have to be complete and that the picture is not explicit even in regard to lipids. Beside cases in which lipids were present in enterocytes bordering on the lumen we found lipids even in the basement membrane and also in the propria in one and the same patient. In our opinion neither the reaction for 'ATP-ase' nor for lipids can distinguish these diseases unequivocally.

2.5.2 Tropical sprue

This disease in contrast to celiac sprue is not influenced by a gluten-free diet. A clinical remission is usually attained by therapy with folic acid, vitamin B_{12} and orally administered antibiotics. The histological picture of the jejunal mucosa is in many respects similar to that encountered in celiac sprue [304, 316, 327, 328, 301, 317].

The activity of enzymes was reported to be slightly decreased in tropical sprue enterocytes [171, 304, 301]. The latter authors describe that the alteration of enzyme activity in the surface epithelium in tropical sprue differs from that in celiac sprue in two respects: the degree of decreased activity is related to the severity of villous alteration and 'ATP-ase' activity was demonstrable even in patients with flat mucosa. Even this picture can be seen in some celiac sprue patients so that so far in our opinion no clear-cut morphological and histochemical distinction between the two entities seems to be possible [47].

In tropical sprue, similarly as in celiac sprue, cytochemical analysis has shown that malabsorption is caused not only by a reduction of the absorptive surface but also by lowered digestive and transport capacities of enterocytes.

2.5.3 *Malabsorption of disaccharides*

The cause of congenital or acquired malabsorption of disaccharides is the defective equipment of the enterocyte with disaccharidases [329, 330, 87]. So far it was chiefly studied by biochemical methods. Only recently were histochemical [36, 111, 112, 38, 39, 114, 48] methods elaborated and applied [36, 42, 303, 46, 47, 38, 39, 48, 87, 331].

As was stated previously [47], beside the constant findings of lowered activity of disaccharidases in gluten enteropathy and in Whipple's disease we did not find in enterobiopsies of our patients with various forms of secondary malabsorption syndrome any constant changes in activities of these enzymes although activities were lower sometimes. A complete defect or a substantial reduction was not encountered.

In our patients with primary deficiency of lactase and sucrase the morphological, cytological and cytochemical pattern was in the limits of the norm, with the exception of the defect of the respective disaccharidase, which could be demonstrated *in situ* and corroborated biochemically [331].

2.5.4 *Glucose-galactose malabsorption*

With the cytochemical methods available for *in situ* analysis no changes were found in patients with glucose-galactose malabsorption in which the transport of actively transported sugars is disturbed, probably owing to some deficiency at the level of the brush border membrane carrier [332]. Similarly in cystinuria, three types of which can be differentiated biochemically [333], no cytochemical changes in enterocytes which could be demonstrated *in situ* were reported, with the exception of one of our cases in which a marked diminution of activities of brush border enzymes in the enterocytes of the upper part of villi was found [47].

2.5.5 Congenital a-β-lipoproteinemia
In congenital a-β-lipoproteinemia (acanthocytosis) fat droplets in enterocytes of fasting patients were described [334], some of which did not have a lipoprotein membrane. Isselbacher *et al.* [334] and Sabesin *et al.* [335] believe that the cause of malabsorption of lipids in this disease is an imperfect transformation of lipids resynthesized in enterocytes into chilomincrons due to a defective synthesis of proteins.

2.5.6 Whipple's disease
Whipple's disease is usually classified in the group of secondary malabsorption syndrome in which the cause lies beyond the absorptive cells of the intestine. In advanced cases of this disease the height of villi is lowered and their width is greater because of the large number of rounded or polygonal macrophages in the propria. These macrophages contain sickle-shaped particles which stain distinctly with PAS reaction. Due to these particles they have been called SPC cells (sickle-form particle-containing cells [336]). These cells were analysed histochemically several times [47]. The stored glycidic material, which arises from the phagocytosed bacteria, which can be demonstrated in the mucous membrane by Gram's staining, is a polysaccharide which is not glycogen. Free carboxylic and sulphate groups are present in trace amounts only. No great amounts of protein amino acids were demonstrated. In SPC cells hydrophobic lipids were also demonstrated. A high activity of acid phosphatase and β-glucuronidase suggests an enhanced phagocytic activity of SPC cells. The reason why a concomitant increase of nonspecific esterase and of its E 600-resistant part does not occur remains unknown [47]. The activity of dehydrogenases in SPC cells is relatively low.

In enterocytes no changes in proteins, polysaccharides and lipids were found [47]. The lysosomes of enterocytes were coarser, however, and displayed a high activity of acid phosphatase. In the brush border was observed a decrease of activities of α-D-glucosidase and alkaline phosphatase [47]. In other cases investigated recently [331] a deficiency of lactase was always present.

Malabsorption in Whipple's disease is caused mainly by obstructed lymphatic drainage through the mesenterial nodes. A decreased digestive and transport function of enterocytes participate also in it. A striking improvement of the pattern was found after therapy with antibiotics similarly as was reported elsewhere [337]. We did not observe a complete normalization in our patients [48].

2.5.7 Other cases of secondary malabsorption syndromes
Among findings in other cases of secondary malabsorption syndrome those in patients with mucoviscidosis will be mentioned first. Some data were given previously (for references see Lojda et al. [47]. Of the 21 patients which were investigated in collaboration with Dr. Jodl from the 1st Children's Clinic, Faculty of Pediatrics, Prague, concomitant celiac sprue was found in three cases and the clinical and cytochemical picture of those patients improved after a gluten-free diet. In the majority of patients the changes were milder and in 5 patients no cytochemical deviation from the norm was found in the bioptic specimens. The prevailing abnormalities were lower activities of disaccharidases, particularly of lactase and trehalase [87]. Enhanced secretion of goblet cell mucus, which adhered to enterocytes and had a greater amount of demonstrable acidic groups than normally [338], was found in 4 patients only. Our data shows that malabsorption in mucoviscidosis is not exclusively of pancreatogenic origin in all cases, and that the decreased functional capacity of enterocytes can also participate in it.

There are some reports on changes in the intestinal mucosa in various forms of secondary malabsorption syndrome. Those after gastrectomy were reported recently by Strukov and Aruin [339]. In evaluating changes in the efferent loop it should be borne in mind that they are dependent on the distance of the taken sample from the stoma [340]. All deviations which are found in the first 20 cm of the afferent loop are of no diagnostical value for malabsorption.

As was pointed out previously [47] the morphological and

cytochemical pattern of the jejunal mucosa in the secondary malabsorption syndrome is not constant.

The most frequent, but not constant, enzymatic defects were observed in the microvillous zone of enterocytes. These defects could be the cause of malabsorption only when a large part of the intestine was afflicted, and this can hardly be decided on the basis of biopsies.

Deviations in the propria were also not constant. In some patients with lymphadenitis mesenterialis tuberculosa droplets of hydrophobic lipids were observed by us in the propria, both intra- and extracellularly, which according to our opinion points to an impeded lymph drainage if they are found in fasting patients. Otherwise in the propria were found at the most signs of a mild nonspecific inflammation regardless of the character of secondary malabsorption. In this case there was a very marked activity of acid phosphatase and β-glucuronidase in macrophages, especially at the tip of villi.

Cytochemical examination did not reveal any unequivocal evidence as to the role of enterocytes in absorption disorders in the investigated forms of secondary malabsorption syndrome.

2.6 CONCLUSIONS

From the survey presented here it would seem at first glance that our knowledge of the biochemical cytology of the intestine is fairly adequate. However, after careful reconsideration it becomes apparent that we are only at the beginning. Systematic and combined electron-microscopical, biochemical and cytochemical investigations carried out with recently elaborated and more sophisticated methods will enable more precise localization on the ultrastructural level. The results, correlated with physiological studies, will lead to a much more profound understanding of the functions of the enterocytes and of other cells in the intestinal mucosa in health and in disease.

REFERENCES

1. M. SHINER, *Lancet*, **1**, 17 (1956).
2. M. SHINER, *Lancet*, **1**, 85 (1956).
3. W. H. CROSBY and H. W. KUGLER, *Am. J. Dig. Dis.*, **2**, 236 (1957).
4. E. L. POSEY, *Gastroenterology*, **37**, 299 (1959).
5. S. J. BAKER and A. HUGHES, *Lancet*, **2**, 686 (1960).
6. A. L. FLICK, W. E. QUINTON and C. E. RUBIN, *Gastroenterology*, **40**, 120 (1961).
7. D. MILLER and R. K. CRANE, *Biochim. Biophys. Acta*, **52**, 293 (1961).
8. H. RUTTLOFF, R. NOACK, R. FRIESE and G. SCHENK, *Biochem. Z.*, **341**, 15 (1964).
9. A. EICHHOLZ and R. K. CRANE, *J. Cell Biol.*, **26**, 687 (1965).
10. G. HÜBSCHER, G. WEST and D. BRINDLEY, *Biochem. J.*, **97**, 629 (1965).
11. J. W. PORTEOUS and B. CLARK, *Biochem. J.*, **96**, 159 (1965).
12. J. W. PORTEOUS and B. CLARK, *Biochem. J.*, **96**, 539 (1965).
13. G. G. FORSTNER, S. M. SABESIN and K. J. ISSELBACHER, *Biochem. J.*, **106**, 381 (1968).
14. J. W. PORTEOUS, *FEBS Letters*, **1**, 46 (1968).
15. T. J. PETERS, *Biochem. J.*, **120**, 195 (1970).
16. S. ITO, *Fed. Proc.*, **28**, 12 (1969).
17. L. HSU and A. TAPPEL, *J. Cell Biol.*, **23**, 233 (1964).
18. L. HSU and A. TAPPEL, *Biochim. Biophys. Acta*, **101**, 83 (1965).
19. J. WRIGGLESWORTH and J. POVER, *Life Sci.*, **5**, 1365 (1966).
20. D. S. HARRER, B. K. STERN and R. W. REILLY, *Nature*, **203**, 319 (1964).
21. A. D. PERRIS, *Canad. J. Biochem.*, **49**, 687 (1966).
22. F. S. SJÖSTRAND, *J. Ultrastr. Res.*, **22**, 424 (1968).
23. D. D. HARRISON and H. L. WEBSTER, *Exptl. Cell Res.*, **55**, 257 (1969).
24. H. L. WEBSTER and D. D. HARRISON, *Exptl. Cell Res.*, **56**, 245 (1969).
25. A. R. IMONDI, M. R. BALLIS and M. LIPKIN, *Exptl. Cell Res.*, **58**, 323 (1969).
26. K. LINDERSTRÖM-LANG and K. R. MOGENSEN, *C.R. Trav. Lab. Carlsberg serie chim.*, **23**, 37 (1938).
27. A. DAHLQVIST and Ch. NORDSTRÖM, *Biochim. Biophys. Acta*. **113**, 624 (1966).
28. F. MOOG and R. F. GREY, *J. Cell Biol.*, **32**, Cl (1967).
29. Ch. NORDSTRÖM, A. DAHLQVIST and L. JOSEFSSON, *J. Histochem. Cytochem.*, **15**, 713 (1968).
30. Ch. NORDSTRÖM, O. KOLDOVSKÝ and A. DAHLQVIST, *J. Histochem. Cytochem.*, **17**, 341 (1969).
31. R. K. CRANE, *Physiol. Rev.*, **40**, 789 (1960).
32. W. O. DOBBINS III, *Am. J. Med. Sci.*, **258**, 150 (1969).
33. A. EICHHOLZ, *Fed. Proc.*, **28**, 30 (1969).

34. W. FISCHER, *Verh. Dtsch. Ges. Path.* (53. Tgg.), 81 (1969).
35. N. J. GREENBERGER, *Am. J. Med. Sci.*, **258**, 144 (1969).
36. Z. LOJDA, *Histochemie*, **5**, 339 (1965).
37. Z. LOJDA, Habilitation Thesis (Charles University, Prague, 1967), p. 138.
38. Z. LOJDA, *Histochemie*, **22**, 347 (1970).
39. Z. LOJDA, *Histochemie*, **23**, 266 (1970).
40. Z. LOJDA, *Histochemie*, **27**, 182 (1971).
41. Z. LOJDA and P. FRIČ, *Histochemie*, **3**, 455 (1964).
42. Z. LOJDA and P. FRIČ, *Csl. Patol.*, **3**, 96 (1967).
43. Z. LOJDA and P. FRIČ, Modern Gastroenterology (O. Gregor and O. Riedl, eds.), p. 853. F. K. Schattauer, Stuttgart, New York (1969).
44. Z. LOJDA and P. FRIČ, *Csl. Pathol.*, **5**, 25 (1969).
45. Z. LOJDA and P. FRIČ, Die heutige Stellung der Morphologie in Biologie und Medizin (G. Kettler, ed.), p. 561. Akademie Verlag, Berlin (1970).
46. Z. LOJDA, P. FRIČ and J. JODL, *Verh. Dtsch. Ges. Path.* (53. Tgg.), 93 (1969).
47. Z. LOJDA, P. FRIČ, J. JODL and V. CHMELIK, *C.T. in Pathology*, **52**, 1 (1970).
48. Z. LOJDA and J. KRAML, *Histochemie*, **25**, 195 (1971).
49. P. FRIČ and Z. LOJDA, *Acta gastro-ent. belg.*, **27**, 526 (1964).
50. A. G. E. PEARSE, Histochemistry, 2nd ed., p. 998. J. & A. Churchill, London (1960).
51. A. G. E. PEARSE, Histochemistry, 3rd ed., vol. 1, p. 759. J. & A. Churchill, London (1968).
52. T. BARKA and P. J. ANDERSON, Histochemistry, p. 660. Harper & Row, New York, Evanston, London (1963).
53. F. VERZÁR and E. J. McDOUGALL, Absorption from the Intestine. Longmans, Green & Co., London (1936).
54. J. S. TRIER, *Fed. Proc.*, **26**, 1391 (1967).
55. J. S. TRIER, Handbook of Physiology, Alimentary Canal (C. F. Code, ed.), p. 1125. Amer. Physiol. Soc., Washington, D.C. (1968).
56. J. S. TRIER and C. E. RUBIN, *Gastroenterology*, **49**, 574 (1965).
57. P. G. TONER, *Int. Rev. Cytol.*, **24**, 233 (1968).
58. H. J. MERKER, *Verh. Dtsch. Ges. Path.* (53. Tgg.), 57 (1969).
59. H. S. BENNETT, *J. Histochem. Cytochem.*, **11**, 14 (1963).
60. S. ITO, *Anat. Rec.*, **148**, 294 (1964).
61. S. ITO, *J. Cell Biol.*, **27**, 475 (1965).
62. S. A. PRATT and L. NAPOLITANO, *Anat. Rec.*, **165**, 197 (1969).
62. T. J. PETERS, *Gut*, **11**, 720 (1970).
63. S. ITO and J. P. REVEL, Gastrointestinal Radiation Injury (M. F. Sulivan, ed.), p. 27. Excerpta Medica Foundation, Amsterdam (1968).
64. R. C. CURRAN, A. E. CLARK and D. LOVELL, *J. Anat*, **99**, 427 (1965).
65. S. S. SPICER, M. W. STALEY, M. G. WETZEL and B. K. WETZEL, *J. Histochem. Cytochem.*, **15**, 225 (1967).

66. S. BRADBURY and P. J. STOWARD, *Histochemie*, **11**, 71 (1967).
67. J. P. REVEL, *J. Microscopie*, **3**, 535 (1964).
68. A. RAMBOURG and C. P. LEBLOND, *J. Cell Biol.*, **32**, 27 (1967).
69. J. D. BERLIN, *Rad. Res.*, **34**, 347 (1968).
69a. J. H. LUFT, *J. Cell Biol.*, **25**, 54A (1964).
70. G. G. FORSTNER, *Am. J. Med. Sci.*, **258**, 172 (1969).
71. L. A. E. ASHWORTH and C. GREEN, *Science*, **151**, 210 (1966).
72. G. G. FORSTNER, K. TANAKA and K. J. ISSELBACHER, *Biochem. J.*, **109**, 51 (1968).
73. P. F. MILLINGTON and D. R. CRITCHLEY, *Life Sci.*, **7**, 839 (1968).
74. A. EICHHOLZ, *Biochim. Biophys. Acta*, **135**, 475 (1967).
75. H. RUTTLOFF, R. NOACK, R. FRIESE, G. SCHENK and J. PROLL, *Acta, Biol. Med. German.*, **19**, 831 (1967).
76. P. MALATHI and R. K. CRANE, *Biochim. Biophys. Acta*, **173**, 245 (1969).
77. A. EICHHOLZ and R. K. CRANE, *Federation Proc.*, **25**, 656 (1966).
78. J. H. HOLT and D. H. MILLER, *Biochim. Biophys. Acta*, **58**, 239 (1962).
79. M. FRIEDRICH, R. NOACK and G. SCHENK, *Biochem. Z.*, **343**, 346 (1965).
80. J. B. RHODES, A. EICHHOLZ and R. K. CRANE, *Biochim. Biophys. Acta*, **135**, 959 (1967).
80a. R. HOLMES and R. W. LOBLEY, *J. Physiol.*, **211**, 50P (1970).
80b. Ch. NORDSTRÖM and A. DAHLQVIST, *Biochim. Biophys. Acta*, **242**, 209 (1971).
81. J. S. K. DAVID, P. MALATHI and J. GANGULY, *Biochem. J.*, **98**, 662 (1966).
82. P. MALATHI, *Gastroenterology*, **52**, 1106 (1967).
83. G. G. FORSTNER, E. M. RILEY, S. J. DANIELS and K. J. ISSELBACHER, *Biochem. Biophys. Res. Comm.*, **21**, 83 (1965).
84. A. NILSSON, *Biochim. Biophys. Acta*, **176**, 339 (1969).
85. G. SEMENZA, Handbook of Physiology, Alimentary Canal (C. F. Code, ed.), p. 2547. Am. Physiol. Soc., Washington, D.C. (1968).
86. E. EGGERMONT and H. G. HERS, *European J. Biochem.*, **9**, 488 (1969).
87. F. MALIŠ, Z. LOJDA, P. FRIČ and J. JODL, *Digestion*, **5**, 40 (1972).
88. G. H. JEFFRIES, E. WESER and M. H. SLEISENGER, *Gastroenterology*, **56**, 777 (1969).
89. R. G. DOELL, G. ROSEN and N. KRETCHMER, *Proc. Nat. Acad. Sci. Wash.*, **54**, 1268 (1965).
90. R. K. CRANE, Handbook of Physiology, Alimentary Canal (C. F. Code, ed.), Sect. 6, Vol. V, p. 2535. Am. Physiol. Soc., Washington, D.C. (1968).
91. R. GITZELMANN, TH. BÄCHI, H. BINZ, J. LINDENMANN and G. SEMENZA, *Biochim. Biophys. Acta*, **196**, 20 (1970).
92. Ch. F. JOHNSON, *Science*, **155**, 1670 (1967).
93. Ch. F. JOHNSON, *Fed. Proc.*, **28**, 26 (1969).

94. Y. NISHI, T. O. YOSHIDA and Y. TAKESUE, *J. Mol. Biol.*, **37**, 441 (1968).
95. T. ODA and S. SEKI, *J. Electr. Micr.*, **14**, 210 (1965).
96. R. B. COHEN, K.-C. TSOU, S. H. RUTENBURG and A. M. SELIGMAN, *J. Biol. Chem.*, **195**, 239 (1952).
97. R. B. COHEN, S. H. RUTENBURG, K.-C. TSOU, M. A. WOODBURY and A. M. SELIGMAN, *J. Biol. Chem.*, **195**, 607 (1952).
98. A. M. RUTENBURG, S. H. RUTENBURG, B. MONIS, R. TEAGUE and A. M. SELIGMAN, *J. Histochem. Cytochem.*, **6**, 122 (1958).
99. A. M. RUTENBURG, J. A. GOLDBARG, S. H. RUTENBURG and T. R. LANG, *J. Histochem. Cytochem.*, **8**, 268 (1960).
100. A. GOTTSCHALK, The Enzymes (J. B. Sumner and K. Myrbäck, eds.), Vol. I, Part I, p. 551. Academic Press, New York (1950).
101. S. VEIBEL, The Enzymes (J. B. Sumner and K. Myrbäck, eds.), Vol. I, p. 551. Academic Press, New York (1950).
102. S. VEIBEL, The Enzymes (J. B. Sumner and K. Myrbäck, eds.), Vol. I, Part I, p. 621. Academic Press, New York (1950).
103. A. DAHLQVIST, B. BULL and D. L. THOMPSON, *Arch. Biochem.*, **109**, 159 (1965).
104. J. KRAML, O. KOLDOVSKÝ, A. HERINGOVÁ, V. JIRSOVÁ, K. KÁCL, M. LEDVINA and H. PELICHOVÁ, *Biochem. J.*, **114**, 621 (1969).
105. K. R. KERRY, *Comp. Biochem. Physiol.*, **29**, 1015 (1969).
106. G. ZOPPI and D. H. SHMERLING, *Comp. Biochem. Physiol.*, **29**, 289 (1969).
107. N.-G. ASP, A. DAHLQVIST and O. KOLDOVSKÝ, *Gastroenterology*, **58**, 591 (1970).
108. A. DAHLQVIST, *Enzym. Biol. Clin.*, **11**, 52 (1970).
108a. N.-G. ASP, Small intestinal β-galactosidases, p. 54. (Dissertation, University of Lund, 1971.)
109. G. M. GRAY and N. A. SANTIAGO, *J. Clin. Invest.*, **48**, 716 (1969).
110. A. DAHLQVIST and A. BRUN, *J. Histochem. Cytochem.*, **10**, 294 (1962).
111. J. JOS, J. FRÉZAL, J. REY and M. LAMY, *Nature*, **213**, 516 (1967).
112. J. JOS, J. FRÉZAL, J. REY, M. LAMY and R. WEGMANN, *Ann. Histochim.*, **12**, 53 (1967).
113. P. FRIČ, Z. LOJDA and J. JODL, *Cs. Gastroent.*, **22**, 235 (1968).
114. Z. LOJDA, Histochemical Demonstration of Enzymes, Part III, p. 28. Czechoslovak Society of Histo- and Cytochemistry, Brno (1970).
115. G. GOMORI, *J. Cell. Comp. Physiol.*, **17**, 71 (1941).
116. A. G. E. PEARSE and E. O. RIECKEN, *Brit. Med. Bull.*, **23**, 217 (1967).
171. G. G. BERG and B. CHAPMAN, *J. Cell. Comp. Physiol.*, **65**, 361 (1965).
118. I. H. ROSENBERG and L. E. ROSENBERG, *Comp. Biochem. Physiol.*, **24**, 975 (1968).

119. K. WATANABE and W. H. FISHMAN, *J. Histochem. Cytochem.*, **12**, 252 (1964).
120. P. FRIČ and Z. LOJDA, *Gastroenterologia*, **106**, 65 (1966).
121. N. K. GHOSH, *Ann. N.Y. Acad. Sci.*, **166**, 604 (1969).
122. H. PELICHOVÁ, O. KOLDOVSKÝ, A. HERINGOVÁ, V. JIRSOVÁ and J. KRAML, *Canad. J. Biochem.*, **45**, 1375 (1967).
123. F. MOOG, M. E. ETZLER and R. F. GREY, *Ann. N.Y. Acad. Sci.*, **166**, 447 (1969).
124. M. E. ETZLER, E. H. BIRKENMEIER and F. MOOG, *Histochemie*, **20**, 99 (1969).
125. H. SCHÜSSLER, *Biochim. Biophys. Acta*, **151**, 383 (1968).
126. W. H. CHASE, *J. Histochem. Cytochem.*, **11**, 96 (1963).
127. S. GOLDFISCHER, E. ESSNER and A. B. NOVIKOFF, *J. Histochem. Cytochem.*, **12**, 72 (1964).
128. J. HUGON and M. BORGERS, *J. Histochem. Cytochem.*, **14**, 629 (1966).
129. J. HUGON and M. BORGERS, *Histochemie*, **6**, 209 (1966).
130. J. HUGON and M. BORGERS, *Histochemie*, **12**, 42 (1968).
131. J. HUGON and M. BORGERS, *Gastroenterology*, **55**, 608 (1968).
132. P. F. MILLINGTON and A. C. BROWN, *Histochemie*, **8**, 109 (1967).
133. A. C. BROWN and P. F. MILLINGTON, *Histochemie*, **12**, 83 (1968).
134. E. REALE and L. LUCIANO, *Histochemie*, **8**, 302 (1967).
135. P. PORTMANN and F. J. LI-CHEN, *Helvet. Physiol. Pharmacol. Acta*, **26**, 400 (1969).
136. A. B. NOVIKOFF, E. ESSNER, S. GOLDFISCHER and M. HEUSS, *Symp. Intl. Soc. Cell Biol.*, **1**, 149 (1962).
137. N. J. INGLIS, M. J. KRANT and W. H. FISHMAN, *Proc. Soc. Exp. Biol. (N.Y.)*, **124**, 699 (1967).
138. U. E. KLEIN, *Dtsch. Med. Wchschr.*, **94**, 526 (1969).
139. U. E. KLEIN, H. Ch. DRUBE and H. Th. HANSEN, *Klin. Wchschr.*, **45**, 95 (1967).
140. M. S. BURSTONE and J. E. FOLK, *J. Histochem. Cytochem.*, **4**, 217 (1956).
141. Z. LOJDA, Histochemical Demonstration of Enzymes, Part II, p. 30. Czechoslovak Society of Histo- and Cytochemistry, Brno (1968).
142. E. D. WACHSMUTH, *Histochemie*, **14**, 282 (1968).
143. V. K. HOPSU-HAVU and T. O. EKFORS, *Histochemie*, **17**, 30 (1969).
144. A. EICHHOLZ, *Biochim. Biophys. Acta*, **163**, 101 (1968).
145. M. I. COHEN, L. M. GARTNER, O. O. BLUMENFELD and I. M. ARIAS, *Pediat. Res.*, **3**, 5 (1969).
146. K. GIBIŃSKI, A. NOWAK and D. KOCHANSKA, *Gastroenterologia*, **108**, 219 (1967).
147. R. K. CRANE, *Gastroenterology*, **50**, 254 (1966).
148. A. M. UGOLEV, *Physiol. Rev.*, **45**, 555 (1965).
149. J. S. TRIER, *Gastroenterology*, **56**, 618 (1969).
150. H. L. MOSES, A. S. ROSENTHAL, D. L. BEAVER and S. S. SCHUFFMAN, *J. Histochem. Cytochem.*, **14**, 702 (1966).

151. J. MALINSKÝ, J. BLATNÝ and Z. KOJECKÝ, *Folia Morphol. (Prague)*, **13**, 300 (1965).
152. E. O. RIECKEN, J. S. STEWART, C. C. BOOTH and A. G. E. PEARSE, *Gut*, **7**, 317 (1966).
153. E. O. RIECKEN, J. S. STEWART and R. H. DOWLING, *Internist (Berlin)*, **7**, 209 (1966).
154. E. O. RIECKEN, H. A. SCHMIDT, K. WÄCHTLER and A. G. E. PEARSE, *Klin. Wchschr.*, **45**, 383 (1967).
155. L. BIEMPICA, A. B. NOVIKOFF, B. KAHN, *J. Histochem. Cytochem.*, **10**, 654 (1062).
156. K. OGAWA, K. MASUTANI and Y. SHINONAGA, *J. Histochem. Cytochem.*, **10**, 228 (1962).
157. T. BARKA, *J. Histochem. Cytochem.*, **12**, 229 (1964).
158. J. HUGON and M. BORGERS, *J. Cell Biol.*, **33**, 212 (1967).
159. J. HUGON and M. BORGERS, *Bull. de l'Assoc. des Anatomistes*, 52e Réunion, 670 (1967).
160. W. STRAUSS, Enzyme Cytology (D. B. Roodyn, ed.), p. 239. Academic Press, London, New York (1967).
161. A. V. HOFFBRAND and T. J. PETERS, *Biochim. Biophys. Acta*, **92**, 479 (1969).
162. T. J. PETERS, *Gut*, **11**, 720 (1970).
163. E. D. KORN, *Fed. Proc.*, **28**, 6 (1969).
164. Z. LOJDA, *Folia Morphol. (Prague)*, **13**, 84 (1965).
165. P. FRIČ and Z. LOJDA, *Clin. Chim. Acta*, **12**, 111 (1965).
165a. Z. LOJDA and P. FRIČ, Enzymes and Isoenzymes (D. Shugar, ed.), p. 185. Academic Press, London and New York (1970).
166. H. R. JERVIS, *J. Histochem. Cytochem.*, **9**, 692 (1963).
167. M. H. FLOCH, S. Van NOORDEN and H. M. SPIRO, *Am. J. Dig. Dis.*, **11**, 804 (1966).
168. M. M. FERGUSON, *Quart. J. Microscop. Sci.*, **106**, 289 (1965).
169. A. H. BAILLIE, M. M. FERGUSON and D. McK. HART, Developments in Steroid Histochemistry, p. 186. Academic Press, London and New York (1966).
170. H. A. PADYKULA, W. E. STRAUSS, A. J. LADMAN and F. H. GARDNER, *Gastroenterology*, **40**, 735 (1961).
171. H. M. SPIRO, M. I. FILIPE, J. S. STEWARD, C. C. BOOTH and A. G. E. PEARSE, *Gut*, **5**, 145 (1964).
172. J. MYRÉN, G. C. LUKETIC, R. CEBALLOS, G. SACHS and B. I. HIRSCHOWITZ, *Amer. J. Dig. Dis.*, **11**, 394 (1966).
173. P. G. BURHOL and J. MYREN, *Scand. J. Gastroent.*, **1**, 314 (1966).
174. Ch. M. PARMENTIER, *Gastroenterology*, **43**, 1 (1962).
175. H. W. DEANE and E. W. DEMPSEY, *Anat. Rec.*, **93**, 401 (1945).
176. A. B. NOVIKOFF, L. KORSON and H. SPATER, *Exptl. Cell Res.*, **3**, 617 (1952).
177. F. MOOG, *Fed. Proc.*, **21**, 51 (1962).
178. M. PETERSON and C. P. LEBLOND, *J. Cell Biol.*, **21**, 143 (1964).
179. A. PRZELECKA, G. EJSMOT, M. G. SARZALA and M. TARACHA, *J. Histochem. Cytochem.*, **10**, 596 (1962).

180. J. S. HUGON, M. BORGERS and D. MAESTRACCI, *J. Histochem. Cytochem.*, **18**, 361 (1970).
180a. Z. LOJDA, *Verh. Anat. Ges.* **66**, Tgg., 19 (1972).
181. D. G. LYGRE and R. C. NORDLIE, *Biochemistry*, **7**, 3219 (1968).
182. B. FRIEDMAN, D. S. STRACHAN and M. M. DEWEY, *J. Histochem. Cytochem.*, **14**, 560 (1966).
183. M. ABE, S. P. KRAMER and A. M. SELIGMAN, *J. Histochem. Cytochem.*, **12**, 364 (1964).
184. L. ARVY, Handbuch der Histochemie (W. Graumann and K. Neumann, eds.), VII/2, p. 154. Gustav Fischer Verlag, Stuttgart (1962).
185. A. A. M. LEWIS and R. L. HUNTER, *J. Histochem. Cytochem.*, **14**, 33 (1966).
186. G. HUBSCHER, B. CLARK and M. E. WEBB, *Biochem. J.*, **84**, 23P (1962).
187. C. P. LEBLOND and B. MESSIER, *Anat. Rec.*, **132**, 247 (1958).
188. B. CREAMER, *Brit. Med. Bull.*, **23**, 226 (1967).
189. M. EDER, *Verh. Dtsch. Ges. Path.*, (50. Tgg.), 75 (1966).
190. H. A. PADYKULA, *Fed. Proc.*, **21**, 873 (1962).
191. R. FORTIN-MAGANA, R. HURWITZ, J. J. HERBST and N. KRETCHMER, *Science*, **167**, 1627 (1970).
192. W. GRAUMANN, Polysaccharide, in Handbuch der Histochemie (W. Graumann and K. Neumann, eds.), p. 743. G. Fischer, Stuttgart (1964).
193. J. LINDNER, *Verh. Dtsch. Ges. Path.*, (53. Tgg.), 111 (1969).
194. M. A. JENINGS and H. W. FLOREY, *Quart. J. exptl. Physiol.*, **41**, 131 (1956).
195. Z. LOJDA, B. VECEREK and H. PELICHOVA, *Histochemie*, **3**, 428 (1964).
196. R. LEV and A. GERARD, *J. Roy. Micr. Soc.*, **87**, 361 (1967).
197. H. W. FLOREY, Lectures on General Pathology, Saunders, Philadelphia (1954).
198. W. D. TROUGHTON and J. S. TRIER, *J. Cell Biol.*, **41**, 251 (1969).
199. G. GEYER and L. NIETZOLD, *Acta Histochem. (Jena)*, **29**, 409 (1968).
200. V. PATZELT, Handb. microskop. Anat. (W. Moellendorf, ed.), **5**, Teil 3, p. 193. Springer, Berlin (1936).
201. E. O. RIECKEN and A. G. E. PEARSE, *Gut*, **7**, 86 (1966).
202. O. MIDORIKAWA and M. EDER, *Histochemie*, **2**, 444 (1962).
202a. A. B. NOVIKOFF, Lysosomes (A. V. S. de Reuck and M. P. Cameron, eds.), p. 36. Little, Brown and Co., Boston (1963).
203. O. BENNKE and H. MOE, *J. Cell Biol.*, **22**, 633 (1964).
204. M. M. de CASTRO, W. da SILVA SASSO and F. A. SAAD, *Acta Anat.*, **38**, 345 (1959).
205. A. D. HALLY, *J. Anat.*, **92**, 268 (1958).
206. H. M. SELZMAN and R. A. LIEBELT, *Anat. Rec.*, **140**, 17 (1961).
207. B. CREAMER, *Lancet*, **1**, 304 (1967).

208. A. J. SPEECE, *J. Histochem. Cytochem.*, **12**, 384 (1964).
209. R. J. DECKX, G. R. VANTRAPPEN and M. M. PAREIN, *Biochim. Biophys. Acta*, **139**, 207 (1967).
210. I. DAWSON, *Histochem. J.*, **2**, 527 (1970).
211. E. SOLCIA and R. SAMPIETRO, *Z. Zellforsch.*, **68**, 689 (1965).
212. E. SOLCIA and R. SAMPIETRO, *Riv. Istoch. Norm. Pat.*, **11**, 265 (1965).
213. E. SOLCIA and R. SAMPIETRO, *Riv. Istoch. Norm. Path.*, **12**, 122 (1966).
214. E. SOLCIA and R. SAMPIETRO, *Nature*, **214**, 196 (1967).
215. E. SOLCIA, G. VASSALLO and C. CAPELLA, *Stain Technol.*, **43**, 257 (1968).
216. C. CAPELLA, E. SOLCIA and G. VASSALLO, *Arch. Histol. Jap.*, **30**, 479 (1969).
217. A. F. CARVALHEIRA, U. WELSH and A. G. E. PEARSE, *Histochemie*, **14**, 33 (1969).
218. A. G. E. PEARSE, I. COULLING, B. WEAWERS and S. FRIESEN, *Gut*, **11**, 649 (1970).
219. W. G. FORSMANN, L. ORCI and Ch. ROUILLER, *Symp. Dtsch. Ges. Endokrinol.*, **14**, 252 (1968).
220. L. ORCI, R. PICTET, W. G. FORSSMANN, A. E. RENOLD and Ch. ROUILLER, *Diabetologia*, **4**, 56 (1968).
220a. W. G. FORSMANN, L. ORCI, R. PICTET, A. E. RENOLD and Ch. ROUILLER, *J. Cell Biol.*, **40**, 692 (1969).
220b. A. G. E. PEARSE, Origin, Chemistry, Physiology and Pathophysiology of the gastrointestinal Hormones (W. Creutzfeld, ed.), p. 95. F. K. Schattauer, Stuttgart—New York (1970).
221. G. VASSALLO, E. SOLCIA and C. CAPELLA, Atti VI. *Congr. Ital. Micr. Elettr.*, 148 (1968).
222. G. VASSALLO, E. SOLCIA and C. CAPELLA, *Z. Zellforsch.*, **98**, 333 (1969).
223. C. DEBRAY, D. CATTAN, C. MARCHE and J. P. JORI, *Bull. Soc. Med. Hop. Paris*, **116**, 1635 (1965).
224. H. W. DEANE, *Anat. Rec.*, **149**, 453 (1964).
225. P. A. CRABBE and J. F. HEREMANS, *Gastroenterology*, **51**, 305 (1966).
226. W. RUBIN, L. L. ROSS, M. H. SLEISENGER and E. WESER, *Lab. Invest.*, **15**, 1720 (1966).
227. A. P. DOUGLAS, P. A. CRABBE and J. R. HOBBS, *Gastroenterology*, **59**, 414 (1970).
228. O. KOLDOVSKÝ, Development of the Functions of the Small Intestine in Mammals and Man, p. 204. S. Karger, Basel and New York (1969).
229. L. VOLLRATH, Uber die Entwicklung des Dunndarms der Ratte. Ergebnisse der Anatomie and Entwicklungsgechichte, 41, H. 2, p. 70. Springer, Berlin, Heidelberg, New York (1969).
230. P. SUNSHINE and N. KRETCHMER, *Science*, **144**, 850 (1964).
231. O. KOLDOVSKÝ, N.-G. ASP and A. DAHLQVIST, *Anal. Biochem.*, **27**, 409 (1969).

232. O. KOLDOVSKÝ, P. SUNSHINE and N. KRETCHMER, *Nature*, **212**, 1389 (1966).
233. G. D. CAIN, P. MOORE, M. PATTERSON and M. A. McELVEEN, *Scand. J. Gastroenterol.*, **4**, 545 (1969).
234. J. VERNE and S. HEBERT, *C.R. Ass. Anat.*, **36**, 402 (1949).
235. F. VERZAR and F. SAITER, *Helv. Physiol. Pharmacol. Acta*, **10**, 247 (1952).
235a. J. HUGON and M. BORGERS, *Histochemie*, **19**, 13 (1969).
236. F. MOOG, *Develop. Biol.*, **3**, 153 (1961).
237. F. MOOG, *Science*, **144**, 414 (1964).
238. F. MOOG, *Adv. in Enzyme Regulation*, **3**, 221 (1965).
239. T. LINDBERG and B. W. KARLSON, *Gastroenterology*, **59**, 247 (1970).
240. R. CORNELL and H. A. PADYKULA, *Am. J. Anat.*, **125**, 291 (1969).
241. F. BIERRING, H. ANDERSEN, J. EGEBERG, M. MATTHEESEN and F. BRO-RASMUSSEN, *Acta Path. Microbiol. Scand.*, **61**, 345 (1964).
242. Z. VACEK, *Csl. Morfol.*, **12**, 292 (1964).
243. J. E. JIRASEK, J. UHER and O. KOLDOVSKÝ, *Acta Histochem. (Jena)*, **22**, 33 (1965).
244. N. BJORKMAN and M. SIBALIN, *Experientia*, **23**, 339 (1967).
245. J. P. KRAEHENBUHL and M. A. CAMPICHE, *J. Cell Biol.*, **42**, 345 (1969).
246. R. M. WILLIAMS and F. BECK, *Histochem. J.*, **1**, 531 (1969).
247. M. J. CONNOCK and W. F. R. POVER, *Histochem. J.*, **2**, 35 (1970).
248. D. L. BAXTER-GRILLO, *Histochemie*, **19**, 31 (1969).
249. W. BUNO, *Acta Anat.*, **60**, 285 (1965).
250. H. PELICHOVÁ, O. KOLDOVSKÝ, A. HERINGOVÁ, V. JIRSOVÁ and J. KRAML, *Canad. J. Biochem.*, **45**, 1375 (1967).
251. K. KUBAT and O. KOLDOVSKÝ, *Acta Histochem. (Jena)*, **33**, 75 (1969).
252. J. T. TOMASINI and W. O. DOBBINS III, *Am. J. Dig. Dis.*, **15**, 226 (1970).
252a. S. L. PALAY and L. KARLIN, *J. Biophys. Biochem. Cytol.*, **5**, 363 (1959).
252b. S. L. PALAY and L. KARLIN, *J. Biophys. Biochem. Cytol.*, **5**, 373 (1959).
252c. M. SHINER and R. A. B. DRURY, *Am. J. Dig. Dis.*, **7**, 744 (1962).
252d. P. C. PHELPS, C. E. RUBIN and J. H. LUFT, *Gastroenterology*, **46**, 134 (1964).
252e. N. VODOVAR and J. FLANZY, *Ann. Biol. Anim. Biochim. Biophys.*, **6**, 315 (1966).
253. Z. VACEK, P. HAHN and O. KOLDOVSKÝ, *Csl. Morfol.*, **10**, 30 (1962).
254. M. DOBIASOVA, P. HAHN and O. KOLDOVSKÝ, *Biochim. Biophys. Acta*, **70**, 711 (1963).
255. M. DOBIASOVA, P. HAHN and O. KOLDOVSKÝ, *Biochim. Biophys. Acta*, **84**, 538 (1964).

256. R. HALLIDAY, *J. Endocrin.*, **18**, 56 (1959).
257. S. L. CLARK, *J. Biophys. Biochem. Cytol.*, **5**, 41 (1959).
258. N. S. ROSENWEIG and R. H. HERMAN, *Am. J. Clin. Nutr.*, **22**, 99 (1969).
259. J. PROSPER, R. MURRAY and F. KERN, *Gastroenterology*, **55**, 223 (1968).
260. B. N. TANDON, P. M. NEWBERN and V. R. YOUNG, *J. Nutr.*, **99**, 519 (1969).
261. K. B. KNUDSEN, E. M. BRADLEY, F. R. LECOCQ, H. M. BELLAMY and J. D. WELSH, *Gastroenterology*, **55**, 46 (1968).
262. I. BERKEL, O. KIRAN and B. SAY, *Acta Paediat. Scand.*, **59**, 58 (1970).
263. M. R. HAUSSLER and L. A. NAGODE, *J. Cell Biol.*, **43**, 51a (1969).
264. E. S. HELDSWORTH, *J. Membrane Biol.*, **3**, 43 (1970).
265. Z. LOJDA and P. FABRY, *Acta Histochem. (Jena)*, **8**, 289 (1959).
266. Z. LOJDA, *Acta Histochem. (Jena)*, **5**, 236 (1958).
267. E. O. RIECKEN, R. H. DOWLING, C. C. BOOTH and A. G. E. PEARSE, *Enzymol. Biol. Clin.*, **5**, 231 (1965).
268. R. J. LEVIN, *J. Endocr.*, **45**, 315 (1969).
269. R. J. LEVIN, 1964, quoted according to Levin, 1969.
270. A. J. WALL, W. R. J. MIDDLETON, A. G. E. PEARSE and C. C. BOOTH, *Virchow's Arch. Abt. B. Zell pathol.*, **6**, 79 (1970).
271. A. SOULAIRAC, *C.R. Soc. Biol.*, **142**, 643 (1948).
272. J. VERNE and S. HERBERT, *C.R. Soc. Biol.*, **142**, 201 (1948).
273. E. LUTHY and F. VERZAR, *Biochem. J.*, **57**, 316 (1954).
274. A. D. CHIQUOINE, *J. Histochem. Cytochem.*, **3**, 471 (1955).
275. R. J. LEVIN, H. NEWEY and D. H. SMYTH, *J. Physiol. Lond.*, **177**, 58 (1965).
276. J. B. ROTGERS, E. M. RILEY, G. D. DRUMMEY and K. J. ISSELBACHER, *Gastroenterology*, **53**, 547 (1967).
277. D. WALSH, *Brit. Med. J.*, **2**, 272 (1897).
278. Y. S. LEWIS, H. QUASTLER and G. SVIHLA, *J. Natl. Canc. Inst.*, **21**, 813 (1958).
279. H. QUASTLER, *Radiology*, **73**, 161 (1959).
280. G. WIERNIK, *J. Path. Bact.*, **91**, 389 (1966).
281. G. WIERNIK and M. PLANT, *C.T. Rad. Res.*, 325 (1970).
282. H. QUASTLER and J. HAMPTON, *Radiat. Res.*, **17**, 914 (1962).
283. J. HUGON and M. BORGERS, *J. Microscopie*, **4**, 643 (1965).
284. J. S. TRIER and T. H. BROWNING, *Clin. Res.*, **13**, 263 (1965).
285. J. S. TRIER and T. H. BROWNING, *J. Clin. Invest.*, **45**, 194 (1966).
286. J. JONEK, S. KOSMIDER and J. KAISER, *Intern. J. Radiation Biol.*, **7**, 411 (1963).
287. P. M. ANSARI, H. EDER and W. NAGELE, *Strahletherapie*, **120**, 275 (1963).
288. H. M. SPIRO and A. G. E. PEARSE, *J. Path. Bact.*, **88**, 55 (1964).
289. P. OTOUPAL and I. KUNSTYR, *Proc. Czech. Soc. Histochem. Cytochem.*, **3**, 359 (1971).
290. J. P. CONCANNON, R. E. SUMMERS, J. KING, G. TCHERKOW, Ch. COLE and E. ROGOW, *Am. J. Roentgen.*, **105**, 126 (1969).

291. G. D. CAIN, E. B. REINER and M. PATTERSON, *Archs. Int. Med.*, 122, 311 (1968).
292. W. O. DOBBINS, B. A. HERRERO and C. M. MANSBACH, *Am. J. Med. Res.*, 255, 63 (1968).
293. D. S. MADGE, *Comp. Biochem. Physiol.*, 36, 467 (1970).
294. R. MARANO, G. PASTORE, O. MANGHISI and O. SCHIRALDI, *Experientia*, 25, 1284 (1969).
295. K. BECKER, *Z. Gastroenterol.*, 6, 181 (1968).
296. J. MYREN, *Acta Histochem. Suppl. IX*, 291 (1971).
297. F. T. RACE, I. C. PAES and W. W. FALOON, *Amer. J. Med. Sci.*, 259, 32 (1970).
298. J. R. McPHERSON and W. H. J. SUMMERSKILL, *Gastroenterology*, 44, 900 (1963).
299. J. R. McPHERSON and R. G. SHORTER, *Amer. J. Dig. Dis.*, 10, 1024 (1965).
300. E. O. RIECKEN, R. ROSENBAUM, R. BLOCH, H. MENGE, E. RITT, M. ASLAN and W. DOELLE, *Klin. Wchschr.*, 47, 202 (1969).
301. E. A. SCHENK, I. M. SAMLOFF and F. A. KLIPSTEIN, *Amer. J. Clin. Nutr.*, 21, 944 (1968).
302. W. RUBIN, A. S. FAUCI, M. H. SLEISENGER and J. H. JEFFRIES, *J. Clin. Invest.*, 44, 475 (1965).
303. J. JOS, J. FREZAL, J. REY and M. LAMY, *Pediat. Res.*, 1, 27 (1967).
304. E. A. SCHENK, I. M. SAMLOFF and F. A. KLIPSTEIN, *Amer. J. Path.*, 47, 765 (1965).
305. F. E. PITTMAN and R. J. POLLITT, *Gut*, 7, 368 (1966).
306. A. P. DOUGLAS and C. C. BOOTH, *Lancet*, 2, 491 (1968).
307. T. LINDBERG, A. NORDEN and L. JOSEFSSON, *Scand. J. Gastroent.*, 3, 177 (1968).
308. E. O. RIECKEN and A. G. E. PEARSE, *Histochemie*, 5, 182 (1965).
309. A. J. WALL, A. P. DOUGLAS, C. C. BOOTH and A. G. E. PEARSE, *Gut*, 11, 7 (1970).
310. W. C. McDONALD, W. O. DOBBINS III and C. E. RUBIN, *New Engl. J. Med.*, 172, 448 (1965).
310a. P. FRIC, Z. LOJDA, J. JODL and F. MALIS, *Digestion*, 2, 35 (1969).
311. H. N. HOFFMAN, E. WOLLAEGER and E. GREENBERG, *Gastroenterology*, 51, 36 (1966).
312. R. R. W. TOWNLEY, M. H. CASS and Ch. M. ANDERSON, *Gut*, 5, 51 (1964).
313. D. L. ABELOFF, quoted according to T. H. Hendrix (1968).
314. R. R. MARONPOT and C. K. WHITEHAIR, *Canad. J. Comp. Med. Veter. Sci.*, 31, 309 (1967).
315. C. E. RUBIN and W. O. DOBBINS III, *Gastroenterology*, 49, 676 (1965).
316. V. L. SWANSON and R. W. THOMASSEN, *Amer. J. Path.*, 46, 511 (1965).
317. N. W. J. ENGLAND, *Amer. J. Clin. Nutr.*, 21, 962 (1968).

318. S. EIDELMAN, R. A. PARKINS and C. E. RUBIN, *Medicine (Baltimore)*, **45**, 111 (1966).
319. A. H. CAMERON, R. ASTLEY, M. HALLOWELL, A. B. RAWSON, C. G. MILLER, J. M. FRENCH and D. V. HUBBLE, *Quart. J. Med.*, **31**, 125 (1962).
320. J. R. COLLINS and K. J. ISSELBACHER, *Gastroenterology*, **49**, 425 (1965).
321. M. SIURALA, K. VARIS and B. A. LAMBERG, *Acta Med. Scand.*, **184**, 53 (1968).
322. H. I. TANKEL, D. H. CLARK and F. D. LEE, *Gut*, **6**, 560 (1965).
323. O. BRUNSER, A. REID, F. MONCKEBERG, A. MACCIONI and I. CONTRERAS, *J. Clin. Nutr.*, **21**, 976 (1968).
324. C. A. HEDBERG, C. S. MELNYK and Ch. F. JOHNSON, *Gastroenterology*, **50**, 796 (1966).
325. T. R. HENDRIX, *Gastroenterology*, **54**, 976 (1968).
326. B. BEDNAR, Malabsorption Syndrome (P. Fric, ed.), p. 100. St. Zd. N., Prague (1969).
327. K. N. JEEJEEBHOY, H. G. DESAI, J. M. NORONHA, F. P. ANTIA and D. V. PAREEKH, *Gastroenterology*, **51**, 333 (1966).
328. F. A. KLIPSTEIN, E. A. SCHENK and I. M. SAMLOFF, *Gastroenterology*, **51**, 317 (1966).
329. A. DAHLQVIST, *Gastroenterology*, **43**, 694 (1962).
330. U. P. HAEMMERLI, H. KISTLER, R. AMMANN, T. MARTHALER, G. SEMENZA, S. AURICCHIO and A. PRADER, *Amer. J. Med.*, **38**, 7 (1965).
331. Z. LODJA, P. FRIČ and J. JODL, *D. Ztschr. Verdaungs. Stoffwechselkr* (in press).
332. G. W. MEEUWISSE and A. DAHLQVIST, *Acta Paediat. Scand.*, **57**, 273 (1968).
333. L. E. ROSENBERG, S. DOWNING, J. L. DURAN and S. SEGAL, *J. Clin. Invest.*, **45**, 365 (1966).
334. K. J. ISSELBACHER, R. SCHEIG, G. R. PLOTKIN and J. B. CAULFIELD, *Medicine (Baltimore)*, **43**, 347 (1964).
335. S. M. SABESIN, G. D. DRUMMEY, D. M. BUDZ and K. J. ISSELBACHER, *J. Clin. Invest.*, **43**, 1281 (1964).
336. J. C. SIERACKI, *Arch. Path.*, **66**, 464 (1958).
337. J. MOPPERT, L. BIANCHI and H. BUHLER, *Virchows Arch. Path. Anat.*, **344**, 307 (1968).
338. D. DUDORKINOVA and Z. LOJDA, *Csl. Patol.*, **8**, (1972).
339. A. I. STRUKOV and L. I. ARUIN, *Verh. Dtsch. Ges. Path.* (53. Tgg), 185 (1969).
340. E. GUDMAND-HOYER, S. JARNUM and H. WORNING, *Gut*, **10**, 451 (1969).

CHAPTER 3

Biological Membranes

D. CHAPMAN

Department of Chemistry,
The University,
Sheffield, England

		Page
3.1	INTRODUCTION	124
3.2	ELECTRON MICROSCOPE STUDIES	125
	3.2.1 *Types of animal-cell membranes*	125
	3.2.2 *Intestinal membranes*	126
	3.2.3 *Fat transport*	130
	3.2.4 *The unit membrane concept*	130
	3.2.5 *The negative staining technique*	135
	3.2.6 *The freeze-etching technique*	135
3.3	MOLECULAR COMPONENTS	137
	3.3.1 *Membrane isolation*	137
	3.3.2 *Lipids*	139
	3.3.3 *Protein*	141
3.4	SOME PHYSICAL PROPERTIES OF MEMBRANE COMPONENTS	142
	3.4.1 *Thermotropic mesomorphism*	142
	3.4.2 *The effects of water*	144
	3.4.3 *Phase polymorphism*	147
3.5	MODEL MEMBRANES AND TRANSPORT PROCESSES	148
	3.5.1 *Model membranes*	149
3.6	THE FUTURE	155
	REFERENCES	155

3.1 INTRODUCTION

The idea of a cell membrane existed for many years almost entirely at a conceptual level; physiological experiments suggested a barrier to the exchange of material between the cell and its surroundings. It was called a membrane because the barrier was thought to be a thin layer completely enclosing the cell contents. The membrane appeared to have definite mechanical and physical properties and, even in the absence of any direct analyses, suggestions were made as to its structure and organization. The introduction of the electron microscope revealed the very many membrane systems present within the cell as well as the outer plasma membrane.

The term 'cell membrane' is used in at least three different senses:

The anatomical sense. In this sense the cell membrane is the external limiting region of the cell, visible occasionally as a darkly staining region in the light microscope and with more certainty in the electron microscope as a layer (or pair of layers).

The biochemical sense. In this case the cell membrane is a 'fraction' of the cell prepared by the techniques of selective disintegration of the whole cell, followed by differential centrifugation. A preparation is obtained which can be analysed chemically and which can be compared by electron microscopy with the 'cell membrane' seen in the whole cell.

The physiological sense. In this case it is a hypothetical structure devised to explain certain data on permeability and which also explains other data on the distribution of metabolites and other molecules between the cell and the fluid in which the cell is immersed.

At the present time there is intense activity aimed to bring these three concepts together so as to explain membrane structure and function in molecular terms. Chemical analysis of the components of cell membranes has now been made possible using chromatographic techniques showing the various lipid and protein components present. Physical studies of the single pure components have begun to reveal the basic physical chemistry involved in cell membrane organization and the ways in which the components can mutually interact and affect each other. Finally, the introduction of model membrane systems has led to a more detailed understanding of the possible ways in which water can be transported through membranes and also the way in which ions can be transported and discriminated. More sophisticated model systems may eventually lead to an understanding of the transport of amino acids and sugars.

3.2 ELECTRON MICROSCOPE STUDIES

3.2.1 Types of animal-cell membranes

Plasma membranes. The plasma membrane interposes a boundary between the cell and its environment by the active transport of ions and nutrients, and it creates and maintains the interior environment of the cell. In certain tissues, e.g. of the intestinal mucosa and kidney, substances are transposed across cellular barriers in the interests of maintaining the internal environment of the whole organism. In the nervous system, the electrochemical potentials to which the transport of Na^+ and K^+ ions give rise are used for the transmission and processing of information. Schwann cells, in the central nervous system, and oligodendrogliocytes, in the central nervous system, insulate the axons of neurones from their neighbours by myelin formation and thereby bring about a speeding up of the impulse transmission by means of saltatory conduction.

Certain types of cells — epithelial cells, cardiac and smooth-muscle cells — make specialized contacts characterized by modifications of the plasma membrane. One type of contact, the desmosome or macula adhaerens, appears to provide anchorage and involves a somewhat wider than normal intercellular space (220—250 Å) filled with a bank of lightly staining material. The tight junction involves membrane-to-membrane apposition without intervening space.

Cytoplasmic membranes
The cytoplasmic membranes (cytomembranes, endoplasmic reticulum) are usually classified as 'rough' and 'smooth'. The roughness is due to the ribosomes that stud their surface and these are numerous in cells that synthesize protein for export.

Both types of membrane form arrays of cisternae of tubules which vary considerably in form and intercommunicate with each other and perhaps with the cell exterior. The Golgi apparatus is a specialized form of smooth membrane, characteristically seen as a somewhat isolated pattern of stacked cisternae, and it is particularly prominent in cells that elaborate secretions rich in carbohydrate.

Organelle membranes

Subcellular organelles are the least homogeneous. The membranes, comprising the nuclear envelope, are thought to be homologous with the endoplasmic reticulum; they are studded with pores about 0.05μ in diameter which may facilitate the exchange of macromolecules (messenger ribonucleic acid, repressors) essential for the genetic control of the cell. The essential structural features of the mitochondrion are the double membrane and the characteristic cristae. As first stressed by Palade [1], the number of cristae and their conformation varies considerably from one type of mitochondrion to another. However, the elements that appear to be common to all mitochondria are an outer membrane, an inner membrane, cristae, mitochondrial 'sap' or 'matrix', dense granules and DNA.

3.2.2 Intestinal membranes

The structure of the intestinal absorptive cell has been discussed in Chapter 1 by Creamer. (See Fig. 3.1.) Many membrane systems are associated with the intestinal cell as revealed by electron microscope techniques. The microvilli of the intestinal absorptive cell has been the subject of many studies because of their location at the interface between the external environment (gut lumen) and the internal environment (the cell cytoplasm). A substance absorbed or excreted from the gut is expected to pass through this cell membrane system.

The plasma membrane covering the microvilli is wider (95 to 115 Å) than the plasma membrane of the lateral and basal portions of the absorptive cells (70 to 80 Å). It is a continuous triple-layered membrane or tram line structure, the outer and inner lines being more electron dense.

Most investigators feel that the membrane is symmetrical but others disagree. Applied to the outer layer of the apical plasma

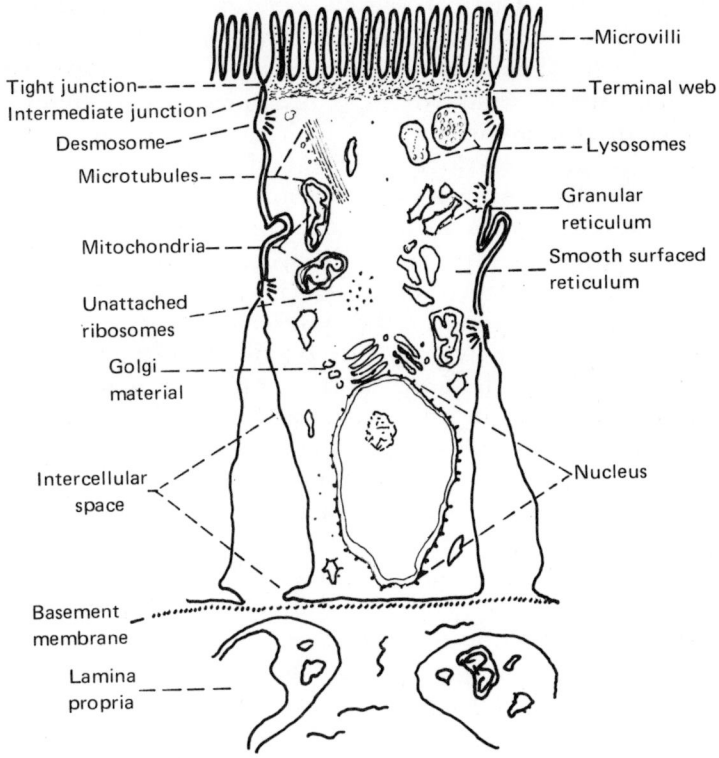

Fig. 3.1. Schematic illustration of an intestinal absorptive cell.

membrane is a coating of very fine filaments oriented perpendicular to the plane of the membrane. This surface coating has been named 'fuzz' by Ito [2]. It is rich in acid mucopolysaccharide. Labelled sulphur and glucose (known mucopolysaccharide precursors) have been traced from absorptive cell cytoplasm into the fuzzy surface layer of the microvilli. This suggests that the fuzz is synthesized by the epithelial cells and is an integral part of the cell, not an extraneous coating of absorbed mucus from the intestinal lumen. This coating may serve as a barrier against noxious substances within the gut lumen and it is seen coating the plasma membranes of many other cell types engaged in active transport.

Cytoplasmic processes of adjacent epithelial cells frequently interdigitate with one another at their lateral borders. The upper halves of the lateral plasma membranes of the cells are generally closely apposed to each other (95 to 100 Å apart). Between the basal portions of the cells, there are relatively large intercellular spaces which are most prominent at the villous tip and which disappear entirely near the base of the villus.

Distinctive specializations of the lateral plasma membranes of adjacent cells have been known for some time [3]. There is a 'tight junction' between adjacent absorptive cells at the level of the upper portion of the terminal web. To form this 'tight junction' the outer layers of the adjacent plasma membranes fuse, completely obliterating the intercellular space [4] for a distance of approximately 0.1 to 0.2 μ. This tight junction completely encircles the upper portion of the web area in belt-like fashion, forming an effective barrier between the intestinal lumen and the intercellular space [4]. Thus, material from the intestinal lumen must first enter and travel through the apical cytoplasm before gaining access to the intercellular space [4] via the lateral plasma membrane. Dewey and Barr [5] have suggested that this tight junction may also be a site of chemical exchange between the adjacent epithelial cells. Immediately below the tight junction the lateral cell membranes diverge and form the 'intermediate junction' which extends for 0.2 to 0.5 μ and which is characterized by an intercellular space approximately 200 Å wide.

The desmosomes are another type of junctional complex which are located at multiple levels along the remainder of the lateral plasma membrane. These button-like zones have a thin, central dense area in the centre of the intercellular space and disc-like accumulations of dense material in the epithelial cytoplasm just lateral to the plasma membranes [6, 7]. Fine filaments of the cytoplasmic matrix appear to attach to all three types of intercellular junctions and desmosomes bind adjacent cells firmly to one another.

The basal surface of the absorptive cell is generally smooth, although occasional small, pseudopod-like extensions of cytoplasm may project into the lamina propria through the basement membrance applied to the basal surface. The basement membrane is approximately 300 Å wide [6] and is composed of homogeneous material of intermediate density. It serves to separate the epithelium from the lamina propria. There is increasing evidence that epithelial basement membranes in general are mucoproteins synthesized by the epithelial cell to which they are applied [3, 8, 9, 10].

Napolitano and Kleinerman [11] have reported that the membranes enclosing the vesicles in this region are of the same dimensions as the plasma membrane covering the microvilli. This is consistent with the idea that some of the vesicles in the terminal web are derived from the apical plasma membrane of the cell [11]. The role of the terminal web is unknown. It has been suggested that it serves a structural function, stiffening and stabilizing the apical cell surface as indicated by its ability to resist osmotic shock which disintegrates the remainder of the absorptive cell cytoplasm [12].

Lysosomes are found in the apical cytoplasm beneath the terminal web of normal absorptive cells. Zetterqvist reported these structures in mouse absorptive cells [6] and they have since been observed in other mammalian species [13]. The lysosomes are enclosed by well defined membranes separating them from the surrounding cytoplasm. They have been shown by histochemical techniques to contain acid phosphatase [13, 14, 15]. There is increasing evidence that these structures, with their lytic enzymes, help cells rid themselves of waste products either by changing them into useful materials which are released into the surrounding cytoplasm or by segregating useless or noxious substances until they can be extruded from the cell [15].

A moderate amount of endoplasmic reticulum is seen in absorptive cells [6, 7, 16, 17]. This is a complex interconnecting system of membrane-bounded tubules, cisterns, and vesicles. Two components are seen; granular reticulum composed of membranes whose outer surface is studded with dense granules rich in ribonucleic acid (ribosomes), and agranular reticulum whose membranes are smooth surfaced and devoid of ribosomal granules. The reticulum found in absorptive cells is not nearly as extensive or as well organized as that of cells with known secretory functions [18, 19, 20].

Goblet cells are abundant in the small intestinal crypts but are also seen as part of the villous epithelium. The fine structure of goblet cells from mouse [21], rat [22, 23] and human [24] small intestinal epithelium has been described. Regardless of their location the general structure of these cells is similar. The typical goblet cell is distended with mucous granules and, as its name implies, is shaped like a brandy goblet. The microvilli on their apical surface are similar to those of the absorptive cells but less abundant and more irregular in shape and length. The lateral and basal plasma membranes and the basement membranes have the same structural features as do those of the villous absorptive cells.

3.2.3 Fat transport

Once fat has entered the absorptive cells, fat particles in rodents and man are enclosed in a membranous envelope within the cytoplasm [25]. As absorption progresses, more fat particles are concentrated in the supranuclear zone and conglomerations of many particles may be enclosed in a single membranous envelope, especially in the Golgi zone. Fat particles then appear in the intercellular spaces between the basal halves of the absorptive cells. Once the fat particles leave the cell, they are divested of their membranous envelopes. It has been suggested that they leave the cell by reverse pinocytosis [25], leaving their envelope behind as part of the plasma membrane. There are interruptions in the basement membrane where the inter-epithelial spaces are continuous with the lamina propria. Fat particles from 0.1 to 1μ in diameter may be seen within these openings in the basement membrane extending into the lamina propria. These openings do not appear to be artifactual and are the same as those through which lymphocytes extend from lamina propria between epithelial cells. As revealed by the electron microscope, there are many membrane systems associated with the intestinal absorptive cell. However, an important question is, does the e.m. data provide information about the molecular organization of the membranes? Robertson [26 27] has discussed this in terms of the unit membrane concept.

3.2.4 The unit membrane concept

The similarity in appearance of the electron micrographs of many different types of cell membranes were considered by Robertson to be evidence for a common structure at the surface of a wide variety of cells. Using a new fixative procedure involving $KMnO_4$, he was able to observe a similar three-layered unit, approximately 75 Å thick, at the surface of a number of different cell types as well as in many different cellular

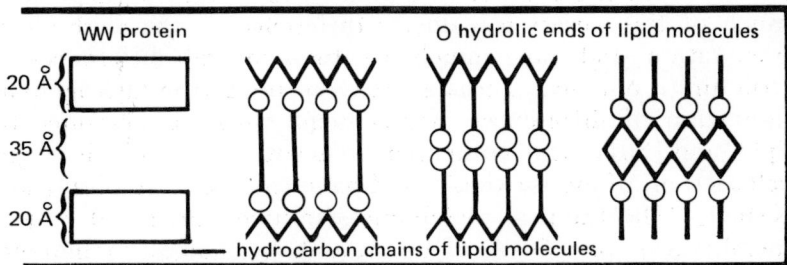

Fig. 3.2. Unit pattern of membranes under the electron microscope appears as at far left. Next to this is shown the classical molecular structure for the membrane. Two right-hand diagrams show alternative structures which could also explain the observed unit pattern. Difficulties in interpreting electron micrographs account for the fact that no one structure is as yet universally accepted.

organelles. This unit appeared as two dense lines about 20 Å wide separated by a lighter space of 35 Å. As we have seen, similar structures are observed with intestinal membranes. Robertson [27] proposed a model of the membrane which he termed a 'unit membrane' (Fig. 3.2). This concept appeared to clarify and unify a wide body of information including other physical data as well as other electron microscope data. By referring to the biological membrane as a unit, he emphasized not only that all three parts of the triple-layered 75 Å structure seen in the electron microscope were part of one membrane, but also that all membranes had a similarity of molecular arrangement and origin. In general with plasma membranes, triple-layered structure are also observed after fixation in $KMnO_4$, and often, but not always, after fixation in OsO_4. The endoplasmic reticulum (cytomembranes), the outer, inner and cristae membranes of mitochondria, chloroplast membranes, the two membranes of the nuclear envelope, and the membranes of bacterial protoplasts and spheroplasts are also revealed as triple-layered structures in electron micrographs.

Despite these general similarities there has been some reluctance (Stoeckenius [29]) to accept this proposal of an identical unit membrane structure for all the membranes in cells and particularly for the mitochondrial membranes.

In general, membranes are thought to be about 75 Å wide.

There are considerable variations in dimensions of membranes. The overall widths of triple-layered plasma membranes appear to vary from about 50 Å to perhaps 130 Å (Elbers [29]). How much of this variation is due to differences in the methods of preparation and how much to fundamental differences in structure is not clear. Perhaps the clearest electron micrographic indication of differences among membranes was obtained by Sjöstrand [30], who compared adjacent membranes in single cell sections of mouse kidney and pancreas fixed with OsO_4 and $KMnO_4$. The thinnest membranes (mitochondrial and α-cytomembranes) were 50 Å to 60 Å, and the thickest membranes (plasma and zymogen granules) were 90 Å to 100 Å. These variations indicate the difficulty of interpreting micrographs in terms of molecular structure.

Several other observations have appeared which seem to disturb the 'unit membrane' picture. Hillier and Hoffman [31] studied the structure of the membrane of erythrocytes by using shadowing techniques which revealed a mosaic structure. They suggested that the erythrocyte envelope was composed of plaques situated on the outside of a fibrous network joined together by lipids. Glaeser and co-workers [32] have studied the membrane structure of OsO_4-fixed erythrocytes viewed 'face on' by electron microscopy. This supports the conclusions of Hillier and Hoffman [31] and shows that the surface of the rat red cell membrane has a 'pebbly' appearance at the level of 400 Å to 500 Å. These authors suggest that these bumps on the surface may be associated with a filamentous structure, the bumps representing the tops of loops of the filaments.

Sjöstrand [30] observed globular subunits in one of the opaque layers of mitochondrial membranes and smooth endoplasmic reticulum, and an asymmetry in the electron opacity of the dense lines in the plasma membrane. Subunits, or cross-linkages bridging the gap between the two opaque bands of the triple-layered structure, have been observed by Robertson [33], who later reinterpreted them as an electron optical artifact derived from a mosaic pattern in the plane of one or both surfaces of the triple layer. In several instances, hexagonal mosaic patterns have been seen on the surfaces of plasma membranes (Benedetti and Emmelot [34]). Strong evidence for membrane subunits is suggested in a paper by Blasie and co-workers [35]. Outer segment membranes of frog retina were isolated and were oriented in ultracentrifugal pellets. Electron microscopic surface views of negatively stained membranes and low-angle X-ray diffraction patterns from unfixed, unstained pellets showed square array of spherical particles. The unit cell size was about 70 Å and the particles had a non-polar core about 40 Å in diameter. When attempts are made to obtain further information about the interpretation of the electron microscope data of fixed sectioned tissues in molecular terms, we find that there are a number of difficulties.

Korn [36] has critically discussed this question in some detail. He points out that Robertson [37] and Fernández-Morán and Finean [38] have questioned whether the manganese atom is responsible for any of the electron opacity in micrographs of $KMnO_4$-fixed cells. He suggests that, if this is not the case, there is then no way to interpret the dense lines in micrographs of membranes fixed with $KMnO_4$ in molecular terms. This is a serious deficiency because it is in such preparations, as we have already seen, that the triple-layered structure is most reproducibly and distinctly seen. There appears to be only one study in which cells were chemically studied during fixation with $KMnO_4$. Korn and Weisman [39] found that the lipids of amoebae were essentially unaffected by fixation with 1% $KMnO_4$ for 1 hour at 0°C. All the neutral lipids and about half of the phsopholipids were extracted from the amoebae during dehydration in ethanol.

Similar triple-layered structures are often observed when OsO_4 is used as a fixative and various experiments have been devised so as to ascertain the site of the fixation process. This fixation was originally thought to take place at the sites of the double bonds in the hydrocarbon chain. Wigglesworth [40] suggested that osmium can cross-link through the ethylenic double bonds and has given evidence to show the occurrence of an insoluble polymeric complex of lipid and osmium which he considers is the basis of cytological fixation. This observation was considered to be consistent with the fact, noted by many authors, that fully saturated phospholipids, such as the phosphatidylethanolamines, do not react easily at room temperature with osmium tetroxide, although suggestions had been made that brominated or hydrogenated phospholipids can take up osmium without fixation occurring. Bahr [41] showed that many amino acids would also react with osmium tetroxide, and so some authors have considered that the dense lines observed in electron micrographs of cell membranes represent protein, whereas the light central line corresponds either to the hydrocarbon chain region of the lipid, or to a gap produced by removal of lipid in the preparative stages for electron microscopy (Robertson [42]).

Stoeckenius [43, 44] studied natural lipids after reaction with osmium tetroxide, and concluded that the dark area seen in electron micrographs corresponds to osmium located at the polar groups of the phospholipid molecules. Finean [45], after an analysis of X-ray data of fixed and unfixed tissue, also considered that the osmium is located between the polar groups of the bimolecular sheets of lipid. Riemersma [46] considers, after chromatographic analysis of the intermediates formed during osmium fixation of unsaturated lecithins, that the initial reaction is with the double bonds, but that there is a subsequent migration of osmium derivatives to the polar groups. Chapman and Fluck [47], in an attempt to clarify the problem, attempted to fix saturated phospholipids with OsO_4, arguing in this case that, as there were no double bonds present, any successful fixation must be at the polar group. This study showed that, whilst fully saturated phosphatidylethanolamine reacts with OsO_4, the fully saturated lecithins do not .

Korn [36] concludes that the dense lines in membranes fixed with osmium tetroxide reveal nothing about the molecular orientation of the phospholipid, in the original membrane. He also points to the fact that triple-layered membranes are seen in osmium-fixed mitochondria from which all the lipid has previously been removed by extraction with acetone (Fleischer and colleagues [48]. [*Escherichia coli* B (van Iterson [49]) also gives a triple-layered membrane despite the fact that this organism contains essentially no unsaturated fatty acids.] If lipids are not necessary to reveal the triple-layered structure, then its explanation in terms of particular molecular configurations cannot be correct.

We see that there are still many doubts concerning a full acceptance of the 'unit membrane'. Many of the supporting arguments for its acceptance depend upon studies of myelin which may be chemically, metabolically and functionally different from all other membranes.

3.2.5 The negative staining technique

The negative staining technique using phosphotungstate has increased in popularity in recent years. In this method the biological membranes are immersed in a pool of electron-dense material (e.g. sodium phosphotungstate) which dries to form an electron-dense glass. Membranes appear as regions of electron transparency against a dark background. The specimen is not sectioned.

The possibility of artifacts being introduced with the negative staining technique has been considered. Bangham and Horne [50] show a number of different phase structures with artificial lipid mixtures, some of which resemble structures seen in negatively stained natural membranes. Whittaker [51] has commented that the negative staining technique only reveals those structures sufficiently tightly packed to exclude the negative stains. Proteins possessing an α-helix are clearly seen, whilst those with an open structure are not. In some instances the fine structures observed can be generated on the electron microscope grid by interaction with negative stains during drying. There is the further possibility of spontaneous or enzyme-catalysed rearrangements occurring when biological organelles are ruptured.

3.2.6 The freeze-etching technique

Recently a new technique has been introduced to electron microscopy (Moor and co-workers [52]) which attempts to overcome the many problems associated with chemical-fixation methods and the production of artifacts associated with these methods. This is the technique of freeze-etching (Fig. 3.3).

Freeze-etching involves six preparational steps: pretreatment of the object, freezing, chipping of the frozen specimen followed by etching and coating and, finally, the cleaning of the replica. One of the most important processes is to get a clean fracture plane through the frozen specimen. This can be done by chipping under high vacuum at a controlled object temperature. In this method there is no chemical treatment of the object during the whole procedure until the replica is formed on the fracture plane from the frozen specimen. By means of snap freezing and/or glycerol impregnation, cells have been frozen so as to preserve the life of the organisms.

Freeze-etching depends upon freezing a very small sample

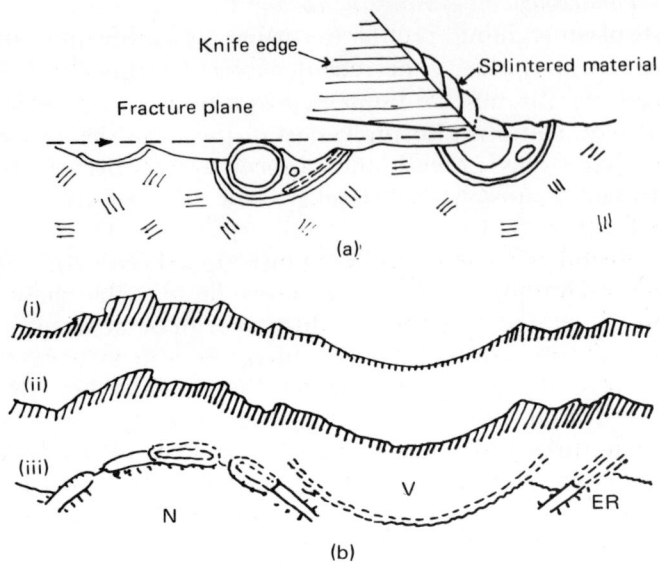

Fig. 3.3. (a) Diagram of the cutting procedure which actually consists of a fine splintering of the deep-frozen object (yeast cells);

(b) Diagram of how the splintering and the etching reveal the fine structure of a frozen object: (i) cross-section through the fracture plane; (ii) etched fracture plane, showing the fine structure; (iii) the reconstructed structural details of the recorded object. N, nucleus, showing a partially removed envelope. V, totally removed vacuole, rendering possible a surface view of the adjacent cytoplasmic ground substance. ER, endoplasmic reticulum, fractured at a low angle. (From Moor and Nuhlethaler [52].)

very rapidly (the sample is frozen in freon cooled with liquid nitrogen), usually in the presence of glycerol, and then putting the rapidly frozen specimen on to the cold stage of a vacuum coating unit. After a very high vacuum has been obtained, say 2 or 3×10^{-6} torr, the surface of the specimen is cut with a cold knife. The temperature of the specimen is then raised to $-100°C$ and a thin layer of ice is sublimed off. The cut surface is therefore etched by vacuum sublimation. Immediately after this the surface of the specimen is shadowed with a heavy metal and a carbon replica is made. This replica is then examined in the electron microscope.

Using this technique, Moor and Mühlethaler [53] reported the results of a study, using the freeze-etching method, on yeast cells, *Saccharomyces cerevisiae*. They show that the cell wall and cytoplasmic membrane are clearly separated and suggest that the fracture plane either follows the surface of the cell wall or penetrates the whole wall perpendicularly. By combining surface views and cross-fracture views, they were able to obtain a three-dimensional image of the invaginations of the yeast cytoplasmic membrane. The invaginations have an average length of 3,000 Å, a width of 200 Å to 300 Å and a depth of 500 A. The cross-fractured membranes give images the same as those shown by membranes in sections of permanganate-fixed material. Thus the membrane consists of three subunits each about 25 Å thick. They suggest that this is in agreement with the structure of the unit membrane suggested by Robertson [26, 27]. Branton [54] suggests that the fracture planes split along the inside of the membrane effectively with a bilayer system along the methyl planes of the hydrocarbon chains.

3.3 MOLECULAR COMPONENTS

3.3.1 *Membrane isolation*

To understand the structure of biological membranes at the molecular level, their various components must be identified and isolated in a pure state. This is by no means easy, particularly because the membrane preparation must lack none of the membrane components themselves. The problem of isolating pure biological membrane material therefore involves establishing a method for destroying cells, isolating the membrane material, demonstrating the material's precise origin, and, finally, estimating the purity of the preparations.

The biological membrane which has received considerable attention is that of the erythrocyte — the red blood cell. When erythrocytes are subjected to certain osmotic pressures or enzymes, they undergo haemolysis — a process in which the cell membrane structure loosens sufficiently to allow the cell contents to escape without itself being ruptured completely. All that remains are the 'ghosts' of the original cells — a collection of erythrocyte membranes.

Recently many other membrane systems have been examined

including other plasma membranes, mitochondria, chloroplast membranes and bacterial membranes [55].

The success of the isolation of the membrane material depends also on its identification. Electron microscopy is sometimes useful. Surface antigens which attach to specific target chemicals have been used as marker materials. A difficulty in this approach is to ensure that the marker molecule is specific to the membrane. Attempts have been made to overcome this by artificially labelling the membrane with a chemical known to be confined to the membrane and easily detectable during the separation of the membranes from the constituents of the disrupted cells.

To be certain which materials are contaminants of this membrane fraction is also rather difficult. With erythrocyte ghosts the question has been raised as to whether haemoglobin is a true component of the membrane or whether it is a contaminant. This question is still unresolved and the significance of the part played by haemoglobin in the membrane structure is somewhat obscure.

Recently modern analytical techniques have been increasingly applied to membrane preparations. The membrane materials which have been analysed appear to contain lipid, protein and carbohydrate material.

The composition of some common membranes in terms of lipid and protein content are given in Table 3.1. It can be seen

TABLE 3.1. Protein and lipid composition of animal and bacterial membranes.

Origin of membrane	Amino acid	Molar ratio phospholipid	Cholesterol	Amino ratio (protein/lipid)*
Myelin	264	111	75	0.43
Erythrocyte	500	31	31	2.5
Bacillus licheniformis	610	31	0	4.8
Micrococcus lysodeikticus	524	29	0	4.3
Bacillus megaterium	520	23	0	5.4
Streptococcus faecalis	441	31	0	3.4
Mycoplasma laidlawii	442	25.2	2.3	4.1

* From Korn, 1966 [36].

that the lipid to protein ratio varies considerably from one type of membrane to another.

3.3.2 *Lipids*

Analysis of cell membranes has shown that a particular membrane may contain one or more different types of lipid. Some of the lipid classes which have been observed are shown in Fig. 3.4.

Associated with each lipid class present in membranes are a range of fatty acids. Usually the saturated fatty acids are found esterified at the 1-position on the phosphoglyceride and unsaturated fatty acids at the 2-position. Typical structures of some of these fatty acids are shown in Fig. 3.5.

Other variations on the fatty acid structure include the presence of a cyclopropane ring; e.g. lactobacillic acid, present in certain bacteria, contains this grouping, and a cyclopentene ring which occurs in hydnocarpic, chaulmoogric and gorlic acids.

Comparisons of the fatty acid analyses of the non-phospholipid and phospholipid fractions of several tissues from a number of mammals (Veerkamp and co-workers [53]) show that the neutral lipid fractions possess some degree of animal specificity, i.e. the fatty acid patterns may differ from animal to animal, but resemble one another for different tissues of one animal. In the rat the fatty acid composition of neutral lipid fractions from several tissues appeared to be rather similar to that of the abdominal depot fat. The phospholipid fractions of the tissues studied were again found to be usually richer in stearic and poorer in palmitic acid when compared with the non-phospholipids. Furthermore, the phospholipids from several tissues of *one* animal species revealed a less characteristic fatty acid pattern than that given by the neutral lipids. By contrast, the fatty acids from phospholipids may exhibit a certain degree of similarity in homologue tissues of different animals. This type of specificity is demonstrated by a comparison of the fatty acid pattern of phospholipids from lung and brain tissue of a number of species.

Sialic acid molecules are also found to be associated with certain membranes, e.g. sialic acids have been found to be associated with erythrocyte and Ehrlich ascites carcinoma cells.

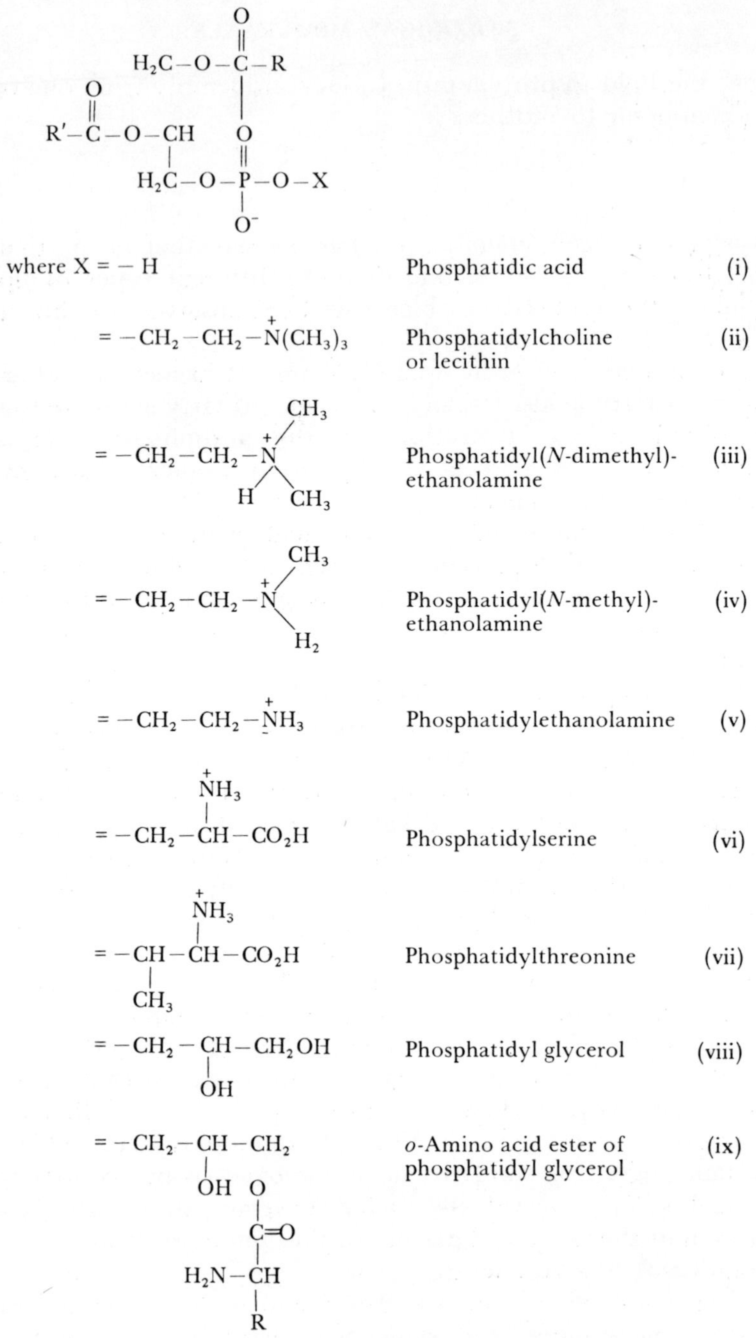

$$\begin{array}{c}
\text{O} \\
\| \\
\text{H}_2\text{C}-\text{O}-\text{C}-\text{R} \\
\text{O}| \\
\|| \\
\text{R}'-\text{C}-\text{O}-\text{CH}\text{O} \\
|\| \\
\text{H}_2\text{C}-\text{O}-\text{P}-\text{O}-\text{X} \\
| \\
\text{O}^-
\end{array}$$

where X = $-$H	Phosphatidic acid	(i)
= $-$CH$_2-$CH$_2-\overset{+}{\text{N}}$(CH$_3$)$_3$	Phosphatidylcholine or lecithin	(ii)
= $-$CH$_2-$CH$_2-\overset{+}{\text{N}}\!\!\begin{array}{l}\diagup\text{CH}_3\\ \diagdown\text{CH}_3\end{array}$ with H	Phosphatidyl(N-dimethyl)-ethanolamine	(iii)
= $-$CH$_2-$CH$_2-\overset{+}{\text{N}}\!\!\begin{array}{l}\diagup\text{CH}_3\\ \diagdown\text{H}_2\end{array}$	Phosphatidyl(N-methyl)-ethanolamine	(iv)
= $-$CH$_2-$CH$_2-\overset{+}{\text{N}}$H$_3$	Phosphatidylethanolamine	(v)
= $-$CH$_2-\underset{\underset{\text{CH}}{\|}}{\overset{+}{\text{NH}_3}}-\text{CO}_2\text{H}$	Phosphatidylserine	(vi)
= $-$CH$-$CH$-$CO$_2$H with $\overset{+}{\text{NH}_3}$ and CH$_3$	Phosphatidylthreonine	(vii)
= $-$CH$_2-$CH$-$CH$_2$OH with OH	Phosphatidyl glycerol	(viii)
= $-$CH$_2-$CH$-$CH$_2$ with OH and O$-$C(=O)$-$CH(NH$_2$)$-$R	o-Amino acid ester of phosphatidyl glycerol	(ix)

Fig. 3.4. The structures of some phosphoglycerides.

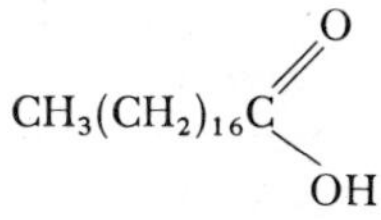

$CH_3(CH_2)_{16}COOH$

Stearic acid 18:0
(Octadecanoic acid)

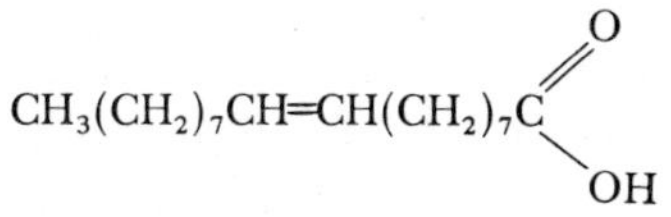

$CH_3(CH_2)_7CH=CH(CH_2)_7COOH$

Oleic acid 18:1
(Octadec-9-enoic acid)

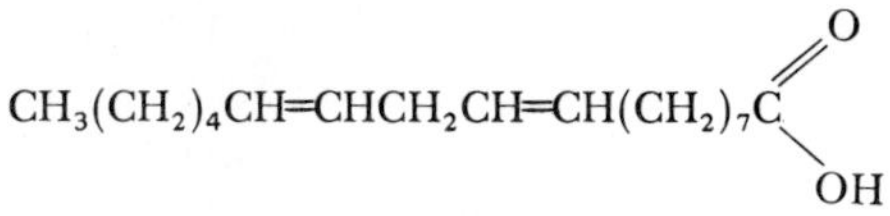

$CH_3(CH_2)_4CH=CHCH_2CH=CH(CH_2)_7COOH$

Linoleic acid 18:2
(Octadeca-9:12-dienoic acid)

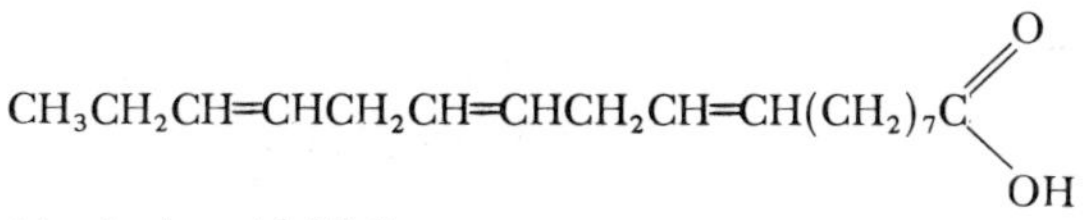

$CH_3CH_2CH=CHCH_2CH=CHCH_2CH=CH(CH_2)_7COOH$

Linolenic acid 18:3
(Octadeca-9:12:15-trienoic acid)

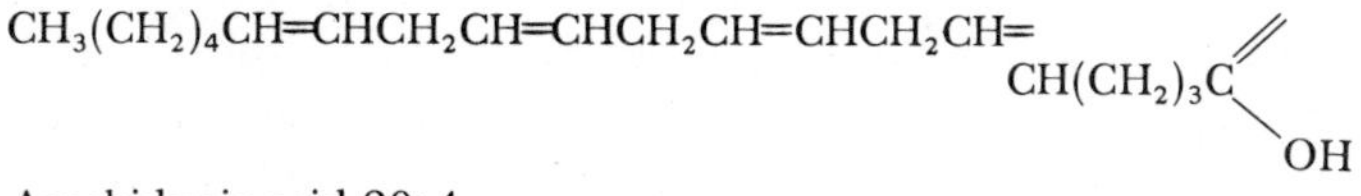

$CH_3(CH_2)_4CH=CHCH_2CH=CHCH_2CH=CHCH_2CH=CH(CH_2)_3COOH$

Arachidonic acid 20:4
(Eicosa-5:8:11:14-tetraenoic acid)

Fig. 3.5. The structures of a range of naturally occurring fatty acids.

3.3.3 Protein

During the past few years a number of studies have been made on the isolation and characterization of proteins from erythrocyte membranes. Maddy [56] has described a method for the solubilization of the protein of ox erythrocyte by n-butanol fractionation. Zwall and van Deenen [57] have described a method for the solubilization of erythrocyte ghosts using n-pentanol. In contrast to the butanol procedure, not only the proteins but also the lipids were recovered in the water layer.

Rosenberg and Guidotti [58] have also carried out studies on the characterization of the protein component of mammalian erythrocyte membranes. Exhaustive lipid extraction of the membrane produced a glycoprotein residue containing greater than 90% of the total membrane protein. The protein was found to be soluble in formic acid and sodium dodecyl sulphate solutions. The protein of this membrane appears to be a heterogeneous collection of proteins, many in the molecular weight range near 50,000. Amino acid composition of the membrane protein shows a high percentage of amino acids with long non-polar side chains. There is a small excess of these amino acids as compared with basic amino acids. Molecular weights of the major protein components fall in the range normal for most globular proteins. The amino acid composition of lipid-extracted membrane protein is shown in Table 3.2.

3.4 SOME PHYSICAL PROPERTIES OF MEMBRANE COMPONENTS

3.4.1 *Thermotropic mesomorphism*

In addition to the capillary melting point, other phase changes have been shown to occur with phospholipids at lower temperatures. For example, when a pure phospholipid, dimyristoylphosphatidylethanolamine, containing two fully saturated chains is heated from room temperature up to the capillary melting point, a number of thermotropic phase changes occur (i.e. phase changes caused by the effect of heat). This was shown by infra-red spectroscopic techniques (Byrne and Chapman [59]), by thermal analysis techniques (Chapman and Collin [60]) and has now been studied by a variety of physical techniques (Chapman and co-workers [61], Chapman and Salsbury [62]).

The main conclusions from these various physical studies are that:

(i) Even with the fully saturated phospholipid at room temperature, some molecular motion occurs in the solid. This is evident from the nuclear magnetic resonance spectra

TABLE 3.2. Amino acid composition of lipid-extracted membrane proteins.

Amino acid	Residues per 100 residues	
Lysine	5.21	
Histidine	2.44	12.18
Arginine	4.53	
Aspartic acid*	8.49	
Threonine	5.86	
Serine	6.26	
Glutamic acid*	12.15	
Proline	4.26	
Glycine	6.73	
Alanine	8.15	
Half-cystine	1.08	
Valine	7.10	
Methionine	2.02	
Isoleucine	5.29	
Leucine	11.34	
Tyrosine	2.41	
Phenylalanine	4.20	
Tryptophan	2.49	

* One mole of NH_3 was formed for every three aspartic and glutamic acids. (From Rosenberg and Guidotti, 1968 [58].)

and from the infra-red spectra taken at liquid nitrogen and at room temperature.

(ii) When the phospholipid is heated to a higher temperature it reaches a transition point, a marked endothermic change occurs, and the hydrocarbon chains in the lipid 'melt' and exhibit a very high degree of molecular motion. This is evident on the appearance of the infra-red spectrum and also in the narrow nuclear magnetic resonance line width. The broad diffuse appearance of the infra-red spectrum is consistent with the chains flexing and twisting and with a 'break-up' of the all-planar *trans* configuration of the chains.* The fact that the phase transition is concerned

* Chapman and Salsbury [62] interpreted this data to indicate that there is a distribution of motion along the lipid chain with greater motion of groups near the methyl end.

primarily with the hydrocarbon chains of the phospholipid is confirmed by the X-ray data. This shows that the space taken up by the glycerol and polar group remains essentially unchanged when this phase transition occurs.

(iii) When phospholipids contain shorter chain lengths, or unsaturated bonds, those marked endothermic phase transitions occur at lower temperatures (Chapman and co-workers [61]). The temperature at which these transitions occur parallels the behaviour of the melting point of the related fatty acids. The transition temperatures are high for the fully saturated long chain phospholipids, lower when there is a *trans* double bond present in one of the chains, and lower still when there is a *cis* double bond present.

(iv) An important conclusion which we obtain from these phase transition studies is that, above the endothermic transition temperature, a given phospholipid will be in a highly mobile condition with its hydrocarbon chains flexing and twisting. This is a *fundamental property* of the phospholipid and we can expect this chain mobility to occur in whatever situation the phospholipid occurs unless, for special reasons, this motion is somehow inhibited. We can envisage that inhibition of chain motion by interaction with other molecules (e.g. water or protein) would provide one of these special reasons. In other circumstances, due to less perfect packing arrangements, we might expect, at a particular temperature, even greater mobility of the chains of the lipid and, indeed, diffusion of the whole lipid molecules.

3.4.2 *The effects of water*

Transition temperatures Small amounts of water can have unusual effects upon the phospholipid mesomorphic behaviour. Thus the diacylphosphatidylcholines (lecithins) exhibit additional liquid crystalline forms between the first transition temperature and the capillary point (Chapman and co-workers [61]). The intermediate liquid crystalline form is found to exhibit X-ray spacings consistent with a cubic phase organization. On the other hand, if all the water is removed from the phospholipid, the lipid no longer exhibits this phase.

When these phospholipids are examined in increasing

amounts of water the various physical techniques, such as microscopy, n.m.r. spectroscopy or differential thermal analysis, show that as the amount of water increases, the marked endothermic transition temperature for a given phospholipid falls (Chapman and co-workers [69]). The transition temperature does not fall indefinitely; it reaches a limiting value independent of the water concentration. We can understand this if we regard the effect of water as leading first to a 'loosening' of the ionic structure of the phospholipid crystals. This, in turn, affects the whole crystal structure and a reduction, up to a certain limit, of the dispersion forces between the hydrocarbon chains. Large amounts of energy are still required to counteract the dispersion forces between the chains and quite high temperatures are still required to cause the chains to melt. The limiting transition temperatures parallel the melting point behaviour of the analogous fatty acids, becoming lower with increasing unsaturation. This further reduction of the endothermic transition temperatures by water means that the natural phospholipids extracted from biological membranes often exhibit this crystalline to liquid crystalline transition many degrees below the biological environmental temperature. At the biological environmental temperature we should expect the phospholipids which contain highly unsaturated chains to be in a highly mobile and fluid condition.

This implies that it is necessary to provide a reason if these unsaturated phospholipids do not occur in a highly mobile condition at this environmental temperature. If they are not in a highly mobile condition in a cell membrane, then we need to have some inhibitory interaction, such as interaction with cholesterol or protein, etc., to explain this. Recent studies have shown that cholesterol can inhibit the chain mobility of some phospholipids (Chapman and Penkett [63], Hubbell and McConnell [64]).

Bound water Some of the water added to the phospholipid appears to be *bounds* to the lipid, e.g. 1,2-dipalmitoylphosphatidylcholine binds about 20% water (Chapman and co-workers [69]). This water does not freeze at 0°C and calorimetric studies made with lipid-water mixtures show that it is only after more than 20% water has been added to the lipid that a peak at 0°C is observed. Some differential scanning calorimetric traces for lipid-water mixtures of different concentrations are shown in Fig. 3.6. This 'bound' water may have considerable relevance to interactions of anaesthetics, drugs and ions with biological membranes. If this 'bound' water varies, either in its properties or in its total amount, dependent upon

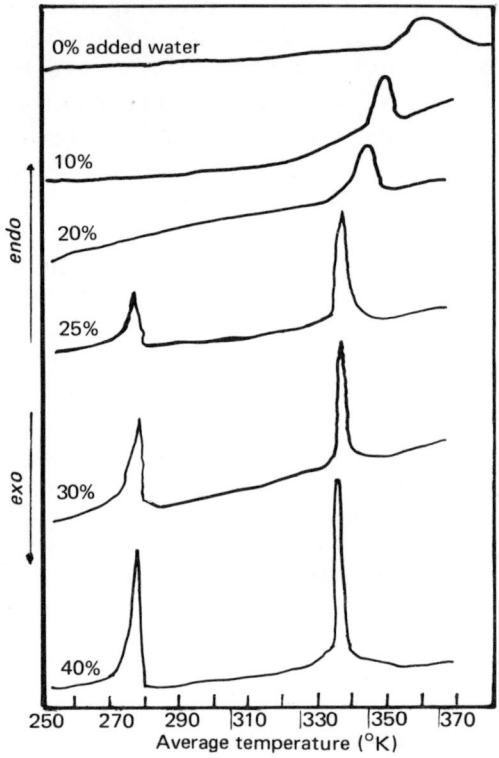

Fig. 3.6. D.s.c. heating curves for 1,2-distearoylphosphatidylcholine in increasing amounts of water (the endothermic peak on the right). This shows that the transition temperature separating the crystalline (gel) phase from the liquid crystalline phase falls with increasing amounts of water but reaches a limiting value. This occurs near the same point as when the added water can form ice. (It can be seen that no melting of ice occurs at 273°K until more than 20% of water is added to the lipid.)

the type of ion or interacting molecule, this in turn may alter transport and diffusion properties across the membrane. The amount of 'bound' water associated with the constituent lipids and proteins will perhaps provide a limit for the amount of water which can be removed from biological membranes before they lose their organization.

Monolayer properties Monolayer studies of phospholipids have been carried out for a considerable number of years. Usually this work has been performed with natural phospholipid mixtures and, in the vast majority of cases, with egg yolk phosphatidylcholine. In recent years a few studies have been made with pure synthetic phospholipids (e.g. van Deenen and co-workers [65]). These show that the fully saturated phospholipids exhibit, at room temperature, monolayers which are more condensed than are the unsaturated phospholipids containing *cis* hydrocarbon chains, i.e. the saturated lipids occupy less area at low surface pressures than do the unsaturated derivatives. These results can be compared with the differential thermal analysis results. The type of monolayer observed is related to the degree of flexing and twisting of the hydrocarbon chains of the lipid (Chapman [66], Phillips and Chapman [67]).

3.4.3 *Phase polymorphism*

Phospholipids in the presence of water can also form different liquid crystalline phases. In some cases, as the concentration of water varies, transitions from lamellar to hexagonal phases occur. These phases have been fully discussed by Luzzati and Husson [68].

In our laboratory recent studies on pure 1,2-diacylphosphatidylcholines in water show that there is no lamellar/hexagonal transition over a wide range of concentration. The presence of impurities, such as ions, can, however, have an appreciable effect upon the amount of water taken up by the lipid (Chapman and co-workers [69]).

The fact that lecithins of different chain length in water adopt a lamellar or bilayer type structure over a wide variation of concentration is interesting. At first sight this may appear to lend some support to the idea that a natural membrane may be built up on this bilayer-type organization. This, however, need not always be the case.* The influence of the membrane protein may be quite considerable on the resultant structure,

---

* X-ray studies by Wilkins *et al* Nature **230**, 72 (1971) support the idea that many membranes contain *some* regions of lipid bilayer.

and the organization of the membrane will depend upon the mode of interaction of the lipid and protein. Both the lipid and protein separately take up stable configurations to minimize hydrophobic and maximize hydrophilic interactions with the surrounding water. The important question for membrane organization is whether the combined lipid-protein complex attempting to balance these interactions will adopt an entirely different arrangement. The way in which membranes are synthesized naturally may also be relevant to this point.

3.5 MODEL MEMBRANES AND TRANSPORT PROCESSES

When we consider membrane function a number of phenomena need to be considered. It has been established that non-electrolytes penetrate cell membranes in direct proportion to their solubility in lipids. However, certain substances do not obey this general rule. Thus glucose enters the human red cell at a rate faster by 5 orders of magnitude than the rate predicted from the lipid solubility rule and from the rate found for the structurally related hexitols whose lipid solubility and molecular dimensions are close to those of glucose. D-glucose enters far more rapidly than L-glucose. The dependence of rate of entry on rate of concentration does not agree with the Fick kinetics expected for simple diffusion.

Three general hypotheses have been advanced to explain these processes: (a) formation of lipid-soluble complexes, (b) aqueous pores, (c) a process of membrane pinocytosis. In the carrier hypothesis the substance to be transported forms a complex with a membrane constituent which is miscible with the lipid of the membrane. The idea of a conducting pore was first proposed by Danielli [70], who suggested that the lipid layer of the membrane is incomplete and interrupted at intervals by infolding of the surface protein layers to form a pore lined by polypeptide chains. Substances would then evade the lipid solubility penetration rule if they could pass readily through these pores. Specificity in this system is provided by the complementary nature of chemical groupings in the permeant as well as molecular groups lining the pore. The idea behind the process of pinocytosis is that external fluid is engulfed into the cell enclosed in an invagination of the plasma membrane and it then forms an intracellular vesicle. In this process materials enter the cell in proportion to their external concentration, but specificity can be introduced by proposing that specific constituents, sugar or electrolytes, might be adsorbed on to active sites in the membrane and carried in as temporary membrane constituents. These authors require that the membrane becomes dissolved after engulfment so that effectively ingestion is intimately connected with the metabolic turnover of the membrane.

The process by which so-called active transport occurs remains obscure. Various theories have been put forward which include: (1) metabolic transformation of the transportant, e.g. glucose is transported as a glucose phosphate ester; (2) a carrier which undergoes a cyclic metabolic change being converted into an active form at one face of the membrane combining with the transportant diffusing to the other face of the membrane and then reverting to an inactive form, by this means discharging the transportant and then diffusing back to the original face of the membrane; (3) a redox-electron transport theory in which a trans-membrane electron gradient is set up by the operation of an oxidation-reduction system which forces ionic transport and perhaps also other transport processes.

3.5.1 Model membranes
In recent years new and improved methods for the study of biological membranes have suggested to the minds of a number of scientists a range of new models. These models have been concerned with either explaining and interpreting the data available on membrane systems, or have been attempts to construct the structure on the basis of the known constituents of the membrane and, in some cases, a model has been determined and built up on the basis of the functions of the membranes under consideration.

Bangham, Standish and Watkins [71] have recently used phospholipids which have been swollen in aqueous environment as a model system for the study of ion permeability. The system consists of the dispersed lipid, e.g. lecithin containing up to 15% of dicetyl phosphoric acid or in some cases phosphatidic acid. The lecithin is known to form lamellar structures of phospholipid bilayers separated by aqueous compartments. The presence of phosphatidic acid of the dicetylphosphoric acid is to give a net charge to the lipid and this appears to cause a separation of adjacent layers. It was found that cations taking part in the original swelling process of the phospholipid are restricted in their specific diffusion out of the lipid. Bangham *et al.* [71] suggest that the amount captured is proportional to some physical property of the lipid structure rather than to any chemical ion-pairing or binding. As the negative charge on the lipid structures was decreased, the diffusion rate for cations leaving the phospholipid decreased. Anions were found to diffuse much more readily than cations and, interestingly, anions appear to be relatively free to diffuse whether fixed charges of either sign were present or not. They conclude that the simple bilayers of phospholipid present a diffusion barrier to cations several orders of magnitude greater than to anions. They suggest that the combination of this property with the permeability to water, and the consequent osmotic sensitivity, makes this system a useful analogue to certain biological membranes.

Lucy [72] has discussed this model system of Bangham *et al.* and has pointed out that there is no direct evidence for the assumption which they make that phospholipid particles, which form when the lipids are dispersed in aqueous salt solutions, are completely closed structures, i.e. that there is really a complete separation between the aqueous environment and the inner concentric aqueous layers inside the phospholipid.

Mueller and co-workers [73] reported formation of single stable bimolecular phospholipid films up to 10 sq. mm in area present in 0.1 M saline solution (Fig. 3.7). These authors pointed out it was possible to form such a membrane by two components filled with saline to study its transverse electric properties. The dimensions of this bilayer were estimated to be between 60 and 90 Å thick with an electric capacity about 1 $\mu F/cm^2$, resistance 10^7 to even greater 10^8 ohms/cm^2. The dielectric breakdown of the film estimated at 2.5×10^5 volts/cm before the membrane. It was also observed to break below pH 5.0 and above pH 9.0. These authors also pointed out that this bimolecular lipid membrane adsorbed molecules obtained from various biological sources and that this led to a drop in resistance from 10^8 to about $10^3 - 10^5$ ohms/cm^2 and that they then became active or electrically excitable. By

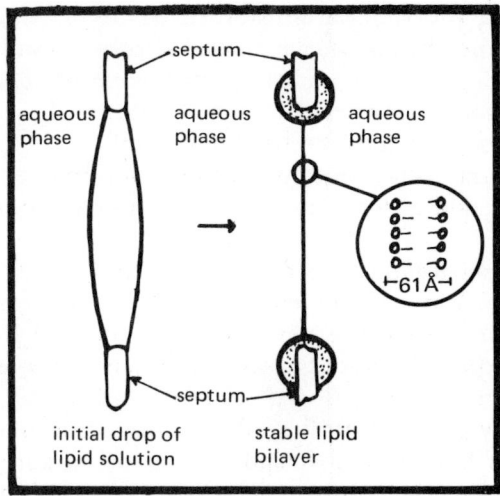

Fig. 3.7. Membrane model can be prepared by painting phospholipid across an orifice. Most of the lipid then drains away (photograph above shows this in process) to leave a thin layer of lipid believed to be bimolecular. The addition of biological protein to the thin layer gives the model an electrical activity which simulates the behaviour of a nerve cell. Certain antibiotics increase the conductance of such layers by a factor of up to 10^3 and affect permeability to sodium and potassium ions.

this they mean that their resistance changes reversibly and regeneratively between two definite values in response to a suprathreshold applied voltage. Hanai, Haydon and Taylor [74] examined black hydrocarbon films stabilized by pure egg lecithin. The films had capacitances which were accurately reproducible and which were independent of frequency and of the nature and concentration of the electrolyte. The thickness of the film was deduced by using a parallel plate condenser formulae. The thickness of the hydrocarbon part of the film was found to be 48 ± 1 Å. This value was compared with twice the average chain length of the lecithin stabilizing molecules. The d.c. conductance of the films were found to be irreproducible and generally less than 10^{-8} mho/cm^2 of film. The films were made by using a hair brush, dipping it into hydrocarbon solution of lecithin and drying it across a hole in a Teflon pot.

Huang and Thompson [75a, b] also made lipid bilayer systems using egg yolk lecithin and N-tetradecane dissolved in chloroform methanol solvent. These authors found a d.c. ohmic resistance of 1×10^6 ohm/cm^2, a dielectric breakdown of 200 millivolts and the surface tension of 0.5 dynes/cm. The thickness of the membrane was estimated to be about 61 Å thick. These authors considered that the membrane, or bilayer, consisted mainly of phospholipid. Cholesterol or one of the cholesterol esters added to the phospholipid in solution also gave stable films. The conductance mechanism over the bilayers was suggested to be electrolytic rather than electronic.

Water permeability. Black lipid films are appreciably permeable to water (Mueller *et al.* [76]). The early measurements suggested that the permeability obtained from tritiated water diffusion experiments was about ten times less than that obtained from osmotic flow [77, 78]. This raised the interesting possibility that a coupling of two unidirectional fluxes was occurring (as had been inferred for diffusion across some cell membranes) and the results might be explicable in terms of, for example, the single-file diffusion model proposed by Hodgkin and Keynes [78]. Alternatively, the water-filled pore model of Pappenheimer *et al.* [80] might have been applicable. However, it has now been shown that there is no real difference between the isotopic and osmotic permeabilities. The discrepancy arose from the presence of unstirred layers of aqueous solution adjacent to the membrane surfaces and from the failure to correct for the fall in concentration of the diffusing species across this region. This boundary layer effect is much more important for isotope diffusion than for osmotic flow. In the former instance it is possible, in unstirred systems, for the boundary layers to constitute almost the total resistance to transfer [81]. Mechanical stirring is not usually able to reduce the boundary layer resistance to a negligible value and indirect means have to be used to find the true membrane permeability [82, 83]. The permeability as measured by osmotic flow usually requires a correction for boundary layer effects of not more than a few percent. This is a consequence of the density gradients and back diffusion of solute which occur during osmotic flow. Both these effects tend to restore the water concentration at the membrane surfaces to its bulk value, the former by causing natural convection in the boundary layers and the latter simply by interdiffusion of solute and solvent [84]. The water permeability depends on the lipid composition of the film [85]. Unfortunately, the precise manner in which the lipid varied in these films is not known. The effect of cholesterol in lowering the permeability of a phosphatidyl choline film has been demonstrated by Finkelstein and Cass [86] but again the precise composition of these films is not known.

The permeability to non-ionic substances other than water. Relatively little work has been carried out on this property of the black films. This is partly, if not entirely, due to technical difficulties. Thus, on the basis of the liquid hydrocarbon slab model the films would not be expected to be very permeable to polar substances. The quantities of material transferred would be small and difficulties of measurement would arise. Substances such as urea and the sugars fall into this category. Urea and sucrose when used in osmotic experiments have reflection coefficients of effectively unity [78] and indeed, by the use of an isotopic method, Vreeman [87] concluded that the permeability of a lipid film to urea was only approximately 4.2×10^{-6} cm/sec. Vreeman also measured permeabilities to glycerol and erythritol by an isotopic method and found values of approximately 4.5 and 0.75×10^{-6} cm/sec, respectively.

The permeability to a number of organic substances has been investigated by Bean and co-workers [88]. The substances chosen were not very lipid soluble, although they permeated much more rapidly than glycerol and erythritol and, in some instances, more rapidly than water. The nature of the membrane lipid had very little influence on the permeabilities. The authors concluded that for unionized molecules the relative permeabilities are as would be expected from a mechanism involving solubility and diffusion in a lipid phase.

Maddy, Huang and Thompson [89] have constructed a bilayer system and added to it a protein derived from beef erythrocyte ghosts. They calculated a composite membrane thickness of about 144 Å and suggested that this composite structure was similar to the Davson-Danielli model. There is, however, no direct evidence that such a structure has been formed. Mueller and Rudin [90] have also adsorbed antibiotics of the valinomycin, actin and enniatin groups to bilayer membranes. They have also shown that such adsorption increases the membrane conductance up to 10^3 times. The single ion conductances were observed to differ by as much as

300 times and were found to increase in the order of lecithin less than sodium less than caesium less than potassium less than rubidin, and resting potentials of up to 150 mV result when 0.1 M solutions of sodium chloride and potassium chloride are placed on opposite sides of the bilayer. These observations were interpreted on the basis that the ring diameters of these molecules permit hydrogen bonding of carbonyl oxygens in the ring to substitute for water molecules of the hydration shells of the cations. Cations with high field strengths and solvation energies show large activation energies and poorer conductances. The ring compound can act as both pores and carrier, depending upon conditions.

The model membrane system has also been applied to the study of uncoupling agents such as dinitrophenol. These uncoupling agents are thought to act as transmembrane proton carriers, thereby preventing the generation of the transmembrane pH gradient necessary for ATP synthesis.

Bielawski, Thompson and Lehninger [91] have studied the effect of 2,4-dinitrophenol on the electric resistance of phospholipid bilayers and shown that this molecule produces a large decrease in their electrical resistance. They suggest that this reduction of resistance is of the right amount sufficient to account for the uncoupling effect of DNP in mitochondria.

Van Zutphen, van Deenen and Kinsky [92] have studied the action of polyene antibiotics on bilayers. From this investigation they suggest that selective toxicity of the polyene antibiotics may be due to interaction with sterol present in the cell membrane. They therefore investigated the effect of these polyene antibiotics with the model bilayer systems. They were able to show that the stability of the bilayers in the presence of the antibiotic paralleled the behaviour of the antibiotic in another system, providing support for the condition that the localization of sterol in a membrane is a necessary prerequisite for polyene sensitivity and that the phospholipid to sterol ratio may be a factor which determines whether or not the membrane is sensitive to polyenes. This conclusion was supported by the observation that a 10-fold increase in the lecithin to cholesterol ratio results in the formation of a bilayer with long survival times in the presence of filipin.

Mueller and Rudin [93] have reported further experiments on the possibilities of obtaining action potential phenomena with bilipid layers. They were able to show properties such as thresholds, delayed rectification, bistable flip-flop kinetics, anodal break excitation and self regenerative changes of membrane potential accompanied by decreases in conductance and spontaneous rhythmic firing, the frequency of which can be precisely controlled by varying the temperature and the applied electrical polarization. These phenomena have also been shown to be reversibly blocked by local anaesthetics such as cocaine at physiological concentrations.

The study of sugars, amino acids and fat transport through model membrane systems of this type may be useful for providing clues as to the nature of these processes in the intestinal cell system.

3.6 THE FUTURE

We are at last beginning to understand the details of the molecular organization of cell membranes and transport processes. In some membranes a bilayer of lipid occurs apparently separating complex lipoprotein regions. Further studies may show similar types of structure with other membrane systems. As our confidence in understanding the structures of cell membranes increases, so will our use of model membranes increase in sophistication. Using these model systems we can hope to understand transport processes at the molecular level. The use of modern physical techniques, such as X-ray methods and nuclear magnetic resonance, electron spin resonance and fluorescence spectroscopy, may aid this understanding.

REFERENCES

1. G. E. PALADE, *Histochem. Cytochem.*, 1, 188 (1953).
2. S. ITO, *Anat. Rec.*, 148, 294 (1964).
3. D. W. FAWCETT, *Circulation*, 26, 1105-1125 (1962).
4. M. G. FARQUHAR and G. E. PALADE, *J. Cell Biol.*, 17, 375-412 (1963).
5. M. M. DEWEY and L. BARR, *J. Cell Biol.*, 23, 553-585 (1964).
6. H. ZETTERQVIST, Monogr. Aktiebolaget. Godvil, Stockholm (1956).

7. S. L. PALAY and L. J. KARLIN, *J. Biophys. Biochem. Cytol.*, **5**, 363-372 (1959).
8. S. M. KURTZ and J. D. FELDMAN, *J. Ultrastruct. Res.*, **6**, 19-27 (1962).
9. G. A. ANDRES, C. MORGAN, K. C. HSU, R. A. RIFKIND and B. C. SEEGAL, *J. Exp. Med.*, **115**, 929-935 (1962).
10. G. B. PIERCE, Jr., T. F. BEALS, J. SRI RAM and A. R. MIDGLEY, Jr., *Amer. J. Path.*, **45**, 929-942 (1964).
11. L. M. NAPOLITANO and J. KLEINERMAN, *J. Cell Biol.*, **23**, 65A (1964).
12. D. MILLER and R. K. CRANE, *Biochim. Biophys. Acta*, **52**, 293-298 (1961).
13. O. BEHNKE, *J. Cell Biol.*, **18**, 251-265 (1963).
14. T. BARKA, *J.A.M.A.*, **183**, 761-764 (1963).
15. A. B. NOVIKOFF, in The Cell (J. Brachet and A. E. Mirsky, eds.), 2, 423-488. Academic Press Inc., New York (1961).
16. H. A. PADYKULA, E. W. STRAUSS, A. J. LADMAN and F. H. GARDNER, *Gastroenterology*, **40**, 735-765 (1961).
17. J. S. TRIER, *Gastroenterology*, **43**, 407-424 (1962).
18. G. PALADE, P. SIEKERVITZ and L. CARO, in The Exocrine Pancreas: Normal and Abnormal Functions, pp. 23-49. Little, Brown and Company, Boston (1961).
19. S. ITO and R. J. WINCHESTER, *J. Cell Biol.*, **16**, 541-572 (1962).
20. A. K. CHRISTENSEN and D. W. FAWCETT, *Anat. Rec.*, **136**, 333 (1960).
21. S. L. PALAY, in Frontiers of Cytology, pp. 305-342. Yale University Press, New Haven (1958).
22. J. A. FREEMAN, *Anat. Rec.*, **144**, 341-358 (1962).
23. F. BIERRING, *Acta Path. Microbiol. Scand.*, **54**, 241-252 (1962).
24. J. S. TRIER, *J. Cell Biol.*, **18**, 599-620 (1963).
25. S. L. PALAY and L. J. KARLIN, *J. Biophys. Biochem. Cytol.*, **5**, 373-384 (1959).
26. J. D. ROBERTSON, *J. Biophys. Biochem. Cytol.*, **3**, 1043 (1957).
27. J. D. ROBERTSON, Intern. Kongr. Elektronenmikroskopie, 4, 259. Springer, Berlin (1958).
28. W. STOECKENIUS, in Principles of Biomolecular Organisation (G. E. W. Wolstenholme and M. O'Conner, eds.), p. 418. J. A. Churchill Ltd., London (1966).
29. P. F. Elbers, *Recent Prog. Surf. Sci.*, **2**, 443 (1964).
30. F. S. SJOSTRAND, *J. Ultrastruct. Res.*, **9**, 561 (1963).
31. J. HILLIER and J. F. HOFFMAN, *J. Cellular Comp. Physiol.*, **43**, 203 (1953).
32. R. M. GLAESER, T. HAYES, H. MEL and C. TOBIAS, *Expl. Cell Res.*, **42**, 467 (1966).
33. J. D. ROBERTSON, *J. Biophys. Biochem. Cytol.*, **19**, 201 (1963).
34. E. L. BENEDETTI and P. EMMELOT, *J. Biophys. Biochem. Cytol.*, **29**, 299 (1965).
35. J. K. BLASIE, M. M. DEWEY, A. E. BLAUROCK and C. R. WORTHINGTON, *J. Mol. Biol.*, **14**, 143 (1965).

36. E. D. KORN, *Science*, 153, 1491 (1966).
37. J. D. ROBERTSON, *Biochem. Soc. Symp.*, 16, 3 (1959).
38. H. FERNANDEZ-MORAN and J. B. FINEAN, *J. Biophys. Biochem. Cytol.*, 3, 725 (1957).
39. E. D. KORN and R. A. WEISMAN, *Biochim. Biophys. Acta*, 116, 309 (1966).
40. V. B. WIGGLESWORTH, *Proc. Roy. Soc. (London) Ser. B*, 147, 185.
41. G. E. BAHR, *Expl. Cell Res.*, 7, 457 (1954).
42. J. D. ROBERTSON, *Prog. Biophys. Biophys. Chem.*, 10, 343 (1960).
43. W. STOECKENIUS, *J. Biophys. Biochem. Cytol.*, 12, 221 (1962).
44. W. STOECKENIUS, *Circulation*, 16, 1066 (1962).
45. J. B. FINEAN, *Circulation*, 26, 1151 (1962).
46. J. C. RIEMERSMA, *J. Histochem. Cytochem.*, 11, 436 (1963).
47. D. CHAPMAN and D. J. FLUCK, *J. Biophys. Biochem. Cytol.*, 30, 1 (1966).
48. S. FLEISCHER, B. FLEISCHER and W. STOECKENIUS, *Federation Proc.*, 24, 296 (1965).
49. W. Van ITERSON, *Bacteriol. Rev.*, 29, 299 (1965).
50. A. D. BANGHAM and R. W. HORNE, *J. Mol. Biol.*, 8, 660 (1964).
51. V. P. WHITTAKER, in Regulation of Metabolic Processes in Mitochondria (J. N. Tager, F. Papa, E. Quagliariello and E. C. Slater, eds.), p. 1. Elsevier, Amsterdam (1966).
52. H. MOOR and K. MUHLETHALER, *J. Biophys. Biochem. Cytol.*, 17, 609 (1963).
53. H. MOOR, K. MUHLETHALER, H. WALDNER and A. FREY-WYSSLING, *J. Biophys. Biochem. Cytol.*, 10, 1 (1961).
54. D. BRANTON, *Proc. Nat. Acad. Sci.*, 55, 1048 (1966).
55. G. ROUSER, G. J. NELSON, S. FLEISCHER and G. SIMON, in Biological Membranes, (D. Chapman, ed.), p. 5, Academic Press, London and New York, (1968).
56. A. H. MADDY, *Biochim. Biophys. Acta*, 117, 193 (1966).
57. R. F. A. ZWAAL and L. L. M. Van DEENEN, *Biochim. Biophys. Acta*, 150, 323 (1968).
58. S. A. ROSENBERG and G. GUIDOTTI, *J. Biol. Chem.*, 243, 1986 (1968).
59. P. BYRNE and D. CHAPMAN, *Nature*, 202, 987 (1964).
60. D. CHAPMAN and D. T. COLLIN, *Nature*, 206, 189 (1965).
61. D. CHAPMAN, P. BYRNE and G. G. SHIPLEY, *Proc. Roy. Soc. (London) Ser. A*, 290, 115 (1966).
62. D. CHAPMAN and N. J. SALSBURY, *Trans. Faraday Soc.*, 62, 2607 (1966).
63. D. CHAPMAN and S. A. PENKETT, *Nature*, 211, 1304 (1966).
64. W. L. HUBBELL and H. M. McCONNELL, *Proc. Natl. Acad. Sci. U.S.*, 61, 12 (1968).
65. L. L. M. Van DEENEN, U. M. T. HOUTSMULLER, G. H. De HAAS and E. MULDER, *J. Pharm. Pharaceut.*, 14, 429 (1962).
66. D. CHAPMAN, *Ann. N.Y. Acad. Sci.*, 137, 745 (1966).
67. M. C. PHILLIPS and D. CHAPMAN, *Biochim. Biophys. Acta*, 163, 301 (1968).

68. V. LUZZATI and F. HUSSON, *J. Cell Biol.*, **12**, 207 (1962).
69. D. CHAPMAN, R. M. WILLIAMS and B. D. LADBROOKE, *Chem. Phys. Lipids*, **1**, 445 (1967).
70. H. DAVSON and J. F. DANIELLI, in The Permeability of Natural Membranes. C.U.P., London (1943).
71. A. D. BANGHAM, M. M. STANDISH and J. C. WATKINS, *J. Mol. Biol.*, **13**, 238 (1965).
72. J. A. LUCY, in Biological Membranes (D. Chapman, ed.), pp. 233-288. Academic Press, London (1968).
73. P. MUELLER, D. O. RUDIN, H. T. TIEN and W. C. WESTCOTT, *Nature*, **194**, 979 (1962).
74. T. HANAI, D. A. HAYDON and J. TAYLOR, *Proc. Roy. Soc.*, **281A**, 377 (1964).
75a. C. HUANG and T. E. THOMPSON, *J. Mol. Biol.*, **13**, 183 (1965).
75b. C. HUANG and T. E. THOMPSON, *J. Mol. Biol.*, **16**, 566, and **15**, 539 (1966).
76. P. MUELLER, D. O. RUDIN, H. T. TIEN and E. C. WESTCOTT, in Recent Progress in Surface Science. (J. F. Danielli, K. C. A. Pankhurst and A. C. Riddiford, eds.), **1**, 379. Academic Press, New York (1964).
77. C. L. HUANG, J. WHEELDON and T. E. THOMPSON, *J. Mol. Biol.*, **8**, 148 (1964).
78. T. HANAI and D. A. HAYDON, *J. Theoret. Biol.*, **11**, 370 (1965).
79. A. L. HODGKIN and R. D. KEYNES, *J. Physiol. (London)*, **128**, 61 (1965).
80. J. R. PAPENHEIMER, E. RENKIN and L. L. BORRERO, *Amer. J. Physiol.*, **167**, 13 (1951).
81. T. HANAI, D. A. HAYDON and W. R. REDWOOD, *Ann. N.Y. Acad. Sci.*, **137**, 731 (1966).
82. A. CASS and A. FINKELSTEIN, *J. Gen. Physiol.*, **50**, 1756 (1967).
83. C. T. EVERITT, W. R. REDWOOD and D. A. HAYDON, *J. Theoret. Biol.*, **22**, 20 (1969).
84. C. T. EVERITT and D. A. HAYDON, *J. Theoret. Biol.*, **22**, 9 (1969).
85. C. HUANG and T. E. THOMPSON, *J. Mol. Biol.*, **15**, 539 (1966).
86. A. FINKELSTEIN and A. CASS, *Nature*, **216**, 717 (1967).
87. H. J. VREEMAN, *Koninkl. Ned. Akad. Wetenschap. Proc. Ser. B.*, **69**, 564 (1966).
88. R. C. BEAN, W. C. SHEPHERD and H. CHAN, *J. Gen. Physiol.*, **52**, 495 (1968).
89. A. H. MADDY, C. HUANG and T. E. THOMPSON, *Fed. Proc.*, **25**, 933 (1966).
90. P. MUELLER and D. O. RUDIN, *Biochem. Biophys. Res. Commun.*, **26**, 398 (1967).
91. J. BIELAWSKI, T. E. THOMPSON and A. L. LEHNINGER, *Biochem. Biophys. Res. Commun.*, **24**, 948 (1966).
92. H. Van ZUTPHEN, L. L. M. Van DEENEN and S. C. KINSKY, *Biochem. Biophys. Res. Commun.*, **22**, 393 (1966).
93. P. MUELLER and D. O. RUDIN, *Nature*, **217**, 713 (1968).

CHAPTER 4

The Passive Permeability of the Small Intestine

ERNEST M. WRIGHT

*Department of Physiology
University of California Medical Center,
Los Angeles, California U.S.A.*

		Page
4.1	INTRODUCTION	159
4.2	NON-ELECTROLYTE PERMEATION	160
	4.2.1 *General principles*	160
	4.2.2 *The permeability of the small intestine to water*	168
	4.2.3 *Permeability to non-electrolytes*	170
4.3	ELECTROLYTE PERMEATION	177
	4.3.1 *General principles*	177
	4.3.2 *Permeability to ions*	183
4.4	PASSIVE PERMEATION AND MEMBRANE STRUCTURE	193
4.5	SUMMARY	194
	ACKNOWLEDGEMENT	195
	REFERENCES	195

4.1 INTRODUCTION

The purpose of the present chapter is to review the passive permeation of molecules across the epithelial cells lining the mucosal surface of the small intestine. However, since intestinal absorption is only one of many physiological processes which involve the transport of molecules across cell membranes I have chosen to review the intestinal literature against a general background of permeability studies. The justification for this approach is that there has been a conspicuous lack of interest in the passive permeability properties of the small intestine.

The transport of a molecule across a membrane is usually described as passive if the rate of transport is directly proportional to the electrochemical gradient. The proportionality constant, the permeability constant, depends upon both the nature of the molecule and the structure of the membrane. Consequently permeability measurements yield information about permeation mechanisms and the structure of the membrane. Furthermore it is possible to begin to analyse these simple transport phenomena in terms of thermodynamics and elementary atomic and molecular forces such as Coulomb's law and hydrogen bonding. The significance of a discussion of the passive permeability of the intestine is that it provides a starting point for the interpretation of the 'active transport' of molecules across the tissue.

The chapter is divided into two main parts; the first deals with the passive permeation of non-electrolytes and the second deals with passive ion transport. In each part there is a brief introduction to some principles of passive transport and a discussion of some practical aspects of permeability measurements. Readers interested in detailed treatments of these aspects of membrane transport are referred to the review by Curran and Schultz [1] and to the monographs by Stein [2] and by Lakshminarayanaiah [3]. Non-electrolyte permeation is discussed within the framework of permeation across epithelial cells in general and the topics include the physical basis for selectivity and the question of permeation via pores. Ion permeation is discussed in the light of recent studies of cation permeation across the rabbit gall bladder, an epithelium with much in common with the intestinal epithelium. Finally some conclusions are drawn about the membranes which control ion and non-electrolyte permeation across epithelial tissues.

4.2 NON-ELECTROLYTE PERMEATION

4.2.1 General principles

Diffusion
A starting point for the consideration of membrane permeability is the diffusion of molecules in solution. The flux, J_i, of

molecules i is defined as the number of molecules of i that cross a unit area in a unit of time,

$$J_i = \frac{1}{A}\frac{dn_i}{dt} = u_i C_i F_i \tag{4.1}$$

where A is the area, n_i the number of moles of i, u_i the mobility of i, t time, C_i the concentration of i and F_i the local force acting on i. The force, F_i, is the negative gradient of local electrochemical potential, $d\bar{\mu}_i/dx$, so that

$$J_i = -u_i C_i \frac{d\bar{\mu}_i}{dx} \tag{4.2}$$

In the absence of temperature and pressure gradients

$$\bar{\mu}_i = \mu_i^0 + RT \ln a_i + z_i F \psi$$

when there is no flow of any other molecular species; μ_i^0 is the standard chemical potential of i, R the gas constant, T the absolute temperature, a_i the activity of i, z_i the valency of i, F the Faraday, and ψ the electrical potential.

Now since
$$\frac{d\bar{\mu}_i}{dx} = RT\frac{d \ln a_i}{dx} + z_i F \frac{d\psi}{dx}$$

it follows that

$$J_i = -RT u_i \left(\frac{C_i\, d \ln a_i}{dx} + \frac{z_i C_i F}{RT}\frac{d\psi}{dx} \right) \tag{4.3}$$

which is known as the general flux equation for i at any point x in the solution.

If the molecule is uncharged, i.e. $z_i = 0$, and if activities can be approximated by concentrations then the general flux equation becomes

$$J_i = -D_i \frac{dC_i}{dx} \tag{4.4}$$

where $u_i RT = D_i$ which is the diffusion coefficient of i. This expression is referred to as the Fick equation.

The units of D are usually given in cm^2/sec and in aqueous solution at infinite dilution they range in magnitude from about 1×10^{-5} to 1×10^{-7} cm^2/sec according to the molecular weight (M) of the diffusing molecule; in simple liquids DM^x = constant where $x = \frac{1}{2}$ or $\frac{1}{3}$ according to the value of M (see Stein [2], figure 3.2).

Consider the application of the Fick equation to the diffusion of molecules within membranes. With a system in a steady state, i.e. the flux of i is identical at all points in the membrane, and assuming that D is a constant, then the flux equation may be integrated across the thickness of the membrane to obtain

$$J_i = -\frac{D_i}{h}\left(C'_i(m) - C''_i(m)\right) \tag{4.5}$$

where h is the thickness of the membrane, $C'_i(m)$ and $C''_i(m)$ are the concentrations of i at side 1 and side 2 of the membrane respectively, and D_i is now the diffusion coefficient of the solute in the membrane phase. If the membrane phase can be represented by a simple liquid then $DM^{1/2}$ or $DM^{1/3}$ should equal a constant but if the membrane is more like a polymer [4] there should be a steeper relationship between D and M, for example in polymethyl acrylate membranes $DM^{3.8}$ = constant. In any case the absolute value of D may be a factor of 100-1000 less than the aqueous solution diffusion coefficients depending on the viscosity of the membrane (see Stein [2] for further discussion).

Practical use of equation 4.5 can be made by relating the concentration of i just within the membrane to the concentration in the external aqueous solutions:
i.e.

$$J_i = -\frac{D_i K_i}{h}\left(C'_i - C''_i\right) = P_i \Delta C_i \tag{4.6}$$

where $-(D_i K_i)/h = P_i$, and K_i is the membrane : water partition coefficient. K is defined as the equilibrium ratio of the solute concentration in the membrane phase to its concentration in the aqueous phase, the two solutions being in contact and mutually immiscible. For this analysis it is assumed that the value of K_i is the same at each membrane interface, and that the concentration of solute in the aqueous solution adjacent to the membrane is identical to the bulk phase concentration.

This relationship, equation 4.6, implies that P_i's can be predicted from the knowledge of the bulk D_i's and K_i's. In at least lipid bilayer membranes this is correct [5]. Furthermore, it shows that the selectivity of membranes may be controlled either by K's or by D's. It is probable however that partition coefficients are more important as they may differ by up to a factor of 10^8 for compounds with similar molecular weights.

Although the K's for the distribution of solutes between biological membranes and aqueous solutions are unknown they are available for a number of model systems. These systems, e.g. olive oil : water and ether : water have the same selectivity [6, 7] (that is, K's are in approximately the same sequence for any given set of solutes) and there is a simple quantitative relationship between the K's in each system. Collander [6] suggested that selectivity was mainly controlled by the degree of hydrogen bonding between the solute and water, and that the quantitative differences between systems were due to the different polar properties of the lipid solvents. Thermodynamic analysis by Diamond and Wright [8] support these suggestions.

P's are usually obtained experimentally by measuring the flux of molecules across the membrane at a known concentration gradient. Fluxes may be determined either by direct chemical analysis or by radioactive tracer techniques.

Permeability measurements and unstirred layers

A tacit assumption about the application of the general flux equation to diffusion of non-electrolytes across membranes is that the external aqueous solutions are perfectly stirred, i.e. that the concentration of i in the aqueous solution adjacent to the membrane is the same as the bulk phase concentration. Alternatively it is assumed that the flux of i across the membrane is slow compared with the aqueous solution diffusion coefficient. In practice it is difficult to achieve 'perfect stirring' and for many solutes the flux across the membrane approaches the rate of diffusion in aqueous solution. Dainty [9] was one of the first to recognize the importance of unstirred layers in membrane phenomena.

Adjacent to the membrane there is a region in which diffusion is the only mechanism of transport. The effective thickness of the unstirred layer (δ) is taken as the width of a uniform concentration gradient equal to the difference in concentration of solute between the bulk phase and the membrane interface. The thickness of unstirred layers in biological systems can be estimated from hydrodynamic or kinetic measurements and they usually range from 50 to 500 microns in well stirred systems [10, 11, 12, 13]. In a variety of *in vitro* intestinal preparations the unstirred layers are not less than 400 microns thick (Wright, unpublished observations).

During the permeation of solutes (or radioactive tracers) across a membrane the concentration of solute at one side of the membrane is reduced below the bulk phase concentration

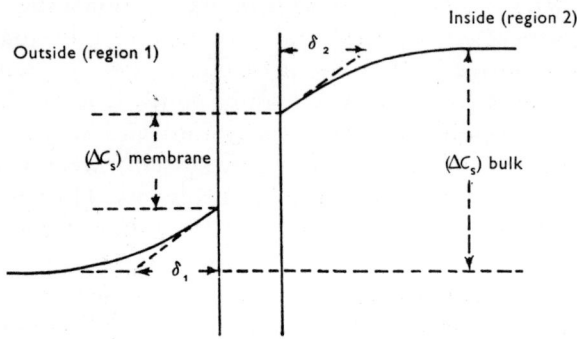

Fig. 4.1. A possible concentration profile for a permeating solute in the solutions adjacent to a membrane. (Taken with kind permission from Dainty and House [10].)

and at the other side the concentration increases above the bulk phase concentration on that side as illustrated in Figure 4.1. In the steady state the concentration gradient is reduced by an amount

$$\phi\left(\frac{\delta_i + \delta_2}{D_1 + D_2}\right) \quad (4.7)$$

where ϕ is the flux of solute across the membrane [9]. Thus the permeability coefficient that is measured experimentally is related to the true membrane permeability coefficient (P_t) by

$$\frac{1}{P} = \frac{1}{P_t} + \frac{\delta_1}{D_1} + \frac{\delta_2}{D_2} \quad (4.8)$$

where D_1 and D_2 are the diffusion coefficients in the unstirred layers δ_1 and δ_2 respectively, and $C'' = 0$. The more general expression, i.e. when $C'' \neq 0$, is

$$\frac{1}{P} = \frac{1}{P_t} + \frac{\delta_1}{D_1} + \frac{\delta_2}{D_2} + \frac{C''}{C'P} \quad (4.9)$$

(Barry, personal communication)

Values of P greater than 10^{-4} cm/sec are of doubtful significance as transport in these cases is probably unstirred layer limited.

Another factor which affects the magnitude of the effective concentration gradient across the membrane is the flow of water. A water flow tends to reduce the effective solute concentration gradient by enhancing the solute concentration on one side of the membrane and by depleting the solute concentration on the other side of the membrane. The solute concentration adjacent to the membrane (C_m) is related to the velocity of water flow (V) by

$$C_m = C_b \exp \pm \frac{V\delta}{D} \qquad (4.10)$$

where C_b is the bulk phase concentration [9]. V is negative for the sweeping away effect and positive for the enhancement effect.

A direct consequence of the presence of unstirred layers in biological systems is that permeability measurements must be made in the absence of water flows (osmotic or solute linked) and under controlled rates of stirring. The absence of such precautions and the omission of unstirred layer measurements, common in studies of intestinal absorption, raises serious questions about the validity of such measurements.

Osmotic measurements of non-electrolyte permeability

Osmotic methods have long been used to measure the permeability of biological membranes to non-electrolytes, for example, in bacteria [14], in algae [15], in mitochondria [16], in erythrocytes [17] and in the intestine [18, 19, 20]. The methods involve the principle that concentration gradients of permeant molecules yield rates of osmotic flow which are less than the flows obtained with impermeant molecules.

Concentration gradients of impermeant molecules produce flows of water across membranes which are directly proportional to the van't Hoff osmotic pressure $RT\Delta C$ and the proportionality constant, L_p, is known as the osmotic permeability or the hydraulic conductivity of the membrane, i.e. the osmotic flow is given by

$$\text{flow} = L_p RT\Delta C \qquad (4.11)$$

and the osmotic pressure is defined as the hydrostatic pressure required to reduce this flow to zero.

In the case of homogeneous non-porous membranes, the diffusional water permeability, P_{H_2O} is related to L_p by

$$P_{H_2O} = \frac{L_p RT}{\overline{V}_\omega} \qquad (4.12)$$

where P_{H_2O} is in cm/sec, L_p in cm/sec/atmos, R the gas constant, T the absolute temperature and $\overline{V}_\omega$ the partial molar volume of water in ml/mole. This relationship has been shown to be valid in lipid bilayer membranes when the unstirred layers are taken into account [5]. In porous membranes the osmotic water permeability should be greater than the diffusional permeability owing to the fact that water molecules move in a cooperative manner during osmosis but not during diffusion. In the artificial membranes doped with nystatin or amphotericin B [21] and in most biological membranes this appears to be true. One exception is the marine algal cell, *Valonia*, where there is no significant difference between osmotic and diffusional water permeability [22]. This is interpreted to mean that there are no pores in *Valonia* membranes.

Concentration gradients of permeant molecules produce volume flows which are less than the expected van't Hoff flow, i.e. the rate of flow is proportional to $\sigma RT\Delta C$ where σ is the reflection or Staverman coefficient [23]. σ is defined as the ratio of the osmotic flows produced by a concentration gradient of the test molecule to that produced by the same concentration gradient of an impermeant molecule. The value of σ depends upon the nature of both the solute and the membrane, = 1 for impermeant molecules and for increasingly permeant molecules σ decreases progressively below 1. Negative values of σ may be obtained, for example in *Valonia* [22] and in the frog choroid plexus [12]. Values of σ less than 1 are due to the fact that the rate of osmotic flow from dilute to concentrated solution is reduced by the volume of the solute moving down the concentration gradient; and the flux of solute may drag along a volume of water from the concentrated to the dilute solution if, and only if, there is frictional interaction between the solute and water in the membrane (see Kedem and Katchalsky [24]). Reflection coefficients may be related quantitatively to permeability coefficients (the precise form of the equation depends somewhat on the model used to describe permeation), for example,

$$\sigma = 1 - \frac{\omega_i \overline{V}_i}{L_p} - \frac{\omega_i f_{im} h}{\phi_w} \quad (4.13)$$

where ω_i is the solute permeability coefficient, $P_i = \omega_i RT$, $\overline{V}_i$ is the partial molar volume of the solute, L_p the hydraulic conductivity of the membrane, f_{im} the frictional coefficients between solute and membrane, h membrane thickness and ϕ_w the volume fraction of water in the membrane [25]. The term $(\omega_i \overline{V}_i)/L_p$ represents the volume flow of solute across the membrane and $(\omega_i f_{im} h)/\phi_w$ represents the frictional interaction term. So, on the basis of irreversible thermodynamics, we can rationalize indirect osmotic measurements of membrane permeability. The validity of osmotic methods has also been established experimentally [7, 26].

Unstirred layer effects in osmotic experiments

Determinations of L_p's and permeability coefficients by osmotic methods are not free of the complications posed by the presence of unstirred layers. The effective concentration gradient across the membrane is reduced by an amount related to the magnitude of the flux of the solute across the membrane and the thickness of the unstirred layers, and by an amount related to the velocity of water flow across the membrane (see above). The former is more important in the case of highly permeable solutes whereas the latter effect is more important for the less permeable solutes. To minimize the contribution of these unstirred layers during osmotic measurements of membrane parameters it is necessary to reduce the thickness of the layers to a minimum and make the measurements close to zero time.

A further complication arises where there are additional solutes present during permeability measurements; this of course is common in biological systems where the integrity of the cell or tissue must be maintained. The flow of water across the membrane affects the concentration of all solutes adjacent to the cell membranes; for the purpose of illustration let us consider a membrane separating two solutions containing equal concentrations of NaCl. Addition of an impermeable non-electrolyte to one side of the membrane generates an osmotic flow of water into the solution containing the non-electrolyte. In the absence of perfect stirring the flow of water will (a) decrease the salt concentration adjacent to the membrane on the same side as the non-electrolyte, and (b) increase the salt concentration close to the membrane on the side containing no other solute, and (c) reduce the concentration of non-electrolyte adjacent to the membrane. The net effect is to produce a salt concentration gradient across the membrane opposite in direction to the non-electrolyte gradient. This reduces the osmotic flow of water across the membrane by an amount proportional to the magnitude of the local salt gradient and the salt reflection coefficient, and leads to an underestimate of the L_p by up to two orders of magnitude. When L_p's are determined by the application of hydrostatic pressures this effect of water flow on the salt concentration will also be present but the error is less owing to the fact that there is one less solute in the system. This may explain, at least in part, the apparent discrepancy between L_p's measured by osmotic and hydrostatic procedures, for example in nerve [27] and in stomach [28].

In the case of ion selective membranes (for example, nerve, gall bladder and intestine) the sweeping away effect on the salt concentrations may produce osmotically induced electrical potentials which may be mistaken for true electrokinetic streaming potentials [29, 30, 31]. In nerve some 25 to 50% of the observed potential may be the true streaming potential and in the gall bladder much or most of the streaming potential is a boundary diffusion potential. Streaming potentials have been observed in the rat small intestine [20] but as yet there is no estimate of the magnitude of the true electrokinetic component.

4.2.2 The permeability of the small intestine to water

Osmotic and flux methods have been used to estimate the water permeability of various intestinal preparations, for example, in frog jejunum [32], in the rat ileum [33], in the rat jejunum [18, 20], and in the human intestine [19, 34]. However, in these studies the unstirred layer effects discussed above have been largely neglected. Thus the values of L_p and P_{H_2O} reported and the conclusions based on them, for example, estimates of pore size, are of doubtful significance. Some indication of the error introduced into these measurements by unstirred layers can be obtained by considering recent L_p measurements in the rabbit gall bladder [26]. The initial L_p, determined within five minutes of applying the osmotic gradient, was found to be an order of magnitude greater than that obtained during the steady state (see Fig. 4.2). Even in these experiments the 'initial' L_p is probably a significant underestimate of the true L_p owing to the fact that the half times for the buildup and decay of streaming potentials are about 10 seconds (an approximate estimate of the time constant for the buildup of the salt concentration gradients in

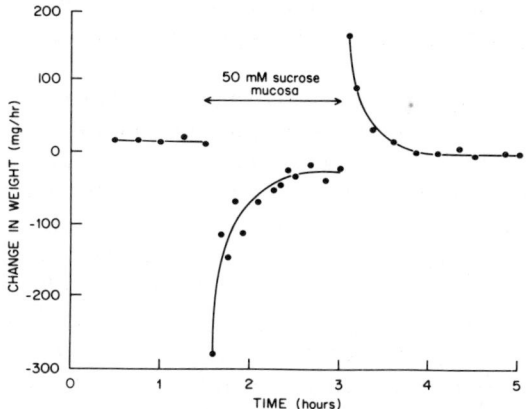

Fig. 4.2. The rate of osmotic water flow across the rabbit everted gall bladder as a function of time. Except for the time indicated there was no osmotic gradient and no significant water flow across the tissue. The mucosal solution was made hypertonic by the addition of 50 mM sucrose at the time indicated on the graph. Initially there was a high rate of water flow into the mucosal solution but this decreased to a flow which was an order of magnitude less than the initial value. This experiment was carried out at 23°C in a bicarbonate free solution. (Taken by permission from the dissertation by Smulders [26].)

the unstirred layers). Consequently it is expected that intestinal L_p's so far reported are orders of magnitude too low. With regard to unidirectional water flux measurements it is predicted that these are unstirred layer limited. Consequently, differences between P_{H_2O} and L_p's in the intestine, and variations in water permeability along the length of the intestine may be in part or wholly explained by variations in unstirred layers rather than in terms of pore size or area.

Asymmetry of water flow

Recently it has been reported by Loeschke, Bentzel and Csáky [32] that there is an asymmetry of osmotic water flow across an *in vitro* preparation of the frog intestine. The osmotic flow from mucosa to serosa was found to be seven times greater than the flow from serosa to mucosa. Furthermore, they observed that this rectification was correlated with changes in the ultrastructure of the tissue; in the absence of net water flow, and during osmotic flow from mucosa to serosa the lateral intercellular spaces were closed but during osmotic flow from serosa to mucosa these spaces were widely dilated. The authors suggested that the pathway for water flow across the tissue is different in the two directions and that asymmetry is a direct consequence of different osmotic permeabilities of the individual barriers.

An alternative explanation is that there is only one pathway for water flow across the tissue, i.e. via the intercellular spaces, and that the higher serosa to mucosa water flux is related to either a lower resistance to flow in the dilated intercellular spaces or an effective increase in the area available for water permeation. Evidence to support this interpretation comes from recent experiments by Smulders [26] on the rabbit gall bladder, which show that dilation of the lateral intercellular spaces is associated with an increase in (i) L_p, (ii) unidirectional non-electrolyte fluxes, and (iii) streaming potentials; and with a decrease in membrane resistance, the magnitude of the resistance change being directly correlated with the change in unidirectional fluxes. *

4.2.3 Permeability to non-electrolytes

There is surprisingly little information available about the passive permeation of non-electrolytes across the intestinal epithelium. What information there is comes mainly from a study by Höber and Höber published in 1937 [35]. These authors studied the absorption of fifteen compounds (polyhydroxy alcohols and aliphatic amides) from the rat small intestine. They concluded that lipid solubility and molecular volume were two major factors controlling the rates of intestinal absorption. Although little beyond this is known about the molecular forces which control the passive permeation across the intestine there have been comprehensive studies on two other epithelial membranes, namely the rabbit gall bladder [36, 37, 38, 26, 8] and the frog choroid plexus [12]. As there are many features common to the function of these epithelia, for instance isotonic water transport, it is advantage-

* See also: *J. Memb. Bid.* 7, 164 (1972) and *J. Memb. Biol.* 7, 198 (1972).

TABLE 4.1. Intestinal σ's. Non-electrolyte reflection coefficients obtained for rat jejunum by a gravimetric method [18] and the streaming potential method [20]. Also included in the Table are the molecular radii. Note the good agreement between the results obtained by the two procedures and the correlation between σ and molecular size.

Compound	σ		Radius
	Gravimetric method	Electrical method	
Formamide	0.22	0.17	2.2
Ethylene glycol	0.27	0.24	2.3
Urea	0.81	0.82	2.3
Erythritol	0.93	0.87	3.2
Mannitol	0.99	1.00	4.0
Sucrose	0.99	(lactose 0.97)	5.4

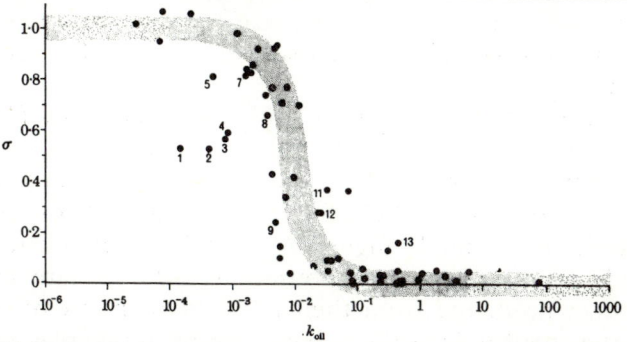

Fig. 4.3. Ordinate, average values of σ for non-electrolytes in rabbit gall bladder at 17 to 20°C; abscissa, oil : water partition coefficient of each non-electrolyte. Points referring to small solutes and branched solutes are numbered: small solutes, 1 = urea, 2 = methyl urea, 3 = formamide, 4 = acetamide, 5 = ethylene glycol, 7 = ethyl urea, 8 = propionamide, 9 = dimethyl formamide; branched solutes, 10 = pinacol, 11 = isovaleramide, 12 = 2-methyl-2,4-pentanediol, 13 = triacetin. The shaded line is drawn to indicate the general pattern of the other points and has no theoretical significance. (Taken from Wright and Diamond [37].)

ous to discuss briefly these results (for more discussions see the original papers).

An osmotic method was used to determine the permeability of both the gall bladder and choroid plexus*; the parameter measured was the reflection coefficient, σ. This was measured by the method first used by Smyth and Wright [20] for the small intestine and is based on the fact that 'streaming potentials' are directly proportional to the rate of osmotic flow. The validity of the electrical procedure has been established experimentally in the gall bladder by directly comparing σ's and

* See also: *J. Membrane Biol.* 10, 93 (1972) and *Aust. J. Biol. Sci.* 25, 931 (1972).

permeability coefficients in the same preparation [26, 39]. In the small intestine there is also good agreement between the σ's obtained by electrical and gravimetric variants of the osmotic procedure (see Table 4.1).

Figure 4.3 is a plot of gall bladder σ's against olive oil partition coefficients. This shows that the gall bladder is virtually impermeable (σ = 1) for solutes with low partition coefficients ($K_{oil} < 10^{-3}$); and that permeability increases (σ decreases) as partition coefficients increase, i.e. permeability of the gall bladder increases as the 'lipid solubility' of the solute increases. Solutes with olive oil partition coefficients greater than 10^{-1} all have reflection coefficients close to o, negative σ's not being observed; this is taken to indicate that diffusion through unstirred layers is rate limiting for these solutes. This explanation has been confirmed by radioactive tracer flux measurements [26].

There are two deviations to the general pattern: (1) small polar solutes (compounds nos. 1-9, $M < 75$, $K < 0.009$) permeate through the gall bladder at a greater rate than predicted from their partition coefficients, i.e. they have σ's expected of compounds with much higher K's; and (2) molecules with branched carbon chains (compounds nos. 11-13) are less permeable than predicted from their K's. The high rate of permeation of the small polar solutes has often been taken as evidence for the existence of aqueous pores in the membrane (see below for a further discussion). Most cell membranes discriminate against branched compounds and this indicates that lipids in membranes are in a more highly ordered configuration than in bulk lipid solvents [8, 37]. Neither the choroid plexus [12] nor the plant cell *Vallisneria* [40] discriminate against branched compounds, which implies that the lipid molecules in these membranes are less rigidly orientated. Similar graphs to that of Figure 4.3 were obtained when σ was plotted against $K_{oil}M^{-1/2}$, K_{ether}, or $K_{ether}M^{-1/2}$.

In the choroid plexus similar curves were obtained and Figure 4.4 shows a plot of σ against K_{ether}. The form of this figure is very similar to the corresponding plot in the gall bladder but there are three differences in detail: (1) σ's for permeant solutes are correspondingly lower in the choroid plexus than in the gall bladder, i.e. the curve is shifted to the left; (2) in the choroid plexus σ is reduced from 0.9 to 0.1 over a 40-fold range in K_{ether} whereas in the gall bladder the same reduction is over a 20-fold range in K_{ether}; and (3) negative σ's were not obtained in the gall bladder. Preliminary results from the frog intestine (Hingson, personal communication) show a σ/K_{ether} curve very similar to that obtained for the choroid plexus (except that negative σ's were not observed and that the intestine discriminates against branched chain compounds).

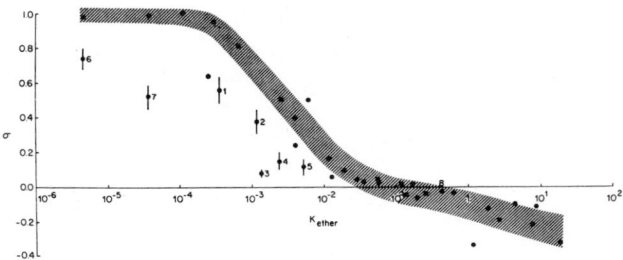

Fig. 4.4. The relation between σ and the ether : water partition coefficient. The ordinate shows the average σ for non-electrolytes in the frog choroid plexus; the abscissa shows the ether : water partition coefficient. The numbered points are, 1 urea, 2 methyl urea, 3 formamide, 4 acetamide, 5 ethylene glycol, 6 D-glucose, 7 L-arabinose, and 8 pinacol; the standard deviations are represented by the bars on each point. The shaded band indicates the general pattern of the other points, and, apart from the fact that the width of the band indicates the standard deviation of the results, the band has no theoretical significance. The anomalous position of hexamethylene tetramine (σ 0.64, K_{ether} 0.00026) may be related to the strong basic nature of this compound. (Taken from Wright and Prather [12].)

These results show that the non-electrolyte selectivity of epithelial membranes is very similar to that in single cell membranes (see Diamond and Wright [8]). In general, non-electrolyte permeation increases with increasing lipid : water partition coefficients as predicted (equation 4.6). So the same solute : water and solute : lipid intermolecular forces which control the distribution of solutes between a bulk lipid phase and water also control the rate of permeation of most solutes through a diverse variety of biological membranes. Thermodynamic analysis [8] of P's and K's show that the observed selectivity patterns are largely controlled by differences in solute : water intermolecular forces; the stronger these forces the lower the P's and K's. The major solute : water intermolecular forces are hydrogen bonds—the term given to the hydrogen bridge between two molecules containing electronegative atoms such as O, N or F which can act as proton acceptors or donors, e.g.

$$H - O - H - - - NH(CH_3)_2$$

water   H bond   urea

Although hydrogen bonds are weak (7 kcals/mole) compared with covalent bonds (50-100 kcals/mole) they are substantially stronger than the van der Waals forces which constitute the major interactions between non-electrolytes and lipid. The rate of permeation is largely controlled by the number and strength of the hydrogen bonds that the solute forms with water; the most effective substituent groups forming hydrogen bonds are

$$-NH-\underset{}{\overset{O}{\overset{\|}{C}}}-NH_2 > -\underset{}{\overset{O}{\overset{\|}{C}}}-NH_2 > -NH_2 > -\underset{}{\overset{O}{\overset{\|}{C}}}-OH,$$

$$-OH > -C\equiv N > -\underset{}{\overset{O}{\overset{\|}{C}}}-, -O-, -\underset{}{\overset{O}{\overset{\|}{C}}}-O-R.$$

However, with more than one such substituent group on the molecule we also have to take into account intramolecular hydrogen bonding and inductive effects. An additional factor controlling permeability is the number of methylene groups, $-CH_2-$, in the molecules; P's are increased by the addition of methylene groups owing to an entropy effect caused by the stabilizing action of the hydrocarbon on the 'ice' structure of water. Undoubtedly it is these same molecular forces which govern non-electrolyte selectivity in the small intestine.

With regard to the magnitude of the selectivity sequences the thermodynamic analysis of P's and K's showed that the quantitative differences between different systems arise entirely from differences in the solute : lipid intermolecular forces; the weaker these forces the greater the spread in the selectivity. The major solute : lipid intermolecular forces are van der Waals forces which are the instantaneous dipole interactions that arise between molecules because the centres of nuclear and electron charges do not coincide at any given instant. Additional bonding between the solute and lipid molecules occurs if the lipids contain hydrogen bonding sites. It is suggested that the ratios of hydrogen bonding sites to methylene groups in the membrane lipid is the parameter which governs the quantitative differences in P's between membranes. This hypothesis is supported by the experiments on liposomes where the permeability to non-electrolytes was measured as a function of lipid composition [41]. It was found that either the introduction of double bonds into the paraffin chains or the reduction in chain length of the membrane lipid enhanced permeation.

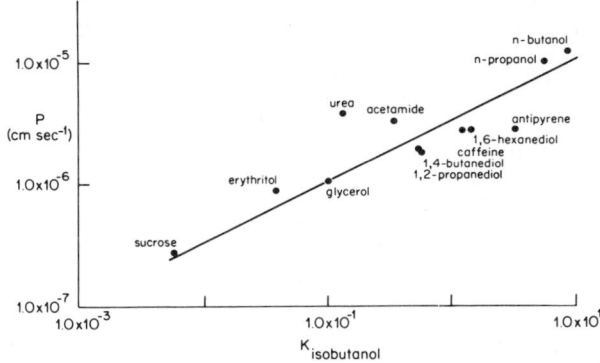

Fig. 4.5. The correlation between gall bladder permeability coefficients and isobutanol partition coefficients. The permeability coefficients were calculated from unidirection fluxes and were corrected for the unstirred layers on both sides of the epithelium. There is a good correlation between P's and K's; P's increasing with increasing K's. (Taken by kind permission from Smulders [26].)

Insight into the solute : membrane interactions can be obtained from direct permeability measurements. In the rabbit gall bladder P's for 16 non-electrolytes have been determined by direct radioactive tracer techniques [26]; and a plot of P against $K_{\text{isobutanol}}$ is shown in Figure 4.5. In general P's increase with the increase in partition coefficients and regression analysis of the results shows that the gall bladder P's increase in proportion to $M^{-0.7}(K_{\text{isobutanol}})^{0.33}$. Two conclusions follow from these experiments: (1) the low selectivity magnitude suggests a high number of hydrogen bonding sites in the membranes controlling non-electrolyte selectivity, i.e. strong solute : membrane intermolecular forces; and (2) the dependence of P on molecular weight is only a little greater than that predicted for diffusion in simple solutions, cf. *Nitella*, $PM^{-1.5}$. This and other evidence to be discussed later suggests that the tight junctions represent a low resistance pathway for the permeation of molecules across the gall bladder epithelium.

Permeation via pores?

Höber and Höber [35] studied the absorption of polyhydroxy alcohols from the rat small intestine and found that neither mannitol nor inositol were absorbed but that for smaller alcohols the rate of absorption increased with decreasing molecular volume; i.e. adonitol < erythritol < glycerol < ethylene glycol. They concluded that molecular volume was a major parameter controlling the absorption from the intestine. Their data is consistent with the permeation of these small polar molecules via pores about 4 Å in radius. The demonstration of solvent drag [42] and reflection coefficient measurements [18, 19, 20, 34] would appear at first hand to confirm both the existence and the dimensions of these intestinal pores. In contrast a pore radius of 30 Å is required to explain the apparent discrepancy between L_p and P_{H_2O} measurements [34]. It is likely, however, that this latter estimate is incorrect owing to the unstirred layer effects discussed above. In this context it is also worth pointing out that variations in σ along the length of the intestine [19, 34] may not reflect variations in membrane structure but variations in unstirred layers.

In the calculation of pore radii it is usually assumed that small polar solutes do not permeate via the membrane lipid. Inspection of the choroid plexus results, Figure 4.4, shows that without proof it is unjustifiable to make this assumption as σ's for urea, acetamide, etc. in the choroid plexus would be well below 1 even in the absence of a polar pathway. Consequently in the determination of equivalent pore radii it is essential to establish that the probing molecules do not permeate via the lipid. Parenthetically, it appears that two of the molecules used to probe the pore size of the intestine, ethylene glycol and formamide (Table 4.1) permeate through the intestine to some extent by virtue of their 'lipid solubility'.

The discussion of pore size in the intestine begs the question as to their actual existence. This problem is part of the more general question about the existence of pores in biological membranes; this has been reviewed in detail elsewhere [43, 8, 37]. Although the anomalous high permeation of polar non-electrolytes across membranes is part of the evidence for the existence of pores, the most impressive evidence comes from the demonstration of frictional interactions between molecules crossing membranes, e.g. L_P greater than P_{H_2O} in red cells [44], electroosmosis in algae [45, 46],

$$\sigma < 1 - \frac{\omega \overline{V}_s}{L_p}$$

in red cell [47] and streaming potentials in squid axio [27].

Recently Lieb and Stein [4] concluded that it is unnecessary to postulate the existence of pores in order to explain the anomalous behaviour of small polar non-electrolytes*. This was based mainly on the analysis of 13 permeability coefficients selected from Collander's study on the alga *Nitella* [7]. They calculated that diffusion within *Nitella* membranes was proportional to $M^{-3.8}$, which is comparable to that obtained for diffusion in polymethyl acrylate membranes. At first sight this would appear to conflict with the conclusions drawn by Collander, i.e. permeability was proportional to $M^{-1.5}$, but closer inspection of the paper by Lieb and Stein shows that they selected 13 compounds with an average molecular weight of 95 (10 with a molecular weight less than 100). In other words, they chose compounds which are likely to permeate via the polar pathway and so their conclusions may only apply to the diffusion of small molecules.

4.3 ELECTROLYTE PERMEATION

4.3.1 General principles

Diffusion

It will be recalled that the general flux equation for the diffusion of molecules in solution is given by

$$J_i = -D_i \left(\frac{dC_i}{dx} + \frac{C_i z_i F}{RT} \frac{d\psi}{dx} \right) \qquad (4.14)$$

where $D_i = u_i RT$ and where activities are approximated by concentrations. Applying this equation to ion permeation across membranes is more complex than for non-electrolytes. If it is assumed that there is a constant electrical potential within the membrane, i.e. we can replace $(d\psi/dx)$ with $(\psi' - \psi'')/h$, it is

* See also ref. 39.

possible to integrate the equation across the membrane in the steady state. The solution takes the form

$$J_i = \frac{\Delta\psi F D_i \left(C'_{i(m)} - C''_{i(m)} e^{-\frac{F\Delta\psi}{RT}} \right)}{RTh \left(1 - e^{-\frac{F\Delta\psi}{RT}} \right)} \quad (4.15)$$

which simplifies to

$$J_i = \frac{\Delta\psi F P_i \left(C' - C'' e^{-\frac{F\Delta\psi}{RT}} \right)}{RT \left(1 - e^{-\frac{F\Delta\psi}{RT}} \right)} \quad (4.16)$$

when $C'_i(m)$ and $C''_i(m)$ are related to the external solution concentrations as before and where $P_i = D_i K_i/h$. This equation has been used to obtain ion permeability coefficients and to predict net fluxes of ions down electrochemical potential gradients, e.g. in the gall bladder [48].

Furthermore in the absence of net current flow across the membrane, equation 4.16 can be solved for the electrical potential difference across the membrane,

$$\Delta\psi = \frac{RT}{F} \ln \frac{P_K(K)' + P_{Na}(Na)' + P_{Cl}(Cl)''}{P_K(K)'' + P_{Na}(Na)'' + P_{Cl}(Cl)'} \quad (4.17)$$

where (K) (Na) and (Cl) are ion activities on side 1 and side 2. This equation is known as the constant field equation or the Goldman, Hodgkin, Katz equation [49, 50]. It is frequently used by biologists, for example to describe membrane potentials or to obtain relative ion permeability coefficients from electrical measurements alone, and in many cases gives a fairly good fit to the experimental data. Nevertheless a 'good fit' does not necessarily validate the constant field assumption used in the derivation, particularly since Sandblom and Eisenman [51] have shown that it is not necessary to make the assumption when the membrane is permeable solely to either cations or anions.

The net flux of i across a membrane can also be defined as the sum of the two undirectional fluxes,

$$J_i = J_{i(12)} - J_{i(21)} \quad (4.18)$$

where $J_{i(1\,2)}$ is the unidirectional flux of i from side 1 to side 2 of the membrane and $J_{i(2\,1)}$ is the unidirectional flux from side 2 to 1. Unidirectional fluxes are determined by the use of radioactive tracer techniques. If it is assumed that $J_{i(1\,2)}$ is independent of the concentration of i on side 2 and likewise $J_{i(2\,1)}$ is independent of the concentration on side 1, it follows from equation 4.16 that

$$J_{i(12)} = \frac{\Delta\psi FP C'}{RT\left(1 - e^{-\frac{F\Delta\psi}{RT}}\right)} \qquad (4.19)$$

and

$$J_{i(21)} = \frac{\Delta\psi FP C'' e^{-\frac{F\Delta\psi}{RT}}}{RT\left(1 - e^{-\frac{\Delta\psi F}{RT}}\right)} \qquad (4.20)$$

By a similar approach it is also possible to relate the unidirectional fluxes to the partial ionic conductances (G_i) e.g.

$$G_i = \frac{J_{i(12)} F^2 e^{-\frac{\Delta\psi F}{RT}}}{RT \ln e^{-\frac{\Delta\psi F}{RT}}} \qquad (4.21)$$

when $C' = C''$, or more correctly $a' = a''$ (see Linderholm [52]).

Although these various equations may be used to obtain either permeability coefficients or unidirectional fluxes across epithelial membranes (see Schultz and Zalusky [53] and Diamond [48]) the drawback is it is difficult to check the validity of the results by independent methods. This difficulty can be circumnavigated to some extent by the following approach. Assuming that there is an independent migration of ions across the membrane it can be shown from equations 4.19 and 4.20 that the ratio of the unidirectional fluxes is given by

$$\frac{J_{i(12)}}{J_{i(21)}} = \frac{C'}{C''} e^{\frac{F\Delta\psi}{RT}} \qquad (4.22)$$

which is commonly referred to as the Ussing Flux ratio equation [54]. The significance of this equation is that the unidirectional flux ratios can be predicted without making any assumptions about the properties of the membrane. In other words ion transport across a membrane is said to be passive when there is no discrepancy between the flux ratios predicted by the equation on the one hand and determined experimentally on the other.

The use of the flux ratio equation is subject to the following conditions: (1) there are no temperature or pressure gradients across the membrane, (2) there are no chemical reactions between the ion and the membrane, (3) the flux of the ion is not influenced by the flux of any other species, i.e. there is no frictional interaction between the ion and any other ion or molecule crossing the membrane, and (4) the system is in a steady state. Consequently deviations from the ratios predicted by the equation do not necessarily imply that transport is not passive. In fact there are five known deviations which do not involve 'active transport', i.e. (a) solvent drag, (b) single file diffusion, (c) non-ionic diffusion, (d) exchange diffusion, and (e) facilitated diffusion. For a further discussion of these phenomena the readers are referred to the review by Curran and Schultz [1].

In summary we have seen that it is possible to distinguish between passive and active ion transport across a membrane by the use of the flux ratio equation. Furthermore, permeability coefficients (absolute or relative) can be obtained from measurements of (a) net fluxes, (b) unidirectional fluxes, and (c) diffusion potentials.

Permeability measurements and unstirred layers

As in the case of non-electrolytes the determination of ion fluxes and permeability coefficients is influenced by the presence of unstirred layers. Thus the effective concentration gradient across the membrane may be less than the difference in the bulk phase concentrations owing to the flux of ion through the membrane and sweeping away effects (see above). The former effect has been shown to influence the measurement of both dilution potentials and bi-ionic potentials across the rabbit gall bladder [55, 56] while the latter effect has been shown to contribute towards the generation of streaming potentials (see above).

Another important unstirred layer effect arises when d.c. currents are passed across ion selective membranes. This is the transport number effect which takes its origin in the difference in the transport numbers in the membrane and in the external solutions. The theory of the effect has been developed and tested experimentally by Barry and Hope [45, 46] in *Nitella* membranes. In cation selective membranes current is carried primarily by cations whereas in the external solutions current is carried by both cations and anions. Consequently when a pulse of d.c. current is passed across such a selective membrane, changes occur in the salt concentration in the unstirred layers with the result that a local salt gradient is generated across the membrane, which in turn causes both an osmotic water flow and a creep in the membrane potential. (Recent experiments in the gall bladder have shown that the voltage creep begins about half a second after applying a constant current pulse [57].) These changes in salt concentration do not continue indefinitely but are opposed by (i) diffusion away from the interface, (ii) sweeping away effects, and (iii) salt diffusion across the membrane. On switching off the current it also takes a finite time for the salt gradients to dissipate and for both the water flow and p.d. to return to baseline.

In *Nitella* membranes [45, 46] up to 50% of the current-induced water flows are due to the transport number effect and in the rabbit gall bladder [31] it accounts for most if not all the current-induced flow. It is also expected that a large fraction of the electro-osmotic flow observed in the rat ileum [58] is due to this effect.

Some practical considerations

In addition to the unstirred layer effects discussed above there are other precautions that should be taken into account either in the design of experiments or in their interpretation, namely junction potentials and solution composition.

Junction potentials: Potential differences across epithelial membranes are usually recorded by connecting two electrodes (usually calomel electrodes) to the solutions bathing the epithelium via salt bridges. To obtain the true membrane potential, at least with asymmetrical solutions on each side of the tissue, it is necessary to know the difference in the liquid junction potentials which arise at the interface between the salt bridges and the solutions. A common practice has been to use agar KCl salt bridges and to neglect the problem of junction potentials. In some cases this procedure proves to be unsatisfactory as the magnitude of the junction potentials may be greater than the true membrane potentials and the junction potentials may be time-dependent (see Barry and Diamond [56]).

Solution composition: Although it is accepted that the composition of the solutions used in experiments on ion permeability have to be well defined, certain precautions are often overlooked. For example, sucrose has been used to vary the osmolarity and to replace salts iso-osmotically in experiments on the intestine, although it is known that sucrose is hydrolysed at the brush border and that the glucose released increases the p.d. across the tissue [59, 60]. In addition sucrose has been found to have a larger effect on electrode standard potentials than smaller non-electrolytes [56]. Other effects of solution composition on epithelial cells include (a) osmotically induced increases in membrane resistance (see *J. Memb. Biol.* 7, 164 (1972)) and (b) slow increases in membrane resistance in KCl, RbCl and CsCl solutions [57]. Both of these effects are probably related to changes in membrane ultrastructure. Readers are referred to the paper by Barry and Diamond [56] for a more detailed discussion of the problems associated with electrical potential measurements in biological membranes.

Although it is advantageous to use *in vitro* techniques to study the permeability of the intestinal epithelium there is one potentially serious drawback, namely, that the lamina propria and the muscle layers may constitute a serious barrier to the diffusion of solutes between the serosal solution and the serosal surface of the epithelial cells. In the hamster intestine about 25% of the resistance of the intact intestinal wall is due to the muscle layers alone [61]. Diffusion potential measurements in both the tortoise and rat intestines also indicate that the muscle layers and the lamina propria offer substantial resistance to the diffusion of ions [59, 62]. Thus considerable care should be exercised while drawing conclusions about the properties of the epithelial cell from data taken from experiments on intact *in vitro* preparations. Specifically this may be important when we attempt to compare diffusion potential measurements with conductance and flux data, as the diffusion potentials originating from changes in the mucosal solution may reflect only the properties of the epithelial membranes.

4.3.2 Permeability to ions

Unidirectional fluxes

Although interest in the transport of ions across the intestine extends back into the last century there have been few modern studies on the passive permeability of this tissue to ions. Early studies were carried out mainly on *in vivo* preparations of the intestine where electrical potential measurements were neglected (see Ussing [63], for a review of the literature prior to 1960). A landmark in the area of intestinal absorption was the publication of two papers by Curran [33, 65] on ion and water transport across the rat ileum. Although these and more recent studies have been directed towards the question of active ion transport these were the first two 'modern' studies of ion transport across the intestine. However, in the *in vitro* study Curran established by the use of the flux ratio equation that sodium and chloride were passively transported across the intestine in the absence of glucose. In the presence of glucose both ions were found to be actively absorbed.

The mechanism of ion transport across the rabbit ileum was also investigated by Schultz and Zalusky [53, 66, 67] using an *in vitro* preparation and radioactive tracers. They concluded that the serosa to mucosa sodium flux was due solely to passive diffusion while the mucosa to serosa flux consisted of both active and passive components. In the case of chloride the unidirectional flux ratios were those predicted by the flux ratio equation in the presence or absence of glucose and even when the p.d. across the tissue was clamped at 25 mV. The partial ionic conductances for sodium and chloride were calculated to be 9.0 and 6.3 mmhos/cm^2 respectively and the sum of the conductances accounted for 91% of the total tissue conductance. Similar conclusions may be drawn from flux measurements in various regions of the rat and rabbit small intestines [68]. On the basis of these results it can be concluded that the intestinal epithelium is permeable to both sodium and chloride ions, the sodium conductance being slightly greater than the chloride conductance, and that no process other than simple diffusion is involved in the passive permeation of these ions.

Diffusion potentials

An alternative approach is to compute relative permeability coefficients from diffusion potentials using the constant field equation. It has been shown for both the rat jejunum and the tortoise intestine [59, 62] that the p.d. across the isolated intestine is linearly related to the logarithm of the sodium chloride concentration in the mucosal fluid when the sodium chloride is replaced with mannitol; the slope yields a P_{Cl}/P_{Na} ratio of 0.1 which indicates that sodium is ten times more permeable than chloride. Similar results have also been obtained in various regions of the rabbit and frog small intestines (Wright, unpublished data). In the rabbit ileum the P_{Cl}/P_{Na} ratio was 0.3 which indicates that the chloride unidirectional fluxes should be less than half those actually observed. In the gall bladder [48] a similar discrepancy between electrical measurements and the unidirectional chloride was taken as evidence for exchange diffusion. However, the rabbit ileum voltage clamp experiments [67] did not reveal any evidence for exchange diffusion. At the present time there is no reasonable explanation for this discrepancy, so further experiments are needed to clarify this issue, e.g. the flux ratio equation should be tested under the same experimental conditions as the diffusion potential measurements.

The polarity of streaming potentials produced by the osmotic flow of water across the intestine, hyperosmotic solution positive, also indicates that the intestine is more permeable to cations than anions [20]. In addition, the variation in the magnitude of the streaming potentials with pH gives information regarding the nature of the membrane negative charges controlling permeation. A pH tritation curve is shown in Figure 4.6 where it can be seen that the intestine behaves as an amphoteric membrane with an isoelectric point of 2.7 and an apparent pK_a of 4.0. Similar titration curves have been obtained for both streaming potentials and diffusion potentials in the gall bladder [69] and for streaming potentials in the choroid plexus [12].

Cation selectivity

Membrane charge should also control the cation selectivity of the small intestine. Cation selectivity has been obtained from measurements of bi-ionic diffusion potentials; bi-ionic potentials are those which arise in the absence of current flow when a membrane separates solutions of two different salts sharing a common anion, or cation, at the same total salt concentration, e.g. 150 mM NaCl vs. 150 mM KCl or 150 mM NaCl vs. 150 mM NaBr. Relative permeability coefficients, e.g. P_K/P_{Na} or P_{Cl}/P_{Br}, are extracted from diffusion potentials by the use of the constant field equation. On the basis of such potential measurements it has been shown that neither the rat nor the tortoise intestines discriminate between sodium and potassium ions, i.e. $P_K/P_{Na} \sim 1$ (Wright [59, 62]). The same conclusion can be drawn from the results obtained by Baillien and Schoffeniels [70] for the tortoise intestine.

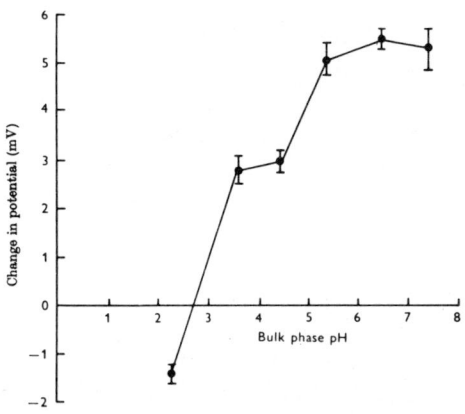

Fig. 4.6. Relation between initial pH of the mucosal and serosal fluids and the change in potential produced by addition of 84 mM mannitol to the mucosal fluid. The abscissa shows the initial pH, and the ordinate the change in potential caused by addition of mannitol. Positive values indicate that the mucosal side becomes less negative to the serosal side on addition of mannitol. The values are the means of five experiments with ±S.E. of the means. (Taken from Smyth and Wright [20].)

This approach has been extended to cover the permeation of all five alkali metal cations (Wright, unpublished). In the rabbit jejunum the selectivity sequence is

$$Cs(1.4) = Rb(1.4) > K(1.3) > Na(1.0) > Li(0.8)$$

and in the frog jejunum it is

$$Rb(1.45) > Cs(1.4) = K(1.4) > Na(1.0) > Li(0.6)$$

In Eisenman's terminology [71] the sequence in the rabbit is either II or III and in the frog either I or II. These are among the eleven sequences which occur in numerous biological and non-living systems such as the glass electrodes and mineral ion-exchangers.

The physical basis of cation selectivity

Although permeation of the 5 alkali metal cations could yield a possible 120 selectivity sequences Eisenman [72] showed that the 11 sequences actually observed could be predicted from a consideration of the cation's hydration energies and their free energies of interaction with membrane negative charges of different field strength. That is, the selectivity of the membrane site controlling permeability for two different cations a and b is governed by the free energy difference

$$\Delta F_{a \text{ site}} - \Delta F_{b \text{ site}} - \Delta F_{a \text{ water}} + \Delta F_{b \text{ water}} \quad (4.23)$$

where $\Delta F_{a \text{ water}}$ and $\Delta F_{b \text{ water}}$ are the free energies of hydration and $\Delta F_{a \text{ site}}$ and $\Delta F_{b \text{ site}}$ are the free energies of interaction between the cation and the negative site. Although the major cation : site interactions are Coulomb forces the negative sites may be either monopolar or multipolar. If the membrane sites have a high field strength relative to the hydration energies, selectivity will be controlled by the differences in the cation : site free energies and the sequence will be Li > Na > K > Rb > Cs (sequence XI), i.e. affinity decreases with increasing ionic radius. On the other hand if the membrane sites have a weak field strength relative to the hydration energies, the selectivity sequence is Cs > Rb > K > Na > Li (sequence I), i.e. the lyotropic series. As the field strength of the site varies between the low and the high values, see Figure 4.7, the selectivity sequences pass through transition sequences (II-X) in a regular fashion which may be calculated from expression 4.23. Thus the different permeability sequences observed in biological membranes may be a reflection of differences in the effective field strength of the sites controlling permeation. The sequences observed in the intestine indicate the presence of relatively weak sites.

In the rabbit gall bladder the cation selectivity sequence is K(2.4) > Rb(1.7) > Na(1.0) > Li(0.9) > Cs(0.8) which is known as sequence VIa in the Eisenman terminology. Essentially the same sequence is obtained from four different kinds of measurements: conductances, the calcium effect, dilution potentials and bi-ionic potentials [57, 73]. Quantitatively there is good agreement between the permeability sequences, and between the permeabilities extracted from the constant field equation and from equations derived for much more physically realistic permeation models. The review by Diamond and Wright [8] should be consulted for more detailed discussions of alkali cation, halide and divalent cation selectivity.

Selectivity magnitudes

Although there is good qualitative agreement between the cation selectivity sequences obtained in the intestine and gall bladder and those obtained in single cells such as nerve, muscle

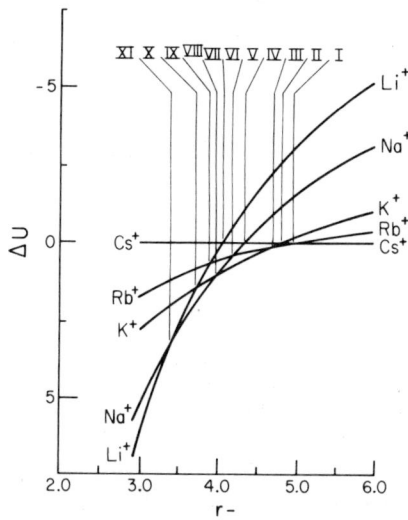

Fig. 4.7. The change in cation selectivity with decreasing field strength of membrane sites. The free energy difference on the ordinate is calculated from expression 4.23 with reference to the free energy of interaction of Cs with the site and with water. The abscissa, r_-^-, is the radius of a monopolar membrane site; the field strength decreases with increasing site radius. The curves were calculated for the hypothetical cation exchange between monopolar anions and a single multipolar water molecule. Note the smooth systematic change in the cation selectivity from sequence XI at the high field strength to sequence I at the low field strength. (Redrawn from Eisenman [71].)

and red cell, there is a striking quantitative difference in the magnitude of the selectivity. Comparison of the selectivity isotherms constructed from p.d. measurements in epithelial tissues (see Figure 4.8) with those constructed by Eisenman from measurements in single cells (reference 74, figure 8) shows that the range in selectivity from the most permeable to the least permeable is an order of magnitude less in the epithelial membranes than in single cells or artificial membranes. Since the

188   ERNEST M. WRIGHT

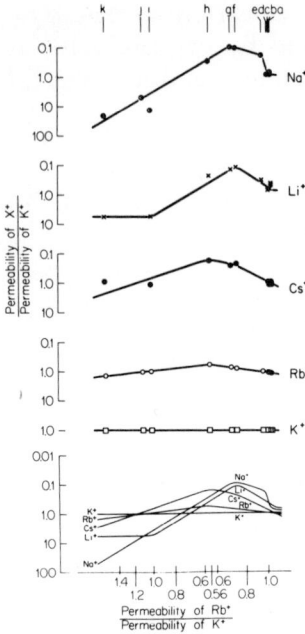

Fig. 4.8. Selectivity isotherms for the alkali cations in epithelia. The isotherm for each cation together with the experimental points on which it is based is plotted separately above, while all five isotherms are replotted together below without the experimental points. Each set of five experimental points arranged vertically above each other represents relative permeability coefficients extracted from transepithelial potential measurements in one epithelium identified by a letter above: (a) frog gall bladder; (b) frog jejunum; (c) rabbit jejunum; (d) frog choroid plexus; (e) serosal surface of toad urinary bladder; (f) inside surface of bullfrog skin; (g) inside surface of leopard frog skin; (h) rabbit gall bladder; (i) outside surface of leopard frog skin; (j) mucosal surface of toad urinary bladder; (k) outside surface of bullfrog skin (points a through d are derived from unpublished measurements by E. M. Wright; e and j from Leb, Hoshiko and Lindley [83]; f, g, i and k from Lindley and Hoshiko [84]; and h from the present study). Permeabilities relative to $P_K = 1$ are plotted logarithmically on the ordinate, and are arranged according to the relative Rb^+ permeability plotted logarithmically on the abscissa. (For instance, permeability coefficients in rabbit gall bladder in the last column of table 4b of this paper, re-expressed relative to K^+, are: $K^+ = 1.0$, $Rb^+ = 0.56$, $Na^+ = 0.28$, $Li^+ = 0.23$, $Cs^+ = 0.17$. These points have therefore been arranged on an imaginary vertical line intersecting the horizontal axis at 0.56. Since the ordinate gives permeability relative to P_K as a function of P_{Rb}/P_K on the abscissa, the K^+ value automatically falls on a horizontal line intersecting the ordinate at 1.0, and the Rb^+ value automatically falls on the line of identity. The Na^+ value of 0.28, the Li^+ value of 0.23, and the Cs^+ value of 0.17 have been used in constructing the empirical isotherms for these ions, drawn by eye through all the points.) The observed regularity means that the same physical factor (probably site field strength or coordination number, increasing from right to left: cf. Diamond and Wright [8]) determining P_{Rb}/P_K determines the permeabilities of the other cations; and that, therefore, once one knows the permeabilities of two cations, one not only can predict qualitatively the whole alkali-cation

range in selectivity depends upon the degree of hydration of the membrane [71, 8] this suggests that the pathway for permeation of cations across epithelial membranes is more hydrated than most other biological and artificial membranes. This conclusion is quite consistent with the finding that the gall bladder membranes controlling non-electrolyte permeation are more hydrophylic than isobutanol (see above).

In general ion selectivity depends upon two factors: equilibrium affinities, such as binding constants between membrane sites and ions; and non-equilibrium factors such as mobilities. In ion-exchange membranes it has been shown that $P_a/P_b = u_a/u_b \cdot K$ where P_a/P_b is the permeability ratio deduced from bio-ionic potentials for the two ions a and b, u_a/u_b is the mobility ratio, and K is the ion-exchange equilibrium constant [74]. Different forces enter into these factors and in fixed ion-exchange membranes the mobility sequence tends to be the inverse of the equilibrium sequence. The exact form of the relationship between equilibrium selectivity and mobility depends on the actual permeation mechanism, and the method used to determine selectivity, e.g. in ion-exchange membranes the permeability sequences determined by either dilution potentials or by conductances are the pure mobility sequences while those determined by bio-ionic potentials are a product of the mobility and equilibrium constants. In the intestine neither the permeation mechanism nor the separation of P's into u's and K's has been resolved.

Ion permeation mechanisms

There are five basic mechanisms by which cations could permeate across lipid membranes [75]: (1) permeation by diffusion through the membrane lipid, which is probably of little significance owing to the low salt partition coefficients; (2) permeation through pores containing negative fixed charges;

---

sequence but also quantitatively the approximate relative permeabilities of the other three cations. The degeneracy of the abscissa and the maxima in the isotherms mean that a given P_{Rb}/P_K may correspond to either of two sequences and relative permeabilities. The intersections of these five experimental isotherms determine transitions between different selectivity sequences. The first four sequences and the eleventh sequence, reading from right to left, are those predicted from a purely Coulomb model of selectivity; sequences 5 through 10 differ from Coulomb predictions only in the higher potency of Li^+; and the last three or 'post-Coulomb' sequences may involve polarization of the largest cations by the sites (see Diamond and Wright [8] for further discussion). (Taken from Barry, Diamond and Wright [73].)

(3) permeation via neutral pores; (4) permeation via mobile negatively charged carriers; and (5) permeation via mobile neutral carriers. Cation selectivity in each case is due to the presence of either charged sites, i.e. monopolar groups such as carboxyl ($\delta + > C = 0\delta$), or 'neutral' sites, i.e. multipolar groups such as carbonyl

$$(\delta + \begin{matrix}>C< \\ >C<\end{matrix}\ 0\ \delta -)$$

The determination of the ion permeation mechanism in epithelial membranes like the intestine is particularly complex owing to the structure of the 'membrane'. The intestinal epithelium consists of a single layer of cells held together at their apical surface by the so-called tight junctions. Consequently ions passing across the intestine either permeate across two cell membranes, at the mucosal and serosal faces of the cells, or alternatively permeate via the tight junctions between the cells. In other epithelial membranes specialized to carry out water transport, e.g. the gall bladder and renal proximal tubule, there is some evidence to indicate that the tight junctions provide a high conductance pathway [56, 73, 76]. Such a high conductance pathway in these epithelia would readily explain the low electrical resistance, the high unidirectional ion fluxes, and the low open-circuit potentials.

However, despite the lack of information about the membrane structure the mechanism of ion permeation may be approached by comparing the characteristics of ion permeation across the epithelia with those in model systems where the permeation mechanisms are well understood [57, 73]. For example, it is possible to distinguish between charged and neutral membranes on the basis of conductance measurements, i.e. the conductance of the membrane is measured as a function of the total external salt concentration. A charged membrane requires that the conductance be independent of the salt concentration below the fixed charge concentration whereas in a neutral membrane the conductance increases linearly with increasing concentration.

Figure 4.9 shows an experiment where the conductance of the rabbit ileum was measured as a function of the sodium concentration. As the sodium chloride concentration was lowered by replacement with mannitol the conductance decreased in a linear fashion which is the behaviour expected of a neutral membrane. Similar experiments in the rabbit gall bladder [37] and in rat jejunum (Wright, unpublished data) show that neutral sites control ion selectivity in these epithelia also. Examples of model membranes with mobile neutral sites include lipid bilayer membranes containing macrocyclic molecules such as monactin [77]. A theoretical treatment of permeation through these membranes was published by Ciani, Eisenman and Szabo [78]. For ion permeation through mem-

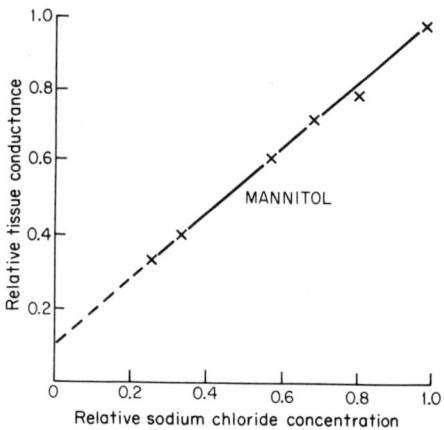

Fig. 4.9. The effect of replacing sodium chloride with mannitol on the tissue conductance of the rabbit ileum. The average conductance of the tissue in normal solution (140 mM NaCl) was 21.6 mmhos/cm^2. The sodium chloride was replaced by mannitol in both the mucosal and serosal solutions. (Taken from Schultz, Curran and Wright [64].)

branes with fixed neutral sites a theory has been presented by Barry and Diamond [79] and an example of such a model membrane is lipid bilayer membranes containing nystatin 'pores' [80].

In the gall bladder the mechanism of cation permeation has been subjected to a systematic investigation and analysis [56, 57, 69, 73]. Preliminary experiments showed that the epithelium behaved as a single membrane with respect to the regulation of passive ion transport across the tissue. Both the variation in membrane conductance with salt concentration and the lack of appreciable effect of concentration on the anion/cation permeability ratio suggested the presence of a neutral rather than a charged membrane. Furthermore the fact that the current-voltage relation was linear in both symmetrical sodium chloride solutions and in the presence of sodium chloride concentration gradients, see Figure 4.10, indicated that the membrane controlling permeation was so thick that microscopic electroneutrality was obeyed, i.e. the membrane is probably much thicker than typical cell membranes. Finally, there was

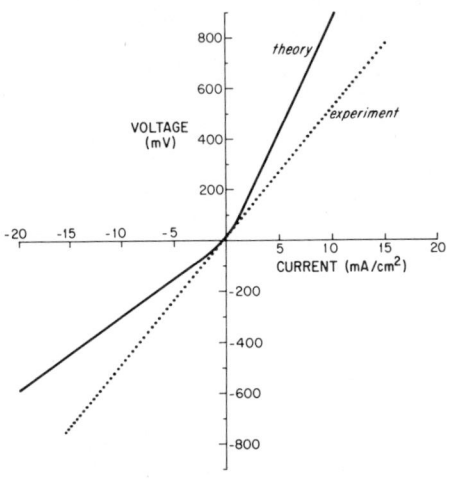

Fig. 4.10. Predicted current-voltage relation in asymmetrical solutions for a thin membrane with fixed neutral sites, compared with the experimental curve for the gall bladder. The curve labelled 'theory' was calculated from eq. 105 of reference 79: choosing values of the parameters to correspond to the situation of the gall bladder separating 30 and 150 mM NaCl solutions ($a'' = 115.3$ mM, $a' = 28.4$ mM, $n = 0.8$, $u_3 = 0$, $B_1 = 1.09$, and taking δ_0 as the average of the two bathing-solution activity coefficients). The curve labelled 'experiment' is the curve actually obtained for the gall bladder. Note that the experimental curve is linear but the theoretical curve is non-linear, so that the membrane controlling permeation in the gall bladder cannot be thin enough for microscopic electroneutrality to be violated. (Taken from Wright, Barry and Diamond [57].)

good qualitative and quantitative agreement between cation selectivity sequences obtained by four independent measurements, namely dilution potentials, bi-ionic potentials, conductances and the calcium effect. On the basis of these experimental results and a theoretical analysis of neutral membranes [79] it was suggested that cations permeate the gall bladder epithelium via channels lined with fixed neutral sites, i.e. cation permeation was via a neutral pore.

4.4 PASSIVE PERMEATION AND MEMBRANE STRUCTURE

The final point for discussion in this chapter concerns the route of passive ion and non-electrolyte permeation across epithelial tissues. Some general morphological features of the columnar epithelial cells lining the small intestine are recalled in Figure 4.11. The features which directly concern this discussion are the membranes which constitute the mucosal and serosal surfaces of the cell, and the so-called 'tight junctions'. Essentially the cell membranes have the basic structure common to all cells and like

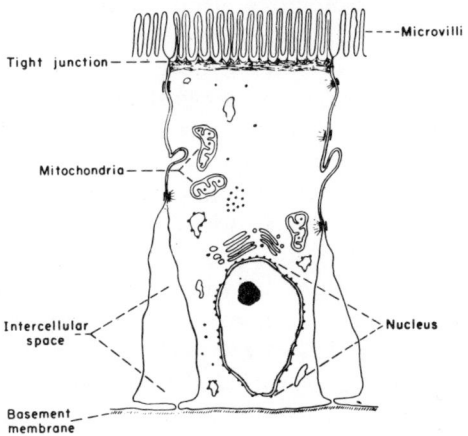

Fig. 4.11. A schematic illustration of an intestinal epithelial cell. (Redrawn from Trier [85].)

most epithelial tissues adjacent cells are joined together at their apical surfaces around the entire circumference of each cell. These tight junctions are formed by the fusion of the lateral plasma membranes over a distance of 1000-2000 Å. There are two possible pathways for the passive permeation of substances from the intestinal lumen into the sub-epithelial spaces, namely through the epithelial cells or through the tight junctions. In the former case substances traverse two cell membranes and the cytoplasm of the cell, and in the latter case substances bypass the cells by permeating into the lateral spaces through the tight junctions.

In epithelial tissues specialized to carry out water transport, e.g. gall bladder, proximal tubule and choroid plexus, there are a number of lines of evidence which suggest, but do not prove, that the principal permeation pathway is the extra-cellular one*: (1) the trans-epithelial resistance of the Necturus proximal tubule is reported to be an order of magnitude less than the sum of the two individual cell membrane resistances [76]; (2) diffusion potentials across the gall bladder have been shown to be dependent on the composition of both the mucosal and serosal solutions but independent of the intracellular composition [56]; (3) linear current-voltage curves across the gall bladder [57, 73] and choroid plexus (Wright, unpublished) suggests that the membranes controlling ion permeation are thick relative to bimolecular lipid membranes; and (4) the magnitude of cation selectivity sequences [73] suggest that the epithelial membranes are much more hydrated than single cell membranes (see above). This suggestion is supported by the analysis of gall bladder non-electrolyte permeability coefficients [26]. Clarkson [58] concluded that an extracellular shunt determined the passive permeability properties of the rat ileum but he suggested that the shunt was the holes left by cells extruded from the epithelium.

The existence of a shunt between epithelial cells would also provide a simple explanation for the low trans-epithelial resistances [81, 48, 53, 76], the high unidirectional fluxes [53, 68], the low value of the trans-epithelial potentials [48, 53, 76, 82], and the symmetry of the permeability characteristics measured from either the mucosal or serosal solutions [48, 55]. Furthermore an increase in the extracellular shunt might explain the increase in permeability of the intestine associated with *Vibrio cholerae* infections [86, 87, 88, 89].

4.5 SUMMARY

Although this review has emphasized the lack of information about the passive properties of the intestinal epithelium it is suggested that the permeation mechanisms are very similar to those in other epithelia which are specialized to carry out water transport. One interpretation of the passive permeability

* More direct evidence has been obtained recently in gall bladder (*Nature*, 235, 9 (1972): *J. Memb. Biol.*, 8, 259 (1972)) and intestine (*J. Gen. Physiol.*, 59, 318 (1972)).

characteristics of these epithelial membranes is that the tight junctions constitute a low resistance extracellular shunt. The test of this interpretation will come when the passive permeability properties of the individual cell membranes are compared with the passive properties of the epithelium as a whole.

ACKNOWLEDGEMENT

I am grateful to my colleagues, Drs. P. H. Barry, J. M. Diamond and A. P. Smulders, for many valuable discussions and their comments on the manuscript.

REFERENCES

1. P. F. CURRAN and S. G. SCHULTZ, in: Handbook of Physiology, Section 6, Volume 3, p. 1217 (Section Ed., C. F. Code, Executive Ed., W. Heidel). Am. Physiol. Soc., Washington (1968).
2. W. D. STEIN, The movement of molecules across cell membranes. Academic Press, New York and London (1967).
3. N. LAKSHMINARAYANAIAH, Transport phenomena in membranes. Academic Press, New York and London (1969).
4. W. R. LIEB and W. D. STEIN, *Nature*, 224, 240 (1969).
5. A. FINKELSTEIN and A. CASS, *J. Gen. Physiol.*, 52, 145s (1968).
6. R. COLLANDER, *Acta Chem. Scand.*, 3, 717 (1949).
7. R. COLLANDER, *Physiol. Plantarum*, 7, 420 (1954).
8. J. M. DIAMOND and E. M. WRIGHT, *Ann. Rev. Physiol.*, 31, 581 (1969).
9. J. DAINTY, *Advan. Botan. Res.*, 1, 279 (1963).
10. J. DAINTY and C. R. HOUSE, *J. Physiol.*, 182, 66 (1965).
11. J. M. DIAMOND, *J. Physiol.*, 183, 83 (1966).
12. E. M. WRIGHT and J. W. PRATHER, *J. Membrane Biol.*, 2, 127 (1970).
13. K. GREEN and T. OTORI, *J. Physiol.*, 207, 93 (1970).
14. W. RUHLAND and C. HOFFMANN, *Planta*, 1, 1 (1925).
15. R. COLLANDER, *Physiol. Plantarum*, 3, 45 (1950).
16. H. TEDESCHI and D. L. HARRIS, *Arch. Biochem. Biophys.*, 58, 52 (1955).
17. D. A. GOLDSTEIN and A. K. SOLOMON, *J. Gen. Physiol.*, 44, 1 (1960).
18. B. LINDEMANN and A. K. SOLOMON, *J. Gen. Physiol.*, 45, 801 (1962).
19. J. S. FORDTRAN, F. C. RECTOR, M. F. EWTON, N. SOTER and J. KINNEY, *J. Clin. Invest.* 44, 1935 (1965).
20. D. H. SMYTH and E. M. WRIGHT, *J. Physiol.*, 182, 591 (1966).
21. R. HOLZ and A. FINKELSTEIN, *J. Gen. Physiol.*, 56, 125 (1970).

22. J. GUTKNECHT, *Biochim. Biophys. Acta*, **163**, 20 (1968).
23. A. J. STAVERMAN, *Trans. Faraday Soc.*, **48**, 176 (1948).
24. O. KEDEM and A. KATCHALSKY, *Biochim. Biophys. Acta*, **27**, 229 (1958).
25. A. KATCHALSKY and O. KEDEM, *Biophys. J.*, **2**, pt. 2, 53 (1962).
26. A. P. SMULDERS, Ph.D. dissertation, University of California at Los Angeles (1970).
27. F. F. VARGAS, *J. Gen. Physiol.*, **51**, 13 (1968).
28. R. P. DURBIN and F. G. MOODY, in: The state and movement of water in living organisms, XIX Symposium of the Society of Experimental Biology, p. 299. University Press, Cambridge (1965).
29. G. SCHMID and H. SCHWARZ, *Z. Electrochem.*, **56**, 35 (1952).
30. F. F. VARGAS, *J. Gen. Physiol.*, **51**, pt. 2, 123s (1968).
31. H. J. WEDNER and J. M. DIAMOND, *J. Membrane Biol.*, **1**, 92 (1969).
32. K. LOESCHKE, C. J. BENTZEL and T. Z. CSAKY, *Am. J. Physiol.*, **218**, 1723 (1970).
33. P. F. CURRAN and A. K. SOLOMON, *J. Gen. Physiol.*, **41**, 143 (1957).
34. K. H. SOERGEL, G. E. WHALEN and J. A. HARRIS, *J. Applied Physiol.*, **24**, 40 (1968).
35. R. HOBER and R. HOBER, *J. Cell. Comp. Physiol.*, **10**, 401 (1937).
36. E. M. WRIGHT and J. M. DIAMOND, *Proc. Roy. Soc.*, B **172**, 203 (1969).
37. E. M. WRIGHT and J. M. DIAMOND, *Proc. Roy. Soc.*, B **172**, 227 (1969).
38. J. M. DIAMOND and E. M. WRIGHT, *Proc. Roy. Soc.*, B **172**, 273 (1969).
39. A. P. SMULDERS and E. M. WRIGHT, *J. Membrane Biol.*, **5**, 297 (1971).
40. R. COLLANDER, *Physiol. Plantarum*, **12**, 139 (1959).
41. J. de GIER, J. G. MANDERSLOOT and L. L. M. van DEENEN, *Biochim, Biophys. Acta*, **150**, 666 (1968).
42. N. LIFSON and A. A. HAKIM, *Am. J. Physiol.*, **211**, 1137 (1966).
43. A. K. SOLOMON, *J. Gen. Physiol.*, **51**, pt. 2, 335s (1968).
44. R. I. SHA'AFI, G. T. RICH, V. W. SIDEL, W. H. BOSSERT and A. K. SOLOMON, *J. Gen. Physiol.*, **50**, 1377 (1967).
45. P. H. BARRY and A. B. HOPE, *Biophys. J.*, **9**, 700 (1969).
46. P. H. BARRY and A. B. HOPE, *Biophys. J.*, **9**, 729 (1969).
47. G. T. RICH, R. I. SHA'AFI, T. C. BARTON and A. K. SOLOMON, *J. Gen. Physiol.*, **50**, 2391 (1967).
48. J. M. DIAMOND, *J. Physiol.*, **161**, 474 (1962).
49. D. E. GOLDMAN, *J. Gen. Physiol.*, **27**, 37 (1943).
50. A. L. HODGKIN and B. KATZ, *J. Physiol.*, **108**, 37 (1949).
51. J. P. SANDBLOM and G. EISENMAN, *Biophys. J.*, **7**, 217 (1967).
52. H. LINDERHOLM, *Acta Physiol. Scand.*, **27**, Suppl., 97 (1952).
53. S. G. SCHULTZ and R. ZALUSKY, *J. Gen. Physiol.*, **47**, 567 (1964).
54. H. H. USSING, *Acta Physiol. Scand.*, **19**, 43 (1949).
55. J. M. DIAMOND and S. C. HARRISON, *J. Physiol.*, **183**, 37 (1966).

56. P. H. BARRY and J. M. DIAMOND, *J. Membrane Biol.*, 3, 93 (1970).
57. E. M. WRIGHT, P. H. BARRY and J. M. DIAMOND, *J. Membrane Biol.*, 4, 331 (1971).
58. T. W. CLARKSON, *J. Gen. Physiol.*, 50, 695 (1967).
59. E. M. WRIGHT, *J. Physiol.*, 185, 486 (1966).
60. P. G. KOHN, D. H. SMYTH and E. M. WRIGHT, *J. Physiol.*, 196, 723 (1968).
61. A. P. SMULDERS and E. M. WRIGHT, *J. Physiol.*, 212, 277 (1971).
62. E. M. WRIGHT, *Nature*, 212, 189 (1966).
63. H. H. USSING, in: The alkali metal ions in biology (H. H. Ussing, P. Kruhfter, J. Hess Thaysen and N. A. Thorn, eds.), p. 1. Springer, Berlin (1960).
64. S. G. SCHULTZ, P. F. CURRAN and E. M. WRIGHT, *Nature*, 214, 509 (1967).
65. P. F. CURRAN, *J. Gen. Physiol.*, 43, 1137 (1960).
66. S. G. SCHULTZ and R. ZALUSKY, *J. Gen. Physiol.*, 47, 1043 (1964).
67. S. G. SCHULTZ, R. ZALUSKY and A. E. GASS, *J. Gen. Physiol.*, 48, 375 (1964).
68. A. E. TAYLOR, E. M. WRIGHT, S. G. SCHULTZ and P. F. CURRAN, *Am. J. Physiol.*, 214, 836 (1968).
69. E. M. WRIGHT and J. M. DIAMOND, *Biochim. Biophys. Acta*, 163, 57 (1968).
70. M. BAILLIEN and E. SCHOFFENIELS, *Biochim. Biophys. Acta*, 53, 537 (1961).
71. G. EISENMAN, in: Symposium on membrane transport and metabolism (A. Kleinzeller and A. Kotyk, eds.), p. 163. Academic Press, New York (1961).
72. G. EISENMAN, *Biophys. J.*, 2, pt. 2, 259 (1962).
73. P. H. BARRY, J. M. DIAMOND and E. M. WRIGHT, *J. Membrane Biol.*, 4, 358 (1971).
74. G. EISENMAN, Proc. 23rd Internat. Congr. Physiol. Sci. Tokyo, p. 489. Excerpta Med. Found., Amsterdam (1965).
75. G. EISENMAN, *Fed. Proc.*, 27, 1249 (1968).
76. E. E. WINDHAGER, E. L. BOULPAEP and G. GIEBISCH, Proc. 3rd Int. Congr. Nephrol., Washington, 1, 35 (1966). Karger, Basel and New York (1967).
77. G. SZABO, G. EISENMAN and S. CIANI, *J. Membrane Biol.*, 1, 346 (1969).
78. S. CIANI, G. EISENMAN and G. SZABO, *J. Membrane Biol.*, 1, 1 (1969).
79. P. H. BARRY and J. M. DIAMOND, *J. Membrane Biol.*, 4, 295 (1971).
80. A. CASS, A. FINKELSTEIN and V. KRESPI, *J. Gen. Physiol.*, 56, 100 (1970).
81. R. J. C. BARRY, D. H. SMYTH and E. M. WRIGHT, *J. Physiol.*, 181, 410 (1965).
82. R. J. C. BARRY, S. DIKSTEIN, J. MATTHEWS, D. H. SMYTH and E. M. WRIGHT, *J. Physiol.*, 171, 316 (1964).
83. D. E. LEB, T. HOSHIKO and B. D. LINDLEY, *J. Gen. Physiol.*, 47, 749 (1965).

84. B. D. LINDLEY and T. HOSHIKO, *J. Gen. Physiol.*, **47**, 749 (1964).
85. J. S. TRIER, in: Handbook of Physiology, Section 6, Volume 3, p. 1125 (Section Ed., C. F. Code, Executive Ed., W. Heidel). Am. Physiol. Soc., Washington (1968).
86. A. H. G. LOVE, *Gut*, **10**, 63 (1969).
87. A. H. G. LOVE, *Gut*, **10**, 105 (1969).
88. D. I. GRAYER, H. A. SEREBRO and T. R. HENDRIX, *Fed. Proc.*, **28**, 718 (1969).
89. H. A. SEREBRO, F. L. IBER, J. H. YARDLEY and T. R. HENDRIX, *Gastroenterology*, **56**, 506 (1969).

CHAPTER 5

Irreversible Thermodynamics

STANLEY G. SCHULTZ*

Department of Physiology,
University of Pittsburgh,
School of Medicine,
Pittsburgh, Pennsylvania, U.S.A.

		Page
5.1	INTRODUCTION	199
5.2	BASIC PRINCIPLES	201
	5.2.1 *Flows, forces and the dissipation function*	201
	5.2.2 *The phenomenological equations and Onsager reciprocity*	203
	5.2.3 *The Kedem-Katchalsky membrane*	206
	5.2.4 *The interpretation of phenomenological coefficients in terms of frictional interactions*	208
	5.2.5 *Composite membranes*	212
5.3	APPLICATIONS OF IRREVERSIBLE THERMODYNAMICS	214
	5.3.1 *The mechanism of isotonic water absorption*	214
	5.3.2 *Solvent drag*	219
	5.3.3 *Streaming potentials and the measurement of* σ	224
	5.3.4 *Solute-solute interactions*	227
5.4	ENERGY CONVERSION IN BIOLOGICAL TRANSPORT PROCESSES	228
5.5	CONCLUDING REMARKS	235
	REFERENCES	236

5.1 INTRODUCTION

Recently, Teorell [1] has described the development of 'membranology' in terms of an inextricable marriage among physiologists, biologists, biochemists, morphologists,

* Research Career Development Awardee, National Institute of Arthritis and Metabolic Diseases, U.S.P.H.S.

biophysicists and physicochemists, all contributing to the forward flow of this field. The primary function of the physical theorist is to provide formalisms that permit systematic quantitation, generalizations and predictions and that decrease our reliance upon highly fallible intuition.

The impact of classical thermodynamics on membranology, through the contributions of Van't Hoff, Nernst, Gibbs, Donnan etc., need not be belaboured. Its unassailable conclusions are dependent only on considerations of the initial and final states of a system and are independent of intervening pathway or mechanism; no discipline could be better suited to a field in which mechanisms and microscopic structure are, for the most part, obscure. Yet, the applicability of classical thermodynamics to living systems is extremely restricted. Its formalisms can only be rigorously applied to systems that are at equilibrium or are undergoing reversible changes (i.e. processes in which energy is not dissipated); for all other conditions, it provides us only with inequalities. It does not deal with flows or rates and thus its usefulness for the description of living systems, in which exchange is a *sine qua non*, is minimal. Indeed, since the rate of a process depends upon pathway, *time* does not play an explicit role in the formalisms of classical thermodynamics. It is by virtue of the explicit consideration of time and flows that irreversible thermodynamics fulfils the need left wanting by classical thermodynamics.

The thermodynamics of irreversible processes, or non-equilibrium thermodynamics, represents an extension of the laws of classical thermodynamics to systems that are slightly displaced from equilibrium. The major impetus for the rigorous development of this field came from the pioneering work of Onsager [2], Prigogine [3], de Groot [4] and Staverman [5]. The application and extension of these new considerations to biological systems was pioneered by Kedem and Katchalsky [6], and their now classic paper 'Thermodynamic analysis of the permeability of biological membranes to non-electrolytes' must be viewed as a turning-point in the formal analysis of transport processes across biological membranes. Briefly, this development makes possible the replacement of the inequalities of classical thermodynamics with equalities and permits a quantitative description of systems involving flows of matter or energy. In addition, it explicitly introduces considerations of *resistance to flow* and thus is operationally cognizant of pathways. However, these important extensions are accompanied by some sacrifice; although the formalisms of irreversible thermodynamics appear to be valid over a wide range of conditions, they do not have the general applicability of the first and second laws of classical thermodynamics.

Clearly, a rigorous development of irreversible thermodynamics in this chapter is neither possible nor warranted. Instead I will limit this discussion to a consideration of some basic principles and attempt to illustrate the application of irreversible thermodynamics to problems of concern to those of us interested in transport across epithelial membranes; it is hoped that both the usefulness of this approach and the potential for misuse will emerge from these considerations. The reader interested in a more detailed or rigorous treatment is referred to the excellent monographs and reviews by Prigogine [3], de Groot [4], Denbigh [7], Katchalsky and Curran [8], Katchalsky and Kedem [9] and Caplan and Mikulecky [10].

5.2 BASIC PRINCIPLES

5.2.1 *Flows, forces and the dissipation function*

The physical properties of a system may be divided into two categories. The first comprises the *extensive* properties, such as mass and volume, which are dependent upon the size and content of the system. The second comprises the *intensive* properties, such as temperature, pressure and electrical potential, which are independent of the size of the system. Thus, if we were to divide a homogenous system into subcompartments, the extensive properties of the entire system would be the sum of the extensive properties of the subcompartments, but the intensive properties of each of the subcompartments are the same as those of the entire system. Each extensive property of a system is associated with a *conjugate* intensive property. Thus, the extensive property *charge* is conjugated with the intensive property *electrical potential*; *volume* is conjugated with *pressure;* and the *amount* of a given substance (in moles), n_i, is conjugated to the *eletrochemical potential* of that substance, $\tilde{\mu}_i$, given by

$$\tilde{\mu}_i = \tilde{\mu}_i^\circ + RT \ln C_i + \bar{v}_i P + z_i \mathcal{F} \psi \qquad (5.1)$$

where P = pressure
R = gas constant
T = absolute temperature
C_i = concentration of the substance i
$\bar{v}_i$ = partial molar volume of i
z_i = the valence of i
$\mathcal{F}$ = the Faraday
$\tilde{\mu}_i^\circ$ = The electrochemical potential in the standard (reference) state (i.e. $\tilde{\mu}_i = \tilde{\mu}_i^\circ$ when $C_i = 1$, $P = 0$ and $\psi = 0$)

Clearly, for an uncharged substance the intensive property $\tilde{u}$ is related to concentration, C_i, and P. For a charged substance, $\tilde{\mu}_i$ is also influenced by the electrical potential of the system. The nature of the conjugate relation becomes apparent when one realizes that *flows* are displacements of the extensive properties and the *driving forces* for these flows are differences in *their* conjugate intensive properties.* Thus, the flow of charge, or current, is driven by a difference in electrical potential; the flow of volume is driven by a difference in pressure and the flow of matter is driven by a difference in electrochemical potential. The products of the intensive and conjugate extensive properties all have units of energy, and the product of the flow of an extensive property times the difference in its conjugate intensive property (the driving force for the flow) has units of energy/time. In other words, if, for example, we permit a flow of matter, dn_i/dt, from a region of high electrochemical potential to a region of lower electrochemical potential, the product

$$(dn_i/dt)\,(\Delta \tilde{\mu}_i)$$

gives the rate at which the free energy represented by $\Delta \tilde{\mu}_i$ is being dissipated. Clearly, flow will cease when $\Delta \tilde{\mu}_i = 0$. From

---

* More rigorously, the driving force is the negative of the gradient of the intensive property, e.g. the driving force for the flow of substance i in the x direction is $-d\tilde{\mu}_i/dx$. When distance is small, the gradient can be approximated by $-\Delta\tilde{\mu}_i/\Delta x$.

these simple considerations we can appreciate the origin of the dissipation function, introduced by Lord Rayleigh, which is defined by

$$\Phi = \sum_i J_i X_i \qquad (5.2)$$

where the J's are the flows of extensive properties and the X's are the conjugate driving forces.

For an isothermal system that is 'close to' equilibrium (i.e. one in which the gradients of intensive properties are sufficiently small so that one can isolate a region of local equilibrium in which properties are defined and homogeneous) it may be shown that

$$\Phi = T\, d_i S/dt = \sum J_i X_i \qquad (5.3)$$

where $d_i S/dt$ is the rate of production of internal entropy.* Thus, under these restricted conditions, the sum of the products of flows and conjugate forces is a measure of the rate at which internal entropy is being produced as the result of ongoing irreversible processes.

5.2.2. *The phenomenological equations and Onsager reciprocity*
It has long been recognized, on empirical grounds, that sufficiently slow flows are linearly related to their driving forces. Fourier's analysis of heat flow, Fick's first law of diffusion and Ohm's law are familiar examples of this relation. Thus, for a single flow and a single driving force we may write

$$J_i = L_i X_i \qquad (5.4)$$

where L_i is the flow per unit driving force and is thus a measure of conductance (e.g. respectively, the thermal conductivity, diffusion coefficient, and electrical conductance for the examples cited above). The essential property of L is that it is strictly a phenomenological coefficient that is dependent upon pathway but is independent of the driving force (X).

It has also been long appreciated that if there are more than one flow and one driving force in a system, the flows are not

---

* Under conditions in which temperature and pressure are constant (e.g. most chemical and biological reactions) $T\, d_i S/dt$ is equal to the rate of decrease or dissipation of the Gibbs free energy ($-dG/dt$). When only temperature is constant, $T\, d_i S/dt$ is equal to the rate of dissipation of the Helmholtz free energy or 'work function' of the system. For a discussion of anisothermal systems see Katchalsky and Curran [8].

solely influenced by their conjugate driving forces but are affected by other flows and, hence, by nonconjugate driving forces. Such a system can be described by the following set of linear phenomenological equations.

$$J_i = L_{ii}X_i + L_{ij}X_j + \ldots L_{in}X_n$$
$$J_j = L_{ji}X_i + L_{jj}X_j + \ldots L_{jn}X_n$$
$$\vdots$$
$$J_n = L_{ni}X_i + L_{nj}X_j + \ldots L_{nn}X_n \tag{5.5}$$

or

$$J_i = L_{ii}X_i + \sum_{\substack{k=1 \\ k \neq i}} L_{ik}X_k \tag{5.6}$$

In these equations, the L_{ii} are referred to as straight coefficients that relate J_i to its conjugate driving force. The L_{ik} ($k \neq i$) are cross coefficients that reflect the effect of the nonconjugate force X_k on J_i.

Equations (5.5) express flows as functions of forces and the phenomenological coefficients are generalized conductances. It is also possible to express the forces as functions of the flows and to relate them by generalized resistances. Thus,

$$X_i = R_{ji}J_i + R_{ij}J_j + \ldots R_{in}J_n$$
$$X_j = R_{ij}J_i + R_{jj}J_j + \ldots R_{jn}J_n$$
$$\vdots$$
$$X_n = R_{ni}J_i + R_{nj}J_j + \ldots R_{nn}J_n \tag{5.7}$$

or

$$X_i = R_{ii}J_i + \sum_{\substack{k=1 \\ k \neq i}} R_{ik}J_k \tag{5.8}$$

Solving equation (5.8) for J_i we obtain

$$J_i = (X_i/R_{ii}) - \Sigma R_{ik}J_k/R_{ii} \tag{5.9}$$

This form more directly conveys one of the important messages that emerges explicitly from this formalism; namely, it subdivides the factors responsible for the flow of any substance into two categories. The first is the conjugate driving force for that flow. The second includes all of the contributions to that flow that derive from coupling to, or interaction with, other flows in the system. It is this latter category that has been too often overlooked in the analysis and classification of biological transport processes, and we will return to this point later.*

It should be stressed that the above equations are strictly phenomenological and represent an extension of numerous empirical observations on systems with two flows and two forces. In 1931, Onsager [2] demonstrated that if the flows and forces are properly chosen so that equation (5.3) is satisfied, then

$$L_{ik} = L_{ki} \qquad (5.10)$$

This law of reciprocity tightly links phenomenology with theory and provides an experimental approach for determining whether a given system obeys equation (5.3). Miller [11] has compiled numerous examples of coupled phenomena in which reciprocity is obeyed. More recently Blumenthal et al. [12] have demonstrated, for the first time, reciprocity for the coupling between the flow of electric current and a chemical reaction; a demonstration that is of particular importance in the consideration of biological transport processes.† Thus the law

---

* Kedem [62] has rewritten equation (5.9) in the following form for the flow of a substance i

$$J_i = -(\Delta\mu_i/R_{ii}) - [\sum_{j \neq i, r} R_{ij} J_j/R_{ii}] - R_{ir} J_r/R_{ii}$$

where the J_j are the flows of other solutes or solvent and J_r is the flow of a chemical reaction. She has suggested that the term 'active transport' be reserved only for flows that are directly coupled to metabolic reactions (i.e. $R_{ir} \neq 0$). This definition differs significantly from that proposed by Rosenberg [63]. For a more detailed discussion of the definition of active transport the reader is referred to Curran and Schultz [55] and Schultz [22].

† Two problems dealing with the coupling between a chemical reaction and the flow of matter are: (i) Unlike the vectorial driving forces for the flows of matter, the driving force of a chemical reaction is a non-directed, scalar quantity. Thus, the phenomenological coupling coefficient must be vectorial and must reflect some anisotropic property of the system. (ii) The flow of a chemical reaction, J_r, will not be a linear function of the driving force unless the reaction is very close to equilibrium (e.g. the free energy change must be less than 600 cal/mole). Thus linearity would not be expected to apply in the case of high energy biological reactions (e.g. ATP hydrolysis) unless the overall reaction is the result of numerous intermediate steps, each of which is close to equilibrium. Prigogine [3] and Katchalsky and Curran [8] have discussed these problems in detail.

of reciprocity provides a test for the range of validity of the formalisms of irreversible thermodynamics; that is, it allows us to determine whether we are sufficiently 'close to' equilibrium to apply the powerful tools of this new discipline.

5.2.3 The Kedem-Katchalsky membrane

In his recent, remarkable treatise 'Membranes, Ions and Impulses' Kenneth Cole [13] refers to the Hodgkin-Huxley axon. In this sense it is not inappropriate to entitle this section 'The Kedem-Katchalsky membrane' to convey the impact of their contributions on the formal analysis of transport processes.

Kedem and Katchalsky [6] considered a two-compartment system, separated by a membrane, in which there was a flow of an uncharged solute s and of solvent w. Thus, the dissipation function is simply

$$T \, d_iS/dt = (\Delta \mu_w)(dn_w/dt) + (\Delta \mu_s)(dn_s/dt) \quad (5.11)$$

where

$$\Delta \mu_s = \bar{v}_s \Delta P + RT \Delta \ln C_s \quad (5.12)$$

and, for dilute solutions (where $C_w \bar{v}_w \cong 1$)

$$\Delta \mu_w = \bar{v}_w \Delta P - RT \, \Delta C_s / C_w \quad (5.13)$$

I will not reproduce the derivations, which are detailed in the original paper, but, by assuming reciprocity and employing a skilful manipulation of flows and forces, the following equations describing volume and solute flow emerge.

$$J_V = L_P(\Delta P - \sigma RT \, \Delta C_s) \quad (5.14)$$

and

$$J_s = (1 - \sigma)\bar{C}_s J_v + \omega RT \, \Delta C_s \quad (5.15)$$

where $\bar{C}_s$ is an average solute concentration across the membrane which, for small concentration differences, can be approximated by the mean concentration.

Let us examine each of these equations individually.

In equation (5.14) J_V is the total volume flow across the membrane and thus is composed of the volume contributed by the flow of solvent (w) and that resulting from the flow of solute (s). Thus

$$J_V = \bar{v}_w J_w + \bar{v}_s J_s \qquad (5.16)$$

L_p is the hydraulic conductivity of the membrane and represents the volume flow per unit pressure. Clearly, when $\Delta C_s = 0$.

$$J_V = L_p \Delta P \qquad (5.17)$$

The significance of σ can be best appreciated by setting $J_V = 0$ so that

$$\Delta P = \sigma RT \Delta C_s \qquad (5.18)$$

Clearly, when $J_V = 0$ (e.g. when volume flow is prevented by rigid walls) ΔP is the osmotic pressure resulting from the presence of a difference in solute concentration on the two sides of the membrane. Staverman [5] demonstrated that the effective osmotic pressure across a membrane that is not ideally impermeable to the solute is given by equation (5.18). σ, which is referred to as the *reflection coefficient,* may vary from 1.0 for a solute to which the membrane is ideally impermeable to 0 for a solute that, as far as the membrane is concerned, is indistinguishable from the solvent. Clearly, when $\sigma = 1$, equation (5.18) reduces to the familiar Van't Hoff equation. Thus equation (5.14) essentially states that volume flow is a consequence of driving forces arising from differences in hydrostatic pressure and differences in *effective* osmotic pressure, and is related to these driving forces by L_p which is simply a property of the barrier. But for the notion of the reflection coefficient, this relation is precisely that proposed by Starling [14] in his classic paper 'On the absorption of fluids from the connective tissue spaces' (1896).

Equation (5.15) can be easily appreciated by first considering the condition when $J_V = 0$

$$J_s = \omega RT \Delta C_s \qquad (5.19)$$

Clearly, ω is a measure of the permeability of the membrane to the solute when $J_V = 0$ and is related to the traditional permeability coefficient, P_s, by $\omega = P_s/RT$. The first term in the equation can be best appreciated when $\Delta C_s = 0$. Then

$$J_s = (1 - \sigma) \bar{C}_s J_V \qquad (5.20)$$

so that J_s is the flow of solute that can be attributed entirely to a flow of volume; in short, this is the flow of solute entirely attributable to 'solvent drag'.

Thus the two flow equations derived by Kedem and Katchalsky [6] invoke three independent properties of the system that can be defined experimentally: L_p, σ and ω. However, before considering the application of these equations to transport processes we will explore the possible physical significance of these parameters.

5.2.4 The interpretation of phenomenological coefficients in terms of frictional interactions

The parameters L_p, σ and ω are phenomenologic coefficients reflecting interactions among solute, solvent and the membrane. However, by themselves they offer little insight into the nature of these interactions. As mentioned above, the conclusions of classical thermodynamics are independent of pathway or mechanism. Although this characteristic permits an analysis to circumvent our ignorance, it also does nothing to diminish our ignorance; that is, information obtained from a classical thermodynamic analysis, though irrefutable, sheds no light on underlying mechanisms.

The incorporation of generalized resistances and conductances in the formalism of irreversible thermodynamics is an explicit recognition of the properties of pathways and has prompted the attempt to provide a physical basis for the phenomenologic coefficients. Spiegler [15] proposed a frictional model to account for the phenomenological coefficients describing flows across ion exchange membranes, and Kedem and Katchalsky [16] have extended these considerations to provide explicit expressions for L_p, σ and ω in terms of frictional forces.

We will illustrate the approach using the simple case of water flow through a membrane driven by a hydrostatic pressure. The driving force is thus

$$X_w = -d\mu_w/dx = -\bar{v}_w\, dP/dx \qquad (5.21)$$

When a steady-state velocity of flow is reached (i.e. acceleration is zero) there must be a balance of forces such that the driving force is equal and opposite to the force generated by the frictional interaction between water and the membrane. We may describe the frictional force between water and the membrane as

$$X_{wm} = f_{wm}(v_w - v_m) \tag{5.22}$$

where f_{wm} is the frictional coefficient between water and the membrane, v_w is the velocity of water and v_m is the velocity of the membrane. Choosing the membrane as the frame of reference, $v_m = 0$ so that

$$X_{wm} = f_{wm} v_w$$

When the steady-state velocity is achieved

$$-\bar{v}_w \, dP/dx = f_{wm} v_w$$

Now, the flow of water through the membrane is

$$J_w = v_w C_{\dot{w}}$$

where $C_{\dot{w}}$ is the concentration of water in the membrane. It follows that

$$-\bar{v}_w \, dP/dx = J_w f_{wm}/C_{\dot{w}} = J_w f_{wm} \bar{v}_w/\phi$$

where ϕ is the volume fraction of water in the membrane. Integrating over the thickness of the membrane, assuming that f_{wm} and ϕ are constant, one obtains

$$J_w \bar{v}_w = J_V = \bar{v}_w \, \phi \Delta P/f_{wm} \Delta x$$

where Δx is the thickness of the membrane. Thus

$$L_p = \bar{v}_w \, \phi/f_{wm} \Delta x \tag{5.23}$$

Thus the phenomenological coefficient L_p can be expressed explicitly in physical terms. Similarly, Kedem and Katchalsky have shown that

$$\omega = K_s/(f_{sw} + f_{sm})\Delta x \tag{5.24}$$

and

$$\sigma = 1 - \omega \bar{v}_s/L_p - K_s f_{sw}/(f_{sw} + f_{sm})\phi \tag{5.25}$$

or

$$\sigma = 1 - \omega \bar{v}_s/L_p - \omega f_{sw} \Delta x/\phi \tag{5.26}$$

In these equations, K_s is the partition coefficient between the aqueous solution and the membrane (i.e. $K_s = C_{memb}/C_{soln}$)

and f_{sw} and f_{sm} are the frictional coefficients between solute and water and solute and membrane respectively.

Several important deductions can be derived from equations (5.25)-(5.26):

(a) As $\omega \to 0$, $\sigma \to 1$ except when $\Delta x \to \infty$. Thus, whereas under most circumstances a low value of ω will be associated with a value of σ close to 1, it is possible to have a low permeability and a low reflection coefficient if the diffusion pathway is long. The significance of this point will become apparent when the model for isotonic water absorption through long narrow intercellular spaces is discussed.

(b) In general, as ω increases, σ decreases and may assume negative values. A negative value for σ corresponds to the condition termed 'anomalous osmosis' in which volume flow is directed from the concentrated solution to the more dilute solution. This situation has been observed for the case of membranes bearing a high density of fixed charge (Grim and Sollner [17]) and has been discussed by Kedem and Leaf [18].

Finally, equation (5.25) has been used to examine the pore hypothesis; that is, the notion that small water-soluble solute may diffuse across biological membranes through water-filled channels or pores. Equation (5.25) may be rewritten as

$$(1 - \sigma) - \omega \bar{v}_s/L_p = K_s f_{sw}/(f_{sw} + f_{sm})$$

Now, if solute and water traverse the membrane by different pathways, $f_{sw} = 0$ and $(1 - \sigma) = \omega \bar{v}_s/L_p$. However, if $f_{sw} > 0$ then $(1 - \sigma) > \omega \bar{v}_s/L_p$, and J_s will be greater when there is a concomitant flow of solvent than can be accounted for by the solute concentration difference alone. The latter situation was termed 'solvent drag' by Andersen and Ussing [19] and was forwarded as evidence for a direct interaction between solutes and bulk flow of water.

It is of interest to examine further the situation where solute and solvent traverse a membrane via non-interacting pathways. For example, consider the case illustrated in Fig. 5.1 where a lipid membrane having water-filled pores of small diameter separates two solutions of a lipid-soluble substance that is too large to traverse the membrane via the pores. Assume further that the solution on the left is more concentrated than that on

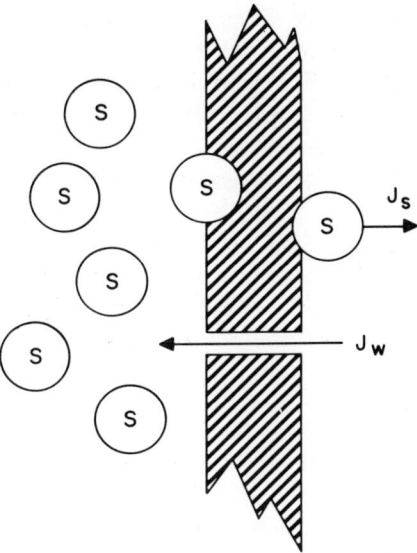

Fig. 5.1. A porous membrane separating two solutions of a solute that cannot traverse the pores but can permeate the membrane following dissolution in the membrane matrix.

the right and that $\Delta P = 0$. Since $f_{sw} = 0$, equation (5.25) reduces to

$$\sigma = 1 - \omega \bar{v}_s / L_p \tag{5.27}$$

From equations (5.14) and (5.16) we obtain

$$J_V = J_w \bar{v}_w - J_s \bar{v}_s = L_p \sigma RT \Delta C_s \tag{5.28}$$

where J_V is the rate of change of volume of the more concentrated solution. Combining equations (5.27) and (5.28) we obtain

$$J_w \bar{v}_w - J_s \bar{v}_s = L_p RT \Delta C_s - \omega \bar{v}_s RT \Delta C_s \tag{5.29}$$

Now, the first term on the right of the equality sign is the *volume flow of water that would be predicted from Van't Hoff's law for an ideally impermeant solute*. The second term on the right of the equality sign is the volume occupied by solute that is diffusing from left to right. Thus, the total volume change is less than that predicted for an ideally permeable membrane (i.e. $\sigma < 1$) only because of a displacement of the volume occupied by the permeable solute. However, because the paths differ, the water flow through the pore is precisely what one would expect for an ideally impermeable solute. If $f_{sw} > 0$, the water flow from right to left is reduced by the drag effect of diffusing molecules moving from left to right through the same pores. Thus, if the permeable molecule interacts with water the reduction of volume flow from right to left, below that expected for an ideal membrane, is the result of both (i) the actual solute volume displaced from left to right and (ii) the *retardation* of water flow by frictional interaction with the oppositely directed flow of solute. This point will be discussed further below.

5.2.5 Composite membranes

Up to this point, all of our considerations have focussed on homogeneous membranes; that is, membranes in which all of the permeation pathways are characterized by the same set of phenomenological coefficients. Clearly, this description cannot apply to epithelial membranes that are bounded by at least two membranes arranged in series that must have different transport properties to effect net transepithelial flows (Schultz and Curran [20]). In addition, most if not all epithelial membranes appear to possess permeation pathways that parallel the absorptive cells. For example, in the intestine, tight intercellular junctions, nonabsorptive cells (e.g. goblet cells) and even denuded areas can provide permeation routes that circumvent the absorptive cells. Kedem and Katchalsky [21] have provided a detailed analysis of the permeability characteristics of composite membranes along the lines developed for homogeneous membranes. A detailed discussion of their analysis is beyond the scope of this chapter. However, several noteworthy points will be summarized.

If two homogeneous membranes, designated α and β, are arranged in series, separated by a homogeneous, well mixed compartment, the steady-state condition implies that the net flows across both membranes will be equal. However, the driving forces across these membranes will, in general, differ. In such a system, the following departures from the behaviour of a simple homogeneous membrane can be observed:

(a) It is well known that ω, in general, is a function of solute concentration. If the concentration dependence of ω_α and ω_β

differ, rectification of solute permeation will be observed. That is, if we initially measure solute permeation with one concentration in the solution bathing membrane α and another concentration in the solution bathing membrane simply reversing the solutions will not yield the same result. Thus, the notion of an 'overall' ω may not be meaningful.

(b) The 'overall' resistance to volume flow $(1/L_p)$ is not equal to the sum of the resistances of the individual membranes. Further, rectification of volume flow may occur since the 'overall' L_p will, in general, depend upon the direction of flow.

(c) As indicated by equations (5.14) and (5.15), σ may be estimated from ultrafiltration experiments in which J_s is determined when $\Delta C_s = 0$ and J_V is produced by the application of hydrostatic pressure. In addition, σ may be evaluated by determining the effective osmotic pressure when $J_V = 0$ but $\Delta C \neq 0$. In general these two methods will not give the same result unless J_V in the ultrafiltration experiment is small and approaches 0.

For a membrane possessing parallel permeation pathways the same driving forces act on all of the flows. The flows through different pathways differ but the total flow is the sum of the individual flows. Again, significant departures from the behaviour of simple homogeneous membranes emerge.

(a) The straight coefficients L_p and ω are non-additive; that is, they are not simply the area-weighted average of the individual coefficients. This indicates a significant circulation of flows among the different parallel pathways.

(b) The 'overall' σ is not simply an area-weighted average of the individual reflection coefficients but may be smaller than the smallest or larger than the largest of the elementary reflection coefficients.

This brief summary does not do justice to the elegance of the original analysis, but it serves to point out the hazards of indiscriminately applying formulations derived for highly simplified, if not idealized, systems to complex systems. An analysis for composite membrane systems comprising both parallel and series arrays has not been published. However, judging from the above considerations, it seems likely that even greater deviations from the behaviour of homogeneous systems (as expressed in equations (5.14) and (5.15)) would emerge. It is often rationalized that although 'overall' phenomenological coefficients may not shed light on the properties of individual elements of a composite membrane system, they nevertheless serve to characterize a given system and can be used to compare one system with another. This argument is acceptable *providing the value of the overall parameter is not a function of the particular experimental approach employed.* If, for example, the evaluation of the overall reflection coefficient in a series array is determined by the experimental approach, so that it can no longer be considered a useful property of the system and could, instead, become the focus of unnecessary dispute.

5.3 APPLICATIONS OF IRREVERSIBLE THERMODYNAMICS

The monumental contribution of irreversible thermodynamics to the analysis of transport processes is that it not only extends the laws of classical thermodynamics to include flows and the element of time, but that it also permits an explicit formal treatment of the *interactions* (or couplings) between flows. Indeed, one of the principle conclusions of the original paper by Kedem and Katchalsky [6] is that the conventional analysis of membrane permeability in terms of solute-permeability coefficients and a water-permeability coefficient alone leads to inconsistencies inasmuch as interactions between the flows of solute and solvent are ignored.

The approach pioneered by Kedem and Katchalsky has been applied to the study of numerous single-cell systems and artificial membranes, and many of these studies are discussed in the text by Katchalsky and Curran [8]. I will restrict the subsequent discussion to some examples of the application of irreversible thermodynamics to transport across epithelial tissues, and, in particular, the small intestine.

5.3.1 The mechanism of isotonic water absorption
In this author's opinion, the high point of the application of the principles of irreversible thermodynamics to transport across epithelial tissues has been in the development of current concepts regarding isotonic water absorption. This subject has been discussed in considerable detail (Schultz and Curran [20],

Schultz [22], Diamond [23]) and I will limit this discussion to illustrating the role of the Kedem-Katchalsky formulations.

By 1962, several characteristics of water absorption by a variety of epithelia appeared to be firmly established. As early as 1892 the brilliant investigations of Reid [24] demonstrated that an isolated segment of small intestine, bathed on both surfaces with identical solutions, will transport water from the mucosal solution to the serosal solution against an applied hydrostatic pressure difference. Much later, Parsons and Wingate [25] demonstrated that water absorption by rat small intestine, *in vitro*, will proceed against a large osmotic pressure difference; that is, the mucosal solution must be made approximately 100 m Osm hypertonic to the serosal solution before water absorption is abolished. In addition it was known that, in the absence of a hydrostatic or osmotic pressure difference, water absorption is dependent upon solute absorption, particularly NaCl, and that the absorbed solution is always essentially isotonic with that bathing the mucosal surface of the tissue. These relations are clearly illustrated in Fig. 5.2. Indeed, as illustrated in Fig. 5.3, the fluid transported across isolated rabbit gall bladder is isotonic with the solution in the lumen over a 6-fold range of osmolarities; this relation can hardly be considered fortuitous.

The problem facing investigators by 1962 was to conceive of a mechanism by which the flow of a single molecule of solute would be accompanied by the flow of several hundred molecules of water so as to produce an isotonic absorbate, and which could operate against a hydrostatic pressure difference.

A model system that could account for these findings was first suggested by Curran [26] and later formally described by Curran and McIntosh [27, 28]. This system is illustrated in Fig. 5.4 and is referred to as the 'double membrane hypothesis'. According to this model, two different membranes, α and β, are arranged in series and are separated by a well stirred compartment (compartment 2). Volume flow across each membrane is given by equation (5.14) as

$$J_{V\alpha} = L_{p\alpha}[(P_1 - P_2) + \sigma_\alpha RT(C_2 - C_1)]$$

and

$$J_{V\beta} = L_{p\beta}[(P_2 - P_3) + \sigma_\beta RT(C_3 - C_2)]$$

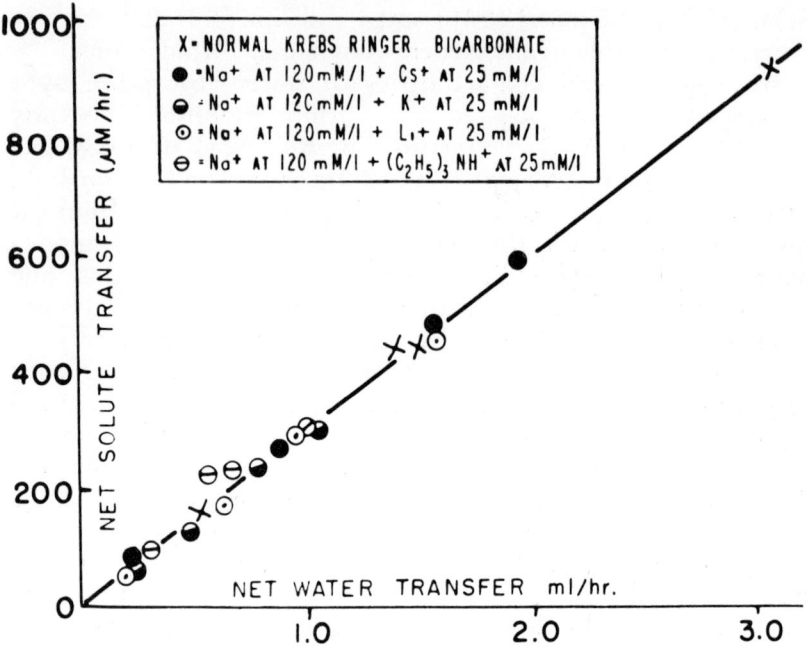

Fig. 5.2. The obligatory relation between water and total solute transfer across *in vitro* rat small intestine. The slope of the line indicates that the osmolarity of the absorbed solution is approximately 300 mOsm, in good agreement with that of the bathing solution (313 mOsm). (From Clarkson and Rothstein [64].)

In the steady-state, $J_{V\alpha} = J_{V\beta} = J_V$ and when $P_1 = P_3$ we obtain

$$J_V = L_p \left[\sigma_\alpha RT(C_2 - C_1) + \sigma_\beta RT(C_3 - C_2) \right] \quad (5.30)$$

where

$$L_p = L_{p\alpha} L_{p\beta} / (L_{p\alpha} + L_{p\beta})$$

Further, when $C_1 = C_3$ we obtain

$$J_V = L_p (\sigma_\alpha - \sigma_\beta) RT(C_2 - C_1) \quad (5.31)$$

Thus, as long as $C_2 > C_1$ and $\sigma_\alpha > \sigma_\beta$ there will be a flow of volume from compartment 1 to compartment 3, even though $P_1 = P_3$ and $C_1 = C_3$. Further, from equation (5.30) we see that there can be volume flow from compartment 1 to compartment

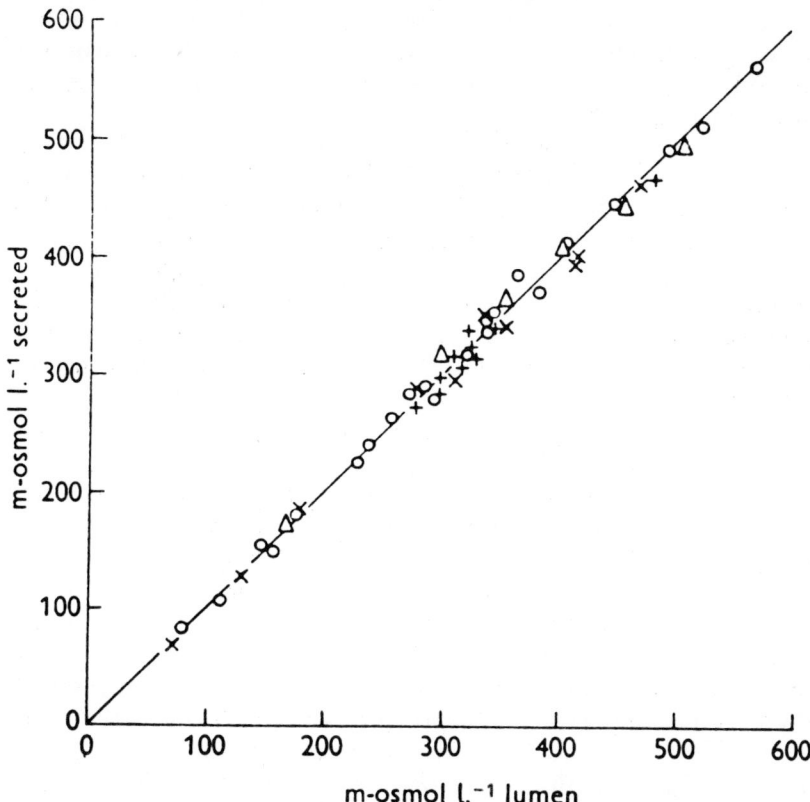

Fig. 5.3. Relation between the osmolarity of the solution transported across an *in vitro* preparation of rabbit gall bladder and the osmolarity of the solution in the lumen. (From Diamond [30] (reproduced by permission of the Rockefeller University Press).)

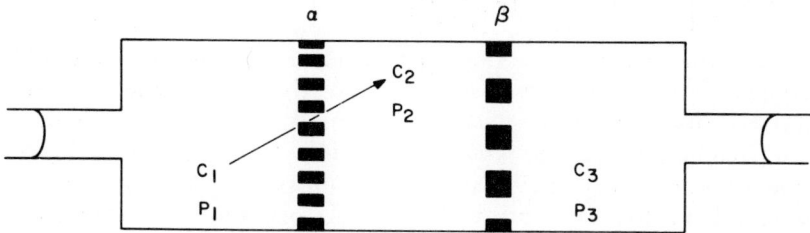

Fig. 5.4. The double membrane model proposed by Curran. C and P designate solute concentration and pressure. The subscript 1, 2 and 3 designate the respective compartments. (From Schultz [22]).

3 even when $C_1 > C_3$ providing σ_α is much greater than σ_β. That is, this model can account for flow against an osmotic pressure difference, and, by extension, a hydrostatic pressure difference.

An intuitive grasp of the workings of this model and its relevance to isotonic water absorption by epithelia can be gained from the following considerations. Assume that membrane contains a mechanism that can transport solute against a concentration difference from compartment 1 to compartment 2 and thereby maintain the condition $C_2 > C_1$. The volume flow from compartment 1 to compartment 2 resulting from the fact that $C_2 > C_1$ is given by (ignoring the hydrostatic pressure for the moment)

$$J_{V\alpha} = L_{p\alpha}\,\sigma_\alpha RT(C_2 - C_1) \tag{5.32}$$

If σ_β is negligible compared to σ_α the fact that $C_2 > C_3$ will not result in a significant volume flow from compartment 3 to compartment 2. Thus the asymmetry with respect to reflection coefficients brings about a polarity of volume flow in an otherwise symmetrical system. If the volume of compartment 2 is constrained, the entrance of fluid from compartment 1 will result in an increase in hydrostatic pressure which in turn drives fluid from compartment 2 into compartment 3 across the more permeable membrane β. Under steady-state conditions P_2 will be slightly greater than P_1 or P_3 so that equation (5.32) is not strictly accurate, but it nonetheless serves the purpose of this intuitive description. A more detailed formal analysis of the properties of this double membrane system has been presented by Patlak et al. [29].

We may now inquire as to the 'desirable' properties of membrane β. Clearly, the model will operate most efficiently if β retards the flow of solute from compartment 2 to compartment 3 and does not permit the rapid dissipation of the concentration difference across α. At the same time, optimal polarization of flow will be obtained when the reflection coefficient of β is very low. Thus, in terms of the phenomenological coefficients, the most desirable properties would be a low ω_β and a low σ_β. As pointed out previously, with reference to equation (5.26), low values of ω can be associated with low values of σ if Δx is large. Thus, from a structural point of view, a long porous barrier (or pathway) could provide the combination of properties that would optimize the operation of the double-membrane model. The lateral intercellular spaces of small intestine could serve admirably.

The model system proposed by Curran provided a firm conceptual foundation for the analysis of isotonic water transport by biological tissue. The subsequent elegant work of Diamond and his collaborators [30, 31, 32] extended and modified some of these idealized concepts to comply with the detailed anatomic structure of several epithelial tissues. Diamond's 'standing osmotic gradient' model for isotonic fluid transport is illustrated in Fig. 5.5. According to this model, solute, principally NaCl, is transported out of the absorptive cells across the lateral membranes into the blind end of the cul-de-sac formed by the lateral membranes and the tight junction. The accumulation of solute in this small region provides the osmotic driving force for the flow of water out of the cell into the lateral space. The resulting increase in volume together with the limited compliance of the surrounding tissues brings about a small elevation in hydrostatic pressure that forces fluid down the intercellular space. This long, slow passage down the intercellular space permits the absorbate to achieve osmotic equilibration with the cell interior, which, in turn, is assumed to be isotonic with the mucosal fluid.

This brief description of Diamond's model does not do justice to this very elegant and important contribution. However, a more detailed discussion does not really fall within the scope of this chapter and the problem of water absorption will be treated elsewhere in this volume. It should be pointed out, however, that our current understanding of the mechanism of isotonic water transport is the result of a long trail of contributions pioneered by Onsager [2], Staverman [5], Kedem and Katchalsky [6, 16, 21], Curran [26] and Diamond [23, 30, 31, 32]; a trail that commences in the depths of theoretical physics and culminates with an elegant correlation of biological structure and function. This is indeed the successful marriage referred to by Teorell [1].

5.3.2 Solvent drag
The possibility that solute may be drawn across a membrane by virtue of being entrained in a solvent stream was clearly

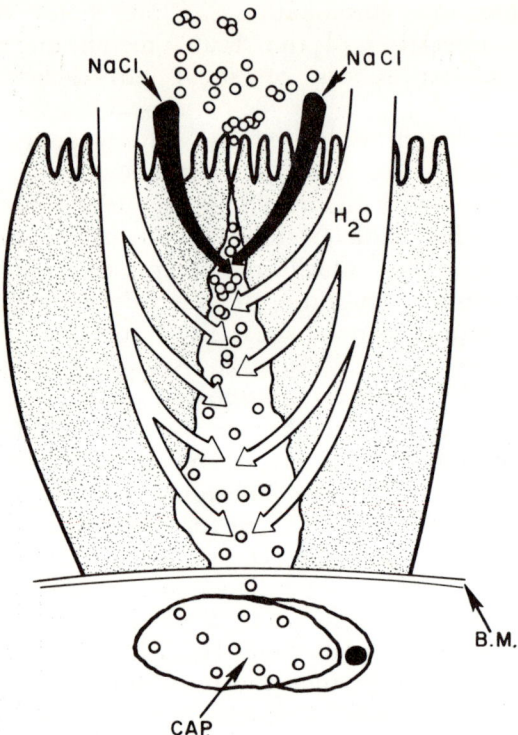

Fig. 5.5. The 'standing osmotic gradient' model for isotonic fluid absorption proposed by Diamond. BM designates the basement membrane and CAP an underlying capillary. (From Schultz [22].)

entertained by Fisher [33] as the result of studies on solute and water transport across rat small intestine. Indeed, this notion formed the basis of the 'fluid circuit' theory postulated much earlier by Ingraham *et al.* [34]. However, the first carefully controlled, explicit experimental demonstration of solvent drag emerged from the studies of Andersen and Ussing [19]. These investigators demonstrated that volume flow across isolated toad skin can bring about the net movement of an uncharged solute in the absence of a concentration difference. This observation gains explicit expression in equation (5.15), where $J_s \neq 0$ when $\Delta C_s = 0$ providing $\sigma_s < 1$ and $J_V > 0$.

The demonstration of solute flow associated with volume flow in the absence of a concentration difference of the solute is frequently cited as evidence for an interaction between solute and water flows in water-filled pores that traverse the membrane. Let us however examine the situation in which $\Delta C_s = 0$ $f_{sw} = 0$ and J_V is the result of a hydrostatic pressure, ΔP. Under these conditions

$$J_s = (1 - \sigma) C_s J_V$$
$$1 - \sigma = \omega \bar{v}_s / L_p$$

and

$$J_V = L_p \Delta P$$

Combining these equations we obtain

$$J_s = \omega C_s \bar{v}_s \Delta P$$

That is, one will still observe a net flow of solute but the driving force is not the result of interaction with water flow but rather the contribution of $\bar{v}_s \Delta P$ to the electrochemical potential difference of the solute (see equation 5.1). Since $\bar{v}_s \Delta P$ generally makes a negligible contribution to the total driving force compared with, say, concentration differences or differences in electrical potential J_s is likewise likely to be small under these conditions.

Recently, Lifson and his collaborators [35, 36] have examined the contribution of solvent drag to solute transport across dog small intestine. Their results are discussed in terms of a 'diffusive-convective' model but, except for terminology and symbolism, this model is identical with that expressed by equation (5.15). These investigators have shown that water flow across *in vivo* as well as *in vitro* dog small intestine can bring about a net flow of urea against a concentration difference. Further, spontaneous water flow from mucosa to serosa, and water flow from serosa to mucosa resulting from the application of a hydrostatic pressure to the serosal solution, exerted approximately the same drag effect on urea. This is illustrated in Fig. 5.6. More recently, this analysis has been applied to the pentoses, arabinose and xylose, and to the actively transported hexose, glucose. The apparent values of σ derived from the *in vivo* studies are: 0.2–0.3 for urea, 0.4 for the pentoses and approximately 0.3–0.4 for glucose. In addition it was observed that convective flow of sugars resulting from spontaneous water absorption was much greater than that resulting from osmotically induced water flow from the plasma into the lumen.

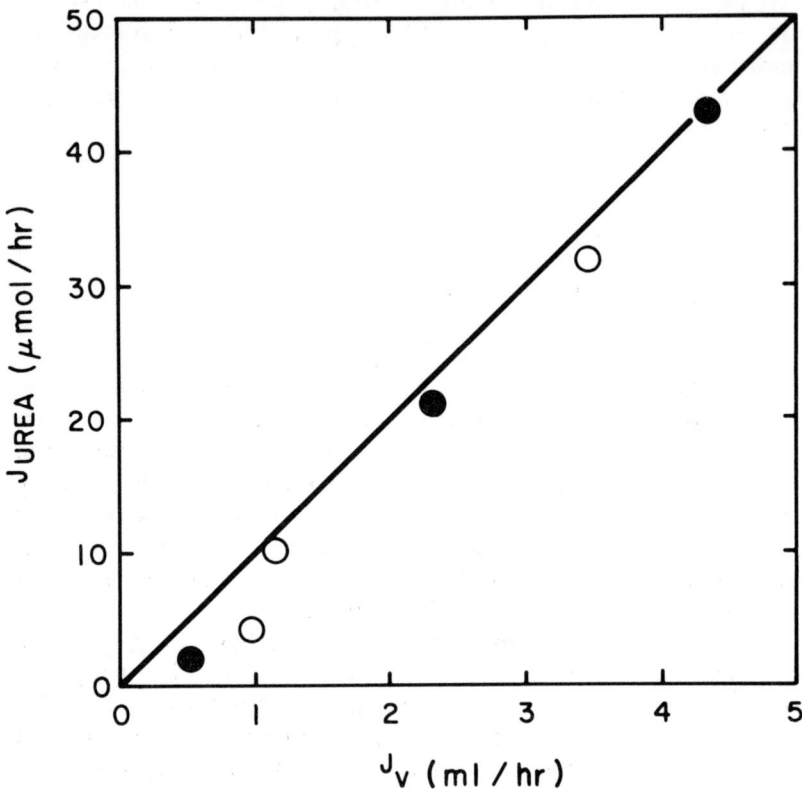

Fig. 5.6. Relation between net volume flow and net movement of urea across *in vitro* dog small intestine. The open circles indicate flow from the mucosal solution to the serosal solution in response to spontaneous water absorption. The closed circles represent flow in the opposite direction produced by application of a hydrostatic pressure to the serosal solution. In both instances the net flow of urea took place from a solution in which the initial concentration was 10 mM to a solution in which its initial concentration was 20 mM. Data from Hakim and Lifson [35]. (From Schultz [22].)

While there is little reason to doubt that Lifson *et al.* [35, 36] have demonstrated solvent-drag effects in dog small intestine, several points should be noted regarding the quantitation of these effects. First, as pointed out in the discussion of composite membranes (section 5.2), 'overall' reflection coefficients in a tissue as complex as intestine may have little quantitative significance. Further, asymmetric effects of the type observed by these investigators are not unexpected, and need not be attributable to the possibility that the pathways of spontaneous water absorption and osmotically induced water flows differ. Finally, one other possibility that could seriously complicate, if not invalidate, the interpretation of these data must be considered: namely, the strong likelihood that water flow alters the transmucosal concentration gradients and that its effect is not entirely attributable to solvent drag. Let us consider the situation in which (i) the concentration of a permeant solute is the same on both sides of the membrane, (ii) the solute and water do not interact within the membrane (i.e. $f_{sw} = 0$), and (iii) water flows across the membrane from side α to side β. At zero time (the moment water flow commences) one would not observe a net flow of solute across the membrane. However, in time, the concentration of solute in the unstirred (or poorly stirred) region adjacent to side α would increase and the concentration of solute in the unstirred layer adjacent to side β of the membrane would decrease. As a result of these simultaneous 'concentrating' and 'washing out' effects a concentration gradient for the solute will be established and diffusion of solute across the membrane will ensue (see Wedner and Diamond [37] for a formal treatment of this problem). It can be easily demonstrated that the magnitude of the concentration difference across the membrane will vary directly with the rate of water flow so that a positive relation between J_s and J_V will be observed; clearly, the interpretation of this relation in terms of solvent drag or convective flow is fallacious.

A similar argument can be made for the case of substances that are transported across membranes by means of carrier processes but which do not interact directly with water flow. For example, in the experiments reported by Levitt *et al.* [36] water flow from mucosal solution to plasma could increase the glucose concentration immediately adjacent to the brush border and decrease the concentration of glucose in the absorptive cells, intercellular spaces and subepithelial tissues. All of these effects should enhance the rate of glucose (or xylose) absorption, and a positive relation between glucose absorption, at constant lumenal glucose concentration, and water absorption would result. Indeed, it would be of great interest to determine whether the convective flow of glucose is affected by phlorizin.

The purpose of this exercise is not to deny the existence of significant solvent-drag effects across small intestine, but to stress that the unequivocal demonstration and precise quantitation of these effects in complex tissues is, at best, difficult. Certainly, the use of reflection coefficients, derived from such experiments, as the basis for speculations regarding pore dimensions or the adequacy of the pore hypothesis is not justified.

5.3.3 Streaming potentials and the measurement of σ

In view of the importance of the reflection coefficient in the characterization of transport processes, several approaches have been employed to evaluate σ's for a variety of epithelial tissues. One such approach is based on the measurement of streaming potentials across the epithelium. The streaming potential itself is an electro-osmotic phenomenon resulting from the frictional drag exerted on ions by water during flow through the membrane. If the mobilities of the ionic species in the membrane differ, as the result of size and/or membrane charge, an electrical potential difference will develop across the membrane such that the side towards which water is flowing assumes the sign of the most permeable ion. This streaming potential is linearly related to water flow over a fairly wide range of flow rates. The rationale for employing streaming potentials to measure reflection coefficients is as follows: If the electrolyte solution bathing the mucosal surface of an epithelial tissue is rendered hypertonic by the addition of an uncharged solute, s, osmotic volume flow directed toward the mucosal solution is given by

$$J_V = L_p \sigma_s RT \Delta C_s$$

and the streaming potential will be proportional to J_V. If one measures the streaming potential produced by a given concentration of an impermeant solute ($\sigma_s = 1$) and then measures the streaming potential produced by the same concentration of a permeant solute ($\sigma_s < 1$), the ratio of the streaming potentials is a measure of the reflection coefficient of the permeant solute. Clearly, for the case of the permeant solute, J_V will be lower than that observed in the presence of the impermeant solute and this, in turn, will result in a lower streaming potential.

Smyth and Wright [38] have employed this method to evaluate reflection coefficients of rat jejunum for a number of water-soluble molecules. They obtained values ranging from 0.2 for ethylene glycol to 0.97 for lactose, when mannitol was used as the reference impermeant solute and assigned a reflection coefficient of unity. More recently Wright and Diamond [39, 40] have employed this method to measure reflection coefficients of rabbit gall bladder to a large number of molecules having a wide range of lipid solubilities. Large lipid-soluble molecules were found to have reflection coefficients close to 0. These findings were interpreted as indicating rapid permeation by virtue of dissolution in the lipoprotein matrix of the membrane, in agreement with the earlier conclusions of Overton [41] and Collander [42].

Although the conclusions and interpretations of the studies of Wright and Diamond are probably correct, the theoretical basis for their experimental observations is unclear. As discussed previously, if a highly permeant lipid-soluble molecule traverses the membrane through a pathway other than that used by water (i.e. f_{sw} =0), osmotic water flow will not be retarded by the drag effect of the diffusing solute and the rate of water flow will be precisely that observed with an ideally impermeant molecule. Since the streaming potential is related to water flow (not volume flow), the large lipid-soluble molecules should have produced streaming potentials close to that observed with the reference molecule, sucrose ($\sigma = 1$). The most likely explanation for the experimental observations of Wright and Diamond invokes an unstirred layer effect. That is, for the case of highly permeable molecules the unstirred layer immediately adjacent to the mucosal membrane is probably rapidly depleted so that the concentration gradient of the test solute across this membrane is much reduced or negligible by the time the streaming potential is recorded. This reasoning could explain why highly permeable molecules were associated with low streaming potentials and why the results of these studies are in general accord with Overton's rule. Certainly, however, the interpretation of these streaming potential measurements in terms of *true reflection coefficients* is not valid. The same criticism applies to the use of streaming potentials to measure reflection coefficients of nonelectrolytes in frog choroid plexus

(Wright and Prather [43] and the use of streaming potentials to measure reflection coefficients of sugars subject to carrier-mediated transport across the choroid plexus (Prather and Wright [44]). By analogy with the above argument, if a solute traverses a membrane by a mediated process for which $f_{sw} = 0$, the streaming potential should be equal to that found for the case of a nontransported molecule. Lower streaming potentials probably reflect depletion of the transported solute in the unstirred layer adjacent to the cell membrane. The quantitative significance that can be attached to such streaming potentials is open to question. In short, the validity of the streaming potential method is probably restricted (as in the study of Smyth and Wright) to molecules that interact with water in their passage through the membrane and partake of no other parallel pathway.

Fortran et al. [45] have attempted to calculate reflections coefficients of human small intestine for a variety of solutes. They first demonstrated that mannitol is not absorbed by human jejunum or ileum and assigned mannitol a reflection coefficient of unity. They then compared osmotic volume flow from plasma to lumen resulting from the instillation of luminal solutions made hypertonic with either mannitol, urea, erythritol or NaCl. Apparent reflection coefficient were calculated from the ratio of osmotic volume flow produced by a solution made hypertonic with a given solute to that produced by a solution made hypertonic with mannitol. In the upper intestine these values were 0.48 for urea, 0.64 for erythritol and 0.58 for NaCl. In the ileum, the osmotic flow ratios were 0.89 for urea, 0.98 for erythritol and approximately unity for NaCl. It should be stressed that although these observations probably have qualitative significance and suggest that jejunum is more permeable to water-soluble molecules than is ileum, the quantitative value of these data is difficult to assess. A large inflow of volume from the plasma undoubtedly reduces the concentration of the test solute in the unstirred region adjacent to the brush border, particularly when a perfusion method that cannot accomplish adequate stirring is employed. Indeed, the concentration of the test solute in this unstirred region will be a function of (i) diffusion of solute into this region from the bulk luminal solution, (ii) penetration of the solute across the brush border, and (iii) the 'washing out' effect. A quantitative evaluation of the overall effect of these processes is a formidable task.

The purpose of this discussion is not to diminish the value of the reflection coefficient, but to put it in its proper perspective. This notion has had an immense impact on the analysis of transport processes and provides the conceptual framework for the interpretation of interactions between flows of solute and solvent. However, it should be apparent that the precise evaluation of reflection coefficients in complex tissues is exceedingly difficult, and the interpretation of overall reflection coefficient in terms of microstructure is hazardous.

5.3.4 Solute-solute interactions
Although the major application of irreversible thermodynamics to the analysis of transport across biological membranes has focussed on solute-solvent interactions, the possibility of interactions between the flows of solutes is expressed explicitly in the phenomenologic equations. Several examples of such interactions will be described.

In 1966, Ussing [46] reported the following observations: A segment of isolated frog skin was mounted between two solutions containing equal concentrations of sucrose. The outer solution was then made hypertonic by the addition of urea. This, not unexpectedly, brought about a flow of volume from the inner solution to the outer solution. However, at the same time, there was a net flow of sucrose from the outer solution to the inner solution in the absence of a concentration difference. This apparent 'active' transport of sucrose took place in the face of an osmotic volume flow in the opposite direction. More recently, Franz and VanBruggen [47] and Biber and Curran [48] have examined this phenomenon in considerable detail and have shown that apparent 'active' transport of an otherwise inert solute occurs only when the outer solution is made hypertonic with a permeant solute. If hypertonicity is established by the use of an impermeant solute, such as raffinose (Franz and Van Bruggen [49]), anomalous transport is not observed. Thus, it appears that Ussing's observation can be explained by a drag effect arising from the diffusion of the hypertonic agent itself from a high concentration in the outer solution to a lower concentration in the inner solution. A formal analysis of this problem has been presented by Biber and Curran [48].

The possibility of solute-solute interactions in flow across biological membranes has also been raised with respect to studies using radioactive tracers and, in particular, the interpretation of flux-ratios. Space does not permit a detailed examination of this problem, but, stated briefly, the flux-ratio equation proposed by Ussing [50] as a criterion for distinguishing between active transport and simple diffusion assumes the absence of isotope interactions. That is, it assumes that there is no interaction between the flows of the abundant species and the tracer used for the measurement of unidirectional fluxes. Hoshiko and Lindley [51], employing this assumption, have shown that the flux-ratio equation can be derived using the phenomenologic equations of irreversible thermodynamics. Kedem and Essig [52] (see also Essig [53], Essig, Kedem and Hill [54]) have re-examined this problem and have explicitly incorporated the effects of possible isotope interactions. Their analysis disclosed that two frequent interpretations of 'abnormal' flux ratios, 'single-file diffusion' and 'exchange diffusion' (for a discussion of these phenomena see Curran and Schultz [55]) could result simply from interactions between the flows of abundant and tracer species so that no specialized biological mechanisms can be deduced from these observations. More recently, de Sousa, Li and Essig (personal communication) have demonstrated that flux ratios across artificial membranes may display abnormalities consistent with the implication of 'exchange diffusion' (see also Kitahara *et al.* [56]). Clearly, the implication of biological carrier mechanisms on the basis of such findings is not justified.

Finally, coupling between the flows of Na and other solutes appears to play an important role in Na-dependent transport of sugars and amino acids by small intestine and kidney and the accumulation of amino acids by a wide variety of animal cells. This subject has been reviewed recently (Schultz and Curran [57]) and will be discussed elsewhere in this volume.

5.4 ENERGY CONVERSION IN BIOLOGICAL TRANSPORT PROCESSES

Up to this point we have focussed on interactions between

flows, and some pertinent biological illustrations have been presented. We have not however explicitly considered the all-important question of energetics, and, in particular, that of energy conversion, a *sine qua non* of life. Many of the examples cited above are obvious instances of energy conversion by coupled systems, e.g. water transport against an osmotic pressure difference, solvent drag of a solute against a chemical potential difference, amino acid accumulation coupled to the asymmetric distribution of Na etc. In general, all coupled processes involve energy conversion since the flow of one substance is affected by a nonconjugate driving force, i.e. a potential or free energy source related to another extensive property of the system.

The principles of energy conversion may be illustrated by considering a simple two flow-two force system illustrated in Fig. 5.7.

The equations that describe this system are

$$T\, d_i S/dt = J_1 X_1 + J_2 X_2 \tag{5.33}$$

$$J_1 = L_{11} X_1 + L_{12} X_2 \tag{5.34}$$

$$J_2 = L_{21} X_1 + L_{22} X_2 \tag{5.35}$$

and, assuming reciprocity,

$$L_{21} = L_{12} \tag{5.36}$$

Since, for any spontaneous process $Td_i S/dt > 0$, it can be shown that $L_{11} L_{22} > (L_{12})^2$.

Now, if there is no coupling between the flows J_1 and J_2, each flow will take place in the direction of its driving force (the spontaneous direction); let us assign these flows positive values. If there is coupling between the flows and if $L_{12} X_2 > L_{11} X_1$ then it is possible for J_1 to be negative; that is, J_1 can be driven in a direction opposite to its spontaneous direction if the effect of X_2 overcomes the effect of X_1 * (note: L_{11} and L_{22} must be positive, but L_{12} can be either positive or

---

* The condition $J_1 X_1 < 0$ is the Rosenberg [63] criterion for an active transport process, i.e. flow against an electrochemical potential difference. According to the Kedem definition (see footnote, p. 00) $J_1 X_1$ can be positive and still be classified as active transport providing $R_{ir} \neq 0$. Under these conditions the investment of metabolic energy serves to accelerate or decelerate the flow, albeit in the spontaneous direction.

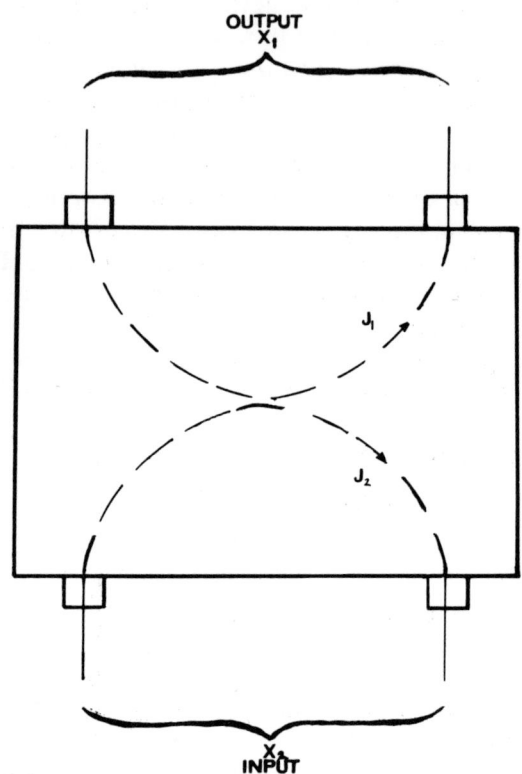

Fig. 5.7. A black box representation of an energy converter in which J_2 and X_2 represent the driving flow and force and J_1 and X_1 the driven flow and force. (From Caplan [59].)

negative). Under these conditions we designate J_2 and X_2 the input or *driving* flows and forces and, conversely, J_1 and X_1 are the output or *driven* flows and forces. That is, when J_1 is negative, $J_2 X_2$ is the rate at which energy is invested into the system and $J_1 X_1$ is the rate at which work is being done. The *efficiency* of such a process is

$$\eta = -J_1 X_1 / J_2 X_2$$

or, substituting into equation (5.33)

$$\eta = 1 - (T \, d_i S/dt)/J_2 X_2 \qquad (5.37)$$

Clearly, when $T\,d_iS/dt = 0$, $\eta = 1$; this is the ideally reversible process. When $J_2 X_2 = T\,d_iS/dt$, the efficiency and $J_1 X_1 = 0$.

Now we may consider three types of steady states:

(a) X_1 and X_2 are constant and restrained at fixed values. This situation is applicable only to infinite systems or when flows are infinitesimally small and has little relevance to biological processes.

(b) If the driving force X_2 is maintained constant, but no restriction is placed on X_1 the system will eventually reach a steady state of minimum entropy production in which $J_1 = 0$ and X_1 is maintained at some constant nonzero value (i.e. $L_{11} X_1 = -L_{12} X_2$). This is the most common form of biological steady state, exemplified by the maintenance of an electrochemical potential gradient for a solute across a membrane with no net flow of that solute (e.g. the maintenance of Na and K asymmetries across most cell membrane). This condition is termed *static head* since energy is invested to *maintain* a gradient against its natural tendency to dissipate.

(c) Alternatively, one may have a condition in which $X_1 = 0$ but $J_1 \neq 0$. Familiar examples are isotonic water transport across intestine or Na transport across short-circuited epithelia. Here, energy is invested to maintain a flow in the absence of a conjugate driving force. This is termed *level flow.*

Clearly, under the conditions of static head and level flow, energy is consumed but useful work is not performed so that the efficiency is zero. Thus, in many biological systems, the notion of efficiency is not very informative.

Kedem and Caplan [58] have addressed themselves to this problem and have introduced a new concept, the 'degree of coupling' that is particularly useful in the characterization of energy conversion processes. The subsequent discussion closely follows their development. From equations (5.33) (5.34) and (5.35) one obtains

$$J_1/J_2 = Z(q + ZX_1/X_2)/(1 + qZX_1/X_2) \qquad (5.38)$$

where

$$Z = (L_{11}/L_{22})^{1/2} \qquad (5.39)$$

and

$$q = L_{12}/(L_{11}L_{22})^{1/2} \qquad (5.40)$$

Since, L_{11} and L_{22} are independent of the J's and X's, Z is simply a dimensionless number that is characteristic of the system. On the other hand, q is restricted by the relation $L_{11}L_{22} > (L_{12})^2$, so that:

$$-1 \leq q \leq 1$$

and is designated the *degree of coupling*. The rationale behind this designation is that when $q = 0$.

$$J_1/J_2 = Z^2 X_1/X_2 = L_{11}X_1/L_{22}X_2 \tag{5.41}$$

Thus, under this condition, the flow ratio is simply the ratio of the products of conjugate forces and conductances. This is characteristic of a system in which there is no coupling or interaction.* On the other hand, when $q = \pm 1$

$$J_1/J_2 = \pm Z \tag{5.42}$$

Under this condition there is a fixed ratio between the flows that is a characteristic of the system but is *independent* of the driving forces. Thus, if we fix one flow, the other flow is uniquely determined, and if one flow is zero the other flow is zero. Clearly this is a tightly coupled system and can be exemplified by two interlocking non-slipping gears.

The degree of coupling, q, may now be related to the efficiency of a process as follows:

$$\eta = -(q - J_1/J_2 Z)/(q - ZJ_2/J_1)$$

or, expressed in terms of forces

$$\eta = -(q + ZX_1/X_2)/(q + X_2/ZX_1)$$

Once more we see that η is not a characteristic of a process since, for any process, flows and forces can be varied without affecting d_iS/dt. On the other hand, since the useful work performed is given by $-J_1X_1$ and since $\eta = 0$ when either $J_1 = 0$ (static head) or $X_1 = 0$ (level flow), it follows that there must be some maximal efficiency in the region where $-J_1X_1 \neq 0$ (the 'driving region'). It can be shown that

$$\eta_{max} = q^2/[1 + (1-q^2)^{1/2}]^2 \tag{5.43}$$

* This is the assumption employed by Ussing [50] in the derivation of the flux-ratio equation. Thus, if J_1 and J_2 represent the bidirectional fluxes of the same species, $L_{11} = L_{22}$ and $J_1/J_2 = X_1/X_2$.

Thus the *maximal efficiency of a process is dependent only on the degree of coupling and is independent of the prevailing flows and forces.*

From equation (5.43) we see that:

(a) $\eta_{max} = 1$ only when $q = \pm 1$. That is, a maximal efficiency of unity can only result from a completely coupled, ideally reversible process.

(b) As shown in Table 5.1, the maximal efficiency of a process decreases rapidly with decreasing degree of coupling. Thus, when the degree of coupling is 0.8, η_{max} is only 0.25. Thus, highly efficient processes must be very tightly coupled.

TABLE 5.1. Relation between degree of coupling and maximal efficiency.*

q	η max
0.1	0.003
0.2	0.010
0.3	0.024
0.4	0.044
0.5	0.072
0.6	0.111
0.7	0.167
0.8	0.250
0.9	0.393
0.95	0.524
0.99	0.753
1.00	1.000

* From Kedem and Caplan [58].

Finally, we may turn to two questions of biological importance. First, what is the energy required to maintain a static head, the steady-state condition characteristic of all biological cells?

Since, the energy expended by the driving reaction is $J_2 X_2$, it can be shown that

$$J_2 X_2 = X_1^2 L_{11} [(1/q^2) - 1] \qquad (5.44)$$

Thus the required energy is (i) directly dependent upon the square of the gradient maintained, (ii) directly dependent upon the conductance or the permeability of the membrane to the solute (this determines the tendency for the gradient to be dissipated), and (iii) indirectly related to the degree of coupling of the process.

Second, what is the energy required to maintain level flow? It can be readily shown that

$$J_2 X_2 = J_1^2/L_{11} q^2 \qquad (5.45)$$

Thus the required energy increases with the square of the flow, and is inversely proportional to both the conductance and the square of the degree of coupling. Now, if a flow is driven by its own conjugate force, the energy expended is $J_1 X_1$ where $X_1 = J_1/L_{11}$. Thus

$$J_1 X_1 = J_1^2/L_{11} \qquad (5.46)$$

Comparison of equations (5.45) and (5.46) indicates that the energy required to drive a level flow by a coupled process is greater than that which would be expended if the flow were driven by its own conjugate force and increases rapidly with decreasing degree of coupling.

These concepts developed by Kedem and Caplan [58] (see also Caplan [59]) have been applied to the coupling of a chemical reaction to current flow across an artificial membrane (Blumenthal et al. [12]) and to the analysis of the energetics of active transport processes (Essig and Caplan [60]). A quantitative analysis of a complex system along these lines is extremely difficult. Nevertheless, these notions provide important insights into the coupling between metabolic reactions and transport processes. One such insight deals with the interpretation of stoichiometric relations between the flow of a metabolic process (e.g. the rate of oxygen consumption, J_{O_2}) and transport (e.g. the rate of Na transport across an epithelial tissue, J_{Na}). Linear relations between J_{Na} and J_{O_2} have been observed for a variety of epithelia; however, in almost all instances J_{Na}/J_{O_2} varies considerably. From equation (5.42) we see that a unique value of J_1/J_2 is characteristic only of a completely coupled system ($q = \pm 1$). Furthermore, as pointed out by Essig and Caplan [60], a linear relation between J_{O_2} and J_{Na} can be observed under a variety of conditions (see equation (5.38)) however the slope is dependent upon q and will not reflect the stoichiometry of the underlying mechanism unless coupling is complete (see also Rottenberg et al. [61].

5.5 CONCLUDING REMARKS

As mentioned at the outset, the purpose of this chapter is to outline some of the basic principles of irreversible thermodynamics and to stress aspects that are particularly relevant to transport across epithelial tissues. To date, a number of relatively simple biological and artificial membrane systems have been systematically analysed in terms of this formalism. That is, the phenomenological coefficients have been evaluated and the behaviour of these membranes can be characterized quantitatively. However, as we have seen, the formal description of composite membranes in these terms is formidable and the additional complexities introduced by the presence of unstirred layers and compartments, as well as multiple series and parallel inhomogeneities, vastly compound these difficulties. Thus, the quantitative description of transport across an epithelial tissue in which there are simultaneous interacting flows of several charged and uncharged solutes, solvent and chemical reactions in meaningful phenomenological terms is not easily realizable. Nevertheless, irreversible thermodynamics has provided the conceptual basis for understanding the behaviour of systems with interacting flows. Capitalizing on relatively simplified models it has provided insight into properties that could be responsible for rectification of solute and solvent flows. And, a foundation has been laid for the analysis of energy conversion in the coupling between metabolic processes and transport. Its contribution in these terms cannot be overestimated.

Since the completion of this chapter, several important observations have been reported that further compound the difficulties of applying the Kedem-Katchalsky formalisms to complex epithelia such as small intestine, renal proximal tubule and gall bladder. Loeschke *et al.* [65, 66] have shown that osmotically induced water flow across *in vitro* frog intestine directed from mucosa to serosa results in an increase in L_p and the permeability to 3-0-methylglucose, whereas osmotically induced water flow from serosa to mucosa results in a decrease in L_p. These findings were attributed to widening of the lateral intercellular spaces when water flow was directed from mucosa to serosa and a collapse of these spaces when water flow was directed from serosa to mucosa. Similar findings have been reported for Necturus proximal tubul [67] and rabbit gall bladder [68]. Thus, in low resistance epithelia where the lateral interspaces appear to provide a major pathway for the flow of water and small solutes, the determination of L_p, the reflection coefficient and solute permeabilities can be markedly influenced by the experimental technique employed. These observations provide an explanation for the asymmetry reported by Levitt *et al.* [36] with respect to the convective flows of sugars elicited by spontaneous water absorption as opposed to osmotically induced water flow from plasma to lumen.

Finally, changes in the dimensions of the lateral interspaces need not be elicited by osmotic gradients but are also associated with changes in the absorptive state of the small intestine. It is well established that the lateral spaces widen in response to enhanced fluid absorption and it has been recently demonstrated that these spaces collapse during choleragenic secretion (DiBona, personal communication). These structural alterations are probably responsible for the recent findings of Lifson *et al.* [69] that during intestinal secretion by canine jejunum elicited by cholera toxin, the reflection coefficient for arabinose increases whereas its permeability coefficient decreases. These changes are in the same direction as those observed during osmotically induced water secretion.

Thus, not only are L_p, ω and σ difficult to define and interpret in complex tissues comprising series and parallel inhomogeneities, but in addition they are not *static* properties of the tissue and will, in general, be influenced by both experimental pertubations as well as the spontaneous functional state of the tissue. A re-examination of published data in the light of these considerations may serve to resolve existing conflicts and, clearly, failure to take these factors into proper account could lead to "apparent" inconsistencies and unnecessary disputes.

REFERENCES

1. T. TEORELL, Problems and perspectives in membranology. In Membrane Models and the Formation of Biological Membranes (L. Bolis and B. A. Pethica, eds.). North-Holland. Pethica, Amsterdam (1968).
2. L. ONSAGER, *Phys. Rev.*, 37, 405; 38, 2265 (1931).
3. I. PRIGOGINE, Introduction to the Thermodynamics of Irreversible Processes. Thomas, Springfield, Illinois (1955).

4. S. R. DE GROOT, Thermodynamics of Irreversible Processes. North-Holland, Amsterdam (1952).
5. A. J. STAVERMAN, *Rec. Trav. Chim.*, **70**, 344 (1951).
6. O. KEDEM and A. KATCHALSKY, *Biochim. Biophys. Acta*, **27**, 229 (1958).
7. K. G. DENBIGH, The Thermodynamics of the Steady State. Methuen, London (1951).
8. A. KATCHALSKY and P. F. CURRAN, Nonequilibrium Thermodynamics in Biophysics. Harvard, Cambridge, Mass. (1965).
9. A. KATCHALSKY and O. KEDEM, *Biophys. J.*, **2**, 53 (1962).
10. S. R. CAPLAN and D. C. MIKULECKY, Transport processes in membranes. In Ion Exchange: A Series of Advances (J. A. Marinsky, ed.), Vol. 1. Dekker, New York (1966).
11. D. MILLER, *Chem. Rev.*, **60**, 15 (1960).
12. R. BLUMENTHAL, S. R. CAPLAN and O. KEDEM, *Biophys. J.*, **7**, 735 (1967).
13. K. C. COLE, Membranes, Ions and Impulses. University of California Press, Berkeley (1968).
14. E. H. STARLING, *J. Physiol.*, **19**, 312 (1896).
15. K. S. SPIEGLER, *Trans. Farad. Soc.*, **54**, 1409 (1958).
16. O. KEDEM and A. KATCHALSKY, *J. Gen. Physiol.*, **45**, 143 (1961).
17. E. GRIM and K. SOLLNER, *J. Gen. Physiol.*, **40**, 887 (1957).
18. O. KEDEM and A. LEAF, *J. Gen. Physiol.*, **49**, 655 (1966).
19. B. ANDERSEN and H. H. USSING, *Acta Physiol. Scand.*, **39**, 228 (1957).
20. S. G. SCHULTZ and P. F. CURRAN, Intestinal absorption of sodium chloride and water. In Handbook of Physiology, Section 6: Alimentary Canal, Vol. III, p. 1245. American Physiological Society, Washington (1968).
21. O. KEDEM and A. KATCHALSKY, *Trans. Farad. Soc.*, **59**, 1919, 1931, 1941 (1963).
22. S. G. SCHULTZ, Mechanisms of absorption. In Biological Membranes (R. M. Dowben, ed.). Little, Brown and Co., Boston (1960).
23. J. M. DIAMOND, Transport mechanisms in the gallbladder. In Handbook of Physiology, Section 6: Alimentary Canal, Vol. V, p. 2451. American Physiological Society, Washington (1968).
24. E. W. REID, *Brit. Med. J.*, 1133 (1892).
25. D. S. PARSONS and D. L. WINGATE, *Biochim. Biophys. Acta*, **46**, 170 (1961).
26. P. F. CURRAN, *J. Gen. Physiol.*, **43**, 1122 (1960).
27. J. T. OGILVIE, J. R. McINTOSH and P. F. CURRAN, *Biochim. Biophys. Acta*, **66**, 441 (1963).
28. P. F. CURRAN and J. R. McINTOSH, *Nature*, **193**, 347 (1962).
29. C. S. PATLAK, D. A. GOLDSTEIN and J. F. HOFFMAN, *J. Theoret. Biol.*, **5**, 426 (1963).
30. J. M. DIAMOND, *J. Gen. Physiol.*, **48**, 15 (1964).
31. J. M. DIAMOND and W. H. BOSSERT, *J. Gen. Physiol.*, **50**, 2061 (1967).

32. J. McD. TORMEY and J. M. DIAMOND, *J. Gen. Physiol.*, **50**, 2031 (1967).
33. R. B. FISHER, *J. Physiol.*, **130**, 655 (1955).
34. R. C. INGRAHAM, H. C. PETERS and M. B. VISSCHER, *J. Phys. Chem.*, **42**, 414 (1938).
35. N. LIFSON and A. A. HAKIM, *Amer. J. Physiol.*, **211**, 1137 (1966).
36. D. G. LEVITT, A. A. HAKIM and N. LIFSON, *Amer. J. Physiol.*, **217**, 777 (1969).
37. H. J. WEDNER and J. M. DIAMOND, *J. Memb. Biol.*, **1**, 92 (1969).
38. D. H. SMYTH and E. M. WRIGHT, *J. Physiol.*, **182**, 591 (1966).
39. E. M. WRIGHT and J. M. DIAMOND, *Proc. Roy. Soc. B.*, **172**, 302 (1969).
40. E. M. WRIGHT and J. M. DIAMOND, *Proc. Roy. Soc. B*, **172**, 227 (1969).
41. E. OVERTON, *Vjschr. Naturforsch. Ges. Zurich*, **41**, 383 (1896).
42. R. COLLANDER, *Physiol. Plantarum*, **2**, 300 (1949).
43. E. M. WRIGHT and J. W. PRATHER, *J. Memb. Biol.*, **2**, 127 (1970).
44. J. W. PRATHER and E. M. WRIGHT, *J. Memb. Biol.*, **2**, 150 (1970).
45. J. S. FORTRAN, F. C. RECTOR, M. R. EWTON, N. SOTER and J. KINNEY, *J. Clin. Invest.*, **44**, 1935 (1965).
46. H. H. USSING, *Ann. N.Y. Acad. Sci.*, **137**, 543 (1966).
47. T. J. FRANZ and J. T. VAN BRUGGEN, *J. Gen. Physiol.*, **50**, 933 (1967).
48. T. U. L. BIBER and P. F. CURRAN, *J. Gen. Physiol.*, **51**, 606 (1968).
49. T. J. FRANZ and J. T. VAN BRUGGEN, *Ann. N.Y. Acad. Sci.*, **141**, 302 (1967).
50. H. H. USSING, *Acta Physiol. Scand.*, **19**, 43 (1949).
51. T. HOSHIKO and B. D. LINDLEY, *Biochim. Biophys. Acta*, **79**, 301 (1964).
52. O. KEDEM and A. ESSIG, *J. Gen. Physiol.*, **48**, 1047 (1965).
53. A. ESSIG, *J. Theoret. Biol.*, **13**, 63 (1966).
54. A. ESSIG, O. KEDEM and T. H. HILL, *J. Theoret. Biol.*, **13**, 72 (1966).
55. P. F. CURRAN and S. G. SCHULTZ, Transport across membranes: general principles. In Handbook of Physiology, Section 6: Alimentary Canal, Vol. III, p. 1217. American Physiological Society, Washington (1968).
56. S. KITAHARA, E. HEINZ and C. STAHLMANN, *Nature*, **208**, 187 (1965).
57. S. G. SCHULTZ and P. F. CURRAN, *Physiol. Rev.*, **50**, 637 (1970).
58. O. KEDEM and S. R. CAPLAN, *Trans. Farad. Soc.*, **61**, 1897 (1965).
59. S. R. CAPLAN, *Biophys. J.*, **8**, 1146 (1968).
60. A. ESSIG and S. R. CAPLAN, *Biophys. J.*, **8**, 1434 (1968).
61. H. ROTTENBERG, S. R. CAPLAN and A. ESSIG, *Nature*, **216**, 610 (1967).
62. O. KEDEM, Criteria of active transport. In Membrane Transport and Metabolism (A. Kleinzeller and A. Kotyk, eds.). Czechoslovak Acad. Sci., Prague (1961).
63. T. ROSENBERG, *Acta Chem. Scand.*, **2**, 14 (1948).

64. T. W. CLARKSON and A. ROTHSTEIN, *Amer. J. Physiol.*, **199**, 898 (1960).
65. K. LOESCHKE, C. J. BENTZEL and T. Z. CSAKY, *Amer. J. Physiol.*, **218**, 1723 (1970).
66. K. LOESCHKE, D. HARE and T. Z. CSAKY, *Pflügers Arch.*, **328**, 1 (1971).
67. C. J. BENTZEL, B. PARSA and D. K. HARE, *Amer. J. Physiol.*, **217**, 570 (1969).
68. A. P. SMULDERS, J. McD. TORMEY and E. M. WRIGHT, *J. Membrane Biol.*, **7**, 164 (1972).
69. N. LIFSON, A. A. HAKIM and E. J. LENDER, *Amer. J. Physiol.*, **223**, 1479 (1972).

CHAPTER 6

Methods of Studying Intestinal Absorption

D. H. SMYTH

*Department of Physiology,
The University, Sheffield, S10 2TN*

		Page
6.1	GENERAL CONSIDERATIONS	241
	6.1.1 *Terminology*	242
	6.1.2 *Balance experiments*	247
	6.1.3 *Marker substances*	249
6.2	METHODS	251
	6.2.1 In vivo *methods*	252
	6.2.2 In vitro *methods*	263
6.3	PARAMETERS FOR MEASUREMENT OF ABSORPTION	273
	6.3.1 *Measurements involving amounts*	274
	6.3.2 *Measurements involving concentration*	276

6.1 GENERAL CONSIDERATIONS

A large number of methods are now available for the study of intestinal absorption varying from the use of the whole animal, including the human subject, to the use of very small pieces of intestine or even cell fractions. No one of these methods gives all the information required, and our knowledge about the absorption of nutrients or drugs must be based on results obtained by a large number of different techniques. The use of different methods also raises important questions about the interpretation of the results, and a number of discrepancies in the literature are due to different concepts of what the absorptive process is, and different criteria for assessing it.

Discussions of these problems and general accounts of methodology and interpretation have appeared in recent years [1, 2, 3, 4, 5, 6, 7]. The account by Parsons [5] is of particular interest from the historical viewpoint.

6.1.1 Terminology

Absorption It is useful to discuss some problems of terminology as much confusion arises from the use of the same

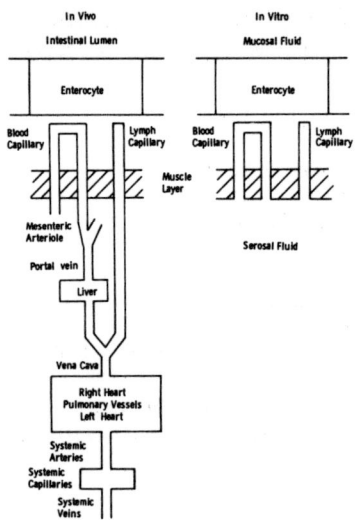

Fig. 6.1. Diagrammatic representation of the stages involved in intestinal transfer *in vivo* (left) and *in vitro* (right). The essential difference is that *in vitro* the products of transfer appear in the serosal fluid instead of the blood stream.

words to mean different things. We can begin by looking at the meaning of the word absorption itself and trying to decide in what way this is going to be used. Figure 6.1 (*left*) shows diagrammatically the series of processes concerned in the movement of nutrients from the intestinal lumen until they are available to the body tissues, i.e. in the arterial systemic blood. It also includes the systemic veins, for in cases where tolerance curves are used as tests of absorption, it is systemic venous blood that is studied. On the other hand when *in vitro* techniques are used, the blood vessels are torn away as they enter the gut wall, and we have the position shown in Fig. 6.1 (*right*).

It is evident that the term intestinal absorption is used in very different senses according to the techniques used. This is indicated in Fig. 6.2 and it is very important to know the precise meaning attached to it in each individual case. It is suggested here that the term absorption should be kept for the whole process of movement of a substance from the lumen of

MEANINGS ATTACHED BY VARIOUS AUTHORS TO ABSORPTION

In vivo	*In vitro*
lumen to enterocyte	mucosal fluid to enterocyte
lumen to mesenteric venous blood	mucosal fluid to gut wall
lumen to systemic venous blood	mucosal fluid to serosal fluid

Fig. 6.2. The word absorption is used by different authors in many different senses and the above indicates some of the meanings which are used in relation to *in vivo* and *in vitro* experiments.

the intestine into the mesenteric venous blood, or into the lymph channels draining the intestine. But even if this is done it must be remembered that the substances may disappear from the lumen in one form but appear in the blood or lymph in a different form; and indeed this is probably the usual pattern for proteins, fats and carbohydrates. It is therefore meaningless to speak of the form in which the substance is absorbed, because, for example, protein may leave the lumen of the intestine as small peptides but enter the blood stream as amino acids.

In vitro *and* in vivo *methods.* The methods for the study of absorption can conveniently be divided into two groups, *in vivo* methods and *in vitro* methods. These terms are very widely used now and it is useful to have a precise definition of their meaning. The term *in vivo* implies that the experiment is carried out in the living animal, while *in vitro* implies that some part of the intestine under study has been removed from the animal. The intestine can however be in the living animal, but not in its physiological relations to other parts of the intestine or to other organs, or it can be connected with the living animal in various ways. The best distinction is probably to define *in vivo* conditions as those where the intestine is supplied with blood through its own systemic blood supply. On the other hand if the intestine is separated from its blood vessels the term *in vitro* should be used. There will be cases where it will be difficult to define whether conditions are *in vivo* or *in vitro,* e.g. if the intestine is perfused with blood through an artificial circulation, but in general the distinction is obvious, and is a useful one. It is also important to appreciate the relation between terms used to describe the results of *in vivo* and *in vitro* techniques.

Polarity of absorption. The terms used to describe absorption must be related to the polarity of movement. Unlike transport processes in many other tissues, e.g. erythrocytes, where transport is a question of entry and exit by the same route, transport in the enterocyte has a definite polarity where the substance being transported enters one side of the cell and leaves at the other. As transport may involve a number of stages it is useful to have a terminology which will include polarity of movement at each stage involved.

It is first essential to have a terminology to describe the polarity of the absorbing cell or enterocyte. The term apical is sometimes used, but the histology of the intestine does not make it quite clear which is the apex of the enterocyte. Functionally there are three borders to the cell—the luminal, the lateral and the basal (Fig. 6.3). The luminal is the part of the cell in contact with the fluid in the lumen of the intestine.

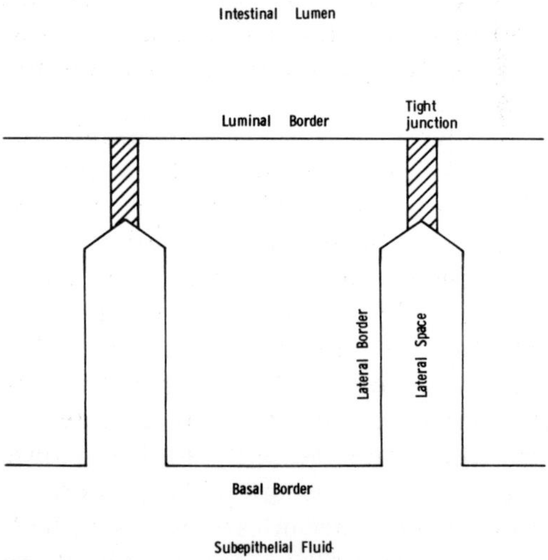

Fig. 6.3. Diagrammatic representation of the three borders of the cell which are distinguished by being in contact with three different kinds of fluid, i.e. the fluid in the lumen of the intestine, the fluid in the lateral space and the subepithelial fluid.

The basal is the part in contact with the main bulk of the subepithelial fluid. This has to be distinguished from the fluid between the cells in the lateral spaces. This fluid is not necessarily in equilibrium with the main bulk of subepithelial fluid, and indeed the standing gradient theory of fluid movement [8] requires that it should not be. The part of the cell bounding the lateral spaces could be called the lateral border. As regards polarity the substances absorbed enter the cell at the luminal borders, and leave by the basal and/or lateral borders.

As regards direction of movement Code [9] has introduced a terminology, in which the prefixes ex, in, ab and entero are used. 'Insorption' is passage of substances from the lumen of the intestine to the blood. 'Exsorption' is movement out of the blood into the intestinal lumen. 'Absorption' is the result when insorption is greater than exsorption, and 'enterosorption' is the result when exsorption is greater than insorption. One of the difficulties of this terminology is the reference points for ex and in. Code uses insorption to indicate the same direction of movement as physiological absorption, and this implies movement into the absorbing cells from the lumen, while exsorption implies movement out of the cells into the lumen. The position becomes more complicated when we think of the various stages in absorption. The physiological direction of absorption is movement into the cells at the luminal border, but movement out of the cells at the other pole. Transport processes in non-polar cells are often described as influx and efflux with obvious meanings, but with polar cells like the enterocyte this can be confusing. Efflux in relation to absorption has been used to mean movement from epithelial cells into intestinal lumen [10], but has also been used for movement in the opposite direction [11]. It would seem better to use the words efflux and influx in relation to the enterocytes, and to specify which pole of the cell we are discussing. One method which avoids the difficulty of describing the direction of movement at different stages is to record movement in the same direction as physiological absorption as positive and in the reverse direction as negative [12, 13, 14, 15]. If we have to discuss two-way movements at various stages absorption will then be the algebraic sum of the positive and negative fluxes.

Stages in absorption. The first stage of absorption *in vivo* is transfer across the luminal membrane and this constitutes luminal transfer. If it is necessary to distinguish between the movement in two directions across the membrane we can speak of luminal fluxes, positive into the cell and negative into the lumen, and the difference between these is net luminal flux or luminal transfer. At the moment we are not able to make sufficiently precise measurements of movement from the enterocyte into the subepithelial fluid but if we were, the term basal transfer might be appropriate. Again, it would be positive in the physiological direction, i.e. out of the cell. Positive and negative fluxes could also be used for two-way movement at the basal pole. If a term is required for transfer out of the cell into the lateral spaces between cells, lateral membrane transfer might be used. This leaves only the final stage, i.e. entry into the capillaries or lymphatics, and perhaps these terms are sufficiently precise.

When we come to *in vitro* studies we may have no intestinal lumen, e.g. with the everted sac preparation. The fluid in which the sacs are suspended is usually called the mucosal fluid and this corresponds with the luminal fluid. Many authors use the term mucosal transfer to indicate movement from the mucosal fluid into the enterocyte, and mucosal transfer *in vitro* therefore corresponds fairly accurately to luminal transfer *in vivo*. Again the term mucosal flux could be used to indicate two-way movement. *In vitro* a special term is necessary for transfer out of the serosal side of the gut and this is usually called serosal transfer. It does not correspond to any parameter *in vivo*. It is discussed later in this paper in relation to expressing results, but certainly it is quite misleading to use the term serosal transfer as a synonym for absorption.

While the above terms are sufficient for most purposes, there is still a possibility of ambiguity or confusion when we consider the movement out of the lumen in more detail. Figure 6.4 shows that there are five different routes of movement from the lumen of the intestine, three of these being transcellular and two extracellular. The transcellular are (1) the lipid route for substances soluble in the cell membrane, (2) the aqueous route for water and small hydrophilic solutes, (3) the carrier route for hydrophilic substances too large to use the aqueous route, e.g. hexoses and amino acids. The intercellular routes are (1) the tight junctions and (2) the spaces left when cells are extruded

METHODS OF STUDYING INTESTINAL ABSORPTION

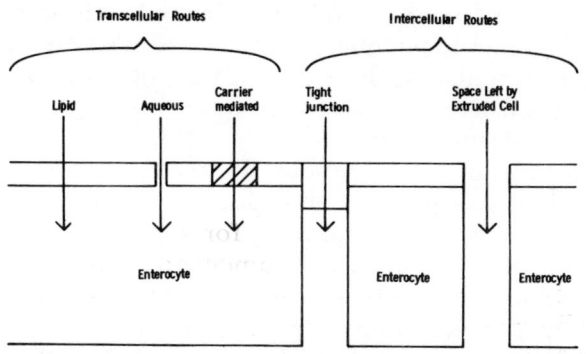

Fig. 6.4. Diagram showing the five different routes by which substances may leave the lumen of the intestine. Three of these are transcellular and two intercellular.

into the lumen of the intestine. In some cases we might require special terms to distinguish between these, but for general purposes the term luminal transfer or mucosal transfer includes movement by all five of these routes. Some further confusion can occur when we think of such problems as the glycocalyx, which according to some authors can play an important part in the final stages of digestion and may form some kind of diffusion barrier. Ugolev [16] associates this with the digestive surface of the cell, and in this region it is easy to have confusion as to what activities are extracellular and what intracellular.

6.1.2 Balance experiments

Whatever experimental technique is used the balance principle is often employed to measure absorption. The general principle is that a certain amount of substance is given by mouth or is placed in the intestine by some other means. After a certain period, the amount not absorbed is recovered and measured, and the difference between these is the amount absorbed. If the amount given is Q, the amount absorbed is A, and the amount recovered R, then

$$A = Q - R$$

If the time elapsing between administration and recovery is t, then the rate of absorption is A/t. This very simple principle becomes more complicated when we consider its practical application. The simplest quantity involved is Q, and this is usually an accurate measurement which is unaffected by the route or method of administration. The determination of R is more complicated, and the first problem is the completeness of recovery. In some techniques where a section of the intestine in an animal is cannulated, this may not present a serious problem. In other cases where the intestine empties itself by physiological means, the difficulties may be very great. Having determined R the question also arises about the interpretation of $Q - R$. Does this really represent absorption, or may it include a fraction of substance disappearing for some other reason, e.g. chemical transformation in the lumen of the intestine due to bacterial metabolism or some other cause? It is also necessary to consider whether the value of $Q - R$ might be affected by some of the substance under discussion being added to the intestinal lumen by some process other than the experimental administration, e.g. by intestinal secretions, haemorrhage, etc. It is also important to remember the recycling of substances which takes place into and out of the alimentary tract. This may be a two-way flux of substances through the same route, or it may be removal by one route and addition by another, e.g. the addition of water, inorganic salts, bile salts and other substances by the pancreatic or bile duct after absorption by the epithelial cells. In some cases it may also be necessary to consider the possibility of coprography as this could affect the interpretation of the results.

The time factor t may raise problems depending on the experimental object in view. If the interest is in knowing whether a substance is absorbed by the body in the normal amounts, e.g. in a fat tolerance test, it does not really matter to the body what the time scale is. The substance is taken by mouth and propelled along the intestine. The time during which it is in contact with the absorbing surface is not under experimental control. When the total amount recovered from the faeces is subtracted from the amount given we have a value for the amount of substance absorbed, and to attempt to explain this formally as a rate (mass/unit time) would not be meaningful. What can be useful in such cases is to express the fraction administered which is absorbed, i.e. $(Q - R)/Q$.

In other cases the time interval can be important, and in such cases it is useful to measure $(Q - R)/t$. This is particularly so when we are trying to describe the absorptive capacity in quantitative terms. It is however always important to have in mind the purpose of the investigation. As regards nutrient substances both from the medical and economic point of view the only thing that matters is that the individual (animal or man) gets enough food in the time available. If we exclude ruminants most mammals spend only a very small amount of their time actually taking in food, and have a long period available for digestion and absorption. There are also probably ample reserves of absorptive surface. Other considerations may arise in the absorption of drugs, where it may be desirable to get high blood concentrations as rapidly as possible. It will be evident that many of the methods discussed subsequently are applications of the balance principle, and the main problem in most cases is the determination of the value of R, i.e. the amount not absorbed.

6.1.3 Marker substances

The difficulty of measuring the amount of substance remaining in the intestine has led to the use of marker substances, and these can be combined with a large number of different techniques. The word marker is unfortunately used in a number of different senses, but as regards the gut the sense used here is the commonest meaning of the term. It perhaps should be called a volume marker, as its basic function is to determine the volume change, where the volume is not susceptible to direct measurement.

The basic principle of the marker is as follows. If we have a marker substance M, and if we dissolve a certain amount (m) in a certain volume (V) then

$$V = \frac{m}{[M]} \tag{6.1}$$

where $[M]$ is the concentration of M.

Suppose a solution of M is present in the lumen of the intestine and both the substance and water are being absorbed. If V_t, m_t and $[M]_t$ are the volume, amount and concentration respectively, at time t, then

$$V_t = \frac{m_t}{[M]_t} \tag{6.2}$$

Hence the volume at any time can be measured if we can determine m_t and $[M]_t$. If the initial values are V_0, m_0 and M_0, the volume absorbed (V_A) can be calculated from the equation

$$V_A = V_0 - V_t = \frac{m_0}{[M]_0} - \frac{m_t}{[M]_t} \qquad (6.3)$$

If the substance M is not absorbed from the intestine but fluid is, the equation becomes

$$V_A = m_0 \left(\frac{1}{[M]_0} - \frac{1}{[M]_t} \right) \qquad (6.4)$$

Hence provided we know the initial amount of M given we can measure volume changes by measuring only the concentration of M. It is clear that it is a great advantage to have an unabsorbed marker.

The marker technique can also be used to estimate the absorption of another solute. Consider the case where a solution containing the marker (M) and another substance (A), whose absorption is being studied, are circulated through the lumen of the intestine by one of the techniques described later in this chapter. Although M is not absorbed its concentration may change due to absorption of fluid, and the concentration of A may change due to its own absorption and also fluid absorption.

Let m be the initial amount of M given, a_0 the initial amount of A given, a_t the amount of A remaining at the end of the period of observation, V_0 the initial volume of fluid circulating and V_t the final volume. Let $[M]_0$, $[M]_t$, $[A]_0$ and $[A]_t$ be the initial and final concentrations and M and A, so that $[M]_0 = m/V_0$, $[M]_t = m/V_t$, $[A]_0 = a_0/V_0$ and $[A]_t = a_t/V_t$.

The amount of A present at any time t is $[A]_t V_t$, and hence the amount of A absorbed (a_A) in time t is given by

$$a_A = [A]_0 v_0 - [A]_t v_t \qquad (6.5)$$

The value of V_0 and V_t can be expressed in terms of m and $[M]$ and hence

$$a_A = [A]_0 \frac{m}{[M]_0} - [A]_t \frac{m}{[M]_t} \qquad (6.6)$$

This can be arearranged so that

$$a_A = m \left(\frac{[A]_0}{[M]_0} - \frac{[A]_t}{[M]_t} \right) \qquad (6.7)$$

It would obviously be possible to determine the amount of A absorbed by using an absorbable marker but in this case the amount of marker (m_t) would also have to be determined. As this raises further difficulties and errors, it is better to find a completely unabsorbable marker. A number of substances have been tried and the most generally used is polyethylene glycol, often called PEG. This was introduced by Sperber *et al.* [17] for use in digestive studies, but was applied to balance studies in the intestine by Borgstrom [18].

As the measurement depends on the difference between concentrations of PEG the accuracy of determination of PEG is an important factor in the measurement. Another important principle is that both marker and substance being studied must be uniformally distributed in the same volume. This will only be so if both have the same partition in the aqueous and lipid phase of the intestinal contents. It will also be so only if there is complete mixing of the luminal content to overcome the effect of absorption of solute producing non-uniformity of luminal concentration. While PEG is the substance most frequently used other substances which have been used are haemoglobin [19, 20], mannitol [21], phenol red [22], amaranth [23], succinylsulfathioazole [24], methyl cellulose [25], chromium sesquioxide [26]. It is important to make sure that in the particular circumstances and with the particular animal species used the marker is reliable.

6.2 METHODS

These can be classifed according to the extent to which the experimental conditions depart from the physiological ones [3]. The most physiological is when the human subject or experimental animal is unanaesthetized, eats an unrestricted diet and subsequently absorbs this free from physiological, psychological or pathological disturbances. Whether these conditions often exist for the human subject in modern conditions must be doubtful, and attempts to impose them might result in serious problems about the length of tea breaks or business lunches. Certainly they can never be achieved in experimental studies on absorption as some conditions necessary for the observer or experimenter must always be imposed. At the other extreme are experiments where the gut is removed from the animal, the cells disintegrated and fractions of these examined. We obtain different kinds of information from these different experiments, and in general it could be said that the less we depart from physiological conditions the less detailed is the amount of information we obtain. In the following description we begin with the most physiological conditions, and progressively go on to the less physiological techniques. It is also convenient to separate these into *in vivo* and *in vitro* methods.

6.2.1 In vivo methods

Techniques applicable to intact unanaesthetized animals including man. There are two different approaches to this problem, i.e. to study the disappearance of substances from the intestinal lumen or to study their arrival in the blood stream.

Disappearance from the lumen. These methods are all variations of the balance experiment discussed above, and a number of different techniques have been used to get over the difficulties involved in determining the amount not absorbed in the conscious animal.

In an attempt to maintain conditions as near to physiological as possible one procedure is to give the nutrient by mouth, and collect the unabsorbed residues in the faeces. In this procedure the main difficulties are the uncertain time required for the passage of the nutrient through the alimentary tract and the uncertainty as to whether the faeces contain all the unabsorbed fraction, or whether some of this might not remain in the lumen of the alimentary tract. Various attempts have been made to get around these difficulties, and hence a number of variations of the technique have been used. In general the faeces are collected over a period during which a substance given by mouth might be expected to appear if not absorbed. Attempts to determine the correct time interval have been made by using marker substances in the food which could be observed in the faeces. (Note that marker substance is used here in quite a different sense from that given above.) For example Cook *et al.* [27] gave carmine capsules on the first day of the test and charcoal on the last day of the test. The faeces were collected from the appearance of carmine to the appearance of charcoal and in this way it was hoped that the collection period would correspond with the time taken for the substance to pass through the gut. On account of the great uncertainty involved in this method it is not in common use.

A variant of the method is to maintain a constant diet over a long period and to assume that a steady state has been set up so that a constant fraction of the substance is being absorbed from the intestine. A study of the amount taken in 24 hours and the amount recovered from the faeces in 24 hours (not necessarily the same 24 hours) could then give a measure of the amount absorbed. This method is in common use for studying defective fat absorption. In this case precise measurements are not required, as differences between the fat recovered from the faeces in the normal person and in a number of clinical conditions are sufficiently large to make the method useful. Different aspects of the problem have been discussed by various authors [28-32]. Another method which has limited application is to give a person a diet for some time which is entirely free from the substance to be studied. One single dose of the substance is then given and the faeces are collected as long as it is considered likely that the substance will still be passing through the alimentary tract. Clearly this method is not applicable to the common food substances, and could be used for the vitamins or other substances which it is not essential to supply every day. Baker and Mollen [33] used this method for the study of absorption of cyanocobalamin and found 4-13 days were required before two successive samples of faeces contained less than 1% of the dose given. Another technique involves use of a marker substance [26, 34] (in the sense discussed on p. 249). In this case the unabsorbed marker substance is given in the food along with the substance to be studied. The ratio of the two substances in the food is known and the ratio is determined in the faeces, and from this the amount absorbed can be calculated. An advantage claimed for this method is that only a sample of faeces is required, and not the total amount passed in a given time.

Intubation Methods. Another form of the balance experiment can be done in the human subject by getting access to the intestinal lumen through swallowed tubes. In 1908 Scheltema [35] introduced a technique for intubation of the intestine in animals and this was also applied to the human subject. It was called permeation or sounding of the gastrointestinal tract. It was however used for introduction of medicinal substances and not for study of intestinal contents. Einhorn [36] also used intestinal intubation and later [37] aspirated intestinal contents from the tube. The first attempts at the use of intubation for serious absorption studies were made by Miller and Abbott [38, 39]. They used either single, double or triple lumened tubes with balloons attached, so that after swallowing the tube the balloons could be inflated thus blocking the lumen of the gut. When two balloons were used a section of gut could be isolated between these, and access to this section was provided by another tube with an opening between the two balloons. Groen [40] considered the two balloons a disadvantage as the isolated segment might become distended and he used only the distal balloon and studied the segment of intestine proximal to this. Nicholson and Chornock [41] considered the presence of balloons could interfere with the absorptive and secretory activities of the gut, and introduced a method of intubation without balloons. The procedure introduced by Blankenhorn [42] and his colleagues forms the basis for many procedures used at present, in which a small-bore plastic tube with a weight attached is swallowed, and the tube moves slowly along the alimentary tract. Samples of contents can be aspirated at any time, and the site of aspiration is determined from knowledge of the length of tube swallowed and the distance between the nose and various parts of the gut (Table 6.1). Borgstrom [18] used a modification in which two

TABLE 6.1. Table for determining the position of swallowed tubes in the human subject. The distances from the nose to the tip of the tube are expressed in centimetres with the mean ± S.E.M. The fractional mean differences are expressed as a percentage of the total nose-anus distance. From Parsons [5].

Nose-pylorus	Nose-ligament of Treitz	Nose-ileocecal Valve	Nose-anus
64 ± 6	83 ± 4	348 ± 17	452 ± 17
Fractional mean distance			
14	18	77	100

openings were made in the tube separated by a solid piece so that aspirations could be made from both ends of the intestine. They also combined this with use of a marker substance and the particular substance used was polyethylene glycol (PEG). Various modifications of these techniques have been used by subsequent workers [43, 44, 45, 46].

Weighing the abdomen. A very different procedure applicable to the human subject is weighing the abdomen and limbs [47]. This gives information about the rate of movement of fluid from the intestine, but is only semi-quantitative.

Appearance in the blood stream. If a substance is given by mouth and subsequently absorbed, its concentration in the blood can be followed by taking blood samples. When a series of blood samples is taken, blood concentration plotted against time constitutes a tolerance curve. The shape of the curve depends on (a) the rate of entry of the substance into the blood stream, (b) the rate of removal from the blood stream. The former is a direct index of intestinal absorption, but without knowledge of the rate of removal the tolerance curve does not give quantitative information about absorption. The removal from the blood stream is due to equilibrium of the substance in a space which will be considerably bigger than the vascular compartment, metabolism by the cells and excretion in the urine or expired air. In general the rate of removal from the blood stream will depend on the concentration in the blood. Without going into a detailed mathematical analysis, it is evident that provided the rate of removal remains proportional to the concentration in the blood, the shape of the tolerance curve will be to some extent dependent on the rate of intestinal absorption. An increased absorption rate will give a higher and more prolonged type of tolerance curve. Hence the shape of a tolerance curve in an individual when compared with tolerance curves in other individuals may give information about differences in the absorption process. Tolerance curves have been used to study absorption of water [48, 49, 50, 51], alcohol [52], methionine, vitamin A and galactose [53], B_{12} [54], sugars [52, 55, 56]. A special example of the tolerance curve is the chylomicron count introduced by Gage and Fish [57] which has been extensively used in the study of fat absorption.

A tolerance curve can also be made following an intravenous administration. In this intestinal absorption makes no contribution to the shape of the curve, which now depends on the rate of removal. If the oral tolerance curve is combined with an intravenous tolerance curve more quantitative information can be obtained about the contribution made by absorption to the shape of the curve [48, 53]. For details of the methods of calculating the results the original papers should be consulted.

The substances absorbed from the intestine pass into the mesenteric veins and hence to the liver. Those that still remain in the blood after passage through the liver enter the systemic circulation. Hence blood changes after absorption will be much smaller in the systemic circulation than in the mesenteric blood. Mesenteric blood is not readily available in human subjects, but use has been made of patients with abnormalities in the portal circulation [58, 59] to obtain samples of mesenteric blood for absorption studies.

Changes in other organs and body fluids. The changes in the blood following absorption will be reflected in changes in various organs and body fluids. Most accessible is the urine, and the amount of substance excreted in the urine provides some evidence of rates of absorption. The radioactivity in the urine after administration of ^{60}Co-labelled vitamin B_{12} was used by Schilling [60]. The radioactivity over the liver was used by Glass et al. [61] for similar purposes, and radioactivity over the thyroid can be used for study of uptake of radio iodine.

An interesting application of changes produced by intestinal absorption is the examination of the expired air as evidence of consumption of alcohol by mouth. This is the basis of the breathalyser, now extensively used as a medico-legal test in connection with driving motor vehicles under the influence of alcohol.

Techniques applicable to intact animals which are subsequently killed. The most important of these is the technique introduced by Cori [62] which is another application of the balance experiment. The animal is weighed, a stomach tube passed and a solution containing the substance to be studied injected into the stomach. In Cori's original experiment various sugars were used. After a certain time the animal was killed, the gastrointestinal tract ligated at both ends and quickly removed, and the content analysed. The amount of hexose recovered was presumed to be the amount not absorbed, and hence the absorbed fraction was obtained. Cori also introduced the term absorption coefficient to express the amount of sugar absorbed per 100 g body weight per hour. Similar techniques were used subsequently by many authors [63, 64, 65]. In using this technique it should be remembered that the amount of substance absorbed depends on the activity of the stomach and intestine as a whole. The supply of nutrient to the intestine depends on the rate of gastric emptying, and this rather than absorptive activity may be the rate-limiting step in the whole process. It is also of interest that the absorption coefficient for sugars as measured by this technique may be very much smaller than that obtained when the glucose uptake is voluntary [66, 67].

This technique was elaborated by Reynold and Spray [22] in a procedure in which a marker substance was also given with the food, and subsequently the intestine was divided into a number of segments so that some information was also obtained about the transit rates of nutrient along the intestine.

Techniques applicable to unanaesthetized animals after modification of the intestine or of its blood or lymph supply

Intestinal fistulas. It is possible to transplant a loop of intestine so that its blood supply is maintained from its natural source but the ends of the loop open on to the abdominal wall, while the continuity of the bowel is restored. This constitutes the Thiry-Vella loop [68, 69] and it would appear to offer a useful approach to studying intestinal absorption in the unanaesthetized animal. The application of the technique is another variant of the balance experiment in which the amount absorbed is the difference between that given and recovered.

The technique of making the loop is well described by Markovitch [70], and loops have been used for absorption studies many times since the first use by Gumilewsky [71]. It must be admitted that these have on the whole been disappointing, and have not made a very significant contribution to our knowledge of absorption. There are two main problems to discuss in relation to this, (1) a technical one—how to achieve water-tight connection between the ends of the loop and an outside circuit, and (2) the more important problem as to whether the loop has the absorptive properties of normal intestine.

As regards the technical problem many devices have been used to make a water-tight junction. In Gumilewsky's original method an inflatable rubber cuff was used, and many authors have used this subsequently. Gregory [72] used a permanent silver cannula, and modifications of this have been used [73]. Other procedures are a suction device against the skin to maintain a connection [74], a nylon cannula [75] and a tube with a rubber cone held against the opening [24]. It is doubtful whether any of these is really satisfactory.

The other problem is whether the isolated loop maintains its normal function. Berger *et al.* [25] have discussed some of the criticisms of the technique, but thought that on the whole the loop was satisfactory. Perhaps one important aspect of the problem is the nutrition of the intestinal epithelium. Smyth [76] has pointed out that the enterocytes differ from most cells in that they have a double source of nutrition, i.e. from the blood supply like other cells, but in addition from the nutrients in the intestinal lumen. Indeed if one considers the various pumps in the enterocyte which remove nutrients from the luminal side of the cell, it is evident that the luminal side of the cell must have some difficulty receiving nutrients from the blood stream. Hence a supply of nutrients in the intestinal lumen may be essential for the maintenance of the enterocytes in physiological conditions. Whether or not this is the cause, the performance of the Thiry-Vella loop as a normally absorbing piece of intestine is still a matter of some uncertainty.

Most of the work with Thiry-Vella loops has been carried out on dogs. There are a few cases where fistulae in human subjects made for therapeutic purposes have been used for absorptive studies [77, 78, 79].

The London cannula. As already mentioned there are advantages in using portal blood rather than systemic blood in studying intestinal absorption. The cannula designed by London [80] is a procedure which enables samples of portal blood to be obtained in conscious animals. The principle is that a metal tube is attached to the portal vein, and the other end opens on the abdominal wall. A needle can be inserted down the tube and into the portal vein for obtaining samples of portal blood. The technique is not simple, and has not been extensively used, but in a few cases [81] has given results of some interest. It enables only concentrations of substances in portal blood to be obtained, and as the portal blood flow is not measured, does not give quantitative information about absorption rates.

More recently Shoemaker and his colleagues [82, 83] have described techniques for cannulation of the hepatic vessels in the dog, which make it possible in the conscious animal to estimate portal blood flow in addition to concentration changes. By this means it was possible to study quantitatively the absorption of glucose and fructose and to describe the metabolic transformations in addition to the amounts absorbed.

Lymph fistulas Since the lymph vessels are the main route of absorption of lipids, the collection of absorbed lipid must be done by lymph fistulas. These have been used for studying absorption in conscious unanaesthetized rats by Bollman *et al.* [84]. A plastic tube of 1-1.5 mm bore is introduced under anaesthetic into a main lymph vessel. After recovery the animal is restrained in a special cage [85], so that it is not able to interfere with the cannula. By this means it is possible to collect up to 20 ml of lymph daily, and since the rate of lymph flow is also measured quantitative absorption studies can be made.

Techniques applicable to the anaesthetized animal without interference with blood or lymph supply. These are the simplest methods available for absorption studies and have been extensively used in many modifications. They measure the amount of substance disappearing from the intestinal lumen, and give no information about the amount appearing in the blood stream. They are all applications of the balance principle. A certain amount of substance is introduced into a loop of intestine, and at the end of the experimental period the amount remaining is estimated, and the difference is taken to be the amount absorbed. In its simplest form an animal is anaesthetized, the abdomen opened, and a loop of gut selected. This is tied off from the rest of the intestine. The tied-off loop is cannulated so that the contents can be washed off, and then a solution introduced containing the substances to be absorbed. At the end of a given time the contents of the loop are withdrawn for analysis. This method was used by Reid [86, 87] and was later used in rats by Verzar and his co-workers [89]. It has since been used many times in various modifications.

In the more recent modifications the loop is not closed, but cannulated so that fluid can be introduced and removed, and this was introduced by Sols and Ponz [89, 90] who cannulated a segment of intestine so that it could be washed out with saline, the test solution run in, and at the end of the experimental period washed out. Horvarth and Wix [19] used a

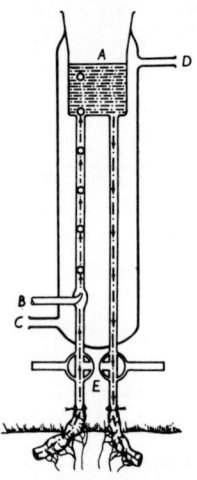

Fig. 6.5. Apparatus used for circulating fluid through a section of the intestine of the anaesthetized animal. The fluid circulates from a reservoir (A) and returns by another tube provided with an oxygen lift (B). Warm water circulated by C and D maintains the fluid at body temperature. The three way stopcocks at E enable rapid washing out of the intestine at the end of the experimental period. (From Jervis, Johnson, Sheff and Smyth [94].)

technique in which fluid was slowly run through the loop, and a comparison of the concentration in the affluent and effluent solutions was used to determine the amount absorbed. In this modification the animal was allowed to recover from the anaesthetic before the absorption studies. Other developments include circulation of fluid through a loop [91, 92, 93]. In these techniques the concentration of solution in contact with the mucosa can be controlled throughout the experimental period. The apparatus used by Sheff and Smyth is seen in Fig. 6.5. The anaesthetized animal is kept warm on a heated table and an important detail, on which stress is laid, is that the rectal temperature should be recorded every 5 minutes. In a modification of this technique [94] two experimental periods of 15 minutes each were used, separated by an interval of 30 minutes. It was found that during the second period the loop absorbed about 90% of the amount absorbed during the first period.

Techniques applicable to anaesthetized animals after interference with blood or lymph supply. The object of these experiments is to study a different aspect of absorption, i.e. the arrival of substances in the blood stream rather than their disappearance from the intestinal lumen. The two parameters can in fact both be studied at the same time, because techniques designed to study the mesenteric blood can usually without much difficulty be combined with study of the contents of the loop [95]. The study of absorption from mesenteric blood analysis presents a number of problems. In the first place, the mesenteric and portal veins are relatively inaccessible, apart from the London cannula or use of polythene cannulae tied into the vessels. Even if this is done, only samples are obtained, and knowledge of absorption must be derived from concentration changes. If we consider the substances absorbed in large amounts, i.e. proteins, fats and carbohydrates, these or their derivatives are already present in the systemic blood, so that comparisons will have to be made of the concentration of substances in mesenteric arterial and venous blood before and after administration of the substances absorbed. On account of rapid mesenteric blood flow considerable amounts of substance could be absorbed with only very small changes in concentration. In the case of glucose the problem is complicated by the fact that the intestine has a high rate of aerobic glycolysis [96], and this glycolysis affects both glucose absorbed from the lumen and glucose delivered by the arterial mesenteric supply. If balances are to be meaningful they should involve lactic acid as well as glucose. In the case of substances absorbed in small quantities, e.g. vitamins, although the amount in the arterial blood is very small, the increase in venous concentration is also very small. Furthermore, in addition to concentrations the total rate of mesenteric flow will be required to make the results quantitative. One way round the difficulty is offered by the technique of Mattthews and Smyth [97] in which the total blood draining one section of the intestine is collected, so that the total amount of substance in the mesenteric blood and not only its concentration can be determined. The procedure is as follows. The dog or cat is anaesthetized, the abdomen is opened, and a suitable loop of intestine is found. The requirements are that the loop of intestine should be drained by a single vein which can be cannulated. This vein is then carefully prepared for cannulation and the loop is washed out. Arrangements must also be made for keeping the loop at body temperature. This can be done by replacing it in the abdominal cavity, if care is taken not to kink the effluent vein. In cases where it is desired to keep the loop under observation, it can be placed on a gauze-covered heated metal plate in close contact with the abdomen, and protected by a Perspex cover. A thermometer is tied inside the lumen of the intestinal loop, and the temperature can be observed throughout the experiment. When everything is completely ready, the vein draining the loop is cannulated and the substance to be studied is introduced into the intestine. What happens now is that the animal is being bled out through the vein draining the loop of intestine. In order to inhibit the glycolysis or other changes, the blood can be cooled down as it leaves the cannula. At the end of the experimental period, the lumen of the loop is washed out, and an estimate made of the amount recovered. The loop itself can be taken for determination of the amount in the intestinal wall. It is possible to control the rate of blood flow through the loop by adjusting the blood pressure. It may be convenient to begin the experiment by removing some blood to reduce the pressure, and later reinjecting this when arterial blood pressure begins to fall. The same principle has been used with the dog and rabbit by Neame and Wiseman [98]. In smaller animals, it was found easier to collect portal blood from the mesenteric venous trunk, and to tie off or remove sections of the intestine other than the loop from which absorption was taking place.

Vascular perfusion of the intestine. This kind of study is on the borderline of *in vivo* and *in vitro* studies as defined above. The intestine has a nutrient supply through its own blood vessels, but the nutrient fluid is supplied through an artificial circulation. The fluid may be blood or some blood substitute. The technique was first used by Salvioli [99] in 1880 and will be recognized as a major technical feat at that time by anyone who has tried vascular perfusion even with modern resources. Since then a number of other attempts have been made but with somewhat limited success [100, 101, 102]. One of the major problems is oedema which occurs very rapidly even with blood perfusion. Recently Parsons and Pritchard [103] have introduced a technique for the perfusion of the frog intestine which appears to be satisfactory.

The word perfusion itself can give rise to ambiguity, as some authors use this to mean circulation of fluid through the blood vessels, while others mean circulation of fluid through the lumen of the intestine. The meaning is usually clear from the context, but ambiguity can be avoided by the terms vascular perfusion and luminal perfusion.

6.2.2 In vitro *methods*

The term *in vitro* is used here to include all cases where the intestine does not have a flow of blood (or other substitute) through its mesenteric vessels. The concept is now new, but it is only since the introduction of an effective method by Fisher and Parsons [104] and perhaps even more the introduction of the everted sac by Wilson and Wiseman [105] that *in vitro* techniques have been extensively used. This very extensive use is largely due to the simplicity of the everted sac technique, and this has some disadvantages, as it is very easy to acquire data without thinking too much about its precise significance. For this reason, the technique and use are dealt with here in some detail.

It is of interest to look first at the historical development of the *in vitro* intestine. Waymouth Reid [106] used a small piece of tissue held in position so that it was separating two fluid compartments, and he found that fluid was transferred from one compartment to the other. The first experiments were done with frog's skin [106] but later a piece of rabbit intestine was used [107]. This work did not arouse much interest at the time, but looking back it marked a very important development in *in vitro* studies, and was really the prototype for much subsequent work on frog skin and intestine, toad bladder, and many other tissues. Reid's work was not followed up, and it was not until 1930 that Magee *et al.* [108] attempted to make an *in vitro* preparation of intestine. Their preparation was unsuccessful because they failed to appreciate the need for oxygenation of the mucosa. It is very interesting to read that they everted the intestine subsequently to make sure the cells were dead to produce a control, whereas they would have had a better chance of achieving success if they had everted it at the beginning. The real break-through came in 1949 when Fisher and Parsons [104] appreciated that the important requirement was oxygenation of the mucosa, and this they achieved by circulating oxygenated saline through the lumen of the intestine.

Isolated intestine with luminal perfusion The original Fisher and Parsons preparation used the whole of the rat small intestine. In order to maintain oxygenation of the mucosa a solution of oxygenated saline was circulated through the lumen while the intestine was still *in situ,* and this time there was no interference with oxygenation of the epithelium when the intestine was removed from the abdomen. It is now known that this precaution is unnecessary. The intestine was then put into a bath of fluid and the circulation through the lumen continued. This preparation can transfer glucose and other substances from the saline inside the lumen to the saline bathing the peritoneal or serosal surface of the intestine. It is now usual to refer to these solutions as the mucosal and serosal fluids respectively. Wiseman [109] used the idea of Fisher and Parsons, but modified this to make a more compact type of apparatus, the principle of which is illustrated in Fig. 6.6. Three short segments (the figure shows one only) are arranged so that they are bathed in serosal fluid, while the mucosal fluid is circulated through them from a reservoir. An oxygen lift supplies at the same time circulation and oxygenation. This technique was modified by Smyth and Taylor [110] by leaving out the serosal fluid, so that the loop of intestine was suspended in air. In this case the fluid transported by the gut appears as small droplets of fluid on the serosal surface of the gut. It runs down the gut and is collected as shown in Fig. 6.6.

The everted sac. The everted sac was introduced by Wilson and Wiseman [105] in 1955 in Sheffield and has been extensively used since, either in the original or modified form. A detailed description of the original procedure has been given by Wiseman [111]. The basic reason for its success is undoubtedly because it offers a preparation with a well-oxygenated intestinal mucosa. When the intestine is everted there is an important difference in the simple geometry of the villi as compared with the situation when the villi are inside the lumen, and this must make a difference in accessibility of oxygen and nutrients. There is the additional advantage that the large volume of fluid outside the sac can contain more O_2 and more nutrient than the smaller volume inside the sac. In the original preparation the

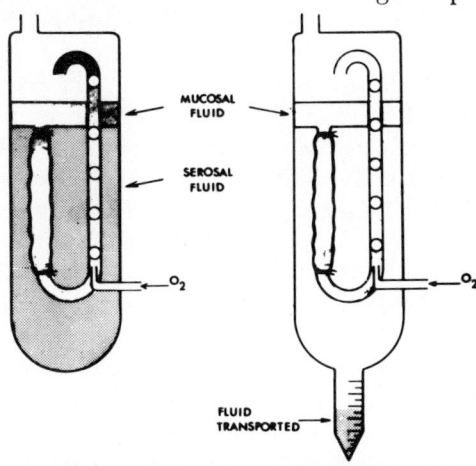

Fig. 6.6. Two different types of *in vitro* experiments. On the left is the technique devised by Wiseman [109] for circulating fluid through loops of intestine. In this case only one loop is shown instead of the three normally used. On the right is a modification by Smyth and Taylor [110] in which no outer fluid is present and the fluid transported is collected. In both cases the circulation is maintained by the oxygen lift.

animal was stunned by a blow on the head; in fact the sac seems to function better if the animal is anaesthetized, a modification used by Barry et al. [112, 113]. This is probably because of damage to the epithelium by anoxia caused by interference with the blood supply as a result of killing. When the animal is anaesthetized the blood supply is maintained until the gut has been removed from the abdominal cavity and washed out, a process which must result in a lower temperature before separation from the blood supply. The original size of sacs were 1-2 cm and for many purposes this is satisfactory. For other purposes larger sacs are used. On account of the difference in intestinal length in different conditions Barry et al. [112, 113] introduced the idea of dividing the intestine into five equal parts, and considered the activity of the different fifths of the intestine. This will be discussed further later.

While different workers have varied the details of their procedures the following description of the technique of dividing the intestine into fifths may be found useful. White rats are anaesthetized with pentobarbitone sodium, and the whole of the jejunum and ileum removed. The duodenojejunal junction is arbitrarily taken as the point where the gut is closely bound to the posterior abdominal wall, and the jejunum was divided as close to this as possible. The combined jejunum and ileum is transferred to a 0.9% NaCl at room temperature and everted. Eversion is done as shown in Fig. 6.7 by tying the gut on to a glass rod with a groove near one end to facilitate fixation. After eversion the intestine is returned to 0.9% saline. In order to facilitate the identification of different parts it may be found useful to place ligatures of different colours on the jejunal and ileal ends. The everted intestine is then transferred to a Perspex trough, 1 m long containing 0.9% NaCl. The bath is constructed by dividing longitudinally a Perspex tube of 1 in (2.5 cm) diameter. The ligatures at the ends of the intestine are fixed to hooks and the intestine stretched until it is just straight. It is then further stretched until the total length in centimetres is the next multiple of five. Usually the extended length is 65-75 cm. With the metre scale alongside, the intestine is now divided by ligatures into five equal parts, and sacs are made from these five parts.

Each sac is treated as follows. The empty sac is weighed (W_1) and this can be done on a spring balance or on a specially modified Petri dish previously weighed. The sac is now tied over a blunt needle, the required volume of serosal fluid injected, the needle withdrawn and the sac securely tied. The sac is weighed again (W_2) and then transferred to a specially designed conical flask (Fig. 6.7) containing a measured volume of mucosal fluid. The glass hook and long looped ligatures on the ends of the sac greatly facilitate rapid handling. The flask is now filled with O_2 or O_2/CO_2 mixture depending on the nature of the saline. It is then incubated at 37°C for the required time by shaking in a Warburg bath or other shaking device. Usually bicarbonate saline (Krebs and Henseleit [114]) is used and in this case a 5% CO_2, 95% O_2 mixture is used. At the end of the experimental period the flask is taken out of the bath, the sac removed, blotted on filter paper and weighed (W_3). One end is then opened, the contents collected and the empty sac weighed again (W_4).

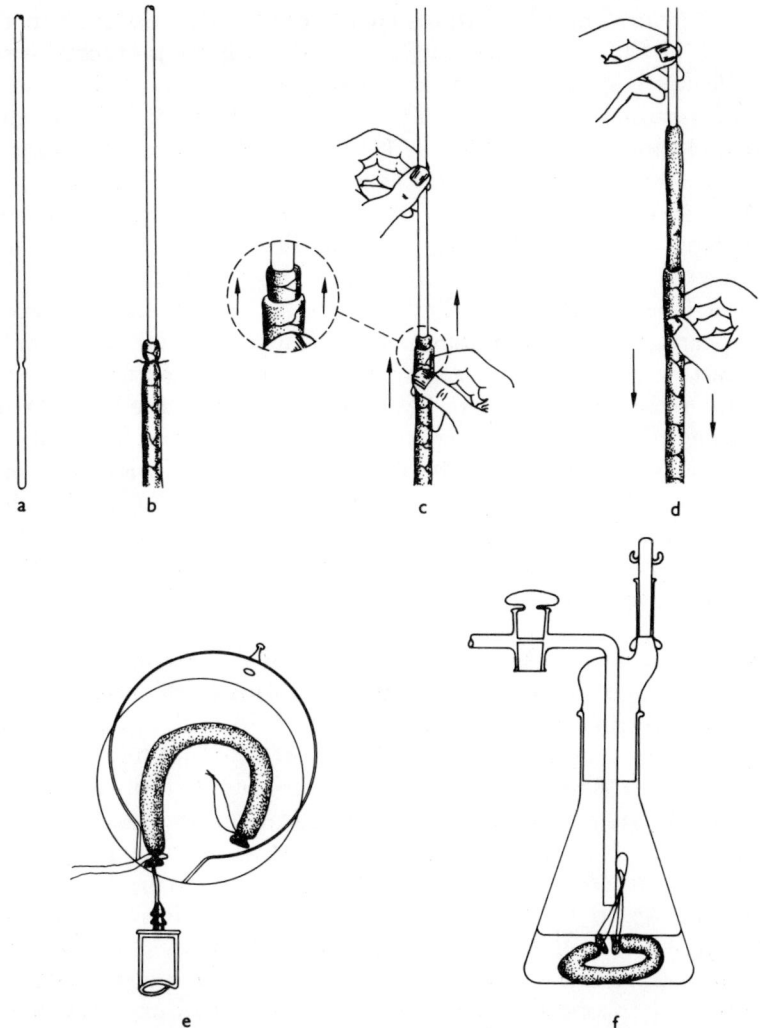

Fig. 6.7. Some stages in making sacs of everted small intestine for absorption studies. *Above:* Eversion of the intestine. A glass rod is used with a groove provided for a ligature (a). This is inserted into the intestine which is then tied close to the end (b). The eversion is commenced by pushing the gut upwards over the place where it is tied (c). The eversion is completed by inverting the rod and pulling the intestine downwards (d). *Below:* (e) Filling the sac, which is lying in a specially modified Petri dish used for weighing the gut. *Below:* (f) The sac in the incubation flask, which is provided with arrangements for gassing the flask with O_2 or an O_2 and CO_2 mixture. The glass hook enables the sac to be lowered into the flask and to be removed again at the end of the experiment. (From Smyth [2].)

Fluid transfer. The fluid movement can be calculated as follows. The initial mucosal volume is the volume measured into the flask. The initial serosal volume is $W_2 - W_1$. The final serosal volume is $W_3 - W_4$. The increase in serosal volume (the serosal fluid transfer) = $W_3 - W_4 - W_2 + W_1$, and the serosal fluid transfer/initial gm wet weight = $(W_3 - W_4 - W_2 + W_1)W_1$. It is found that W_4 is always considerably greater than W_1 as all the fluid taken up by the gut is not transferred to the serosal side. The volume remaining in the gut (the gut fluid uptake) is $W_4 - W_1$. The volume of fluid disappearing from the mucosal fluid (mucosal fluid transfer) is the sum of the amounts transferred to the serosal fluid plus the amount retained in the gut wall, or more simply the increase in the weight of the filled sac during the course of incubation $(W_3 - W_2)$. The final volume of the mucosal fluid is the initial volume $-(W_3 - W_2)$. The assumption of a S.G. of 1.00 introduces only a negligible error. In carrying out these weights it is necessary to consider whether the ligature is wet or dry at the time of weighing.

Solute transfer. The concentrations of the particular solutes are determined in the initial and final fluids both mucosal and serosal, and this information combined with the fluid movement described above will give the transfers of solute. As in the case of fluid there will be three amounts to consider: the amount leaving the mucosal fluid (mucosal transfer), the amount entering the serosal fluid (serosal transfer) and the amount retained in the gut wall (gut uptake). An important consideration in the case of solutes is the question of whether the particular substance is metabolized (e.g. glucose) or produced (e.g. lactic acid) by the intestine. The question of the selection of the best parameter to use and the interpretation of these is discussed later. The calculations of most of these parameters is based on four weighings and four concentrations, and it is easy to prepare a suitable programme for a digital computer so that all the possible parameters are printed out from the eight measurements.

The use of the everted sac enables movement of solute or fluid to be determined only at the end of the experimental period. To enable measurements to be made during the course of the experiment Crane and Wilson [115] introduced the open-ended sac and a neat modification of this was made by Clarkson and Rothstein [116] in which the open-ended sac was supported by a cork so that it floated in the mucosal fluid.

Metabolic experiments. The everted sac was used by Wilson and Wiseman [117] to measure metabolic changes as well as transfer activities. For this purpose the small sacs shaken in standard Warburg flasks as originally described are most convenient. By standard manometric techniques O_2 consumption and CO_2 production can be measured, and chemical transformations can also be studied.

Other modifications of isolated intestine. Another preparation of isolated intestine is the small ring cut from either the everted or non-everted intestine. This was introduced by Agar *et al.* [118] and used subsequently in various modifications [119, 120]. This has one disadvantage, that the separation of mucosal and serosal compartments is lost, as these now form one common pool out of which substances are taken and into which they are again returned. The assessment used here is the accumulation in the gut wall.

Reid's original isolated preparation [106] was a small diaphragm of frog skin separating two compartments. This idea was very successfully exploited by Ussing [121] and Ussing's techniques have been applied to the intestine by Schultz and Zalusky [122] and used by many subsequent workers.

Isolated fractions of the intestinal wall. One of the difficulties of the isolated intestine preparation is that the muscle layers separate the serosal fluid from the enterocytes, and the composition of the fluid in the lamina propria is unknown. For this reason attempts have been made to make preparations of sheets of mucosa separated from the rest of the intestinal wall. Hakim *et al.* [123] used a preparation of dog mucosa as a diaphragm separating two compartments. There is considerable species difference in the ease with which the mucosa can be separated from the rest of the intestinal wall and preparations of mucosa from the Greek tortoise have been successfully used [124, 125, 126, 127].

Smaller fractions of the intestinal wall have been used by

many workers. Scrapings of the mucosa were used by Dickens and Weil-Malherbe [96] to study the metabolism but not for transport studies. Small pieces of the intestine can be sliced off and Crane and Mandelstam [119] used a preparation of isolated villi. Miller and Crane [128] introduced the isolated brush border preparation, which is of great interest because of the location in this region of the cell of some important mechanism in transport. Recently it has been found possible to disintegrate the mucosa and make preparations of isolated enterocytes [129]. While the isolated enterocyte is a useful preparation for studying the isolated cell, its use loses one important feature of the enterocyte, i.e. its polarity. For example it would not readily reveal the different properties of the hexose transport mechanisms at the two sides of the cell [130] or the fact that substances do not only move in and out of the cell but also across the cell.

Under this heading shall also be included the technique introduced by Shiner [131, 132] for obtaining small pieces of duodenal or jejunal mucosa from the human intestine for biopsy. This technique has been expanded and developed by subsequent workers and gives valuable information about the condition of the mucosa of the intestine in health and disease. As it is not directly related to absorption studies it will not be discussed in detail here, but useful references will be found in the review by Duthie [7].

Cell fractionation. Like other cells the enterocyte can be broken up and various fractions separated. By examination of these fractions information can be obtained about the location of the enzymes and other constituents concerned in the transfer mechanisms or in energy supply. Many of these techniques together with the results obtained are discussed in Chapter 2 of this volume, and will not be further discussed here.

Measurement of electrical changes. Most tissues transferring ions show electrical changes which can be related to the transfer mechanisms, and Ussing [121] introduced the measurement of the short circuit current as a useful way of measuring ion transfer. Barry *et al.* [133] showed that the intestine differs from other tissues, in that a potential is associated with the transfer of non-electrolytes, indicating that ion movement is involved in hexoses and amino-acid transfer. They also measured the short-circuit current [134], and showed that it was increased when hexoses were transferred. Since then the measurement of electrical changes has become an important technique for the study of intestinal transfer and has been much used in particular by Schultz and his colleagues [122]. As it is discussed in detail in Chapter 18, together with the methods involved, it will not be further discussed here.

Interpretation of in vitro *techniques*
As *in vitro* techniques are so widely used it is important to examine them critically with a view to establishing how far these preparations give information about what happens physiologically. The essentials of the transfer process can be described under the three headings: (1) the actual physiological route by which substances move from the intestinal lumen to the blood stream, (2) the osmotic work involved in this movement which will depend on the concentration gradient against which movement takes place, (3) the source of the energy for this work and how it is coupled to the transfer processes. It is useful to examine whether there are any differences in these three respects *in vivo* and *in vitro*.

Route of absorption. A diagrammatic representation of the intestine, both *in vivo* and *in vitro*, has been presented earlier (Fig. 6.1). In both cases the first stages are uptake by the epithelial layer and transfer into the lamina propria or subepithelial fluid. As regards this stage the route is identical *in vivo* and *in vitro,* and this is important since we consider the main features of absorption depend on the enterocytes. There is however an important difference in the subsequent route. In physiological absorption the absorbate is removed by the blood vessels and appears in the mesenteric blood, while *in vitro* the absorbate passes through the thickness of the gut wall and appears in the serosal fluid. It has however been pointed out [2] that the difference is less than appears at first sight. *In vitro* it is observed that the first fluid to come through is bloodstained, and this suggests that the fluid is pushing blood out of the capillaries into the serosal fluid. It seems therefore that at least one route *in vitro* is the physiological one, through the capillaries. It is possible that there is also some diffusion through the muscular wall, but in general the difference between *in vivo* and *in vitro* is not very great as regards the actual route.

Concentration gradients. When the absorbate reaches the subepithelial fluid the only forces available for its further movement are probably hydrostatic pressure for fluid and diffusion for solutes. The hydrostatic pressure and the concentration necessary for diffusion are built up by the enterocytes. As regards hydrostatic pressure there is probably no great difference between the conditions *in vivo* and *in vitro,* or at least such differences are restricted to the different capacities of the epithelial cells to generate a hydrostatic pressure *in vivo* and *in vitro.* There is however an important difference as regards concentration gradient for solute movement. *In vivo* the substances enter the capillaries and are rapidly carried away, so that no concentration builds up. *In vitro,* and particularly in the everted sac where the volume of serosal fluid is small, the concentrations must build up fairly rapidly, so that *in vitro* the transport process is working in less favourable conditions. This constitutes a major difference between *in vivo* and *in vitro* conditions, but ways of avoiding it are discussed below.

Source of energy. An important difference between *in vivo* and *in vitro* conditions is the source of oxygen and nutrient. *In vivo* these are obtained from the blood stream and also from the intestinal lumen. *In vitro* no blood stream exists, so that one important source is missing, and the enterocytes *in vitro* have to rely on the mucosal fluid for oxygen and nutrient. This is certainly true of oxygen, but is not quite true for nutrient, as it is known that substances present initially in the serosal fluid can serve as a source of energy for the metabolizing enterocytes. They are not however able to get to the subepithelial fluid as easily as substances carried by the blood stream. Although as pointed out above the capillary route is available for penetrating the outer layers of the gut wall, the movement of fluid in the capillaries which assists substances in the subepithelial fluid to reach the serosal fluid, will hinder the movement of substances from the serosal fluid to the subepithelial fluid. Hence we must conclude that the enterocyte is less well provided with oxygen and nutrients *in vitro* than *in vivo.*

Conditions of the epithelial cells. In view of the diminished oxygen supply the question must be asked whether the epithelial layer remains in good condition *in vitro* as histological studies have suggested gross deterioration of the mucosa. It must be assumed with all *in vitro* preparations of all tissues that progressive deterioration of the tissue begins as soon as it is separated from its blood supply. It is however possible that the histological procedures cause more damage to the tissues maintained some time *in vitro* than they do to normal tissues. Certainly even after one hour the gut wall *in vitro* is able to maintain a diffusion barrier as shown by large differences in concentration across the wall, and this could hardly happen if the epithelial layer was really in the condition sometimes found by the histologists [135].

Methods of improving in vitro *techniques.* There is one experimental condition particularly relevant to all these difficulties and that is the duration of the experiment. Many *in vitro* experiments have been carried out for durations of 1 hour and even longer. There is a good case for reducing this time dramatically as this will affect the building up of concentration gradients and the amount of deterioration of the tissues. By the use of labelled substances it is possible to use very short experimental periods, and as low as 5 sec has been used [136]. This certainly answers most of the criticism. Another important condition in deterioration of the intestine is the absence of metabolizable hexose. If the isolated intestine is incubated in the presence and absence of glucose it will be found that deterioration is greater when glucose is absent. Experiments are often done in which the tissue is incubated for a preliminary period of up to 30 min, e.g. to reduce the Na content, and if glucose is absent during this period it might be doubtful whether the epithelium is in good condition.

6.3 PARAMETERS FOR MEASUREMENT OF ABSORPTION

The parameters used depend partly on what is meant by absorption (see p. 243). In *in vivo* conditions there are two kinds of measurement, i.e. disappearance from the lumen and arrival in the blood stream. In the human subject neither of these is likely to be a very precise measurement, and even if the quantity involved could be measured with some accuracy it is difficult to know how much intestine is involved in the process. The disappearance from the lumen is probably easier to measure quantitatively as arrival in the blood stream depends on tolerance curves and is only a rough guide as to what is happening. Measurements in the human subject are made mainly for diagnostic purposes and a very crude measurement may be adequate, e.g. the diagnosis of steatorrhoea. In animals where loops of intestine can be isolated and studied for definite experimental periods, disappearance from the lumen can be quite a precise measurement although the question still remains, and is discussed further below, what the rate of absorption should be related to — gut length, weight, or absorptive area.

Since so much work has been done *in vitro* and since other parameters have been used, this requires more detailed consideration. If the problem is considered in physico-chemical terms as the work done by the cell, and if we simplify this by considering osmotic work only, this is given by the formula

$$W = 2.3\ NRT \log \frac{C_1}{C_2}$$

where N is the number of moles transferred, R is the gas constant, T the absolute temperature, and C_1 and C_2 the initial and final concentrations of the solution. It is in fact not possible to evaluate the work precisely, but the equation emphasizes two different kinds of measurements which are made, and these are amounts of substance moved and the concentration changes produced. All measurements are based on some aspect of amount moved or concentration change produced.

6.3.1 *Measurements involving amounts*

There are three obvious measurements involving amounts which can be made *in vitro*, i.e. the amount leaving the mucosal fluid (mucosal transfer), the amount entering the serosal fluid (serosal transfer) and the amount accumulating in the gut wall (gut uptake). Of these the one which corresponds most to an *in vivo* measurement is mucosal transfer which corresponds to luminal disappearance. There is therefore much to be said for using mucosal transfer in assessing *in vitro* performance, or at least to give mucosal transfer in addition to other parameters selected. If serosal transfers are used, as they frequently are, their significance should be appreciated. The epithelial layer takes up a certain amount (mucosal transfer) either of fluid or solute. The epithelial cells may metabolize or retain part of this, the rest passes to the subepithelial fluid. Part of this is retained in one or other layers of the gut wall, and the rest enters the serosal fluid. Serosal transfer is therefore an unknown fraction of what is transferred by the enterocytes. It will be dependent on the duration of the experiment, the capacity of the gut to metabolize the particular substance and the capacity of the gut wall to retain fluid and solutes. If the duration of the experiment is very short, there may be no serosal transfer at all, as this will presumably only begin when the capacity of the gut wall has been reached. The relation between serosal transfer and mucosal transfer may also vary in different parts of the gut in the same animal, and in the same part of the gut in different animals, in different experimental conditions, or in different species. The serosal transfer is therefore a less satisfactory measure of absorption than the mucosal transfer; and while it gives some useful information, if possible the mucosal transfer should always be given as well.

As regards the actual measurement of mucosal transfer, this can be done in two ways, as shown by the following consideration. The mucosal transfer can be regarded as (1) the difference between the initial mucosal amount and the final mucosal amount, or (2) the sum of the gut uptake and serosal transfer. The first of these statements applies to all substances, the second only to substances not metabolized by the gut. The first is the balance principle and has the advantages and disadvantages of this. The main disadvantage is that it may be a small difference between two large amounts and therefore subject to the error inherent in such a measurement. If we are dealing with a substance not present initially in the gut wall or bathing fluid and not metabolized, the measurement of mucosal transfer as the sum of the gut uptake and serosal transfer will probably be much more accurate [137]. If the duration of the experiment is very short, the amount in the serosal fluid may be negligible, and if this is first established by experiment, then the amount in the gut wall will be the best measure of transfer capacity.

Having decided which measurement we are going to use we

then have to decide how to relate this to the amount of gut involved, and the three possibilities are weight, length or surface area. The surface area is perhaps the most logical but is ruled out by practical difficulties. The length is somewhat uncertain, as a very small tension on the gut can change the length considerably and the most practical measurement is the weight of the gut. If the weight is chosen it could be the initial wet weight, the final wet weight, the dry weight or the de-fatted dry weight, as all of these have been used. The simplest is the initial wet weight, which is automatically recorded in doing the sac experiments. A complication arises with the weight, if we are testing the effect on absorptive capacity of some condition such as starvation. In the starved animal the weight of the intestine is reduced and this loss of weight is due to change in all layers of the gut wall and not only the epithelium. Hence if we express the absorptive capacity as mucosal transfer per unit wet weight and if starvation reduces gut weight without affecting absorption, the expression of the results per unit weight would indicate an increase in absorption capacity. There is no really satisfactory solution to the problem, and it is best to give a number of parameters, making clear what the significance of these is, and making judgements based on taking all possible factors into consideration. One approach to the difficulty is to express the activity in relation to the fractional length of the gut. If the whole gut is divided into five parts, we can talk about the transfer per fifth of the intestine but again we should specify which fifth this is, as transfer capacity varies considerably in different parts of the intestine. These problems have been discussed by various workers [4, 6, 138, 139].

6.3.2 Measurements involving concentration

The *in vitro* preparation is a very convenient one for measuring concentration changes, which are a very useful index to absorptive capacity. There are, however, a number of different concentrations which can be measured. The most obvious are the initial and final concentrations in the mucosal fluid, the initial and final concentrations in the serosal fluid and the final concentration in the gut wall. Having determined these the concentration changes could be expressed in a number of ways. Final concentration ratio is the final serosal concentration/final mucosal concentration. The final concentration difference is the serosal concentration minus the mucosal concentration. Concentrating activity on the part of the gut will be indicated by a final concentration ratio greater than unity or a positive value for the final concentration difference. In both cases it should be remembered that concentration change can also be caused by fluid movement and hence concentrating capacity cannot be assumed if there is fluid movement in the opposite direction. Both these parameters have the objection that concentration changes might be caused by a metabolism of substance from one side of the intestine, and a stricter criterion of transfer capacity would be final serosal concentration/initial mucosal concentration [140]. Similarly, if differences are required, the final serosal concentration minus the initial mucosal concentration could be used.

Some authors have expressed transfer capacity in terms of the concentration in the gut wall [118, 141, 142]. In the simplest form this is the total amount of substance recovered from the gut wall divided by the volume of fluid in the gut wall at the end of the experiment. One method of assessing this is to assume the initial volume of fluid in the gut wall as 0.80 the initial weight, and add this to the increase in weight in the gut wall during the course of the experiment.

The most complicated parameter is the final concentration in the enterocyte. This has been used by a number of authors [129, 143, 144], but its significance should be considered carefully as it involves a number of assumptions. The principle of measuring is as follows: A marker substance is used which will diffuse into the extracellular space but not into the enterocyte, and it is assumed that when equilibrium is established the concentration of this substance is the same in the extracellular fluid and in the bathing medium. If the amount of marker substance in the gut wall is determined, this divided by the concentration of marker in the bathing fluid will give the volume of the extracellular fluid. It is also assumed that the concentration of absorbate is the same in the bathing fluid and in the extracellular fluid, although this assumption must be to some extent erroneous. However, if it is accepted as correct then the solute concentration in the bathing fluid multiplied by the volume of extracellular fluid will give the amount of solute in the extracellular fluid. If this is subtracted from the total amount of absorbate in the gut wall it will give the intracellular amount. The total volume of fluid in the cells is the wet weight of the gut minus the dry weight of the gut minus the volume of extracellular fluid. From these calculations the intracellular concentrations can be calculated. The calculation can be summarized as follows:

$$\text{Intracellular solute concentration} = \frac{\text{total amount of solute in gut} - \left(\dfrac{\text{amount of marker in gut}}{\text{concentration of marker in bathing fluid}} \times \text{solute concentration in bathing fluid} \right)}{\text{wet weight of gut} - \text{dry weight of gut} - \left(\dfrac{\text{amount of marker in gut}}{\text{concentration of marker in bathing fluid}} \right)}$$

As pointed out above, one assumption which must be incorrect is that absorbate concentration in the extracellular fluid is the same as that in the bathing medium, as this would be inconsistent with movement of substance into or out of the cells. Another assumption which is made is that the concentration in the enterocyte is the same as the average concentration in all the cells in the preparation. If the preparation is isolated enterocytes [129] the assumption is justified; however, if it contains other tissues, lymph cells, connective tissues, muscle etc., it is likely to be erroneous. There are still other considerations in using intracellular concentration as an index for absorptive activity. There is a possibility of compartmentation inside the cell; and the concentration in all the compartments is not necessarily the same. Furthermore, the effect of transfer activities on the concentrations in a compartment will depend on the topographical relation of the transfer mechanism to the compartment. If the substance is being pumped into the compartment, increased pumping activity will result in increased concentrations. If the substance is being pumped out of the compartment, increased pumping activity will result in decreased concentrations in the compartment. The problems have been discussed in detail by various authors [5, 145].

REFERENCES

1. T. H. WILSON, Intestinal Absorption. Saunders & Co., Philadelphia (1962).
2. D. H. SMYTH in Recent Advances in Physiology (R. Creese, ed.), Chapter 00. Intestinal Absorption. Churchill, London (1963).

3. D. H. SMYTH in Methods in Medical Research (J. H. Quastel, ed.), Vol. 9. Methods for study of intestinal absorption *in vivo*. Year Book. Medical Publishers Inc.
4. H. NEWEY and D. H. SMYTH, *Proc. Nutr. Soc.*, **26**, 5 (1967).
5. D. S. PARSONS in Handbook of Physiology (C. F. Code and W. Heidel, eds.), Section 6. Alimentary Canal, Vol. III. Intestinal Absorption, p. 1177. Williams and Wilkins Co. Baltimore (1968).
6. R. J. LEVIN, *Brit. Med. Bull.*, **23**, 209 (1967).
7. H. L. DUTHIE, *Brit. Med. Bull.*, **23**, 213 (1967).
8. J. M. DIAMOND and W. H. BOSSERT, *J. Gen. Physiol.*, **50**, 2061 (1967).
9. C. F. CODE, *Perspectives Biol. Med.*, **3**, 560 (1960).
10. D. NATHANS, D. F. TAPLEY and J. E. ROSS, *Biochem. Biophys. Acta*, **41**, 271 (0000).
11. P. F. CURRAN and A. K. SOLOMON. *J. Gen. Physiol.*, **41**, 143 (0000).
12. M. B. VISSCHER, R. H. VARCO, C. W. CARR, R. B. DEAN and D. ERICKSON, *Am. J. Physiol.*, **141**, 488 (1944).
13. B. J. PARSONS, D. H. SMYTH and C. B. TAYLOR, *J. Physiol.*, **144**, 387 (1958).
14. H. NEWEY, B. J. PARSONS and D. H. SMYTH, *J. Physiol.*, **148**, 83 (1959).
15. B. A. BARRY, J. MATTHEWS and D. H. SMYTH, *J. Physiol.*, **157**, 279 (1961).
16. A. M. UGOLEV, *Physiol. Rev.*, **45**, 555 (1965).
17. I. SPERBER, S. HYDEN and N. J. EKMAN, *Ann. Agr. Coll. Sweden*, **20**, 337 (1953).
18. B. BORGSTROM, A. DAHLQVIST, G. LUNDH and J. SJOVALL, *J. Clin. Invest.*, **36**, 1521 (1957).
19. I. HORVARTH and G. WIX, *Acta Physiol. Acad. Sc. Hungaricae*, **2**, 435 (1951).
20. P. F. CURRAN and A. K. SOLOMON, *J. Gen. Physiol.*, **41**, 143 (1957).
21. C. M. PAINE, H. J. NEWMAN and M. W. TAYLOR, *Am. J. Physiol.*, **197**, 9 (1959).
22. P. C. REYNELL and G. H. SPRAY, *J. Physiol.*, **131**, 000 (1956).
23. J. E. TREHERNE, *J. Exp. Biol.*, **35**, 611 (1958).
24. J. W. PEARSON, *J. Appl. Physiol.*, **13**, 313 (1958).
25. E. Y. BERGER, G. KANZAKI, M. A. HOMER and J. M. STEELE, *Am. J. Physiol.*, **196**, 74 (1959).
26. A. F. SCHURCH, L. E. LLOYD and E. W. CRAMPTON, *J. Nutr.*, **41**, 629 (1950).
27. W. T. COOKE, J. J. ELKES, A. C. FRAZER, J. PARKES, A. L. P. PEANEY, H. G. SAMMONS and G. THOMAS, *Quart. J. Med.*, **15**, 141 (1946).
28. A. C. FRAZER in Modern Trends in Gastroenterology (A. V. Jones, ed.), p. 477. London (1952).
29. W. T. COOKE in Modern Trends in Gastroenterology (A. V. Jones, ed.), p. 495. London (1952).

30. W. T. COOKE and J. M. FRENCH in Modern Trends in Gastroenterology (A. V. Jones, ed.), 2nd Series, p. 218. New York.
31. M. W. COMFORT, E. E. WALLAEGER and A. B. TAYLOR, *Gastroenterology*, **23**, 155 (1953).
32. J. M. BEAZELL, C. R. SCHMIDT and A. C. IVY, *J. Am. Med. Assoc.*, **116**, 2735 (1941).
33. S. J. BAKER and D. L. MOLLEN, *Brit. J. Haematol.*, **1**, 46 (1955).
34. E. C. OWEN, R. A. DARROCH and R. PROUDFOOT, *Brit. J. Nutr.*, **13**, 26 (1959).
35. G. SCHELTEMA, *Arch. Roentg. Ray & Allied Phenomena*, **14**, 144 (1908–9).
36. M. EiINHORN, *New York Med. J.*, **110**, 456 (1919).
37. M. EINHORN, *Amer. J. Med. Sci.*, **161**, 546 (1921).
38. T. G. MILLER and W. O. ABBOTT, *Amer. J. Med. Sci.*, **187**, 595 (1934).
39. T. G. MILLER, *Gastroenterology*, **3**, 141 (1944).
40. J. GROEN, *J. Clin. Invest.*, **16**, 245 (1937).
41. J. T. L. NICHOLSON and F. W. CHORNOCK, *J. Clin. Invest.*, **21**, 505 (1942).
42. D. H. BLANKENHORN, J. HIRSCH and E. H. AHRENS, *Proc. Soc. Exp. Biol. Med.*, **88**, 356 (1955).
43. H. P. SCHEDL and J. A. CLIFTON, *J. Clin. Invest.*, **40**, 1079 (1961).
44. C. D. HOLDSWORTH and A. M. DAWSON, *Clin. Sci.*, **27**, 371 (1964).
45. H. COOPER, R. LEVITAN, J. S. FORTRAN and F. J. INGELFINGER, *Gastroenterology*, **50**, 1 (1966).
46. R. SHIELDS and J. B. MILES, *Postgraduate Medical Journal*, **41**, 435 (1965).
47. F. H. SMIRK, *J. Physiol.*, **78**, 113 (1933).
48. J. F. SCHOLER and C. F. CODE, *Gastroenterology*, **27**, 565 (1954).
49. F. H. SMIRK, *J. Physiol.*, **78**, 127 (1933).
50. P. R. SCHLOERB, B. J. FRIIS-HANSEN, I. S. EDELMAN, A. K. SOLOMON and F. D. MOORE, *J. Clin. Invest.*, **29**, 1296 (1950).
51. I. M. LONDON and D. RITTENBERG, *J. Biol. Chem.*, **184**, 687 (1950).
52. H. A. SALVESON and A. KOLBERG, *Acta Med. Scandinav.*, **161**, 135 (1958).
53. T. L. ALTHAUSEN, K. UYEYAMA and R. G. SIMPSON, *Gastroenterology*, **12**, 795 (1949).
54. C. C. BOOTH and D. L. MOLLIN, *Brit. J. Haematol.*, **2**, 223 (1956).
55. K. N. JEEJEEBHOY, H. G. DESAI and R. V. VERGHESE, *Lancet*, **2**, 666 (1964).
56. P. CUATRECASAS, D. H. LOCKWOOD and J. R. CALDWELL, *Lancet*, **1**, 14 (1965).
57. S. H. GAGE and P. A. FISH, *Amer. J. Anat.*, **34**, 1 (1924).
58. S. SHERLOCK and V. WALSH, *Clin. Sci.*, **6**, 113 (1948).
59. W. B. BEAN, M. FRANKLIN, J. F. EMBICK and K. DAUM, *J. Clin. Invest.*, **30**, 263 (1951).

60. R. F. SCHILLING, *J. Lab. & Clin. Med.*, **42**, 860 (1953).
61. G. B. J. GLASS, L. J. BOYD, G. A. GILLIN and L. STEPHENSON, *Arch. Biochem.*, **51**, 251 (1954).
62. C. F. CORI, *J. Biol. Chem.*, **66**, 691 (1925).
63. H. HELLER and F. H. SMIRK, *J. Physiol.*, **76**, 1 (1932).
64. R. H. WILSON and H. B. LEWIS, *J. Biol. Chem.*, **84**, 511 (1929).
65. H. C. TREMBLE and B. W. CAREY, *J. Biol. Chem.*, **100**, 125 (1933).
66. E. M. MACKAY and W. G. CLARK, *Amer. J. Physiol.*, **135**, 187 (1942).
67. F. PAULS and D. R. DRURY, *Am. J. Physiol.*, **137**, 242 (1942).
68. L. THIRY, *Sitzber. Akad. Wiss. Wein. Math-Naturw. Kl.*, **50**, 77-96 (1864).
69. L. VELLA, *Moleshott's Untersuch. Naturl. Mensch. Thiere.*, **13**, 40 (1888).
70. J. MARKOVITCH, Experimental Surgery. Williams and Wilkins Co., Baltimore (1949).
71. D. GUMILEWSKI, *Pfleuger Archiv. Ges. Physiol.*, **39**, 556 (1886).
72. R. A. GREGORY, *J. Physiol.*, **111**, 119 (1950).
73. D. H. P. STREETON and E. M. VAUGHAN WILLIAMS, *J. Physiol.*, **112**, 1 (1951).
74. E. W. CLARKE and D. H. SMYTH, *J. Physiol.*, **112**, 45P (1951).
75. H. T. NEWMAN and M. W. TAYLOR, *Am. J. Vet. Res.*, **19**, 473 (1953).
76. D. H. SMYTH, *Gastroenterology*, **42**, 76 (1962).
77. A. U. ORTEN, *Fed. Proc.*, **20**, 2 (1961).
78. W. C. de GRAAF and W. NOLEN, *Ned. Tijdschr. Geneesk.*, **10**, 113 (1921).
79. P. R. SCHLOERB and B. L. LUKERT, *Arch. Surgery*, **86**, 356 (1963).
80. E. S. LONDON, *Harvey Lecture Series*, **23**, 208 (1928).
81. C. E. DENT and J. A. SCHILLING, *Biochem. J.*, **44**, 318 (1949).
82. W. C. SHOEMAKER, W. F. WALKER, T. B. van ITALLIE and F. D. MOORE, *Am. J. Physiol.*, **196**, 311 (1959).
83. W. C. SHOEMAKER, H. M. YANOF, L. N. TURK and T. H. WILSON, *Gastroenterology*, **44**, 654 (1963).
84. J. L. BOLLMAN, J. C. CAIN and J. H. GRINDLEY, *J. Lab. Clin. Med.*, **33**, 1349 (1948).
85. J. L. BOLLMAN, *J. Lab. Clin. Med.*, **33**, 1348 (1948).
86. E. W. REID, *Phil. Trans. Roy. Soc. Lond. Ser. B*, **102**, 211 (1900).
87. E. W. REID, *J. Physiol.*, **26**, 427 (1900–1).
88. F. VERZAR and E. J. MACDOUGALL, Absorption from the Intestine. Longmans Green, London (1936).
89. A. SOLS and F. PONZ, *Rev. Espan. Fisiol.*, **2**, 283 (1946).
90. A. SOLS and F. PONZ, *Rev. Espan. Fisiol.*, **3**, 207 (1947).
91. M. F. SHEFF and D. H. SMYTH, *J. Physiol.*, **128**, 67P (1955).
92. P. M. FULLERTON and D. S. PARSONS, *Quart. J. Exp. Physiol.*, **41**, 387 (1956).
93. F. A. JACOBS and M. LUPER, *J. Appl. Physiol.*, **11**, 136 (1957).

94. E. L. JERVIS, F. R. JOHNSON, M. F. SHEFF and D. H. SMYTH, *J. Physiol.*, **134**, 675 (1956).
95. R. M. ATKINSON, B. J. PARSONS and D. H. SMYTH, *J. Physiol.*, **135**, 581 (1957).
96. F. DICKENS and H. WEIL-MALHERBE, *Biochem. J.*, **35**, 7 (1941).
97. D. M. MATTHEWS and D. H. SMYTH, *J. Physiol.*, **126**, 96 (1954).
98. K. D. NEAME and G. WISEMAN, *J. Physiol.*, **140**, 148 (1958).
99. G. SALVIOLI, *Arch. Physiol. Leipzig Suppl.*, **95**, 000 (1880).
100. R. OHNELL, *J. Cell. Comp. Physiol.*, **13**, 155 (1939).
101. R. MOND, *Arch. Ges. Physiol.*, **206**, 172 (1924).
102. E. GELLHORN and D. NORTHUP, *Am. J. Physiol.*, **103**, 382 (1933).
103. D. S. PARSONS and J. S. PRITCHARD, *Nature*, **208**, 1097 (1965).
104. R. B. FISHER and D. S. PARSONS, *J. Physiol.*, **110**, 36 (1949).
105. T. H. WILSON and G. WISEMAN, *J. Physiol.*, **123**, 116 (1954).
106. E. W. REID, *Brit. Med. J.*, **1**, 323 (1892).
107. E. W. REID, *Brit. Med. J.*, **1**, 1133 (1892).
108. D. W. AUCHINACHIE, J. J. R. MACLEOD and H. E. MAGGEE, *J. Physiol.*, **69**, 612 (1930).
109. G. WISEMAN, *J. Physiol.*, **120**, 63 (1953).
110. D. H. SMYTH and C. B. TAYLOR, *J. Physiol.*, **136**, 632 (1957).
111. G. WISEMAN in Methods and Medical Research (J. H. Quastel, ed.), Vol. 9, 287 (1961).
112. B. A. BARRY, J. MATTHEWS and D. H. SMYTH, *J. Physiol.*, **149**, 78P (1959).
113. B. A. BARRY, J. MATTHEWS and D. H. SMYTH, *J. Physiol.*, **157**, 279 (1961).
114. H. A. KREBS and K. HENSELEIT, *Z. Physiol. Chem.*, **210**, 33 (1932).
115. R. K. CRANE and T. H. WILSON, *J. Appl. Physiol.*, **12**, 145 (1958).
116. T. W. CLARKSON and A. ROTHSTEIN, *Am. J. Physiol.*, **199**, 898 (1960).
117. T. H. WILSON and G. WISEMAN, *J. Physiol.*, **123**, 126 (1954).
118. W. T. AGAR, F. J. R. HIRD and G. S. SIDHU, *Biochem. Biophys. Acta*, **14**, 80 (1954).
119. R. K. CRANE and P. MANDELSTAM, *Biochem. Biophys. Acta*, **45**, 460 (1960).
120. A. BOASS and T. H. WILSON, *Fed. Proc.*, **21**, 469 (1962).
121. H. H. USSING and K. ZERAHN, *Acta Physiol. Scand.*, **23**, 110 (1951).
122. S. G. SCHULTZ and R. ZALUSKY, *J. Gen. Physiol.*, **47**, 567 (1964).
123. A. HAKIM, R. G. LESTER and N. LIFSON, *J. Appl. Physiol.*, **18**, 409 (1963).
124. M. BAILLIEN and E. SCHOFFENIELS, *Biochim. Biophys. Acta*, **53**, 521 (1961).
125. M. BAILLIEN and E. SCHOFFENIELS, *Biochim. Biophys. Acta*, **53**, 537 (1961).

126. M. BAILLIEN and E. SCHOFFENIELS, *Archs. Int. Physiol. Biochem.*, **70**, 140 (1962).
127. E. M. WRIGHT, *J. Physiol.*, **185**, 486 (1966).
128. D. MILLER and R. K. CRANE, *Biochim. Biophys. Acta*, **52**, 281 (1961).
129. G. A. KIMMICH, *Biochemistry*, **19**, 3669 (1970).
130. D. H. SMYTH in Intestinal Transport of Electrolytes, Amino Acids and Sugars (W. M. Armstrong and A. S. Nunn, eds.). Thomas, Springfield, Illinois.
131. M. SHINER, *Lancet*, **1**, 17 (1956).
132. M. SHINER, *Lancet*, **1**, 85 (1956).
133. R. J. C. BARRY, S. DIKSTEIN, J. MATTHEWS and D. H. SMYTH, *J. Physiol.*, **155**, 17P (1960).
134. R. J. C. BARRY, D. H. SMYTH and E. M. WRIGHT, *J. Physiol.*, **168**, 50P (1963).
135. R. R. LEVINE, W. F. McNARY, P. J. KORNGUT and R. LEBLANC, *European J. Pharmacol.*, **9**, 211 (1970).
136. H. NEWEY, A. J. RAMPONE and D. H. SMYTH, *J. Physiol.*, **211**, 539 (1970).
137. H. NEWEY and D. H. SMYTH, *J. Physiol.*, **164**, 527 (1964).
138. R. J. LEVIN, H. NEWEY and D. H. SMYTH, *J. Physiol.*, **177**, 58 (1965).
139. R. J. LEVIN and D. H. SMYTH, *J. Physiol.*, **169**, 755 (1963).
140. P. A. SANFORD and D. H. SMYTH, *J. Physiol.*, **215**, 769 (1971).
141. J. W. L. ROBINSON and J. P. FELBER, *Gastroenterologia*, **104**, 335 (1965).
142. J. R. BRONK and D. S. PARSONS, *J. Physiol.*, **184**, 942 (1966).
143. S. J. SAUNDERS and K. J. ISSELBACHER, *Biochem. Biophys. Acta*, **102**, 397 (1965).
144. S. G. SCHULTZ, R. E. FUISZ and P. F. CURRAN, *J. Gen. Physiol.*, **49**, 849 (1966).
145. H. NEWEY and D. H. SMYTH in The Biological Basis of Medicine. (E. E. Bittar and N. Bittar, eds.), Vol. 5, p. 347. Academic Press, London and New York (1969).

CHAPTER 7

Membrane (Contact) Digestion

A. M. UGOLEV

*I. P. Pavlov Institute of Physiology,
Academy of Sciences of the U.S.S.R.,
Leningrad, U.S.S.R.*

		Page
7.1	BASIC TYPES OF DIGESTION	286
7.2	ROLE AND PLACE OF MEMBRANE DIGESTION IN THE FUNCTIONING OF THE ALIMENTARY TRACT IN MAN AND HIGHER ANIMALS	289
7.3	POSSIBLE MECHANISMS OF FINAL STAGES OF HYDROLYSIS. APICAL INTRACELLULAR OR MEMBRANE DIGESTION?	294
	7.3.1 *The concept of Crane and Miller*	294
7.4	THE ENZYME APPARATUS ENSURING MEMBRANE DIGESTION	306
	7.4.1 *The cellular enzymes associated with the membrane surface*	307
	7.4.2 *Adsorbed enzymes*	311
	7.4.3 *The characteristics of membrane enzymatic apparatus regarded as a whole*	316
7.5	PHYSICO-CHEMICAL ASPECTS OF MEMBRANE DIGESTION	317
7.6	STERILITY OF MEMBRANE DIGESTION	321
7.7	FACTORS FAVOURING THE ENTRANCE OF SUBSTRATES INTO THE BRUSH BORDER	323
7.8	RELATIONSHIPS BETWEEN CAVITAL AND MEMBRANE DIGESTION	327
7.9	INTENSIFICATION OF ENZYMATIC PROCESSES IN THE CAVITY AS INFLUENCED BY MEMBRANE DIGESTION	328
7.10	MEMBRANE DIGESTION AND ABSORPTION	329
	7.10.1 *Membrane hydrolysis and transmembrane transport*	329
	7.10.2 *Absorption of monomers either preformed or released during membrane hydrolysis*	331

	7.10.3	*Possible coupling mechanism of membrane hydrolysis and transport*	334
	7.10.4	*Digestive-transport conveyors and digestive-transport ensembles*	337
	7.10.5	*Coupling of membrane hydrolysis and transport as regulated parameters*	339
7.11		ORGANIZATION LEVELS AND REGULATION OF SYSTEMS PERFORMING MEMBRANE DIGESTION	342
7.12		EVOLUTIONARY ASPECTS OF MEMBRANE DIGESTION	342
7.13		CONCLUSION	344
		REFERENCES	345

During the past decade it has been established that in addition to the two classical types of digestion, extra- and intracellular, a third exists—membrane digestion. It occurs when food substrates come into contact with cell surfaces on the external side of the membranes to which enzymes are fixed. About the same time the essential characteristic features of membrane digestion were described. The discovery of membrane digestion increased our understanding of a number of important aspects of the functioning of the alimentary tract, in particular: (1) the mechanism of the final stages of hydrolysis of food substances; (2) the surprisingly high rates of food digestion which cannot be reproduced *in vitro*; (3) the highly effective coupling of digestive and transport processes; (4) the nature of many disorders in the digestive-transport functions of the alimentary tract.

7.1 BASIC TYPES OF DIGESTION

It is proposed to recognize three types of digestion: cavital, intracellular and membrane. Each of the three types is represented schematically in Fig. 7.1. As seen in Fig. 7.1 cavital or extracellular digestion is caused by enzymes secreted by cells and acting outside cells, and is best described as extracellular distant digestion [reviews : 1-17]. At the molecular level cavital digestion can be characterized as follows. Enzymes are dissolved in the liquid phase, their motion being determined by the laws

MEMBRANE (CONTACT) DIGESTION

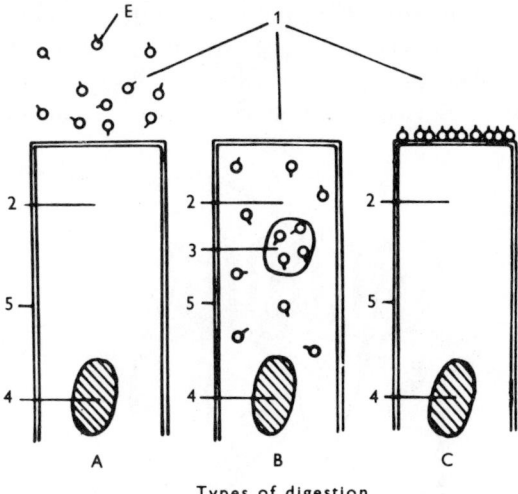

Fig. 7.1. Localization of hydrolysis of food during various types of digestion (according to Ugolev [8]).
A—extracellular distant; B—intracellular; C—membrane (contact). 1—extracellular fluid; 2—intracellular fluid; 3—intracellular vacuole; 4—nucleus; 5—cell membrane; E—enzymes.

of thermal agitation. Hence any orientation of their active centres is possible. The structural organization of the enzyme systems is either limited or impossible. Cavital digestion is responsible for the initial stages of hydrolysis and in particular for the hydrolysis of supermolecular aggregations and large molecules which are unable to penetrate the brush border zone.

Intracellular digestion [reviews 5-10, 18, 19, 20, 21, 22, 23, 24, 25, 26, 27, 11, 28, 29, 15, 16] causes cleavage of substrates penetrating the cell and thus is restricted by the permeability of membranes and by such specific functions as phagocytosis and pinocytosis [30, 31, 32, 33, 34, 35, 36, 37].

Membrane digestion [38, 39, 40, 5, 6, 7, 8, 9, 10, 41] is carried out by enzymes localized on the surface of the cell membrane and may be characterized spatially as intermediate in relation to extra- and intracellular digestion. There are physicochemical analogies between membrane digestion and heterogeneous catalysis on non-homogeneous surfaces. Since membrane digestion is accomplished (at least in the majority of higher animals) on the surface of the brush border this mechanism is not sufficiently effective for the hydrolysis of large molecules, but on the other hand is the main mechanism of the intermediate and final stages of hydrolysis. In addition, membrane digestion is likely to be the basic mechanism ensuring coupling of digestive and transport processes. Since

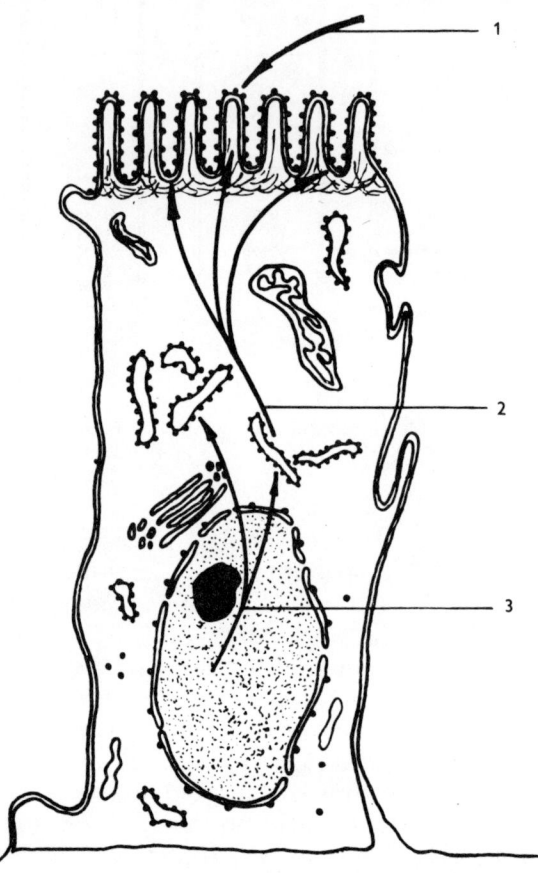

Fig. 7.2. The ways of formation of an enzymatic apparatus accomplishing membrane digestion.
Adsorption of enzymes from chyme (1); synthesis (3) and translocation (2) of enteric enzymes.

hydrolysis is completed and transport starts on the same surface, both these processes are brought close together in space and time. This is due to a special organization of the digestive and transport functions of the external cell membrane.

In the past few years the concept of the membrane structure

as a three-layered lipoprotein complex changed significantly [42, 43, 44, 45, 46, 47, 48, 49, 50, 51, 52, 53, 54, 55, 56, 58, 59, 60, 61, 62, 63, 64, 65, 66, 67, 68, 69, 70, 71, 72, 73, 74, 75, 76, 77, 78, 79, 80, 81, 82, 83, 84]. At present the concept of the so-called 'thick' membrane predominates. It has been suggested, however, that membrane digestion occurs not only on the external surface of the lipoprotein membrane but also in the glycocalyx [65, 66, 68, 73; reviews 77, 78, 9, 10, 85, 86, 87, 88, 89].

Enzymes responsible for membrane digestion may have two origins: (1) enzymes produced by enterocytes and structurally associated with the external surface of the brush border membrane and (2) enzymes adsorbed from chyme (Fig. 7.2). The former group are referred to subsequently as cellular enzymes.

7.2 ROLE AND PLACE OF MEMBRANE DIGESTION IN THE FUNCTIONING OF THE ALIMENTARY TRACT IN MAN AND HIGHER ANIMALS

This section presents a general viewpoint of the alimentary tract functioning in higher animals and gives a proper evaluation of the part played by membrane digestion. A more detailed characteristic of the latter will be given below.

The food substances in higher animals and in many lower animals undergo a three-stage assimilation, (1) cavital digestion, (2) membrane digestion and (3) absorption (Figs. 7.3, 7.4).

The first stage occurs in the digestive cavities (cavital digestion); with break-down of the complex tissue and cellular food structures, desaggregation of physico-chemical and chemical complexes in the food substances, and rupture of primary bonds in biopolymer molecules. The products of intermediate hydrolysis penetrate into the brush border zone where the next stage of hydrolysis (membrane digestion) takes place. The final products of digestion (mainly monomers) are transferred from hydrolytic systems to transport systems of the membrane by means of mechanisms discussed below.

The distribution of endohydrolases (concerned with the initial stages of hydrolysis) and exohydrolases (concerned with

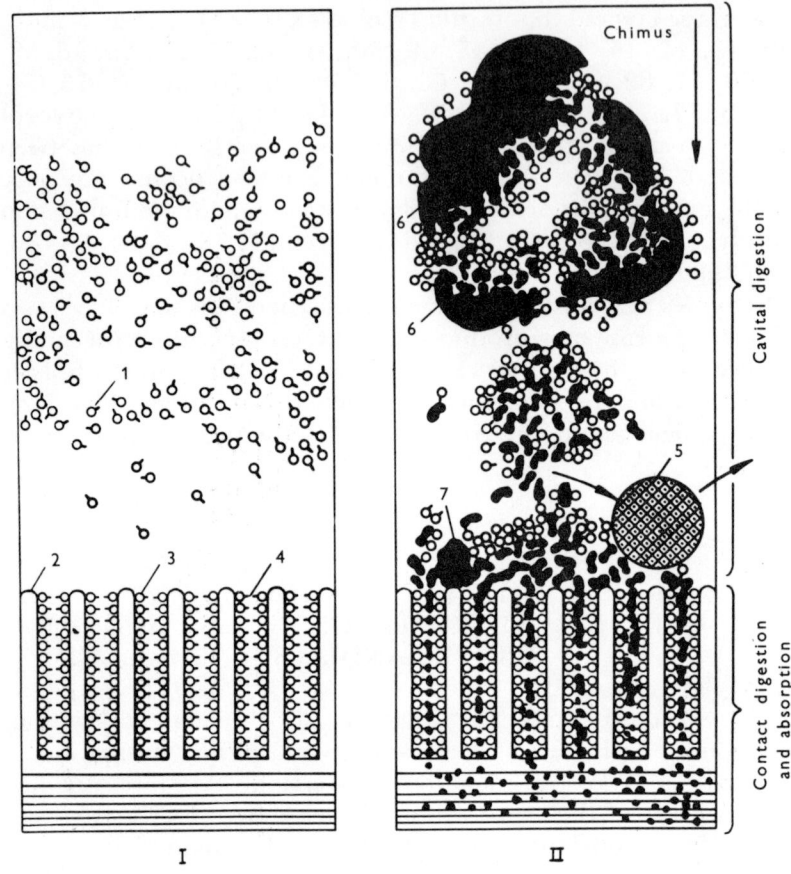

Fig. 7.3. Detailed scheme of interrelationships between cavital and membrane digestion in absence of food substances (I) and in their presence (II) (according to Ugolev [5]).

1—enzymes in the intestinal cavity; 2—microvilli; 3—enzymes on the microvilli surface; 4—brush border pores; 5—microorganisms; 6, 7—food at the different stages of hydrolysis.

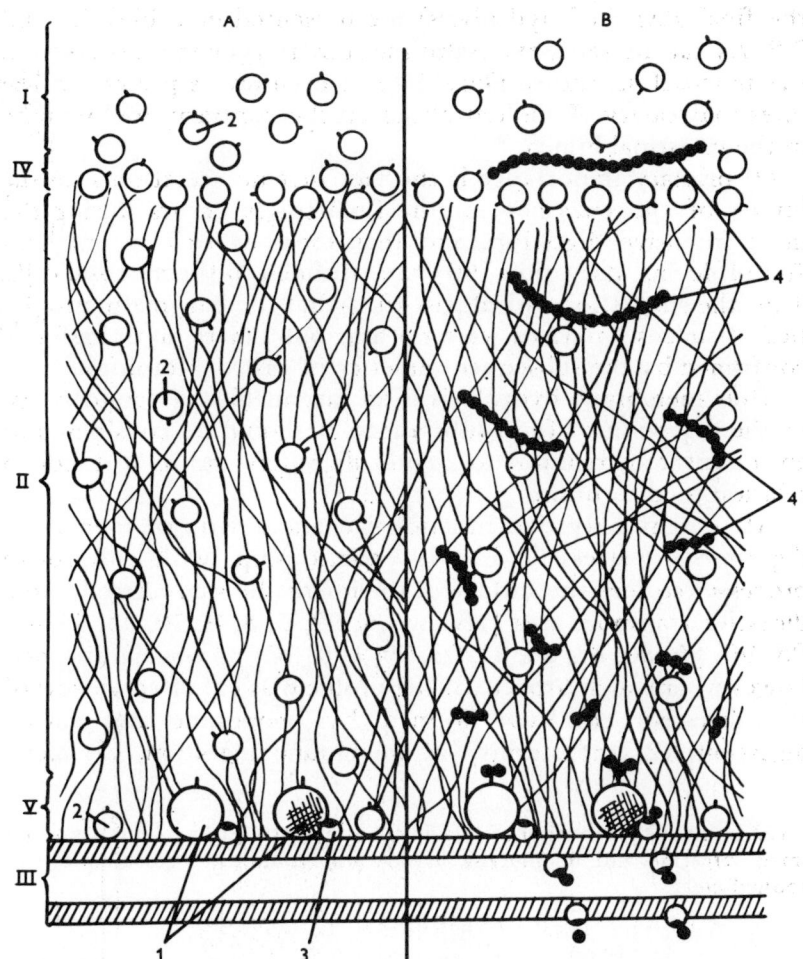

Fig. 7.4. Enteric and adsorbed enzymes during membrane digestion (schematic representation of a luminal surface fragment of the microvillus).

A—distribution of enzymes; B—relationship of enzymes, carriers and substrates. I—cavity; II—glycocalix; III—three-layered membrane; IV—luminal surface of glycocalix; V—luminal surface of the three-layered membrane. 1—enteric enzymes; 2—adsorbed enzymes; 3—carriers; 4—substrates.

the final stages of hydrolysis) are presented in Tables 7.1 and 7.2. As can be seen, exohydrolase activity is mainly observed in the mucosal strutures. Only 10% or even less is present in the intestinal cavity. Endohydrolases on the contrary predominate in the intestinal lumen.

Of primary importance is the fact that no secretion into the lumen of the enzymes for the final stages occurs during the most intensive digestion, and that these as well as enzymes found during starvation [90] are confined to the mucosal cells. Thus the suggestion of Dalqvist [91] that the enzymatic properties of succus entericus were nothing but a mere myth has been confirmed by the subsequent practice of gastroenterology.

Thus membrane hydrolysis is an intermediate (second) stage in the digestion of food substances. It should be considered as an acceptor in relation to cavital digestion and as a donor in relation to absorption.

At present only one functional characteristic of membrane digestion is stressed, i.e. its relationship with adsorptive processes so that the surface of the microvilli is converted into a digestive transport area [reviews, 6, 92, 7, 8, 9, 10, 93, 94, 41, 95, 96, 85, 86, 87, 88, 97, 98, 99, 100]. Of no less significance, however, are such aspects of the problem as (1) effectiveness of the digestive conveyor formed by membrane and cavital digestion, (2) the sterility of membrane digestion, (3) mem-

TABLE 7.1. Distribution of certain enzymes between the contents of the small intestine and its mucosa in rats two hours after consumption of cooked meat.

Enzymes investigated	Object of study	
	Contents	Mucosa
Amylase	70	30
Invertase	11	89
Glycyl-l-leucine-dipeptidase	9	91
Tributyrinase	52	48
Monoglyceridelipase	21	79

Note: Enzyme activity is given in percent in relation to total activity of mucosa and small intestine contents taken as 100%.

TABLE 7.2. Distribution of certain digestive enzymes between the contents of the small intestine and its mucosa in dogs three hours after consumption of cooked meat.*

No. of animal	Object of study	Amylase	Invertase	Glycyl-leucine-dipeptidase	Tributyri-nase	Monoglyceride lipase
1	Contents	1071.1 (52.5)	920.5 (6.4)	153.9 (10.8)	645.86 (73.2)	1.866 (25.4)
	Mucosa	967.9 (47.5)	13485.7 (93.6)	1272.1 (89.2)	234.7 (26.8)	5.48 (74.6)
		2039.9 (100)	14406.2 (100)	1426.0 (100)	880.56 (100)	7.346 (100)
2	Contents	—	594.3 (5.3)	4.8 (0.3)	467.3 (48)	0.874 (10.3)
	Mucosa	—	10693.5 (94.7)	1805.2 (99.7)	508.3 (52)	7.567 (89.7)
		—	11287.8 (100)	1810.0 (100)	975.6 (100)	8.441 (100)
3	Contents	8994.0 (69.6)	786.5 (4.1)	15.8 (1.04)	—	—
	Mucosa	3924.5 (30.4)	18158.6 (95.9)	1513.5 (98.06)	—	—
		12918.5 (100)	18945.1 (100)	1529.3 (100)	—	—

* The activities were determined in absolute units—in milligrams per minute and in percent (in parentheses). Absolute activity was assayed: for amylase by amount of hydrolysed starch; for invertase by formation of glucose; for glycyl-leucine-dipeptidase by formation of glycine; and for tributyrinase and monoglyceride lipase by amount of cleaved tributyrine (Iezuitova et al., 1967 [90]).

brane digestion as an important separatory and barrier mechanism. All this will be discussed below.

Moreover, membrane digestion is the function of cell membranes, and should not be necessarily associated with the presence of the brush border. There is strong evidence in support of the view that in the process of evolution membrane digestion appeared long before the brush border and that it already exists even on 'smooth' membrane surfaces of yeasts and bacteria [101, 102, 103, 104, 105, 106, 107; reviews 6, 9, 10].

7.3 POSSIBLE MECHANISMS OF FINAL STAGES OF HYDROLYSIS. APICAL INTRACELLULAR OR MEMBRANE DIGESTION?

Bearing in mind the characteristics of the hydrolytic and transport processes, it is important to consider in detail the possible location of the final stages of hydrolysis. It may occur on the external surface of membranes and thus the formation of the end products will precede penetration of these substances through the membrane. Digestion may be localized inside the membrane in the microvilli structures, and finally, it may occur in different parts or structures of the enterocyte cytoplasm (Fig. 7.5). It was shown by numerous investigations that the final stages of hydrolysis and the initial stages of transport are closely interrelated. Due to this fact our ideas on localization of the initial stages of active transport should be dependent on the localization of the final stages of digestion [23, 24, 25; reviews 6, 92, 7, 8, 9, 10, 93, 94, 108, 109, 110, 41, 95, 96, 85, 86, 87, 88, 111, 97, 98, 99]. In this connection, it is extremely important to differentiate between the two concepts of the digestive transport functions of the brush border that emerged almost simultaneously.

7.3.1 *The concept of Crane and Miller* [23, 24, 25]
According to this concept the final stages of hydrolysis and initial stages of transport are accomplished by the brush border. The special experiments of these authors [23, 24, 25], however, showed hydrolysis to take place after the oligomers had passed through a diffusion barrier, which in this case was the cell

membrane. Let us see how the authors themselves regard their own results.

'... The data presented that *in vitro* uptake by intestinal tissue of a monosaccharide formed by hydrolysis of a disaccharide or of glucose-l-phosphate does not depend upon the liberation of the monosaccharide in the incubation medium at

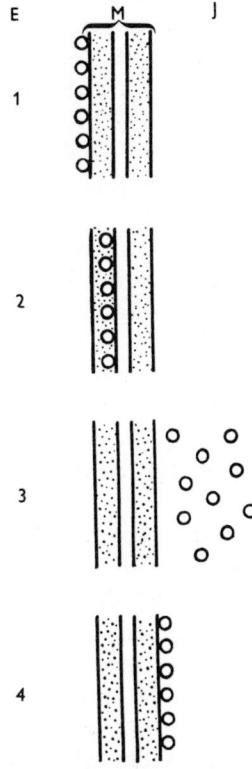

Fig. 7.5. Possible localization of enteric enzymes in relation to the microvilli membranes according to Crane (Crane [95]) (1-3) with addition (4).

the free surface of the mucosa; hydrolysis occurs at a site from which diffusion into the tissue occurs more rapidly than diffusion into the medium. The only conclusion that it seems possible to draw is that the hydrolases are intracellular enzymes [23].

'... The fact that here the tissue levels do exceed those in the medium, at least initially, is strong evidence against the theory that glucose is formed by hydrolysis outside the tissue.... In this situation if digestive hydrolysis with the formation of glucose takes place in the medium or at the cell surface, glucose should be oxidized as it is formed, and the amount taken up by the tissue should be decreased. This did not happen. . . .

'In all the experiments essentially similar results were obtained using disaccharides, sucrose and maltose, and the sugar phosphate ester, glucose-1-phosphate. Tissue uptake of the monosaccharides resulting from digestive hydrolysis does not depend on the liberation of these sugars in the incubation medium or at the free surface of the tissue. Hydrolysis takes place at a site from which the diffusion of the sugar into the tissue occurs more readily than its diffusion into the medium. The only conclusion that seems possible is that these hydrolases are intracellular enzymes.' [25]

The presence of different enzymes in the brush border was demonstrated by the numerous investigators by means of histochemical methods [112, 113, 114, 115, 116, 117, 118, 30, 119, 120, 121, 122, 123, 124, 125, 126, 127, 128, 129, 130, 131, 132, 133]. There is concentration of alkaline phosphatase, peptidases, disaccharidases in the brush border [24, 25, 134, 135, 136, 137; reviews 77, 78, 138, 139, 140]. As will be shown, these data are entirely consistent with the concept of membrane digestion.

In accordance with our view-point (membrane digestion) hydrolysis is performed by enzymes associated with the cell structures but localized on the external surface of the membrane. This concept was first announced at the congress of physiologists in 1959, and later published [38, 39, 40, 5]. Thus, oligomers were assumed to be hydrolized by enzymes (some adsorbed and some produced by the enterocyte) associated with the external membrane surface, which only monomers penetrate.

The finding that the external surface of the brush border is a morphological surface responsible for a particular and previously unknown type of digestion suggested a study of possible adsorption of pancreatic enzymes on the intestine surface of the mucosa [38, 39, 40, 141; reviews 5, 6]. This fact made some investigators identify membrane digestion with the function of enzymes adsorbed from chyme. In practice, this was the first case that enzymes associated with the external structures of the cell membrane accomplished digestion on principles different from those for cavital and intracellular hydrolyses. Due to their large size, amylase molecules (molecular weight about 45,000) [142, 143] could not penetrate the cell, and hence a possibility of the enzyme being rapidly and completely adsorbed was envisaged. Adsorbed amylase was shown to hydrolyse intensively soluble (not colloid) starch [review 6] and later it was shown that the intestinal enzymes proper were involved in membrane digestion but not in intracellular digestion [reviews 6, 92, 7, 8, 9, 10, 93, 94, 144, 145, 146, 147, 148, 149, 150, 151, 152, 153, 154, 155, 156, 157, 158, 159].

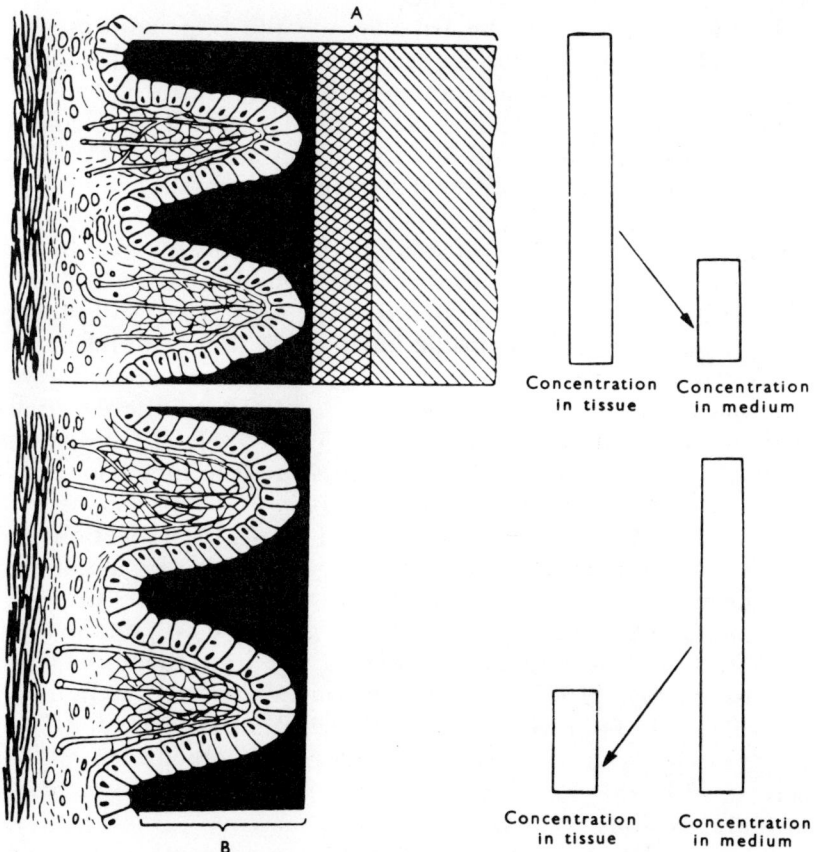

Fig. 7.6. The scheme of experiments of Miller and Crane [23, 24, 25] (A—average values) and Iezuitova, De Laey and Ugolev (B—membrane tests). The distribution of hexoses, released during sucrose hydrolysis, between extra- and intracellular fluid (according to Iezuitova, De Laey and Ugolev [151, 221]).

The principal difference between these two concepts is quite obvious. Digestion follows transport through the membrane in the first case and precedes it in the second. According to the first concept, there are a variety of intracellular enzymes (mainly in the apical region), while the other view suggests that hydrolysis is extracellular and at the cell surface (Fig. 7.5). In the first case only small molecules are thought to be hydrolysed, the size of these molecules being limited by the membrane permeability. In the second case, limitation of hydrolysis by molecular size depends on the microvilli pores and by the size of glycocalyx network.

In the first half of the past decade the first concept was universally accepted, but later ideas were developed which made it possible to differentiate between the apical intracellular and membrane digestion [reviews 7, 9, 10, 93, 94]. It was demonstrated that peptidases and disaccharidases are localized on the external and not on the internal surface of membrane (Tables 7.3 and 7.4). Further strong evidence in support of this hypothesis was measurement of the effective pore radius of the membrane for water soluble substances which was found to be 4-6 Å [160, 161, 162]. These data suggested that the majority of substances must be hydrolysed not inside but outside the membrane.

Special experiments showed at the same time that different disaccharidases and dipeptidases are localized on the external side of the membrane (see Tables 7.3 and 7.4). The investigations of rate limiting processes in multistage reactions also speak in favour of this idea [reviews 6, 7, 9, 10].

It is well known that the level of multistage reactions is usually determined by the slowest stage: the limiting factor for intracellular hydrolysis will be the passage of oligomers through the membrane; while for membrane digestion it will be the rate of diffusion of substances to be hydrolysed from the bulk phase to the surface [163]. It was demonstrated that membrane destruction [144; reviews 92, 7, 9, 10, 93, 94, 159] does not affect hydrolysis of oligosaccharides and dipeptides which agrees with the membranous (or superficial) but not with intramicrovillous localization of these enzymes. This idea was supported also by the fact that the movement of fluid in relation to the intestinal surface results in rapid increase in hydrolysis of disaccharides and dipeptides [141, 144, 146, 152, 153, 150, 154, 155; reviews 6, 7, 9, 10, 93].

TABLE 7.3. Experimental criteria employed in the determination of enteric enzymes localization in the intestinal cells.

Ord. nos.	Criterion	References
1	Histochemical data on localization of enzymes in enterocytes	113, 115, 425, 130
2	Electron microscopy and histochemistry	426, 32, 33, 126, 53
3	Blocking of the cell membrane surface in combination with electron microscopy	427, 428
4	Combination of morphological and immunological approaches	429
5	Combination of electron microscope technique with immunological approaches	185
6	The negative contrasting of microvillous membrane surfaces	165, 167, 168
7	Destruction of cell membranes	144, 92, 428, 430, 159, 431, 378
8	Influence of convective flows on the reaction rate	152, 153, 150, 315, 316, 154, 155, 92, 146, 398
9	Ratio of concentrations of reaction products in the incubation medium and intracellular liquid	104, 23, 24, 25
10	The precise concentration criterion	151, 145, 159
11	,, ,, ,, ,,	147
12	Effect of glucose oxidase on glucose accumulation in the intestinal tissue during hydrolysis of olygosaccharides	23, 24, 25
13	The precise glucose oxidase criterion	164
14	Distribution of amino acids formed during hydrolysis of dipeptides with different molecular weight in intra- and extracellular liquid	156, 158
15	The effective pore radius	160, 161, 162
16	Fractioning of subcellular mucosal structures and, in particular, preparative isolation of the brush border	432, 171, 172, 179, 66
17	Enzyme sedimentation together with the microsome fraction	433, 434
18	The preparative separation of microvillous stroma from membranes	137, 435, 138

TABLE 7.4. The localization of various enteric enzymes in the intestinal cell structures and their possible relation to the intracellular and membrane digestion.

Enzyme	Substrate	Criterion used (see Table 7.3)	Cytoplasm	Brush border	Intracellular digestion	Membrane digestion	Reference
1	2	3	4	5	6	7	8
invertase	sucrose	7				+	144
,,	,,	9			+		23
,,	,,	16		+			432
,,	,,	8				+	152
,,	,,	8				+	150
,,	,,	4				+	429
,,	,,	6				+	165
,,	,,	7				+	92
,,	,,	8				+	315
,,	,,	8				+	316
,,	,,	8				+	146
,,	,,	8				+	92
,,	,,	10				+	145
,,	,,	10				+	151
,,	,,	3				+	427
,,	,,	3				+	428
,,	,,	7				+	428
,,	,,	8				+	153

1	2	3	4	5	6	7	8
invertase	sucrose	6				+	167
,,	,,	11				+	147
,,	,,	7				+	430
,,	,,	6				+	168
,,	,,	5				+	185
,,	,,	7				+	431
,,	,,	7					378
maltase	maltose	17	+				433
,,	,,	9			+		23
,,	,,	12			+		24
,,	,,	9			+		25
,,	,,	12					25
,,	,,	13					135
,,	,,	1		+			130
,,	,,	6				+	167
,,	,,	11				+	147
,,	,,	7				+	430
,,	,,	6				+	168
,,	,,	7				+	431
,,	,,	7				+	147
lactase	lactose	4				+	429
,,	,,	13	+	+		+	135
,,	,,	13				+	135
,,	,,	4				+	436
oligo-1, 6-glucosidase		1					130
trehalase	trehalose	17		+			433
		1		+			130

1	2	3	4	5	6	7	8
peptidases	glycyl-l-leucine	7				+	144
,,	,,	8				+	154
,,	,,	7				+	6
,,	,,	7				+	92
,,	,,	8				+	146
,,	,,	8				+	92
,,	,,	8				+	155
,,	,,	14				+	156
,,	,,	14				+	158
,,	,,	16				+	209
,,	,,	11				+	147
,,	,,	7				+	430
,,	,,	8				+	398
,,	,,	7				+	431
,,	,,	7				+	378
,,	glycyl-dl-leucin	8				+	315
,,	,,	8				+	316
,,	alanyl-glycin	18				+	435
,,	,,	18				+	138
,,	glycyl-dl-alanin	14				+	156
,,	,,	14				+	158
,,	glycyl-l-valin	16				+	209
,,	glycyl-glycin	11		+			147
,,	,,	18		+			435
,,	,,	18		+			138
,,	glycyl-glycyl-glycin	16		+			171
,,	,,	16		+			172

1	2	3	4	5	6	7	8
"	glycyl-leucyl-tyrosine	18				+	435
"	"	18				+	138
"	glycyl-tryptophane	14				+	156
"	"	14				+	158
"	leucil-glycine	18				+	138
"	"	18				+	176
"	leucil-glycyl-glycine	18				+	176
"	"	18				+	138
"	leucil-glycyl-leucine	18				+	435
"	"	18				+	138
"	leucil-leucine	18				+	435
"	"	18				+	138
"	l-alanyl-l-glutamine acid	16				+	209
"	dl-alanyl-dl-serine	8				+	211
"	"	8				+	92
"	"	8				+	146
"	leucynamide	18				+	435
"	"	18				+	138
"	l-leucyl-4-metoxi-2-naphtilamide	1	+				425
"	l-leucil-dl-naphtilamide	16		+			171
"	"	16		+			432
"	"	16		+			172
alkaline phosphatase	β-glycerophosphate	1		+			112
"	"	1		+			113
"	"	1	+	+			437
"	"	1		+			114

1	2	3	4	5	6	7	8
,,	,,	1	+	+			115
,,	,,	1	+	+			116
,,	,,	1	+				438
,,	,,	7				+	144
,,	,,	1		+			439
,,	,,	16		+			129
,,	,,	16		+			128
,,	glucose-1-phosphate	1	+	+			332
,,	,,	1	+	+			114
,,	,,	1	+	+		+	115
,,	,,	2		+			440
,,	,,	9			+		119
,,	,,	9			+		23
,,	glucose-6-phosphate	1	+	+			25
,,	,,	1	+	+			114
,,	,,	16	+				115
,,	ATP	2		+			441
,,	fructose-di-phosphate	1	+	+			122
,,	,,	1	+	+			114
,,	p-nitrophenylphosphate	16		+			115
,,	,,	16		+			432
,,	1-L-naphtilphosphate	2					133
,,	,,	1				+	53
,,	naphtol AS-TR-phosphate	1		+			439
,,	,,	1		+			439
,,	glycerophosphate lecytin	1		+			442
,,	naphtol AS-BI-phosphate	1		+			115
cholesterol-esterhydrolase	cholesterine esters	16		+			439
monoglyceridelipase	tri-, di-, monoglycerides	17	+	—			179
,,	tributyrin	10				+	434
							159

The preliminary data of Miller and Crane [23, 24, 25] obtained from experiments with glucose oxidase were due to the fact that imperfectly purified enzymatic preparations could not penetrate the glycocalyx space. When crystalline glucose oxidase was used, transfer of the formed glucose inside the cell was disturbed, which thus proves the localization of the final stages of hydrolysis of oligo- and disaccharides to be on the external side of the membrane [164]. Computer models showed some inevitable errors in investigation of gradients in the extra-/intracellular fluid and showed that the real gradients corresponded with those previously described [reviews 7, 8, 9, 10, 149, 151, 145, 147] (Figs. 7.6, 7.7).

After a successful attempt to separate microvillous membrane from stroma of 'isolated brush border' [136, 137] Crane departed from his primary concept and proposed another one illustrated in Fig. 7.5 [95]. Other work [165, 166, 167, 168,

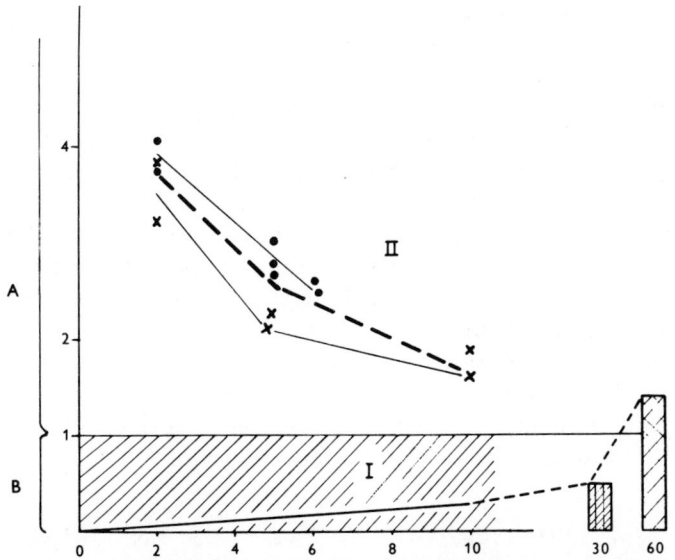

Fig. 7.7. Relationship of concentrations of hexoses formed during hydrolysis of sucrose, in the incubation medium and in the intracellular fluid (according to Ugolev [9]).

Abscissa—time of incubation (in minutes); ordinate—ratio of the concentration of hexoses in the incubation medium to those in the intracellular fluid. I—during the determination of hexoses in the large volume of sucrose solution (according to Miller and Crane [23]); II—during determination of hexoses in the immediate vicinity of the surface of the cells. A—zone typical of intracellular digestion; B—zone typical of extracellular digestion.

169] made it evident that at least disaccharidase activities are localised in the granular structures of 50-60 Å in diameter bound to the external surface of membrane. At present the majority of enzymatic activities are agreed to be localized on the external surface of membrane and this supports the view that membrane digestion is the basic site of the final stages of hydrolysis (Table 7.4). Such localization of the final stages of hydrolysis is likely to provide some functional advantages mentioned already above, which are the subject of further more detailed discussion in a later section.

However, the following facts must be kept in mind. Glycyl-glycine-dipeptidase was shown to be localized intracellularly, from which it follows that it accomplishes the intracellular hydrolysis of the appropriate dipeptides [170, 171, 172, 173, 147; review 15]. Furthermore, many amino acids, sugars and fatty acids may be formed due to the effect of certain enzymes acting on the intestinal cavity (at least in small amounts). Thus there may be some special cases which do not entirely fit in with the general concept of membrane digestion. On the other hand, the very commonly used terms 'enzymes of the brush border' or 'digestion in the brush border' are too general and include the two quite different mechanisms of membrane and intracellular digestion. Along with the experimental data a number of purely theoretical considerations support the idea of intestinal enzymes responsible for digestion (such as oligo- and disaccharide hydrolase, oligo- and dipeptide hydrolase, monobutyrin hydrolase monophospho-ester-hydrolase) being localized on the external surface of the membrane. This idea agrees with the well-established localization of transport systems performing active transport of monomers across the cell membrane.

7.4 THE ENZYME APPARATUS ENSURING MEMBRANE DIGESTION

If we consider membrane digestion in general the origin of the enzymes responsible has no particular importance. The minimal requirement is the presence of some enzymes connected with the structures of the outer membrane surface which result in formation of the end products directly on the resorptive surface.

In practice, however, membrane digestion needs both the adsorbed and cellular enzymes as will be shown below.

7.4.1 The cellular enzymes associated with the membrane surface

A large number of enzymes produced by enterocytes accomplish their functions on the external surface of microvilli. Among these are di- and tri-saccharides (α- and β-glucosidases) and γ-amylase. A very useful characterization of these enzymes exists in a number of reviews [174, 97, 15, 16, 175]. Here were also included tetra-, tri- and dipeptidases, amino-peptidases, and probably, carboxypeptidase [176]. Hydrolysis of phosphoesters is accomplished by the different isoenzymes of alkaline phosphatase [177, 178]. Participation in membrane digestion of intestinal lipases was also demonstrated [157, 159]. Of theoretical and clinical interest are the early data on the presence of enzymes on the brush border surface which hydrolyse cholesterol esters [179].

Thus if enzymes acting in the intestinal lumen are mostly endohydrolases and polymerhydrolases, the cellular surface enzymes are mainly represented by exohydrolases splitting predominantly oligomers and dimers with formation of the products to be transported.

The electron-microscope investigations by Oda *et al.* and Johnson [165, 166, 167, 168, 169] made the spatial relationships between membrane and enzymes, in particular disaccharidases, more clear. Disaccharidase activities are located on the external surface of membrane as 'knobs' 40-60 Å in diameter. The latter can be separated from the membrane by papain treatment. (The molecular weight of disaccharidases is about 200 000 [180, 181, 183, 184]. It was convincingly shown that disaccharidase activities are associated with the solubilized 'knobs' in various preparative procedures, and not with the fragments of smooth membranes.

It is not quite clear at present whether all the other cellular enzymes (peptide hydrolases, alkaline phosphatases, esterases etc.) are also present as 'knobs' in direct contact with the external surface of membrane. For instance, Ito [66] assumes that alkaline phosphatase may be bound to glycocalyx throughout its thickness.

The close connection between digestive hydrolases and entrance into the transport systems will be shown below, and makes one expect the main localization of the cellular enzymes to be on the membrane surface. This location, however, requires special examination for every individual case, as Gitzelman *et al.* [185], combining electron-microscope and immunochemical approaches, showed that the invertase molecule may be separated from the surface of the membrane proper by at least 120 Å.

How are the cellular enzymes involved in membrane digestion connected with the external surface of the three-layered membrane? The nature of the connection in the majority of cases is unknown. Nevertheless, a peculiar interest is their permanence. In contrast to the adsorbed enzymes the cellular enzymes are not subject to spontaneous desorption (except for some dipeptidases). This was the reason for speaking of the structural bonds between enzymes and membranes (the term is not used in a chemical sense).

The separation of enzymes from membrane has been most thoroughly investigated with disaccharidases and can be achieved in several ways. In the majority of cases disaccharidases can be solubilized by papain treatment in potassium-phosphate buffer [186, 187, 188, 189, 190, 191, 192, 193, 182, 194, 195, 196, 197, 198]. Treatment by trypsin [199, 200, 201, 202, 203, 186, 187, 188, 204, 198] and ficin, autolysis and use of such detergents as triton X-100 [182, 194, 205] and to a lesser extent sodium desoxicholate [199, 206, 203] is also very efficient. Ultrasonics also yields partial solubilization [203, 207].

These data allow us to assume that the enzyme membrane bonds do not involve Van der Waals forces or ionic bonds. Bearing in mind the effectiveness of the proteolytic enzymes as solubilizing agents one can suggest the presence of peptide bridges. On the other hand, the effectiveness of detergents, particularly of the non-polar detergent triton X-100, suggested an important role of hydrophobic interactions. The latter are of great importance for the attachment of membrane components since the membrane can be divided into fine fragments with the help of detergents only [208].

The bonds in different animals are unlikely to be identical. This is shown by the fact that trypsin effectively solubilizes disaccharidases in the pig and produces a slight effect on disaccharidases in man and rat [199, 186]. Of no less interest is the effect of papain and triton X-100 which equally and completely solubilize the majority of α- and β-glucosidases in human small intestine [182, 194].

The complete solubilization of disaccharidases both by papain and triton X-100 was achieved in rats [205]. Thus here the enzyme membrane bridges must contain two different bonds, one of which is sensitive to the action of proteases and the other to detergents. Their order and functional significance are still unknown. The bonds of various enzyme groups and individual enzymes with the membrane are not likely to be identical or at least not entirely identical. The data of Crane and Eichholz support this suggestion. It was found that papain solubilized first disaccharidases, then alkaline phosphatase and finally aminopeptidases which were the most resistant [reviews 85, 139].

Unlike aminopeptidases, which are bound to the membrane very tightly, other peptidases are easily solubilized. This was first demonstrated by Josefsson and Sjostrom [209] and later confirmed in our laboratory [205, 210]. In addition, an extremely interesting and mysterious phenomenon of sensitivity of dipeptidase-membrane bridges to oxygen was revealed. It was shown that solubilization and transfer of peptidases to the small intestinal lumen does not take place while the normal blood circulation is preserved. Interference with the blood supply leads to enhanced solubilization of peptidase activities.

Thus the nature of the bonds between the intestinal enzymes and membrane is largely unknown, not only because of the lack of proper information about their nature but also because of their great diversity. While some bonds are extremely resistant and probably not spontaneously destroyed in physiological conditions, others on the contrary are very liable.

The synthesis of cellular enzymes is accomplished by the general mechanisms controlling this process, DNA, RNA and ribosome apparatus. Numerous data on the influence of different inhibitors of proteosynthesis, and on the formation of digestive enzymes by enterocytes support this fact. Moog, however [177, 178], described an interesting possibility of the formation of some isoenzymes of alkaline phosphatase when proteosynthesis is blocked at the transcription and translation stages.

The synthesis of enzymes is a necessary condition, although not the only one, for the cellular enzymes in membrane hydrolysis. Experiments with X-ray treatment [211, 212, 9, 10, 213] showed that after irradiation invertase synthesis stops though synthesis of glycyl-L-leucin depeptidase proceeds normally. The latter, however, remains inside the cells and is not included into the microvilli surface. This was experimentally revealed after

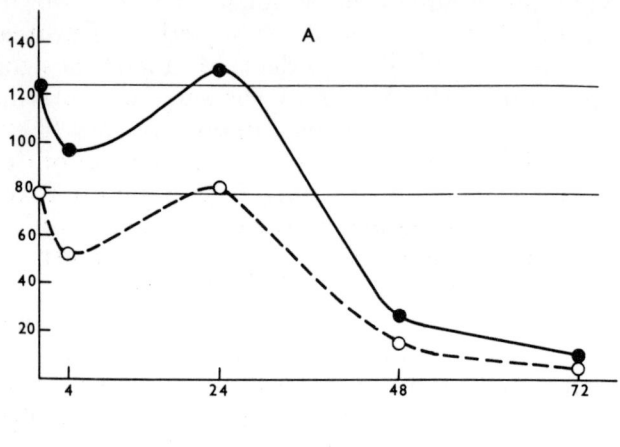

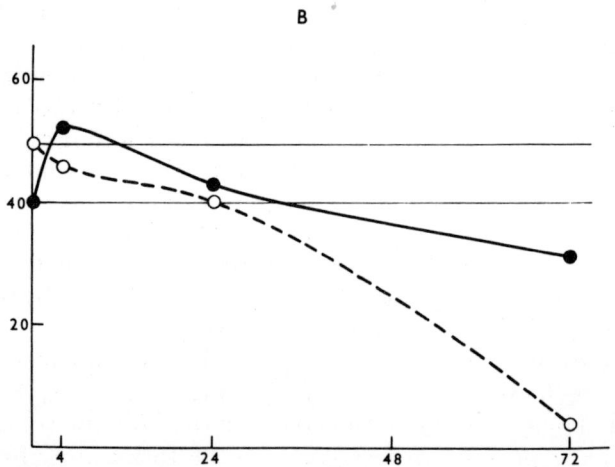

Fig. 7.8. Effect of irradiation on hydrolysis of sucrose (A) and glycyl-L-leucine (B) (according to Ugolev et al. [212]).
Abscissa: time after irradiation (in hours); ordinate: enzymatic activity in arbitrary units. Continuous line—homogenates; interrupted line—everted intestinal sacs.

comparing activities of homogenates and intact cells. It was found that despite the high activity of homogenates the intact cells were unable to hydrolyse glycyl-L-leucine (Fig. 7.8).

Thus two stages may be distinguished: (1) synthesis, including all the mechanisms of protein formation; (2) translocation including the poorly investigated processes of attachment of enzymes to the membrane structure and transition to the external surface of microvilli.

In the majority of cases these processes are highly coordinated. But the more recent investigations showed that on inducing invertase by hydrocortisone both effects become partially separated in time [214]. Moreover, the action of some stressers induced sharp disturbance in the translocation of dipeptidases without producing noticeable effect on the synthesis (Fig. 7.9 C, D) [215, 216, 217, 218]. The mechanism and site of enzyme membrane complex formation are not entirely clear. A. M. Ugolev [reviews 9, 10] suggested that the membrane fragments are coupled with the enzymes inside the cells and built into the microvilli as functionally and morphologically complete units. This process may be based on reverse pinocytosis. It is not excluded, however, that incorporation of enzymes may occur later [219].

7.4.2 Adsorbed enzymes

The various enzymes adsorbed on the surface of the brush border are mostly of pancreatic origin: α-amylase, pancreatic lipases and proteases [38, 39, 40, 141, 5, 6, 7, 8, 9, 10, 144, 220, 148, 149, 150, 221, 152, 153, 222, 223, 224, 225, 226, 227, 228, 229, 230, 231, 232, 233, 234, 235, 14]. More detailed study revealed the presence of trypsin, chymotrypsin [236] and elastase.

Four methods were used to investigate adsorbed enzymes: (1) analysis of the adsorbed properties of carefully prewashed mucosa; (2) study of desorption kinetics; (3) comparison of the digestive properties of mucosa in the presence or absence of adsorbed enzymes (on the basis of one of the above described methods); (4) specific identification of cellular and exogenous enzymes in relation to enterocyte origin.

The importance of adsorbed enzymes (especially the very thoroughly investigated α-amylase) for digestion *in vivo* is shown by comparison of starch hydrolysis in the intestine at various stages of desorption of amylase bound to the surface (Fig. 7.10). As seen from this figure the effectiveness of starch hydrolysis diminishes repeatedly with the desorption of enzymes from the surface; while readsorption of enzymes on the mucosal surface restores the initial effects [reviews 6, 221],

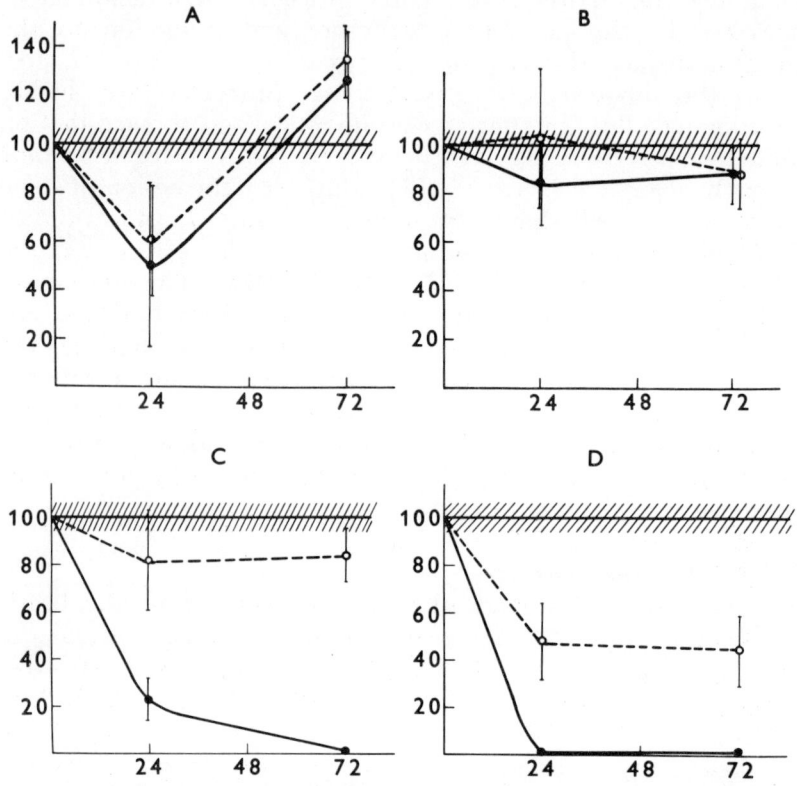

Fig. 7.9. Effect of immobilized stress on disaccharidase (A, B), dipeptidase (C) and tripeptidase (D) activities in small intestine of white rats.

Abscissa: time after stress (in hours); ordinate: enzymatic activity in relation to control taken for 100%. Continuous line—intact mucosa activity; interrupted line—homogenate activity.

Some authors [237, 238, 239] showed that disturbances in adsorption is very important evidence of poor adsorption and digestion of starch clinically. In experiments on rats it was demonstrated that 3-4 days after the cessation of pancreatic juice entrance into the intestine, the absorption of glucose formed during starch hydrolysis (technique of everted sacs) decreases 6-10 times in comparison with that in controls [241, 41]. The mechanism of adsorbed enzyme participation in the processes of polymer utilization will be considered in detail. Though the significance of enzyme adsorption as well as the reality of this process were confirmed both in experimental animals and in clinical experience, the mechanism of enzyme absorption from the intestinal surface and the nature of adsorptive forces are still insufficiently studied. All the more, the analysis of precise localization of enzymes in the brush border structure is badly needed.

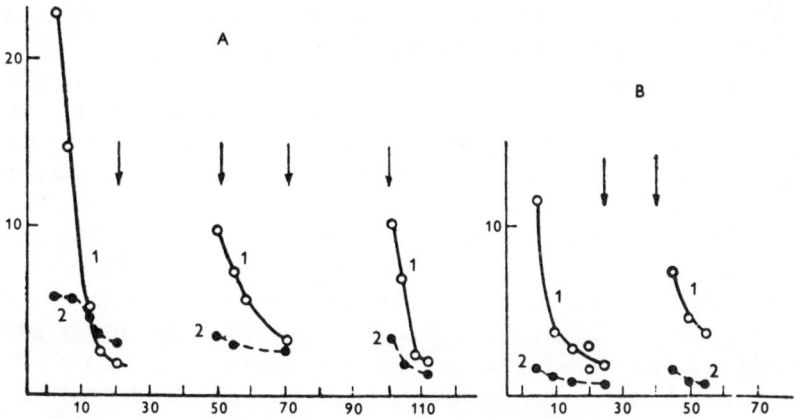

Fig. 7.10. Changes in amylolytic activity during perfusion of rat duodenum and jejunum (entry of duodenal juice into the intestine is preserved) (according to data of P. De Laey (A) and ours (B)). Determination after 3 min by the modified method of Smith-Roy. During interruptions (arrows) perfusion was stopped.
Abscissa: time (in minutes); ordinate: amylolytic activity (in mg of hydrolysed starch/ml/min). 1—*in vivo*; 2—*in vitro*.

It was clear from the very beginning [38, 39, 40] that these enzymes are adsorbed towards the outside of the three-layered membrane. These data, however, were obtained before it was known that the glycocalyx is an important element of microvillous membranes. The investigations by Ito and others made it clear [65, 66, 67, 68, 69, 73, 77, 78] that not only the cell membrane but particularly the glycocalyx may serve as an adsorptive surface (adsorbent), the more so as De Laey [230] showed adsorption of enzymes on mucopolysaccharides.

The glycocalyx is a true component and product of the membrane, forming a layer of 100-1,000 Å (in some cases even more) thick on the external surface. The fact that the glycocalyx structure contains filaments is beyond doubt. Nevertheless, one group of investigators is inclined to think that glycocalyx consists of a large number of filaments [65, 66, 68, 77, 78] while another believes it to be rather a three-dimensioned network, pores of which may be formed by permanent or labile bridges [89]. Whichever is the case, the glycocalyx forms an ultraporosity of the second order in relation to the brush border and thus as already mentioned can be both the site where enzymes are adsorbed and also the site for resistance to diffusion for molecules of large size. The true correlation between the size of glycocalyx and enzymes adsorbed suggests a multilayered adsorption illustrated by Fig. 7.4.

It is not known at present whether different layers of the glycocalyx are identical as regards number and composition of adsorbed enzymes, or where there are gradients of enzyme adsorption in different layers. It can be supposed that according to spatial distribution three fractions exist: (1) the external-on the glycocalyx-chyme border; (2) the intraglycocalyx fraction; (3) the fraction associated with the three-layered membrane.

In the physico-chemistry of adsorption processes it is customary to differentiate between physical adsorption and chemosorption. A kinetic analysis suggested the important role of physical adsorption of enzymes on the brush border surface; at the same time other experiments show that the bulk of enzymes are fixed with the help of chemical bonds [234, 235, 241]. In fact such an enzyme as bacterial glucoamylase is incapable of being adsorbed on the surface of small intestine [221].

It was shown that amylase adsorption depends on temperature, requires activation energy, and appears to be a chemical rather than a physical process [241]. De Laey [233, 234, 235], in a number of ingenious investigations, analysed this problem in detail. This author studied the influence of different cationic detergents (in particular cetrimide and tetraethyl-ammonium-bromide) and anionic detergents (lauryl-sulphate and aerosole OT) on the exchange of amylase between the luminal solution and the surface of the small intestine (Fig. 7.11). The competition for adsorption between lauryl-sulphate and amylase on the same site of binding suggests the ionic character of the bonds. Besides, the author comes to a conclusion about the importance

MEMBRANE (CONTACT) DIGESTION

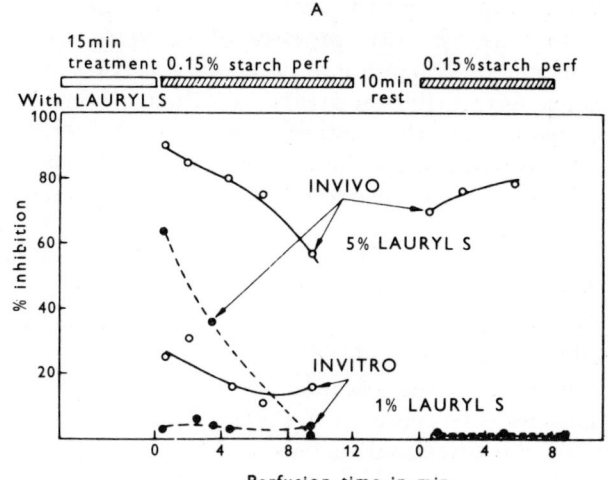

Fig. 7.11. A—effect of lauryl-sulphate on starch digestion *in vivo* (cavital and membrane digestion) and *in vitro* (cavital only) (according to De Laey [235]).

Abscissa: time of perfusion (in minutes); ordinate: percent of inhibition of starch digestion after treatment with lauryl-sulphate (in contrast to control taken as 100%). Continuous line—5% lauryl-sulphate (the upper curve—digestion *in vivo*; the lower curve—digestion *in vitro*); interrupted line—1% lauryl-sulphate (the upper curve—digestion *in vivo*; the lower curve—digestion *in vitro*).

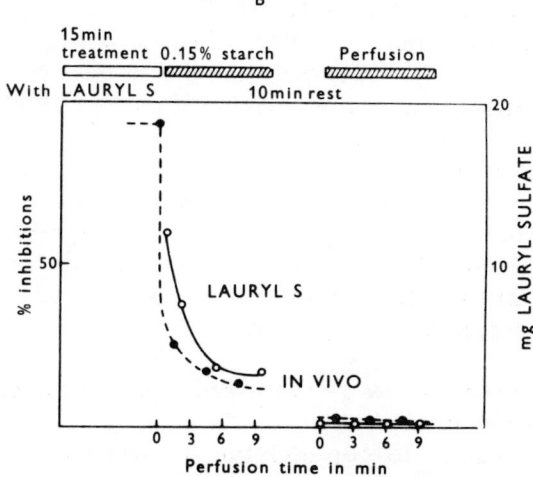

B—displacement of lauryl-sulphate and regeneration of starch membrane hydrolysis during perfusion.

Abscissa: time of perfusion (in minutes); ordinate: percent of inhibition of starch digestion *in vivo* and mg of adsorbed lauryl-sulphate. Continuous line—starch digestion *in vivo*; interrupted line—adsorbed lauryl-sulphate.

of the main groups of phospholipids of membrane and polysaccharide complexes in the process of amylase binding. The activating effect of cetrimide in concentrations less than critical micellar ones shows the importance of fluctuations of the local surface tension for the activity of adsorbed enzymes and stability of adsorption. In the process of adsorption (at least of some enzymes) the stereospecificity of the absorbate are likely to play a certain role. So A. M. Ugolev and E. M. Ustinkova [241] showed that intestinal mucosa in rats binds amylase from dogs more strongly than rat amylase. The alteration of the kinetic characteristics of adsorbed enzymes will be considered in the section 'Physico-chemical and structural aspects of membrane digestion.'

In most cases enzymes are reversibly associated with the structures of the intestinal surface and can be released easily. In this connection, the differentiation between the cellular enzymes and adsorbed enzymes as regards the stability of their bonds with the surface is most important. This is, however, not a very reliable feature. Actually, Goldberg et al. [236] showed that the binding of trypsin and chymotrypsin by intact intestinal cells in man was fairly stable, and a number of agents used caused practically no desorption (freezing and thawing, treatment by H_2SO_4, NaCl, KCl and also by p-chloromercuribenzoate). In a number of cases stability of enzyme binding (α-amylase) depended on the animal species. So a number of investigators [144, 152, 153, 6, 9, 10] showed that desorption of amylase in rats proceeds more readily than in guinea pigs, while hamsters and white mice take an intermediate position. Amylase adsorption is a regulatory process. De Laey and Rakhimov [225, 232] with different experimental techniques independently showed that in fed animals the amount of adsorbed amylase was greater and the desorption rates lower than in fasting animals.

7.4.3 The characteristics of membrane enzymatic apparatus regarded as a whole

It is now possible to consider the interaction of the cellular enzymes and adsorbed enzymes on the process of membrane digestion. Mixing of the chyme brings its small molecules into contact with the paramembrane surface of the glycocalyx, which is not a diffusion barrier for these small molecules. At the same time when the larger molecules come into contact with the glycocalyx surface they are unable to penetrate (we showed that membrane digestion is not effective for the large molecules of glycogen) [6]. In addition, molecules of intermediate size come into contact with enzymes adsorbed on the free surface, and are hydrolysed, thus acquiring the ability to penetrate more or less inside the glycocalyx.

When the molecules have passed inside the glycocalyx space, they are further depolymerized by enzymes localized there. Some additional hydrolysis with a help of adsorbed enzymes is possible on the membrane outer surface. In addition the cellular enzymes localized there are available, which are mainly tri- and dimerhydrolases though some other enzymes such as aminopeptidase and γ-amylase are capable of hydrolysing more highly polymerized products. It is of interest that after the complete desorption of α-amylase from the mucosa, starch hydrolysis by γ-amylase does not occur. This confirms the idea above postulated of the importance of enzymes in the glycocalyx. Another role of the glycocalyx is a separating one, since passage of only some polymers is facilitated. The polymers which pass through must find the appropriate enzymes at the cell surface.

Thus the cellular enzymes and adsorbed ones (Fig. 7.4) functionally complement each other. It will be apparent that there must be considerable overlapping and duplication of the hydrolytic functions, and this is likely to increase the reliability of this system. This interesting and significant question is discussed by ourselves elsewhere.

7.5 PHYSICO-CHEMICAL ASPECTS OF MEMBRANE DIGESTION

Fundamental differences exist between the processes in the bulk phase and interphase as viewed from the standpoint of physical chemistry. These differences are important for understanding membrane digestion, as the latter in many aspects is a special case of catalysis on a non-homogenous surface. The peculiarities of processes in the interphase were characterized in detail for biological interphases in general [242, 243, 244, 245, 246, 59, 247, 248, 249, 250] and for membrane digestion in particular [6, 7, 9, 10].

The most important properties of membrane digestion distinguishing it from cavital digestion are as follows: (1) decrease in the number of degrees of freedom for enzymes bound to cell structures in contrast to those dissolved; (2) important differences in concentrations of organic and inorganic substances which might modify enzyme action; (3) pH of the medium; (4) energy of molecules; (5) form and conformation of molecules; (6) in general, a definite orientation of the molecules in the interphase in contrast to the bulk phase where they are in constant thermal agitation.

Each of these factors is quite sufficient to change essentially the enzymatic activity. For example, the binding of enzyme to the membrane structure renders impossible movement and orientation of the enzyme to the substrate. The substrate must move to the enzyme and orientate itself to the active centre. On the other hand, the fixation of enzymes provides a possibility for the spatial organization of enzymatic chains, grouping of enzymes in definite points and so on.

The hypothesis of organized membrane was advanced by a number of investigators [6, 9, 10, 136, 137, 138, 139]. On the average, the molecules at the surface possess higher energy than those in the bulk phase. Under certain conditions this fact is responsible for higher reactive capacity. The significance of pH for enzymatic reactions is well known. The pH of the surface is determined by the properties of its charged groups and depends on the surface potential. In case of a negative surface charge cations are attracted and the relation between pH and the surface and bulk phase is given by the formula:

$$pH_{surface} = pH_{bulk\ plane} + \frac{\psi}{57}$$

where ψ is the potential of the cell surface [251, 247, 252].

A more detailed analysis of this problem can be found in the literature already mentioned. We would like to dwell here on those peculiarities of membrane hydrolysis that depend on the alteration of the properties of enzymes due to formation of complexes with membrane structures. It concerns the cases of both adsorption and incorporation of the cellular enzymes into membrane.

Many cellular enzymes incorporated into the membrane structure preserve only some of their initial properties when separated from the enzyme-membrane complexes. These changes may be more or less definite and concern only the kinetic characteristics of the enzyme and its stability, or, what is very important, they may alter the intensity or even the character of regulating factors. This question has been insufficiently discussed. Data are available, however, to show that solubilization may result in the alteration of such important indicators as Km, Vmax, pH, temperature dependence, activation energy, to such an extent that it must be of serious physiological importance.

Dalqvist and Thomson [188] found that Km for invertase was significantly higher when investigated on intact tissue than on isolated enzyme. These data were interpreted by Crane [95] as an evidence of the functional separation of disaccharidases from the medium, and an additional proof that free access to the active enzymatic site is limited. On the other hand, this author believes the Km shift is a property dependent on the association of disaccharidases with the surface [253], and accumulation at the membrane surface of glucose in high local concentration induces an inhibitory effect. Since the complex geometry of intact mucosa can fundamentally affect the course of the enzymatic processes, it would be preferable to analyse the membrane influence upon the kinetic properties of enzymes in more simple systems. In our laboratory we investigated some kinetic characteristics of invertase in isolated cells, homogenates and solubilized enzymes. The separation of enzymes from membrane (enzymatic activities of homogenates and solubilized enzymes are compared) leads to statistically significant lowering of Vmax and increase of Km [254].

The alteration of pH function connected with the membrane and solubilized intestinal invertase was demonstrated by G. G. Lapteva and A. M. Ugolev [256]. When enzymes are separated from the brush border membrane, not only their kinetic but also their regulatory characteristics change. Specifically, changes in the regulatory effects of tributyrin and peptides on alkaline phosphatase of small intestine have been shown [255].

Unfortunately, we must admit that while many important properties of the cellular enzymes have been poorly investigated, still less is known about the natural enzyme-membrane complexes. Nevertheless, data are at present available that permit the assumption that in the evolutionary process not only the properties of isolated enzymes but also those of enzyme-membrane complexes undergo selection. The enzyme-membrane complex is likely to differ from the solubilized enzyme not only because of change in the protein conformation but also in a number of properties of the appropriate site of the membrane. G. G. Lapteva and A. M. Ugolev [256] observed that the variability of various characteristics of the enzyme associated with the membrane is higher than that of the solubilized enzyme. These data are in accord with the above postulated viewpoint.

Some changes in the characteristics of enzymes during the coupling of the latter with water-insoluble or hydrophobic carriers were studied in detail in model experiments [257, 258, 259, 260, 261]. During the coupling of trypsin and chymotrypsin with cellulose derivatives the amylase activity can increase to 30-50 times that of the initial enzyme preparation, while the proteolytic activity does not change significantly. A considerable increase of chymotrypsin and ribonuclease was observed after their binding with DNA [262, 263].

Of interest are the changes in the properties of pancreatic enzymes that occur in the course of their adsorption on the surface of small intestine. The adsorbed amylase was shown to possess a higher resistance to heat and acid denaturation than dissolved enzyme. Moreover, the activity of adsorbed amylase also proved to be higher [6]. De Laey [233, 235] found that the activation energy for adsorbed amylase was lower than that for dissolved amylase. He also established that there are significant differences in effects produced by a number of modifying agents on this enzyme in both solubilized and adsorbed states. The examples described illustrate the significance of the problem concerning the interaction between enzymes and cell membrane structures, in spite of the fact that many of its most important aspects remain unknown. It may be assumed that the understanding of the interaction between enzymes and cell membrane structures is important for analysis of the physiology of membrane digestion and for its pathology as well. Among other things, we can find description of disturbances in digestion that may be caused by reduced enzyme adsorption [237, 238] as well as by its abnormal adsorption [212, 9, 10].

7.6 STERILITY OF MEMBRANE DIGESTION

The early publications on membrane digestion pointed out that one of the most important functional characteristics of this mechanism was sterility of the final stages of hydrolysis and initial stages of transport. According to this hypothesis the final stages of digestion resulting in the formation of monomers occur in the zone inaccessible to microorganisms. Comparison of the dimensions of bacteria in the small intestine with the size of the brush border pores and particularly with the size of glycocalyx pores showed (Fig. 7.12) that the brush border presents a bacterial filter by means of which the final stages of hydrolysis are separated from the bacteria populated lumen.

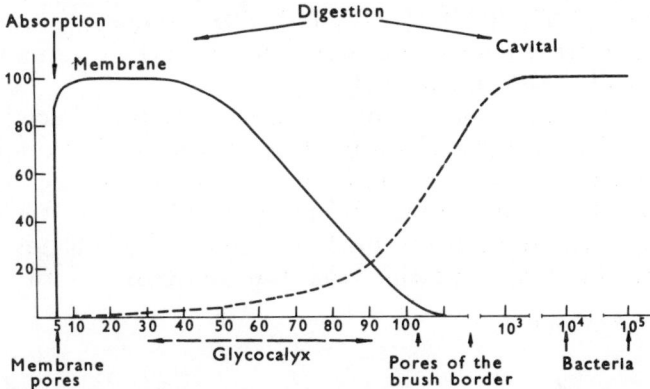

Fig. 7.12. Scheme illustrating the interrelationships of basic processes in the small intestine in relation to size of hydrolysable substrates (water-soluble) (according to Ugolev [9]).

Abscissa: sizes in Å; ordinate: digestion intensity in percent in contrast to maximal digestion taken as 100%.

Thus the sterility of the hydrolysis final stages can be regarded as rather an acquired adaptation of the intestinal cells to the coexistence with the intestinal bacterial flora.

Unfortunately, this aspect of the problem has been poorly investigated experimentally, since the basic hypothesis was in contradiction to the classical idea that the small intestine of monogastric animals and man does not contain any considerable numbers of bacteria. The rapid development of gnotobiology helped to clarify the importance of bacterial flora in the small intestine for the structure and functions of this organ [264,

265, 266, 267, 268, 269, 270, 271, 272, 273]. Comparison of germ-free and conventional animals revealed a number of most essential effects of bacterial flora on the structure and function of small intestine [264, 274, 266, 269, 271, 270, 275] including the following:

(1) Alteration of villous structures [276, 269]. (2) Change of the structure and ultrastructure of enterocytes and among other things the ultrastructure of the brush border [277, 278, 279, 280, 281, 282, 283, 284, 285, 286]. (3) Change of the enzymatic [287, 288, 289] and transport [290, 291, 292, 278, 293] functions of enterocytes. (4) Direct participation of bacteria in metabolism of different food components [294, 295, 296, 297, 298, 266, 299, 300]. (5) Direct participation of bacteria in metabolism of the constituents of digestive secretions [301, 302, 303, 304, 306, 307, 308, 309, 310]. Moreover, bacteria were found to be producers of various vitamins and amino acids [311, 9, 10, 312].

Many other important functions of intestinal bacteria will become clear in the near future. Unfortunately, the comparison of germ-free and conventional animals, however, in the majority of cases does not permit differentiation between the effects caused by the flora in the small intestine, on one hand, and by that in the large intestine and fore-stomach, on the other.

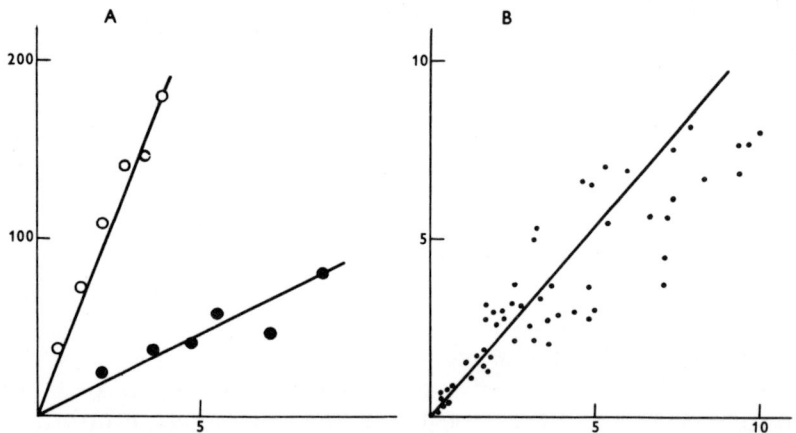

Fig. 7.13. Effect of the perfusion rate on the hydrolysis of sucrose (A) and starch (B) (according to Iezuitova, Ugolev and Feushina [150]).

Abscissa: volume outlay in ml/min to capacity of the intestine (in ml); ordinate: A—the hydrolysis rate of sucrose (in arbitrary units/min); B—relationship of starch hydrolysis *in vivo* to hydrolysis *in vitro*.

Nevertheless, the question arises, how the enterocytes, under ordinary conditions, can compete effectively with the bacterial flora for available food substances, and how the extremely high reproduction of bacteria in small intestine found under many pathological conditions is prevented from occurring normally. As note above, one of the ways is the limitation of nutrients to the bacteria due to the intensive hydrolysis of food in the intestinal lumen with the simultaneous removal of the products into the sterile zone of membrane digestion. The assumption may be made that if the final stages of hydrolysis occurred in the intestinal lumen, then not only the loss of nutrients and rapid reproduction of bacteria would occur, but also formation of many toxic metabolites [311, 9, 10, 312].

7.7 FACTORS FAVOURING THE ENTRANCE OF SUBSTRATES INTO THE BRUSH BORDER

The rate of catalytic reactions on the surface of heterogenous catalysts is determined mainly by: (a) rate of transfer of substances from the bulk phase to the surface, (b) rate of surface reaction and (c) rate of transition of reaction products from the surface into the bulk phase or into the depths of the porous structure [313, 314, 163, 6, 7, 9, 10]. When catalysis on the surface is proceeding rapidly it will be limited by the rate of transfer of substances from the bulk phase to the surface. This is true for the majority of heterogenous catalytic processes in general and for membrane digestion in particular. The role of the factors intensifying the transfer of substances from the intestinal cavity to the surface was not understood immediately. Shortly after the first experiments by Ugolev [38, 39, 40, 141, 5], who demonstrated the digestive functions of the intestinal surface, Dahlqvist and Borgstrom [22] conducted a number of very interesting tests and were forced to conclude that membrane digestion does not play an essential role in hydrolysis of starch. The analysis of these contradictions confirmed the results obtained by De Laey and Iezuitova [220]. It was found, in addition, that the intensity of membrane hydrolysis of soluble starch was a function of liquid flow rate through the small intestinal cavity [144, 220, 152, 153, 149, 150, 221, 154; reviews 6, 7, 9, 10]. The intensity of membrane digestion, as apparent from Fig. 7.13, over a wide range is a linear function of liquidflow rate in relation to the surface. A simple interpretation of these results was proposed: at low rate of liquid conduction through a tube (in this case through the gut) flow becomes laminar with a stagnant substrate-impoverished layer near the surface. With an increase of perfusion rate flow becomes turbulent, thus improving contact of solutes with the surface. These considerations are most important for membrane digestion. The efficiency of membrane digestion is determined not only by the potential power of the enzymatic layer on the surface of the intestine but also by the rate of substrate exchange between chyme and surface.

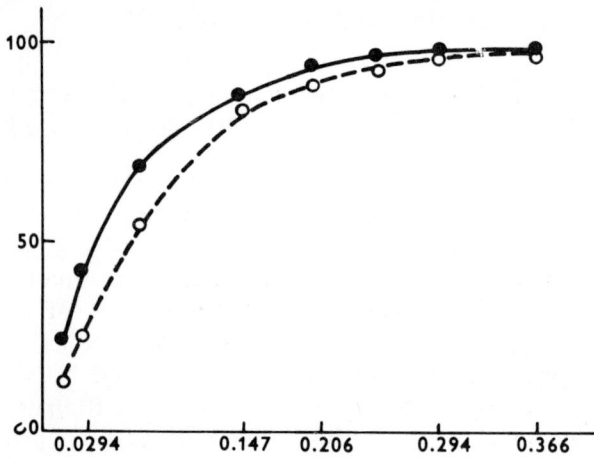

Fig. 7.14. Effect of mixing on sucrose hydrolysis by intact small intestine of white rats depending on substrate concentration.
Abscissa: concentration of sucrose in M/l; ordinate: hydrolysis rate in percent of maximal rate in each experiment take as 100%. Continuous line—without mixing; interrupted line—with mixing.

Later on this idea was confirmed by numerous investigators. It was shown among other things, that hydrolysis of various peptides and oligosaccharides increases proportionally with the intensity of convective flows of fluid containing substrates [6, 9, 10, 152, 153, 149, 150, 315, 316, 154, 155, 146].

The effective transition of substances from the intestinal cavity into the brush border zone is likely to be affected by peristalsis of the intestinal tract. Numerous studies on the dependence of absorption upon a motility function [317, 318] provide support for this view. Let us consider in detail the role of the fluid movement in relation to the intestinal surface. It may be supposed that owing to such movement, the concentration of substrates in the lumen and in the membrane digestion zone is accompanied not only by intensification of membrane digestion, but also of cavital digestion. The importance of motility (it is clear from theoretical considerations) will be greatest at low substrate concentrations which was demonstrated by G. G. Lapteva and A. M. Ugolev [256] (Fig. 7.14).

One must consider two essential facts that are difficult of interpretation.

(1) Fluid movement in the glycocalyx space does not seem to depend on convection in the lumen (particularly fluid in the pores of the brush border). Hence, from the physico-chemical standpoint, there exists a stationary (or autonomous) paramembrane layer.

(2) The convective flow ensures both enrichment of the luminal surface of the glycocalyx with substrate and the removal of the reaction products, and the latter could adversely effect adsorption.

Actually, the experiments of V. Kirse and A. M. Ugolev on everted sacs of sm all intestine in rats showed that the movement of liquid in relation to the surface causes not only intensification of membrane hydrolysis of sucrose, but also sharp reduction of passive transmucosal transfer (Fig. 7.15). Under natural conditions, however, this factor does not affect absorption in this negative manner. The untreated intestinal patterns under the same conditions showed that liquid movement increased both digestion and absorption. This is due to two more important factors. (1) Factors favouring the transfer of substrates from the luminal to the membrane surface of the glycocalyx space. (2) Mechanisms coupling the digestive and transport systems of the enterocytes. In this section the first item will be discussed. The other one will be analysed in the section entitled 'Membrane digestion and adsorption'.

The repeated increase of hydrolysis by convection of fluid in relation to the glycocalyx surface proves that the movement of substances in the microvilli pores and glycocalyx space is not the result of simple diffusion but is connected with the presence of an accelerating mechanism. An attempt made in our laboratory to find out this mechanism in experimental conditions was unsuccessful. However, we may assume that under natural conditions the movement of substances across autonomous paramembrane fluid layers is enhanced due to the intensive water flow inside the intestinal cell [319, 320, 321].

Simple calculations show that the movement of fluid is (1) relatively independent of the molecular weight of substances

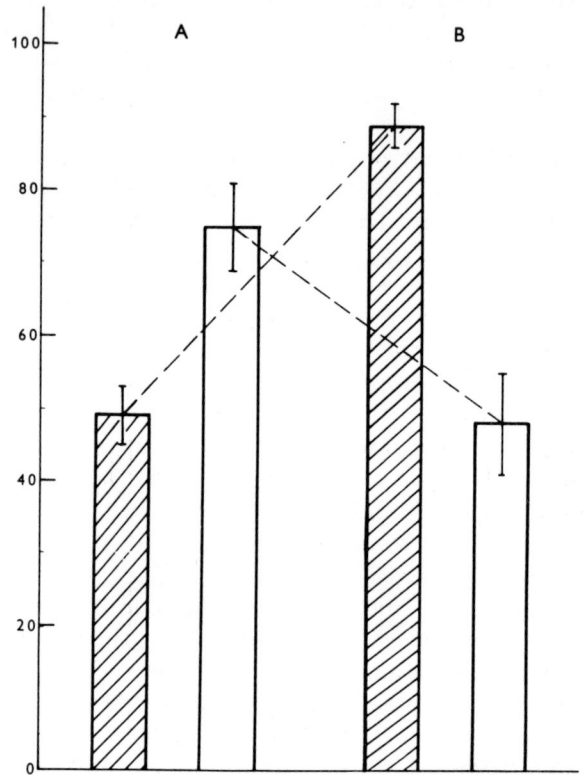

Fig. 7.15. Effect of mixing on the hydrolysis of sucrose (shaded columns) and passive transport of hydrolysis products (light columns).
Ordinate: hydrolysis and passive transport intensity in arbitrary units. A—without mixing; B—with mixing.

and (2) much more effective than simple diffusion. Thus absorption of water may be regarded as a mechanism ensuring not only effective movement of substances towards the membrane in the paramembrane layers, but also regulating concentration of substances in the membrane layer. From this point of view the data on concentration of substances in the membrane layer are of great interest [151, 145, 147, 322, 173].

Thus there are important mechanisms ensuring intensive contact of substrates with the membrane surface. Two of these mechanisms — convection in relation to the surface and transport of substrates in the liquid flow — have been considered. It is important to keep in mind that both normal functioning and disturbances of membrane hydrolysis and transport may be dependent not only on the state of the enzymatic layer on the intestinal surface but also on the state of the mechanisms ensuring transfer of substrates from the bulk phase to the surface.

7.8 RELATIONSHIPS BETWEEN CAVITAL AND MEMBRANE DIGESTION

The size of the brush border pores and, even more, the size of the glycocalyx pores make it impossible for large molecules and supermolecular aggregations to penetrate the zone of membrane digestion. (The adsorption of enzymes is a comparatively slow process.) The maximum size of substrates able to pass from the gastrointestinal tract into the zone of membrane digestion is still unknown. The approximate data obtained by Ugolev and Marauska [223] permit the assumption that possible dimensions of dextrines, which can penetrate the membrane digestion zone amounts to 10-20 glucoside residues. Thus preliminary treatment of food substances in the lumen is a necessary condition for such subsequent stages as membrane hydrolysis and transport [6, 92, 7, 8, 9, 10, 93, 94, 108, 109, 110, 41]. The greater the sizes of the molecule the greater is the relative role of cavital digestion. Such facts as the non-effectiveness of membrane digestion for substrates of large size (for example, amylose and glycogen) [6] provide support of this viewpoint. On the other hand, the more the food products are broken down and the less the polymerization of molecules constituting these substances the less important is the role of cavital digestion. Under certain conditions membrane digestion is either the only or dominating mechanism of hydrolysis. This is in part observed in mammals during the suckling period [323, 324, 315, 316, 222, 223, 224, 226, 227; reviews 7, 9, 10]. Cavital digestion in mammals rapidly develops at the period when animals pass from milk feeding to solid definitive nutrition [325, 326, 327, 328, 329, 330, 331; reviews 6, 9, 10, 332]. It is interesting to note that various parasites have well developed membrane digestion and poor cavital digestion. It means that these parasites utilize food substances obtained both by host and also utilize the host's mechanisms for their preliminary treatment. These and other biological aspects of the problem are discussed more in detail elsewhere [5, 6, 9, 10].

Since in the course of cavital hydrolysis the extent of depolymerization is sufficient for transition to membrane hydrolysis, the reduction of enzymatic processes in the cavity may happen to be the limiting step in assimilation of food substances in general. Here also one should observe the shift of membrane hydrolysis from the proximal portion of the small intestine towards the distal region. The enhancement of starch assimilation in the distal region and its reduction in the proximal zone was found in rats after ligation of the pancreatic duct [333, 240, 41].

Thus transition from cavital hydrolysis to membrane hydrolysis is determined by a number of factors and, in particular, by the size of the brush border filters. It would be tempting to assume that conditions of nutrition could alter the size of glycocalyx pores and the density of the 'packing' of microvilli in the brush border. The numerous investigations, however, demonstrating variability of the above parameters in animals of various species [6, 9, 10, 77, 78] as well as in one species do not prove the presence of adaptive regulation in transition from cavital to membrane digestion.

7.9 INTENSIFICATION OF ENZYMATIC PROCESSES IN THE CAVITY AS INFLUENCED BY MEMBRANE DIGESTION

Cavital and membrane digestion thus form an intimately interacting system. So far we have considered only direct links between these processes. But it has been demonstrated that there exist also links which could be expressed in terms of cybernetics as positive feedback.

At the beginning of the 1960's it was found in our laboratory [38, 39, 40, 141, 5, 6] that when digestion is occurring in the fluid in contact with the mucosal cells the initial rate of hydrolysis of proteins, fats and carbohydrates is soon greatly increased. This was an important discovery in relation to the nature of cavital digestion, as it has been widely believed that most of the digestive activities demonstrated *in vitro* represent what happens physiologically. This is in fact only true for gastric and crop digestion as here membrane digestion does not exist. Reasons for the increased rate of hydrolysis during the course of digestion were advanced by Ugolev [6, 7, 9, 10].

It was assumed that, physiologically, there is excess of substrate so that the rate of hydrolysis initially is proportional to the enzyme concentration. *In vitro*, however, the enzyme concentration may be much less, and enzymes may be bound not only to the initial substrates but also to the hydrolysis products. As digestion proceeds these hydrolysis products and intermediate products immobilize much of the active enzyme. Membrane digestion therefore deals with the products of hydrolysis and thus frees the enzyme again causing stimulation of the early stages of digestion.

7.10 MEMBRANE DIGESTION AND ABSORPTION

7.10.1 *Membrane hydrolysis and transmembrane transport*

It should be stated that membrane digestion is the main mechanism of the final stages of hydrolysis, and in this sense, it affects directly the transport processes, being the donor of the majority of transported substrates. But to understand completely the advantages that can be gained from membrane digestion, one should compare absorption rates when the formation of transported substances occurs as a result of (1) cavital digestion, (2) intracellular digestion, and (3) membrane digestion.

According to the classical concept, the formation of end products of digestion takes place in the intestinal lumen, some time being required for the monomers to reach the intestinal surface [334, 1, 335, 2, 336, 19, 4, 11, 29]. In this case, there is also a very small probability that a given monomer molecule will be absorbed by an enterocyte. Other probabilities are that it could be absorbed by the intestinal bacteria or pass on to the distal region. Furthermore, the concentration of reaction products in the lumen may be low at any given moment due to the large bulk of chyme. From this standpoint, membrane digestion has some significant advantages. Hydrolysis and transport are performed by the cell membrane which simultaneously plays the role of digestive and absorptive area. The structural connections between enzymes and transport systems with the membrane enable us to define the optimal forms of their spatial distribution and functional organization, and this is provided by a number of hypotheses discussed below. The probability of monomer absorption increases and even reaches its summit if we take into account some additional mechanisms.

Intracellular digestion as a mechanism coupling hydrolytic and transport processes is likely to be advantageous for unicellular and lower multicellular organisms where the absorbing cell is providing mainly for its own nutritive needs. In higher animals the ratio between the numbers of absorbing cells and the total number of cells in the body is within the range $1:10^2$ to $1:10^5$. In other words, each absorbing cell satisfies the nutritive needs of 100-100 000 other cells. Pinocytosis and phagocytosis processes, due to their slow rate, are not fit for such a high transport activity. Actually, pinocytosis though active in new-born animals, not only does not increase but falls rapidly with the growth of the organism [120, 34, 36, 337, 78, 332].

Let us consider potentialities of intracellular hydrolysis independent of phago- and pinocytosis. It is likely that higher animals have no selective systems accomplishing oligomer and polymer transport, and their passive permeability is limited by the pores with an effective radius of about 4-6Å [160, 162, 161]. Thus the hypothesis that the final stages of hydrolysis are performed after oligomers have penetrated enterocyte membrane seems unlikely. In this connection it is most important to note that in case of intracellular hydrolysis of oligomers, transport systems of the apical membrane in the intestinal cell are not important since they are not involved in the oligomer transport.

In this instance also membrane digestion occurring on the external surface of membrane has indisputable advantages. This process is not limited by phago- and pinocytosis. Within a wide range it does not depend on molecule size if this does not exceed the size of the glycocalyx pore. The formation of actively transported monomers occurs on the external border of the main diffusion resistance, which can be overcome by special transport systems.

It is thus scarcely surprising therefore, that the triad: cavital digestion, membrane digestion and absorption have been so recognized as the main principle of digestive system activity.

7.10.2   Absorption of monomers either preformed or released during membrane hydrolysis

Since the microvillous membrane ensures both digestive and transport functions, the final stages of hydrolysis and the initial stages of absorption are closely drawn together in space and time. As a result the coupling system is formed. It could be defined as a digestive-transport conveyor. In technology one of the basic characteristics of the conveyor process is the coordination between the preceding and following stages which makes each of these stages maximally effective and reduces to a minimum the loss of material and time. The comparison of transport rates on administration of monomers (e.g. glucose and amino acids) on one hand, and polymers and oligomers (from which membrane hydrolysis releases the substances to be transported) on the other, can be used as an index of effectiveness of the digestive-transport conveyor [41].

It should be readily apparent that a conveyor will be ideal when the rates of both processes are equal. In other words, the transport of substances requiring the preliminary hydrolysis should proceed at the same rate as that of the equal mixtures of monomers. As will be shown below comparisons of such kind were made for a large number of monomers and polymers. The results proved to be interesting and unexpected so that they deserve special consideration.

The very first evidence that the transport of monomers released during hydrolysis of poly- and oligomers occurs without any loss of velocity and practically does not differ from that of equivalent monomer solutions were obtained at the end of the 1950's—beginning of the 1960's. In this field, a major contribution was made by Wilson and Vincent [338] who showed that the glucose and fructose released during sucrose hydrolysis are transferred through the intestinal wall as fast as the mixture of monosaccharides constituting saccharose. Chain *et al.* [339] investigated transport of a number of labelled mono-, di- and polysaccharides and revealed the same phenomenon. These authors began a large series of investigations on carbohydrate transport. The pioneer works by Gupta et al. [340], Crane [341], Smyth [342] contributed very much to our knowledge of transport of amino acids and their oligo- and polymers.

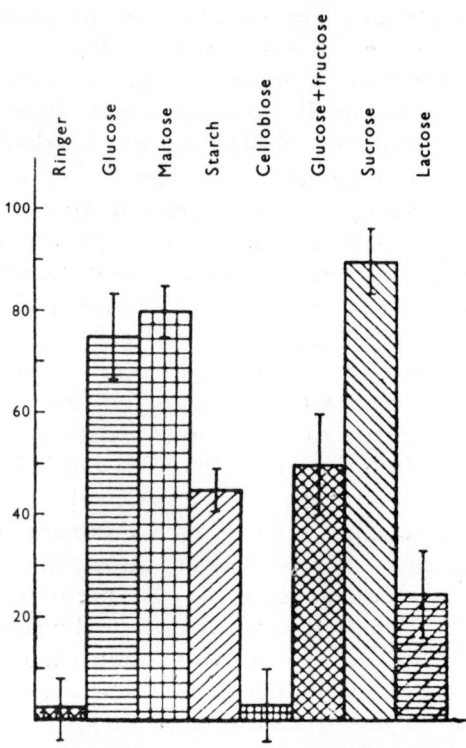

Fig. 7.16. Enhancement of hexose concentration (in mg %) in blood of white rats 15 min after the administration into the small intestine of various mono-, di- and polysaccharides (equivalent to 200 mg of glucose) (according to Ugolev [9]).

During the 1960's, the comparison of absorption rates of starch, maltose, sucrose, trehalose, lactose and equivalent solutes of their constituting monosaccharides was conducted. A great number of animal species (man, rat, hamster, rabbit, frog and toad) were studied. The experiments performed were both *in situ* and *in vitro* [338, 339, 23, 187, 188, 6, 92, 7, 8, 9, 10, 343, 344, 345, 346, 347, 348, 349, 350, 351, 352, 353, 354, 240, 355, 16, 356].

Only some of the authors observed absorption of monomer mixture to occur more rapidly than that of the equivalent amounts of oligomers. In the overwhelming majority of cases oligomers were absorbed as fast as monomers and in some cases even faster. The latter seems quite unexpected and paradoxical. But however that may be, the paradox is well substantiated.

The typical relationship between absorption of disaccharides and monosaccharides are illustrated in Figs. 7.16 and 7.17. Fig 7.17 represents mucosa-serosa transport of glucose during incubation of everted slices of the small intestine in equivalent solutions of maltose and glucose. As can be clearly seen, in all

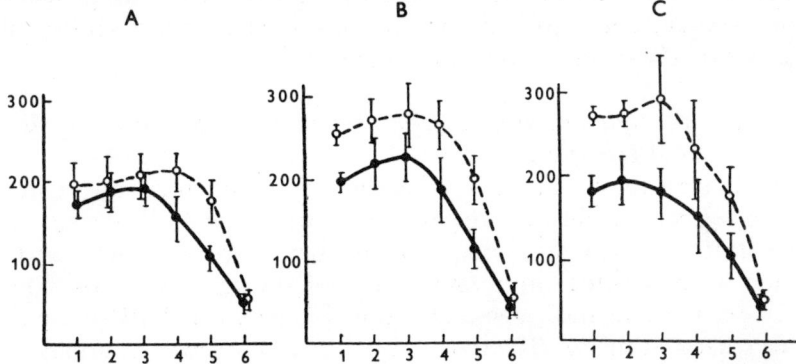

Fig. 7.17. Transport of glucose (continuous line) and maltose (interrupted line) by various segments of the small intestine in white rats during three subsequent incubations: 0-30 min (A), 30-60 min (B), 60-90 min (C) (according to Ugolev [41]).
Abscissa: intestinal segments from jejunum beginning up to ileum end; ordinate: concentration of glucose transferred into 'serosal' solution (in mg %).

intestinal segments save for caudal segments lacking active transport, more rapid absorption of glucose formed during maltose hydrolysis occurs. The relative (as compared with the absorption from monomer solution) effectiveness of so-called maltose transport increases as incubation proceeds, especially in most active proximal segments. The phenomenon of transport facilitation during hydrolysis attracted attention of many investigators and turned out the basis for a number of interesting hypotheses. It appears that many of these theories which will be considered below have much in common.

At the present time high absorption rates have been revealed not only for oligosaccharides as opposed to equal mixtures of their constituting monosaccharide, but also for different peptides as compared to amino acids [357, 358, 359, 360, 361, 362, 351, 363, 364, 318, 15, 365, 366]. Finally, most important was the demonstration of a more rapid absorption of fatty acids during triglyceride hydrolysis in comparison with free fatty acids [367].

Thus at present quite reliable data are available to suggest that the coupling of the final stages of digestion and transport are accomplished efficiently with minimum loss of time and reaction products. Furthermore, monomers released during hydrolysis in many cases though not always are absorbed more rapidly than if they were initially present as monomers. Whatever the explanation may be, these concepts are of primary importance, and significantly change our idea of the biological nature of the digestive-transport conveyor.

7.10.3 Possible coupling mechanism of membrane hydrolysis and transport

The mathematical analysis of mono- and oligomer movement from mucosal fluid into the brush border zone and subsequent penetration through the apical membrane showed that facilitated transport of monomers released during hydrolysis does not occur at all the stages. The transfer from the bulk phase to the surface can be effective only if the molecular volumes or molecular radii of oligo- and polymers is of such a size as to make penetration of the glycocalyx possible.

As compared to monomers, the penetration of oligomers through membrane, in case of passive transport, shows a significant decrease of rate due to small effective radii of pores. In the case of active transport the differences must be even greater as systems of selective transport, at least in higher animals, are efficient only in relation to monomers.

These considerations lead us back to the concept that hydrolysis of food substances must occur at the outside of the apical membrane, and makes us conclude that this membrane digestion is the most effective coupling mechanism of digestive and transport processes.

When oligomers and equal mixtures of monomers are initially present in the mucosal fluid, the monomers which are transported through the membrane are absolutely identical in their characteristics [97, 98, 99]. The high effectiveness of the digestive-transport conveyor must be connected with the stage of transfer of reaction products from enzyme to carrier. This poorly investigated stage where digestive and resorptive functions are closely integrated is one of the immediate central problems of gastroenterology. At present few phenomena are known which can shed light upon the mechanisms of interaction between the final enzymes and entrance into the transport system.

(1) In order to gain some insight into the interaction between membrane digestion and transport systems one must investigate kinetic constants characterizing absorption of oligomers and monomers. A study of such kind was first begun by Prichard and Parsons. Such works are very few. The bulk of investigations concern maltose and glucose [346, 353, 354, 347, 355] and some peptides [348, 349, 351, 366, 368, 364].

(2) In some cases a striking similarity was revealed between the effects of various agents on the final enzyme and entrance into the transport system. So Semenza *et al.* showed that activation kinetics of glucose carrier and brush border invertase have much in common, and this may reflect either physico-chemical similarity of sodium sites on the enzyme and carrier, or spatial coupling of invertase and entrance into the transport system [369, 111, 97, 98, 99, 183].

(3) High sensitivity to phlorizin of so-called maltose transport to glucose transport is well established [240, 41, 370].

(4) Maltose transport is greater than equivalent transport of glucose only if transfer occurs down an electro-chemical gradient. In case of transport against a gradient glucose transport is always higher than that of maltose.

Table 7.5 presents information for the transport of mono- and disaccharides in frogs and toads. As apparent from this table, transport of liberated glucose released by hydrolysis in the small intestine of the frog has a lower Km than any other Km value obtained. The Vmax for glucose released from maltose in the toad is slightly smaller than for glucose initially present. Assuming at the first approximation that Km as a measure of affinity of the acceptor site for substrate, it seems that the affinity for released glucose is higher, though in both cases the nature of substrate seems to be identical. Alterations in Km and Vmax were also observed by comparison of transport of some amino acids released during peptide hydrolysis and equal amino acid mixtures [363, 364].

TABLE 7.5. Kinetic parameters of processes of maltose and glucose transport in *Rana pipiens* and *Bufo vulgaris* (Parsons, Prichard, 1968b, p. 144 [354]).

	Maltase activity	Glucose transport	
		From maltose	From glucose
Rana pipiens			
K_m (mM)	1.89 ± 0.26 (3)	0.38 ± 0.08 (3)	0.45 ± 0.13 (4)
V_{max} (MM glucose/hr/g)	308.7 ± 63.5 (3)	86.7 ± 2.08 (3)	137.5 ± 35.5 (4)
Bufo vulgaris			
K_m (mM)	1.90 (2)	0.45 ± 0.098 (3)	0.54 (1)
V_{max} (MM glucose/hr/g)	210.0 (2)	70.3 ± 9.53 (3)	76.0 (1)

All the data presented above (and some others) permit the suggestion that transport of substances resulting from membrane hydrolysis in some instances appears to be facilitated and more effective. The most convincing data on this matter were obtained by the experiments on the effect of phlorizin upon glucose-maltose transport and assessment of kinetic constants. The majority of investigators concerned with this problem believe that the coupling of digestive and transport functions is the result of direct transition of monomers released during hydrolysis from enzyme to the transport system, and diffusion and dispersion into the fluid phase does not occur (Fig. 7.18). Such concept is a strong argument in favour of the localization of final hydrolysis at the external membrane surface.

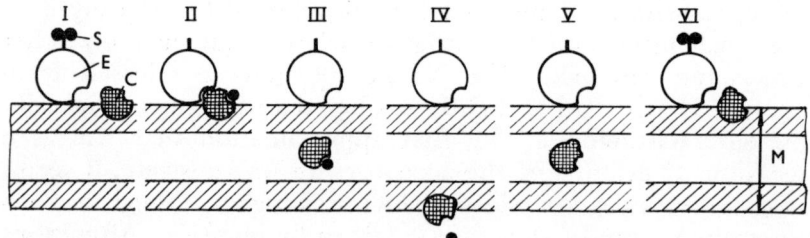

Fig. 7.18. The stages of interrelationship between an end enzyme and carrier in the digestive-transport conveyer.
E—enzyme; C—carrier; S—substrate; M—membrane.

7.10.4 Digestive-transport conveyors and digestive-transport ensembles

Direct transference of reaction products from the final enzyme to a transport system can be achieved only by contact between enzyme and carrier. Such contact is feasible in cases where the enzyme and transport system form a complex (e.g. the quaternary structure) which may be either permanent or intermittent [92, 111, 97, 98, 99, 183, 240, 371, 41].

The data presented above and other evidence demonstrating the transference function of the final enzymes involved in membrane digestion serve as an accessory argument in favour of this hypothesis [111, 97, 98, 99]. Figs. 7.18, 7.20 and 7.21 show enzyme transport pools on the membrane surface. However, the model proposed here is not the only possible one, since it would be premature to claim a definitive model while there is still uncertainty about some of the basic processes. Hence we hesitate in choosing between two hypotheses: (1) hypothesis of mobile carrier and (2) hypothesis of relay race [372]. General analysis of the interaction processes between membrane hydrolysis and transport showed that the both hypotheses are acceptable. To account for the effect of facilitation during transference of substrate from enzyme to the transport system the direct contact between the latter or contact through an intermediate particle (adapter) is required.

The designing of the scheme of a mobile carrier is a rather difficult task, which we shall discuss in detail below. The main characteristics of transport with the help of mobile carriers are thoroughly discussed in a number of reviews [373, 374, 95, 85, 86, 88, 321]. The carriers appear to be proteins with the molecular weight of 10 000-70 000 and most often 30 000 [373].

The carriers are supposed to be bound to the transported molecules and further pass from the external surface of the membrane to the internal one where these molecules are released. The mechanism of this transfer is not known. The operation of the carriers is thus cyclic and at least at one of the stages needs free energy. If this idea is correct the periodically free carriers will form complexes with the products issuing from the hydrolytic system, and these complexes dissociate after transported molecules have passed through the membrane (probably Na^+ involved in case of amino acids and sugars). It may be assumed that the end enzymes and carriers interact allosterically. This interaction facilitates the binding of substrates to the contact or accepting sites of the carrier [41].

This concept is illustrated by Fig. 7.18 which summarizes current ideas on membrane hydrolysis, the highly effective digestive-transport conveyor and mobile carrier. If the relay race hypothesis should prove correct, the same mechanism may function effectively, but the idea of 'shuttle' movement of the carrier must be substituted for the hypothesis of conformation changes occurring at the entrance to the transport system. The latter necessarily comes into contact with either the end enzyme or issuing products. However, we should acknowledge that all attempts to reveal the presence of an enzyme-transport ensemble have so far failed. The attempts made by the Semenza group and in our laboratory to affect the invertase activity by phlorizin were also unsuccessful. However, it is clear that if there exists a contact between the glucose carrier and disaccharidases (for example, invertase), then the administration of phlorizin which forms the complex with a glucose carrier on the external membrane surface should affect the activity of the enzyme with which the carrier interacts. Had phlorizin changed the enzyme activity it would have been possible to suggest not only direct contact of enzyme and entrance into the transport system but also some possibility for carriers to be regulators of activity and intensity of membrane digestion.

It should be noted that all previous attempts to influence invertase activity by phlorizin were made an enterocyte homogenates and solubilized enzymes. Recently, Mitushova, Gozite and Ugolev [375, 376, 377] have developed a new technique by means of which it is possible to obtain a significant amount of surviving isolated intestinal epithelium. Another attempt was made by ourselves to affect invertase activity by phlorizin [254, 41]. As seen from Fig. 7.19, phlorizin does not change invertase activity in homogenates and solubilized invertase. However, in case of intact cells under oxygenation a significant though slight decrease of invertase activity was observed.

Thus with the digestive-transport mechanism of well preserved enterocytes we succeeded in demonstrating the interaction between carrier and enzyme. These data permit the following interpretation. (1) Phlorizin is bound to be the transport component of the enzyme-carrier complex, changing its conformation and thus altering the enzyme activity allosterically. (2) Phlorizin is bound to a carrier preventing formation of the invertase carrier complex. The enzyme association with the carrier shows higher enzymatic activity.

Since the final enzymes and transport systems are functionally integrated it seems reasonable to introduce such terms as 'digestive-transport ensembles' or 'coupling of digestive-transport systems' of the enterocyte membrane. The degree of coupling between the digestive and transport processes may be to a certain extent characterized in different quantitative ways and thus the idea of coupling, in a sense, becomes a problem for investigation.

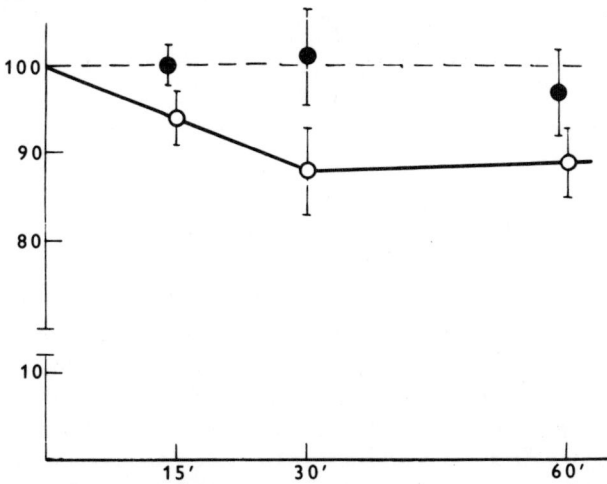

Fig. 7.19. Effect of phlorizin on the invertase activity of intact cells (continuous line) and their homogenates (interrupted line).

Abscissa: time of incubation (in minutes); ordinate: invertase activity in percent (the enzymatic activity in absence of phlorizin taken as 100%).

It should be pointed out that in the modern physiology of this field, besides two extensive fields of research, i.e. digestion and absorption proper, there appeared a new one, i.e. the coupling (integration) mechanisms and phenomena of digestive and transport processes.

7.10.5 *Coupling of membrane hydrolysis and transport as regulated parameter*

The detailed description of digestive-transport ensembles is difficult at present as there is no answer to such fundamental

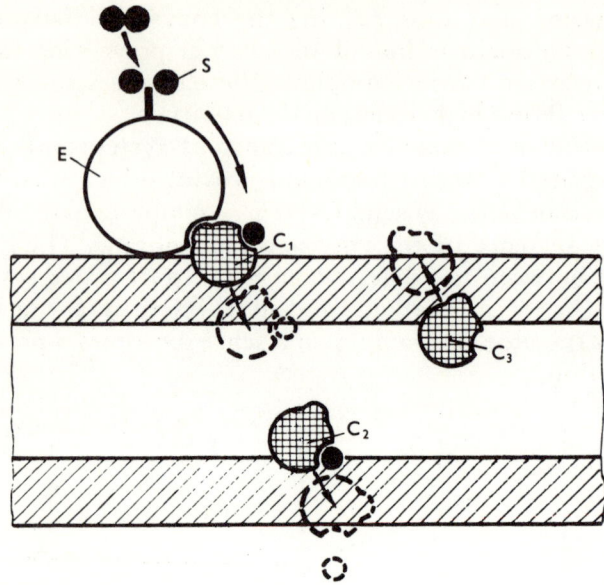

Fig. 7.20. Enzyme-carrier pool.
E—enzyme; C_1, C_2, C_3—carrier forming a digestive conveyer. S—substrate.

question as how many molecules of enzymes and carriers interact within the ensemble. Figs. 7.21 and 7.22 illustrate some of the possibilities. The enzymes of a maltose transport ensemble seem to be served by at least two carriers which are required for the acceptance of two formed molecules of glucose. One may assume also that a larger number of carriers

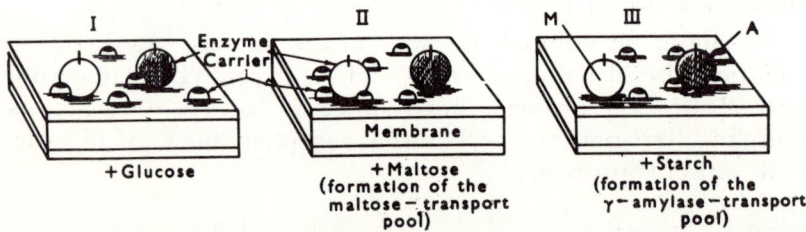

Fig. 7.21. Integrating inductions (scheme) (according to Ugolev [41]).
Hypothetic distribution of carriers on the external membrane surface. I—initial distribution (in absence of substrates); II—formation of maltase-transport pool; III-formation of γ-amylase-transport pool. M—maltase; A—γ-amylase.

may compose this ensemble, and possibly several carriers can adhere in sequence to the same contact enzymatic site, thus forming the relay race of mobile carriers.

To interpret the specific increase of so-called maltose transport in the process of continuous incubation of intestine, provided the maltose activity is unchanged, Ugolev [41] suggested the following hypothesis of integrating inductions. Amino acid and glucose carriers should be included into various enzyme transport ensembles. In addition, one may suggest the presence of a number of free (not associated with enzymes) carriers (Fig. 7.21). In this case the assumption must be made

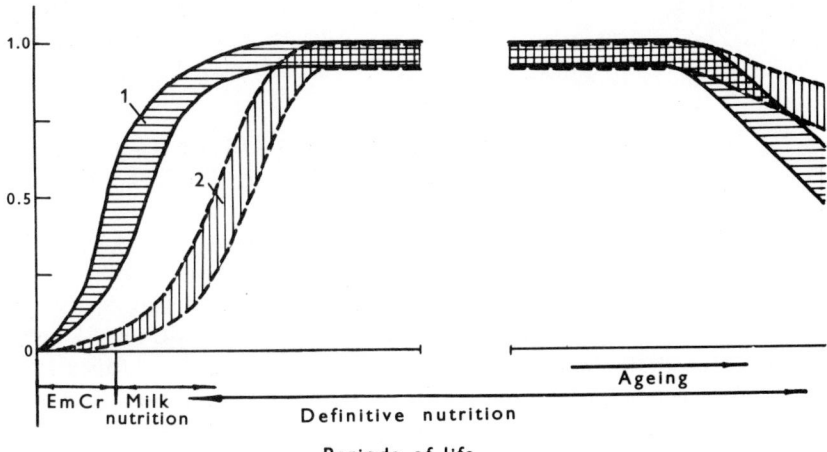

Fig. 7.22. The scheme illustrating interrelationships of membrane and cavital digestion in various periods of life in mammals.
1—membrane digestion; 2—cavital digestion. Abscissa: life periods; ordinate: activity of cavital and membrane digestion (maximal activity is taken for 1 unit).

transport of glucose itself, the whole system of glucose carriers begins to act. Such an organization is not economic and we may assume that an enterocyte is capable of producing all the necessary and effective ensembles. These ensembles may have comparatively rigid construction and can be included into the membrane composition. On the other hand, their structure may be more plastic. Let us imagine the model composed of two enzymes—maltase and glucoamylase, of two types of glucose that on maltose hydrolysis glucose transport will be accomplished not by all the carriers but some of them. During sucrose hydrolysis, other carriers will function and only in case of carriers (M-carrier associated with maltase, A-carrier associated with glucoamylase) a number of free glucose carriers and carriers associated with other enzymes not considered here. Assuming that certain transformations may occur during absorptive processes, we suggest that depending on which of the hydrolytic systems is the most effective at a given moment, carriers can use either one or other of the digestive-transport conveyors. In both cases, not only alterations in the number of definite enzymes and transport activities occur, which might be expected, but also definite transport metabolic chains are formed. As a result of such inductions, the systems are integrated. Hence these inductions were defined as integrating or organising [41]. The alterations in coupling of digestive and transport processes may occur under the influence of various diets.

7.11 ORGANIZATION LEVELS AND REGULATION OF SYSTEMS PERFORMING MEMBRANE DIGESTION

While considering membrane digestion attention is mainly drawn to various levels of organization as: (1) molecular level; (2) enzyme-membrane complex levels at which organization of digestive and digestive-transport conveyors is possible; (3) level of the brush border as a total structure. Membrane digestion, however, is one of the basic functions of the small intestine and in this connection, higher levels of organization should also be considered for understanding the essential functioning of membrane digestion under natural conditions. For this reason, the following levels are also to be stressed: (4) enterocyte; (5) crypt-villous system; (6) intestine as a total organ.

The limitations of the present review give us no possibility of discussing higher levels of organization. That is why problems of regulation and autoregulation of membrane digestion are not considered here. Most of them as well as some published evidence were included in special reviews [6, 7, 9, 10, 41, 378, 217].

7.12 EVOLUTIONARY ASPECTS OF MEMBRANE DIGESTION

This problem has been already considered in the previous reviews [6, 7, 9, 10]. As indicated, membrane digestion is a widespread and, probably, universal mechanism of the final stages of hydrolysis and coupling of digestive and transport processes. Membrane digestion is found in various mammals [6, 7, 8, 9, 10] including ruminants and pigs [379, 380, 381, 382, 383, 384, 385, 386]. It was detected also in birds [9, 10, 387, 388, 389, 390, 391, 392], in amphibians, fish [393, 394, 395, 396, 397, 388, 398, 391], in cyclostomata as well as in a number of invertebrates such as crustaceans, molluscs and worms. Recently Dr. Y. Eeckhout (Louvain, Belgium) has informed me that he observed membrane digestion in a representative of Protozoa. This evidence is most important for understanding the evolution of basic types of digestion. These questions are discussed in detail in a special review containing available data on membrane digestion in bacteria, yeasts and plant roots [9, 10].

Of interest is the fact that in invertebrates as well as in mammals the initial stages of hydrolysis occur in the cavity. The final stages are brought about by means of membrane digestion (Table 7.6) [6, 9, 10, 399, 400, 401, 402, 403, 404, 405].

TABLE 7.6. The distribution of enzymatic activities in the digestive cavity and hepatopancreas in crab

Enzyme	Goitre juice (cavital digestion)	Hepatopancreas (membrane digestion)
Amylase	+++	+
Maltase	+++	++
Invertase	−	−
Lactase	−	−
Proteinases	+++	+
Dipeptidase	−	+++
Lipase	+++	−
Monoglvceride lipase	−	+++

Many of the animals utilizing food consisting of low-molecular substances are characterized by a well-developed membrane digestion and reduced cavital digestion [9, 10].

The relationships between cavital and membrane digestion can essentially change in the process of ontogenesis. In mammals membrane digestion is a dominating mechanism in the period of milk nourishment [7, 9, 10]. The intensive development of cavital digestion coincides with transition to adult feeding [406, 407, 408, 409, 410, 411, 412, 413, 326, 414, 415, 327, 416, 328, 417, 418, 34, 419, 420, 421, 422, 423, 424, 323, 324, 7, 9, 10, 332]. Fig. 7.22 represents the simplified relationships between membrane and cavital digestion in mammals during lifetime.

7.13 CONCLUSION

Membrane digestion was discovered little more than ten years ago, and at that time a term to describe this new and very important type of digestion was proposed. At present membrane digestion is a rapidly developing scientific problem which presents interest for many biological and medical sciences. This progress is the result of mutual efforts of numerous investigators engaged in different branches of science all over the world.

This review begins with a description of some unique properties and limitations distinguishing membrane digestion from other basic types of digestion. Then the details of this mechanism functioning were described, some of them being very specific. In most cases these principles are insufficiently investigated.

The study of membrane digestion as a mechanism of the final and intermediate stages of hydrolysis and coupling of digestive and transport processes proper allow the understanding of two important aspects:

(1) In the process of evolution membrane digestion could not replace cavital and intracellular digestion, but also the latter could not displace the membrane digestion.

(2) The well developed digestive system in higher animals and man, in particular, is an integration of the three basic stages: cavital digestion, membrane digestion and absorption. It has been shown that at the stage of membrane digestion not only transition from initial hydrolysis to absorption occurs but splitting of 50-80% and more of the primary bonds in food biopolymers as well [6, 7, 9, 10].

The functioning of the digestive system is thus based (in language of ancient legends) on three whales: cavital digestion, membrane digestion and absorption. All of them are equal in importance and interact accordingly. These whales seem to understand that they need each other. From this standpoint, the presence of the chapter on membrane digestion in the book dedicated to the fundamental problems of absorption becomes explicable.

REFERENCES

1. I. P. PAVLOV (1897), Lectures on the Work of Principal Digestive Glands, Complete Works, 2, Book 2. Izd. AN SSSR, Moscow and Leningrad (1951).
2. E. S. LONDON, Physiology and Pathology of Digestion. Petrograd (1916).
3. B. P. BABKIN, External Secretion of Digestive Glands, Gosizdat, Moscow and Leningrad (1927).
4. W. BUDDENBROCK, Vergleichend Physiologie, Bd. III. Basel and Stuttgart (1956).
5. A. M. UGOLEV, Digestion and its Adaptation Evolution. Izd. Vysshaya Shkola, Moscow (1961a).
6. A. M. UGOLEV, Membrane (Contact) Digestion. Izd. AN SSSR, Moscow and Leningrad (1963).
7. A. M. UGOLEV, Membrane (Contact) Digestion, *Physiol. Rev.*, 45 555 (1965a).
8. A. M. UGOLEV, *Die Nahrung*, 10, 483 (1966a).
9. A. M. UGOLEV, Physiology and Pathology of Membrane Digestion. Izd. 'Nauka', Leningrad (1967).
10. A. M. UGOLEV, Physiology and Pathology of Membrane Digestion. Plenum Press, New York (1968a).
11. H. L. BOCKUS, Gastroenterology, 2. Philadelphia and London (1964).
12. P. DESNUELLE, in: Handbook of Physiology, Sec. 6, Alimentary Canal, 5, p. 2629. Washington (1968).
13. P. KELLER, in: Handbook of Physiology, Sec. 6, Alimentary Canal, 5, p. 2605. Washington (1968).
14. G. SEMENZA, in: Handbook of Physiology, Sec. 6, Alimentary Canal, 5, Ch. 124, p. 2637. Washington (1968b).
15. K. TAUFEL, *Die Nahrung*, 13, 559 (1969a).
16. K. TAUFEL, *Die Nahrung*, 13, 271 (1969b).
17. Handbook of Physiology, Sec. 6, Alimentary canal, (C. F. Code, ed.), Washington (1968).
18. G. IORDAN, Practicum on the Comparative Physiology. Biomedizd., Moscow and Leningrad (1934).
19. Kh. S. KOSHTOYANTS, Fundamentals of Comparative Physiology, 1. Izd. AN SSSR, Moscow and Leningrad (1950).
20. B. BORGSTROM, A. DAHLQVIST, G. LUNDH, J. SJOVALL, *J. Clin. Invest.*, 36, 1521 (1957).
21. E. M. HEISIN, in: Manual on Cytology, p. 166. Izd. 'Nauka', Moscow and Leningrad (1965).
22. A. DAHLQVIST and B. BORGSTROM, *Nutr. Rev.*, 20, 203 (1962).

23. D. MILLER and R. K. CRANE, *Biochim. Biophys. Acta*, 52, 281 (1961a).
24. D. MILLER and R. K. CRANE, *Biochim. Biophys. Acta*, 52, 293 (1961b).
25. D. MILLER and R. K. CRANE, *Amer. J. Clin. Nutrit.*, 12, 220 (1963).
26. A. B. NOVIKOFF, in: J. Brachet, A. E. Mirsky, The Cell, II. New York and London (1961).
27. M. MULLER, P. ROHLICH, J. TOTH and I. TORO, Ciba Found. Sympos. on Lysosomes, p. 201 (1963).
28. G. K. SHLYGIN, *Vestnik Akad. Nauk SSSR*, 5, 21 (1964).
29. G. K. SHLYGIN, Enzymes of the Intestine in Norm and Pathology. Moscow and Leningrad (1967).
30. R. J. BARRNETT, *Exper. Cell Research*, Suppl. 7, 65 (1959).
31. S. L. CLARK, *J. Biophys. Biochem. Cytol.*, 5, 41 (1959).
32. S. L. PALAY and L. J. KARLIN, *J. Biophys. Biochem. Cytol.*, 5, 363 (1959a).
33. S. L. PALAY and L. J. KARLIN, *J. Biophys. Biochem. Cytol.*, 5, 373 (1959b).
34. B. S. PLATT, *Fed. Proc.*, 20, 188 (1961).
35. E. SANDERS and C. T. ASHWORTH, *Exper. Cell. Res.*, 22, 137 (1961).
36. T. H. WILSON, Intestinal Absorption. Philadelphia and London (1962).
37. P. F. MILLINGTON and J. B. FINEAN, in: Biochem. Problems of Lipids, p. 116. Amsterdam (1963).
38. A. M. UGOLEV, *Bull. Experim. Biologii i Med.*, 49, 12 (1960a).
39. A. M. UGOLEV, Works of Scient. Confer. on the Problem of Physiol. and Pathol. of Digestion dedicated to the Memory of Acad. K. M. Bykov, p. 829. Ivanovo (1960b).
40. A. M. UGOLEV, *Nature*, 188, 588 (1960c).
41. A. M. UGOLEV, *Fiziologitcheskij Zhurnal SSSR*, 56, 651 (1970).
42. J. F. DANIELLI and H. DAVSON, *J. Cell. Comp. Physiol.*, 5, 495 (1935).
43. J. F. DANIELLI and E. N. HARVEY, *J. Cell. Comp. Physiol.*, 5, 483 (1935).
44. S. ELKES and J. B. FINEAN, *Exper. Cell. Res.*, 4, 69 (1953).
45. J. B. FINEAN, *Exper. Cell. Res.*, 5, 202 (1953).
46. J. B. FINEAN, in: The Metabolism of the Nervous System (D. Richter, ed.). Pergamon Press, London (1957).
47. J. F. DANIELLI, in: Surface Phenomena in Chemistry and Biology, p. 246. Pergamon Press (1958).
48. J. D. ROBERTSON, in: The Structure and Function of Subcellular Components, *Bioch. Soc. Symp.*, 16, p. 3. Cambridge (1959).
49. J. D. ROBERTSON, in: Cellular Membranes in Development (M. Locke, ed.), 1. Academic Press Inc., New York (1964).
50. J. ROBERTSON, *Progr. Biophys.*, 10, 343 (1960).
51. J. ROBERTSON, *Protoplasma*, 63, 218 (1967).
52. F. S. SJOSTRAND, *Radiation Res.*, Suppl. 2, 349 (1960).

53. F. S. SJOSTRAND, *J. Ultrastruct. Res.*, **8**, 517 (1963).
54. F. S. SJOSTRAND, *Protoplasma*, **63**, 248 (1967).
55. F. S. SJOSTRAND, in: Protides of the Biol. Fluids (H. Peeters, ed.), p. 15. Amsterdam, London and New York (1968).
56. H. DAVSON, *Circulation*, **26**, 1022 (1962).
57. H. FERNANDEZ-MORAN, *Circulation*, **26**, 1039 (1962).
58. J. L. KAVANAU, *Nature*, **198**, 525 (1963).
59. J. L. KAVANAU, Structure and Function in Biological Membranes. Holden-Day, San Francisco, London and Amsterdam (1965).
60. J. L. KAVANAU, in: Papers on Biological Membrane Structure (select. by D. Branton and R. B. Park) (1968).
61. P. A. VANDENHEUVEL, *Ann. N.Y. Acad. Sci.*, **122**, 57 (1963).
62. A. A. BENSON, *Ann. Rev. Plant. Physiol.*, **15**, 1 (1964).
63. A. A. BENSON, *J. Amer. Oil Chemists Assoc.*, **43**, 265 (1965).
64. A. A. BENSON, 7th Intern. Congr. Bioch. (Tokyo III), 525 abstr. (1967).
65. S. ITO, *Anat. Record*, **148**, 294 (1964).
66. S. ITO, *J. Cell. Biol.*, **27**, 475 (1965).
67. S. ITO, in: The Specificity of Cell Surfaces (B. D. Davis and L. Warren, eds.), p. 211. Prentice Hall, Englewood Cliffs, New York (1967).
68. S. ITO, *Fed. Proc.*, **28**, 12 (1969).
69. S. ITO and J. P. REVEL, *J. Cell. Biol.*, **23**, 44A (1964).
70. J. A. LUCY, *J. Theor. Biol.*, **7**, 360 (1964).
71. J. A. LUCY, *Brit. Med. Bull.*, **24** (2), 127 (1968).
72. J. A. LUCY and A. M. GLAUERT, *J. Molec. Biol.*, **727** (1964).
73. D. W. FAWCETT, *J. Histochem. Cytochem.*, **13**, 75 (1965).
74. W. STOCKENIUS, in: Principles of Bimolecular Organization. Ciba Foundation Symposium, p. 357. Boston (1966).
75. V. L. BOROVYLAGIN, *Citologiya*, **5**, 505 (1967).
76. D. E. GREEN and R. F. GOLDBERGER, Molecular Insights into the Living Process. Academic Press, New York and London (1967).
77. J. S. TRIER, *Fed. Proc.*, **26**, 1391 (1967).
78. J. S. TRIER, in: Handbook of Physiology, Sec. 6, Alimentary Canal, 3, Ch. 63, p. 1125. Washington (1968).
79. A. D. BANGHAM and D. A. HAYDON, *Brit. Med. Bull.*, **24**, 124 (1968).
80. Iu. M. VASILYEV and A. G. MALENKOV, Cell Surface and Reactions of the Cell. Izd. 'Meditsina', Leningrad (1968).
81. J. B. FINEAN, R. COLEMAN, S. KNUTTON, A. R. LIMBRICK and J. E. THOMPSON, *J. Gen. Physiol.*, **51**, 19 (1968).
82. E. D. KORN, *J. Gen. Physiol.*, **52**, 257 (1968).
83. W. D. McELROY and C. P. SWANSON. Modern Cell Biology. Prentice-Hall, New York (1968).
84. J. STEMPAK and M. LAURENCIN, *J. Microscop.*, **9**, 465 (1970).
85. R. K. CRANE, in: Handbook of Physiology, Sec. 6, Alimentary Canal, 3, p. 1323 (1968a).
86. R. K. CRANE, *Ann. Rev. Med.*, **19**, 57 (1968b).
87. R. K. CRANE, *Amer. J. Clin. Nutr.*, **22**, 242 (1969a).

88. R. K. CRANE, *Fed. Proc.*, **28**, 5 (1969b).
89. Ya. Yu. KOMISSARCHIK and A. M. UGOLEV, *Dokl. Akad. Nauk SSSR*, **194**, 731 (1970).
90. N. N. IEZUITOVA, N. M. TIMOFEEVA, M. Yu. CHERNYA-KHOVSKAYA, E. K. ZABELINSKII and A. M. UGOLEV, *Dokl. Akad. Nauk SSSR*, **173**, 475 (1967).
91. A. DAHLQVIST, *Gastroenterology*, **43**, 694 (1962).
92. A. M. UGOLEV, Symposium 'Physiology and Pathology of Absorption in the Gastrointestinal Tract' (Odessa), p. 7 (1964).
93. A. M. UGOLEV, *Ernahrungsforschung*, **13**, 119 (1968b).
94. A. M. UGOLEV, in: Protides of the Biol. Fluids (H. Peeters, ed.), p. 149. Amsterdam, London and New York (1968c).
95. R. K. CRANE, in: Intracellular Transport (K. B. Warren, ed.), p. 71, New York and London (1966a).
96. R. K. CRANE, *Gastroenterology*, **50**, 254 (1966b).
97. G. SEMENZA, in: Handbook of Physiology, Sec. 6, Alimentary Canal, **5**, Ch. 119, p. 2543. Washington (1968a).
98. G. SEMENZA, *Modern Probl. Pediatrics*, **11**, 32 (1968c).
99. G. SEMENZA, in: Protides of the Biol. Fluids, **15** (H. Peeters, ed.), p. 201. Amsterdam, London and New York (1968d).
100. N. J. GREENBERGER, *Am. J. Med. Sci.*, **258**, 144 (1969).
101. K. MYRBACK and E. VASSEUR, *Hoppe-Seyler's Ztschr. f. physiol. Chemie*, **277**, 171 (1943).
102. A. ROTHSTEIN and R. MEIER, *J. Cell. Compar. Physiol.*, **32**, 77 (1948).
103. D. J. DEMIS, A. ROTHSTEIN and R. MEIER, *Arch. Biochem. Biophys.*, **48**, 55 (1954).
104. A. ROTHSTEIN, in: Handbuch der Protoplasmaforsch., Bd. II, Cytoplasma, 1 (1954).
105. J. B. BEST, *J. Cell. Compar. Physiol.*, **46**, 29 (1955).
106. G. DE LA FUENTE and A. SOLS, *Biochim. Biophys. Acta*, **56**, 49 (1962).
107. D. D. SUTTON and J. O. LAMPEN, *Biochim. Biophys. Acta*, **56**, 303 (1962).
108. A. M. UGOLEV, Congressus Gastroenterologiae Internationalis, VIII, p. 357. Prague (1968d).
109. A. M. UGOLEV, Medizin und Naturwissenschaften, S. 815 (1969a).
110. A. M. UGOLEV, Modern Gastroenterology (VIII Intern. Congr. of Gastroenterol., July 7-13, 1968, Prague), p. 1017. Stuttgart and New York (1969b).
111. G. SEMENZA, *Caries Res.*, **1**, 187 (1967).
112. G. GOMORI, *Proc. Soc. Exptl. Biol. Med.*, **42**, 23 (1939).
113. G. GOMORI, *J. Cell. Compar. Physiol.*, **17**, 71 (1941).
114. H. W. DEANE and E. W. DEMPSEY, *Anat. Record*, **93**, 401 (1945).
115. E. W. DEMPSEY and H. W. DEANE, *J. Cell. Comp. Physiol.*, **27**, 159 (1946).
116. V. M. EMMEL, *Anat. Record*, **95**, 159 (1946).

117. B. F. MARTIN and F. JACOBY, *J. Anat.*,83, 351 (1949).
118. J. R. G. BRADFIELD, *Biol. Reviews*, 25, 113 (1950).
119. H. ZETTERQVIST and T. R. HENDRIX, *Bull. Johns Hopkins Hosp.*, 106,240 (1960).
120. S. L. CLARK, *Anat. Record*, 139, 216 (1961).
121. C. T. ASHWORTH, W. C. CHEARS Jr., E. SANDERS and M. B. PEARCE, *Arch. Pathol.*, 71, 13 (1961).
122. C. T. ASHWORTH, F. J. LUIBEL and S. C. STEWART, *J. Cell. Biol.*, 17, 1 (1963).
123. A. PRZELECKA, G. EJSMOND, E. G. SARZALA and M. TOROCHA, Vth Intern. Congr. of Biochem., 1, p. 399. Izd. AN SSSR, Moscow (1961).
124. R. A. BRODSKII, *Arch. Anat., Gistol. i Embriol.*, 42, 92 (1962).
125. R. A. BRODSKII, Symposium, Physiology and Pathology of Absorption in Gastrointestinal Tract, p. 150. Odessa (1964).
126. H. A. PADYKULA, *Fed. Proc.*, 21, 873 (1962).
127. J. DAWSON and J. PRYSE-DAVIES, *Gastroenterology*, 44, 745 (1963).
128. R. NOACK and G. SCHENK, *Biochem. Ztschr.*, 343, 139 (1965).
129. R. NOACK, G. SCHENK, J. PROLL, P. HAHN and O. KOLDOVSKY, *Biochem. Ztschr.*, 343, 146 (1965).
130. J. JOS, J. FREZAL, J. REY and M. LAMY, *Nature*, 213, 516 (1967).
131. F. MOOG and R. GREY, *J. Cell. Biol.*, 32, C1 (1967).
132. M. S. LAASPERE and L. A. VILLAKO, Ann. Tartu University, fasc. 215, Works on Medicine, XVIII, Gastroenterology, p. 237 (1968).
133. N. K. GHOSH and L. KOTOWITZ, *Enzymologia*, 36,54 (1969).
134. L. L. GALLO and C. R. TREADWELL, *Proc. Soc. Exptl. Biol. Med.*, 114, 69 (1963).
135. H. RUTTLOFF, R. NOACK, R. FRIESE and G. SCHENK, *Biochem. Ztschr.*, 341, 15 (1964).
136. A. EICHHOLZ and R. K. CRANE, *J. Cell. Biol.*, 26, 687 (1965).
137. J. OVERTON, A. EICHHOLZ and R. K. CRANE, *J. Cell. Biol.*, 26, 693 (1965).
138. A. EICHHOLZ, *Biochim. Biophys. Acta*, 163, 101 (1968).
139. A. EICHHOLZ, *Fed. Proc.*, 28, 30 (1969).
140. Y. TAKESUE and R. SATO, *J. Biochem.*, 64, 873 (1968).
141. A. M. UGOLEV, *Bull. Experim. Biologii i Med.*, 52, 8 (1961b).
142. M. DIXON and E. WEBB, Enzymes (2nd ed.). Longmans (1964).
143. T. E. BARMAN, Enzyme Handbook, 1. Berlin-Heidelberg, New York (1969).
144. A. M. UGOLEV, N. N. IEZUITOVA, N. A. KASHINA, B. I. SABSAI, T. I. SPIRIDENKOVA, N. M. TIMOFEEVA, I. N. SCHTCHELOVANOVA, T. V. SHALYGINA and O. E. SHERSTOBITOV, Symposium 'Physiology and Pathology of Absorption in the Gastrointestinal Tract', Abstracts and Reports, p. 240. Odessa (1961).

145. A. M. UGOLEV, N. N. IEZUITOVA and P. DE LAEY, *Nature*, **203**, 879 (1964a).
146. A. M. UGOLEV, N. N. IEZUITOVA, N. M. TIMOFEEVA and I. N. FEDJUSHINA, *Nature*, **202**, 807 (1964b).
147. A. M. UGOLEV, N. N. IEZUITOVA, N. M. TIMOFEEVA and R. I. KUSHAK, *Die Nahrung*, **11**, 595 (1967a).
148. N. N. IEZUITOVA, N. M. TIMOFEEVA, A. M. UGOLEV and I. N. FEDUSHINA, Data of the Scientific Conference on Physiology and Pathology of Intestine, p. 16. Moscow (1962).
149. N. N. IEZUITOVA, P. DE LAEY, M. MARAUSKA, V. R. MEBOLDT, N. M. TIMOFEEVA, A. M. UGOLEV, I. N. FEDUCHINA and K. I. KHALIMOV, in: Physiology and Pathology of Digestive System (Reports). Data of the Scientific Conference in honour of I. P. Rasenkov's seventy-fifth birthday, p. 59. Moscow (1963d).
150. N. N. IEZUITOVA, A. M. UGOLEV and I. N. FEDUSHINA, *Dokl. Akad. Nauk SSSR*, **149**, 746 (1963b).
151. N. N. IEZUITOVA, P. DE LAEY and A. M. UGOLEV, *Dokl. Akad. Nauk SSSR*, **159**, 1191 (1964a).
152. N. N. IEZUITOVA, First All-Union Biochemical Conference, Abstracts, 2, Supplementary (sectional) Conference 1-9, p. 43. Izd. AN SSSR, Moscow and Leningrad (1963).
153. N. N. IEZUITOVA, Data of Scientific Conference on the Problem: 'Physiology and Pathology of Cortico-Visceral Interrelationships and Functional Systems of the Organism', p. 415. Ivanovo (1965).
154. N. M. TIMOFEEVA, First All-Union Biochem. Confer., Abstracts, 2, Supplementary (sectional) Confer., 1-9, p. 50. Izd. AN SSSR, Moscow and Leningrad (1963).
155. N. M. TIMOFEEVA, Data of the Scientific Conference on the Problem 'Physiology and Pathology of Cortico-Visceral Interrelationships and Functional Systems of the Organism', 2, p. 343. Ivanovo (1965).
156. R. I. KUSHAK and A. M. UGOLEV, *Dokl. Akad. Nauk SSSR*, **168**, 477 (1966).
157. A. M. UGOLEV and M. Yu. CHERNYAKHOVSKAYA, Materials on the XVIth Scient. Session of Inst. of Nutrition AMN SSSR, p. 146. Moscow (1966).
158. A. M. UGOLEV and R. I. KUSHAK, *Nature*, **212**, 859 (1966).
159. M. Yu CHERNYAKHOVSKAYA and A. M. UGOLEV, *Dokl. Akad. Nauk SSSR*, **187**, 701 (1969).
160. B. LINDEMAN and A. K. SOLOMON, *J. Gen. Physiol.*, **45**, 801 (1962).
161. D. H. SMYTH and E. M. WRIGHT, *J. Physiol.*, **182**, 591 (1966).
162. A. K. SOLOMON, *J. Gen. Physiol.*, **51**, 335 (1968).
163. J. E. GERMAIN, Catalyse heterogene. Paris (1959).
164. H. RUTLOFF, R. FRIESE and K. TAUFEL, *Hoppe Seyler's Ztschr. f. physiol. Chemie*, **341**, 134 (1965).
165. T. ODA, S. SEKI, J. TAKESUE and R. SATO, Kosokagaku Shinposium, **17**, 389 (1964).

166. T. ODA and S. SEKI, 6th Intern. Congr. Electron Microscopy, p. 387. Kyoto (1966).
167. C. F. JOHNSON, *Science*, **155**, 1670 (1967).
168. C. F. JOHNSON, *Fed. Proc.*, **28**, 26 (1969).
169. Y. NISHI, T. O. YOSHIDA and Y. TAKESUE, *J. Molec. Biol.*, **37**, 441 (1968).
170. M. FRIEDRICH, R. NOACK and G. SCHENK, *Biochem. Ztschr.*, **343**, 346 (1965).
171. R. NOACK, Data on the XVIth Scientific Session of the Inst. of Nutrition AMN SSSR, **1**, p. 147. Moscow (1966).
172. R. NOACK, Biochemische, ernahrungsphysiologische und klinisch-chemische Untersuchungen uber Verdauungsenzyme der Dunndarmmucosa. Habilitationsschrift Humboldt-Universitat, Berlin (1969).
173. N. M. TIMOFEEVA and R. I. KUSHAK, Materials on the XVIth Scientific Session of Inst. of Nutrition AMN SSSR, **1**, p. 150. Moscow (1966).
174. A. DAHLQVIST, Hog Intestinal α-Glucosidases. Solubilization, Separation and Characterization. Lund (1960).
175. K. TAUFEL, *Die Nahrung*, **13**, 631 (1969c).
176. J. B. RHODES, in: Handbook of Physiology, Sec. 6, Alimentary Canal, **5**, p. 2589. Washington (1968).
177. F. MOOG, in: Bioch. Animal Development, p. 307 (Academic Press, New York and London (1965).
178. F. MOOG, M. E. EYLER and R. D. GREY, *Ann. of the N.Y. Acad. Sci.*, **166**, 447 (1969).
179. J. S. K. DAVID, P. MALATHI and J. GANGULY, *Biochem. J.*, **98**, 662 (1966).
180. Y. TAKESUE, T. ODA and T. KASHIWAGI, Ann Rep. of Res. Inst. Environment. Med. Nagoya Univ., **18**, 145 (1966).
181. Y. TAKESUE, T. KASHIWAGI and T. O. YOSHIDA, 7th Int. Congress of Biochemistry, Abstr., **V**, 1-29, 920 (1967).
182. E. EGGERMONT, The Biochemical Defects in Sucrase Intolerance and in Glucose-galactose Malabsorption. Universite catholique de Louvain, Belgium (1968).
183. G. SEMENZA and J. KOLINSKA, in: Protides of the Biol. Fluids, **15**, p. 581 (H. Peeters, ed.). Amsterdam, London and New York (1968).
184. J. W. HEATON, *Gastroenterology*, **58**, 1044 (1970).
185. R. GITZELMANN, Th. BACHI, H. BINZ, J. LINDEMANN and G. SEMENZA, *Biochim. Biophys. Acta*, **196**, 20 (1970).
186. S. AURICCHIO, A. DAHLQVIST and G. SEMENZA, *Biochim. Biophys. Acta*, **73**, 582 (1963).
187. A. DAHLQVIST and D. L. THOMSON, *Acta Physiol. Scand.*, **59**, 111 (1963a).
188. A. DAHLQVIST and D. L. THOMSON, *J. Physiol.*, **167**, 193 (1963b).
189. A. DAHLQVIST, B. BULL and D. L. THOMSON, *Arch. Biochem. Biophys.*, **109**, 159 (1965b).

190. G. SEMENZA, S. AURICCHIO and A. RUBINO, *Biochim. Biophys. Acta*, **96**, 487 (1965).
191. D. L. THOMSON, *Gastroenterology*, **48**, 854 (1965).
192. A. E. EICHHOLZ and R. K. CRANE, *Fed. Proc.*, **25**, 656 (1966).
193. D. I. I. HSIA, M. MAKLER, G. SEMENZA and A. PRADER, *Biochim. Biophys. Acta*, **113**, 390 (1966).
194. E. EGGERMONT, *Europ. J. Biochem.*, **9**, 483 (1969).
195. D. H. ALPERS, *J. Biol. Chem.*, **244**, 1238 (1969).
196. H. G. ASP, A. DAHLQVIST and O. KOLDOVSKY, *Biochem. J.*, **114**, 351 (1969).
197. A. DAHLQVIST and U. TELENIUS, *Biochem. J.*, **111**, 139 (1969).
198. Y. TAKESUE and T. KASHIWAGI, *J. Biochem.*, **65**, 427 (1969).
199. B. BORGSTROM and A. DAHLQVIST, *Acta Chem. Scand.*, **12**, 1997 (1958).
200. A. DAHLQVIST, *Acta Chem. Scand.*, **13**, 945 (1959a).
202. A. DAHLQVIST, *Biochem. J.*, **86**, 72 (1963).
203. J. A. CARNIE and J. W. PORTEOUS, *Biochem. J.*, **85**, 620 (1962b).
204. Y. TAKESUE, *J. Biochem.*, **65**, 545 (1969).
205. A. A. GRUZDKOV, V. V. EGOROVA, N. N. IEZUITOVA, N. M. TIMOFEEVA, E. Kh. TULYGANOVA, M. Yu. CHERNYAKHOVSKAYA and A. M. UGOLEV, Materials of the All-Union Conference on Gastroenterology, p. 53. Riga (1970).
206. J. A. CARNIE and J. W. PORTEOUS, *Biochem. J.*, **85**, 450 (1962a).
207. R, GITZELMANN, E. DAVIDSON and J. OSINCHAK, *Biochim. Biophys. Acta*, **85**, 69 (1964).
208. N. S. HELMAN, *Uspekhi Sovrem. Biologii*, **68**, 3 (1969).
209. L. JOSEFSSON and H. SJOSTROM, *Acta Physiol. Scand.*, **67**, 27 (1966).
210. R. I. KUSHAK, D. R. GIGURE and G. G. SCHERBAKOV, Materials of the All-Union Conference on Gastroenterology, p. 64. Riga (1970).
211. N. M. TIMOFEEVA, N. N. IEZUITOVA, T. Ya NADIROVA and A. M. UGOLEV, Brief Account on Reports of the Scientific Conference, p. 272. Lvov (1965).
212. A. M. UGOLEV, N. N. IEZUITOVA, T. Ya. NADIROVA and N. M. TIMOFEEVA, *Dokl. Akad. Nauk SSSR*, **166**, 472 (1966).
213. R. I. GAVRILOV and V. Yu. DAMANSKII, *Radiobiologiya*, **6**, 394 (1966).
214. A. M. UGOLEV, N. N. IEZUITOVA, N. M. TIMOFEEVA and N. V. TOROPOVA, Materials from the Conference on Physiol. and Biochem. of the Function Systems in Organism, p. 146. Kiev (1968b).
215. M. P. DOMBROVSKAYA, N. N. IEZUITOVA, N. M. TIMOFEEVA and A. M. UGOLEV, VIIIth Scientific Conference on Corticovisceral Relationships in Physiol., Med., and Biology on Occasion of the 50th Anniv. of October Revol., p. 52. Theses of Reports, I (1967).

216. A. M. UGOLEV, N. M. TIMOFEEVA, N. N. IEZUITOVA, M. P. DOMBROVSKAYA and K. R. RAKHIMOV, *Dokl. Akad. Nauk SSSR*, **188**, 489 (1969b).
217. A. M. UGOLEV, N. M. TIMOFEEVA, N. N. IEZUITOVA and K. R. RAKHIMOV, Symposium. 'Adaptation of Organism of Man and Animals to Extreme Natural Factors of Surroundings', p. 75. Novosibirsk (1970b).
218. N. M. TIMOFEEVA, N. N. IEZUITOVA, K. R. RAKHIMOV and A. M. UGOLEV, IVth Scientific Conference on Biochem. and Pharmac. West. Siberian Union, 1, Physiology, p. 467. Krasnoyarsk (1969).
219. B. F. POGLAZOV, Assembling of Biological Structures. Izd. 'Nauka', Moscow (1970).
220. P. DE LAEY and N. N. IEZUITOVA, *Dokl. Akad. Nauk SSSR*, **146**, 731 (1962).
221. N. N. IEZUITOVA, P. DE LAEY and A. M. UGOLEV, *Biochim. Biophys. Acta*, **86**, 205 (1964b).
222. N. V. TOROPOVA, in: The Realistic Problems of Obstetrics and Pediatry. Collective Scientific Works of Kirghiz Scientific Research Inst. for Care of Motherhood and Childhood, vip. 2, p. 158. Frunze (1964).
223. A. M. UGOLEV and M. K. MARAUSKA, *Bull. Experim. Biologii i Med.*, **57**, 16 (1964).
224. A. M. UGOLEV and I. K. SALENIETSE, *Bull. Experim. Biologii i Med.*, **58**, 15 (1964).
225. K. R. RAKHIMOV, in: Physiology and Pathology of Digestion, p. 247. Lvov (1965).
226. N. V. TOROPOVA, O. T. TOKTORBAYEVA and Yu. M. TOROPOV, Reports of the 7th Scientific Conference of the problem of Age Morphol., Physiol., Biochem., p. 464. Moscow (1965a).
227. N. V. TOROPOVA, Yu. M. TOROPOV and O. T. TOKTOBAYEVA, in: Metabolism in Animals and Plants, p. 162. Frunze (1965b).
228. P. DE LAEY, *Nature*, **212**, 78 (1966a).
229. P. DE LAEY, *Die Nahrung*, **10**, 641 (1966b).
230. P. DE LAEY, *Die Nahrung*, **10**, 649 (1966c).
231. P. DE LAEY, *Die Nahrung*, **10**, 655 (1966d).
232. P. DE LAEY, *Die Nahrung*, **11**, 1 (1967a).
233. P. DE LAEY, *Die Nahrung*, **11**, 9 (1967b).
234. P. DE LAEY, *Die Nahrung*, **11**, 17 (1967c).
235. P. DE LAEY, in: Protides of the Biological fluids (H. Peeters, ed.), p. 159. Amsterdam, London and New York (1968).
236. D. M. GOLDBERG, R. CAMPBELL and A. D. ROY, *Biochim. Biophys. Acta*, **167**, 613 (1968).
237. C. HOOFT, J. Van HAUWAERT, P. DE LAEY and K. ADRIAENSSENS, *Helv. Pediatri. Acta*, **18**, 502 (1963).
238. C. G. MASEVICH, A. M. UGOLEV and E. K. ZABELINSKII, *Terapeuticheskij Arkhiv*, **39**, 53 (1967).
239. A. M. UGOLEV, in: A. M. Ugolev, N. N. Iezuitova, C. G. Masevich,

T. Ya. Nadirova and N. M. Timofeeva, Investigation of the Digestive Apparatus in Man, p. 196. Izd. 'Nauka', Moscow and Leningrad (1969c).
240. G. I. LOGINOV and A. M. UGOLEV, IVth Scientific Conference of Physiologists, Histologists and Anatomists of Pedag. Inst. RSFSR. Yaroslavl (1969).
241. A. M. UGOLEV and E. M. USTINKOVA, IXth Confer. on Physiology of Digestion. Abstr. of Reports, Part II, p. 122. Odessa (1967).
242. N. K. ADAMS, The Physics and Chemistry of Surface (3rd ed.). Oxford (1947).
243. G. A. DEBORIN and L. B. GORBACHEVA, *Biokhimya*, 18, 618 (1953).
244. A. D. McLAREN and K. L. BABCOCK, in: Subcellular Particles, p. 23. New York (1959).
245. A. KATCHALSKY, in: Membrane Transport and Metabolism, (A. Kleinzeller, A. Kotyk, eds.) p. 69. Academic Press, New York (1961).
246. A. KATCHALSKY, *Biophys. J.*, 4, 9 (1964).
247. A. S. CURTIS, The Cell Surface; Its Molecular Role in Morphogenesis. Logos Press and Academic Press (1967).
248. S. Ya. DAVYDOVA, *Cytologiya*, 9, 1248 (1967).
249. O. M. POLTORAK, *Zhurnal Fizich. Khimiyi*, 41, 2544 (1967).
250. H. TIEN, *J. Gen. Physiol.*, 52, 125 (1968).
251. J. T. DAVIES and E. RIDEAL, Interfacial Phenomena. Academic Press, New York (1961).
252. R. M. C. DAWSON, in: Biological Membranes. Physical Fact and Function (D. Chapman, ed.). Academic Press, London and New York (1968).
253. A. D. McLAREN, *Science*, 125, 697 (1957).
254. I. K. GOZITE, Materials of the All-Union Conference on Gastroenterology, p. 50. Riga (1970).
255. V. V. EGOROVA, E. Kh. TULYAGANOVA, G. G. SCHERBAKOV and A. M. UGOLEV, Materials of the IIIrd Conference on 'Importance of Fat in Nutrition and Enlargement of Nutritive Products Range with Use of Plant Oils', part II, p. 40. Moscow (1969).
256. G. G. LAPTEVA and A. M. UGOLEV, in ref. 9.
257. A. BAR-ELI and E. KATCHALSKI, *Nature*, 188, 856 (1960).
258. A. BAR-ELI and E. KATCHALSKI, *J. Biol. Chem.*, 238, 1690 (1963).
259. J. J. CEBRA, D. GIVOL, H. J. SILMAN and E. KATCHALSKI, *J. Biol. Chem.*, 236, 1720 (1961).
260. M. A. MITZ and L. J. SUMMARIA, *Nature*, 189, 576 (1961).
261. B. P. SURINOV, Highly Active Water-Insoluble Compounds of Trypsin and Chymotrypsin with Cellulose. Diss., Leningrad (1965).
262. B. S. DISKINA, *Biokhimya*, 25, 43 (1960).
263. T. Yu. UGAROVA and B. S. DISKINA, *Biokhimiya*, 29, 914 (1964).

264. H. W. SMITH and W. E. CRABB, *J. Pathol. Bacteriol.*, **82**, 53 (1961).
265. T. ROSEBURY, Microorganisms Indigenous to Man, p. 435. Blakiston, New York (1962).
266. R. M. DONALDSON Jr., *New Engl. J. Med.*, **270**, 938, 994, 1050 (1964).
267. R. M. DONALDSON Jr., in: Handbook of Physiology, Sec. 6, Alimentary Canal, 5, p. 2807. Washington (1968).
268. T. D. LUCKEY, Confer. on Nutrition on Space and Waste Problems, p. 227. Florida (1964).
269. R. W. SCHAEDLER, R. DUBOS and R. COSTELLO, *J. Exptl. Med.*, **122**, 59 (1965a).
270. R. W. SCHAEDLER, R. DUBOS and R. COSTELLO, *J. Exptl. Med.*, **122**, 77 (1965b).
271. H. W. SMITH, *J. Pathol. Bacteriol.*, **90**, 495 (1965).
272. M. E. COATES, H. A. GORDON and B. S. WOSTMAN, The Germfree Animals in Research. Academic Press, London and New York (1968).
273. B. S. DRASAR and M. SHINER, *Gut*, **10**, 812 (1969).
274. R. J. DUBOS and R. W. SCHAEDLER, *J. Exptl. Med.*, **115**, 1161 (1962).
275. A. CANAS-RODRIGUEZ, and H. W. SMITH, *Biochem. J.*, **100**, 79 (1966).
276. J. C. HAMPTON and B. ROSARIO, *Lab. Invest.*, **14**, 1464 (1965).
277. H. A. GORDON, *Am. J. Digest. Diseases*, **5**, 841 (1960).
278. H. A. GORDON and E. BRUCKNER-KARDOSS, *Am. J. Physiol.*, **201**, 175 (1961).
279. H. SPRINZ, D. W. KUNDEL, G. J. DAMMIN, R. E. HOROWITZ, H. SCHNEIDER and S. B. FORMAL, *Am. J. Pathol.*, **39**, 681 (1961).
280. H. SPRINZ, *Feder. Proc.*, **21**, 57 (1962).
281. H. SPRINZ, Proc. 16th Intern. Congr. Zool., 3, 155 (1963).
282. G. D. ABRAMS, H. BAUER and H. SPRINZ, *Lab. Invest.*, **12**, 355 (1963).
283. S. LESHER, H. E. WALBURG and G. A. SACHER, *Nature*, **202**, 884 (1964).
284. G. D. ABRAMS and J. E. BISHOP, *Arch. Pathol.*, **79**, 213 (1965).
285. R. B. FISHER and D. S. PARSONS, *J. Anat.*, **84**, 272 (1950).
286. G. L. BRODY, J. E. BISHOP and G. D. ABRAMS, *Arch. Pathol.*, **81**, 258 (1966).
287. H. R. JERVIS and D. C. BIGGERS, *Anat. Record*, **148**, 591 (1964).
288. A. DAHLQVIST, B. BULL and B. E. GUSTAFSSON, *Arch. Biochem.*, **109**, 150 (1965a).
289. B. S. REDDY and B. S. WOSTMAN, *Arch. Biochem. Biophys.*, **113**, 609 (1966).
290. B. B. MIGICOVSKY, A. M. NIELSON, M. GLUCK and R. BURGESS, *Arch. Biochem. Biophys.*, **34**, 479 (1951).
291. R. W. CARROLL, G. W. HENSLEY, C. L. SITTLER, E. L. WILCOX and W. R. GRAHAM, *Arch. Biochem. Biophys.*, **45**, 260 (1953).
292. H. H. DRAPER, *J. Nutr.*, **64**, 33 (1958).

293. J. B. HENEGHAN, *Amer. J. Physiol.*, **205**, 412 (1963).
294. A. C. FRAZER, *Fed. Proc.*, **20**, 417 (1961).
295. B. E. GUSTAFSSON and A. NORMAN, *Proc. Soc. Exptl. Biol. Med.*, **110**, 387 (1962).
296. A. NORMAN and M. S. SHORB, *Proc. Soc. Exptl. Biol. Med.*, **110**, 552 (1962).
297. O. W. PORTMAN, *Fed. Proc.*, **21**, 896 (1962).
298. J. P. W. WEBB, A. T. JAMES and T. D. KELLOCK, *Gut*, **4**, 37 (1963).
299. E. EVRARD, P. P. HOET and H. EYSSEN, *Brit. J. Exptl. Pathol.*, **45**, 409 (1964).
300. B. E. GUSTAFSSON, T. MIDTVEDT AND A. NORMAN, *J. Exptl. Med.*, **123**, 413 (1966).
301. B. BORGSTROM, A. DAHLQVIST, B. E. GUSTAFSSON, G. LUNDH and J. MALMQUIST, *Proc. Soc. Exptl. Biol. Med.*, **102**, 154 (1959).
302. R. M. DONALDSON Jr., H. A. DOLCINI and S. J. GRAY, *Amer. J. Physiol.*, **200**, 794 (1961).
303. E. W. STRAUSS, R. M. DONALDSON Jr. and F. H. GARDNER, *Lancet*, **2**, 736 (1961).
304. R. M. DONALDSON Jr., *Amer. J. Physiol.*, **202**, 289 (1962).
305. H. BOSTROM, B. E. GUSTAFSSON and B. WENGLE, *Proc. Soc. Exptl. Biol. Med.*, **114**, 742 (1963).
306. B. J. HAVERBACK, B. J. DYCE, P. J. GUTENTAG and D. W. MONTGOMERY, *Gastroenterology*, **44**, 588 (1963).
307. S. LEPKOVSKY, M. WAGNER, F. FURUTA, K. OZONE and T. KOIKE, *Poultry Sci.*, **43**, 722 (1964).
308. D. CARTER, A. EINHEBER, H. BAUER, *Surg. Forum*, **16**, 79 (1965).
309. G. LINDSTEDT, S. LINDSTEDT and B. E. GUSTAVSSON, *J. Exptl. Med.*, **121**, 201 (1965).
310. I. J. KAHN, G. H. JEFFRIES and M. H. SLEISENGER, *New England J. Med.*, **274**, 1339 (1966).
311. A. M. UGOLEV, *Vestnik Akad. Nauk SSSR*, **7**, 43 (1966b).
312. A. R. VALDMAN, Materials of the All-Union Conference on Gastroenterology, p. 309. Riga (1970).
313. J. H. De BOER, The Dynamic Nature of Adsorption. IIL, Moscow (1962). English edition: Oxford University Press (1953).
314. J. H. De BOER, in: Catalysis. Certain Aspects of Theory and Technology of Organic Reactions, p. 18. IIL Moscow (1959).
315. N. N. IEZUITOVA, N. M. TIMOFEEVA, O. K. KOLDOVSKII, Ya. Ya. NURKS and A. M. UGOLEV, *Dokl. Akad. Nauk SSSR*, **154**, 990 (1964c).
316. N. N. IEZUITOVA, N. M. TIMOFEEVA, O. K. KOLDOVSKII, Ya. Ya. NURKS and A. M. UGOLEV, Annals of Kirghiz Scientific Research Institute for Care of Motherhood and Childhood, p. 114. Frunze (1964d).
317. G. WISEMAN, Absorption from the Intestine. Academic Press, New York (1964).

318. G. WISEMAN, in: Handbook of Physiology, Sec. 6, Alimentary Canal, 3, p. 1277. Washington (1968).
319. S. G. SCHULTZ and P. F. CURRAN, in: Handbook of Physiology, Sec. 6, Alimentary Canal, 3, p. 1245. Washington (1968).
320. J. S. FORDTRAN and F. J. INGELFINGER, in: Handbook of Physiology, Sec. 6, Alimentary Canal, 3, p. 1457. Washington (1968).
321. H. NEWEY and D. H. SMYTH, in: Alimentary Tract, Part III, Ch. 10, p. 347 (1970).
322. Ya. P. SKLYAROV, The Absorptive Capacity of Small Intestine. Izd. 'Zdorovye', Kiev (1966).
323. M. K. MARAUSKA, I. SALENIETSE and A. M. UGOLEV, Data of VIth Scientific Conference on Age Morphology, Physiology, and Biochemistry, p. 402. Izd. APN RSFSR, Moscow (1963).
324. I. K. SALENIETSE, A. T. STEPANOVA and A. M. UGOLEV, Data and Abstracts of Reports of Conference, p. 151. Gagra, Tbilisi (1963).
325. C. W. BAILEY, W. D. KITTS and A. J. WOOD, *Canad. J. Agric. Sci.*, **36**, 51 (1956).
326. K. J. HILL, *Quart. J. Exper. Physiol.*, **41**, 421 (1956).
327. B. MOSINGER, Z. PLACER and O. KOLDOVSKY, *Nature*, **184**, 1245 (1959).
328. A. T. STEPANOVA, VIth Conference of Young Scientists Inst. of Normal and Pathol. Physiol. AMN SSSR, Abstr. of Reports, p. 81. Moscow (1960).
329. A. ALVAREZ and J. SAS, *Nature*, **190**, 826 (1961).
330. R. G. DOELL and N. KRETCHMER, *Biochim. Biophys. Acta*, **62**, 353 (1962).
331. R. G. DOELL and N. KRETCHMER, *Fed. Proc.*, **22**, 495 (1963).
332. O. KOLDOVSKY, Development of the Functions of the Small Intestine in Mammals and Man. Basel and New York (1969).
333. G. I. LOGINOV, Materials on Republic Conference, p. 124. Izd. 'Zdorovye', Kiev (1969).
334. R. MALI, in: L. Herman, Physiology of the Mammals, 5. St Petersburg (1886).
335. E. H. STARLING, Mercers' Company Lectures on Recent Advances in the Physiology of Digestion. London (1906).
336. K. M. BYCKOV, Lectures on the Physiology of Digestion. Leningrad (1940).
337. J. J. DEREN, in: Handbook of Physiology, Sec. 6, Alimentary Canal, 3, p. 1099. Washington (1968).
338. T. H. WILSON and T. N. VINCENT, *J. Biol. Chem.*, **216**, 851 (1955).
339. E. B. CHAIN, K. R. L. MANSFORD and F. POCCHIARI, *J. Physiol.*, **154**, 39 (1960).
340. J. D. GUPTA, A. M. DAKROURY and A. E. HARPER, *J. Nutr.*, **64**, 447 (1958).
341. C. W. CRANE and A. NEUBERGER, *Biochem. J.*, **74**, 313 (1960).
342. D. H. SMYTH, *Proc. Roy. Soc. Med.*, **54**, 769 (1961).

343. C. D. HOLDSWORTH and A. M. DAWSON, *Clin. Sci.*, **27**, 371 (1964).
344. G. M. GRAY and F. J. INGELFINGER, *J. Clin. Invest.*, **44**, 390 (1965).
345. N. N. IEZUITOVA, T. Ya. NADIROVA, N. V. TOROPOVA and A. M. UGOLEV, in: Physiology and Pathology of Digestion. Brief Communications of Scientific Conference 24-28 September, p. 102. Lvov (1965).
346. D. S. PARSONS and J. S. PRICHARD, *Nature*, **208**, 1097 (1965).
347. H. B. McMICHAEL, J. WEBB and A. M. DAWSON. *Clin. Sci.*, **33**, 235 (1966).
348. D. M. MATTHEWS, *Hospital Medicine*, September, 1382 (1968a).
349. D. M. MATTHEWS, *Lancet*, **49**, 6 July (1968b).
350. D. M. MATTHEWS, *Lancet*, **401**, 17 August (1968c).
351. D. M. MATTHEWS, I. L. CRAFT and R. F. CRAMPTON, *Lancet*, **49**, 6 July (1968a).
352. I. M. McDONALD and L. F. TURNER, *Lancet*, **2**, 841 (1968).
353. D. S. PARSONS and J. S. PRICHARD, *J. Physiol.*, **198**, 405 (1968a).
354. D. S. PARSONS and J. S. PRICHARD, *J. Physiol.*, **199**, 137 (1968b).
355. J. S. PRICHARD, *Nature*, **221**, 369 (1969).
356. A. G. NOIM, Ya. Ya. NURKS and A. M. UGOLEV, Materials of the All-Union Conference of Gastroenterology, p. 78. Riga (1970).
357. H. NEWEY and D. H. SMYTH, *J. Physiol.*, **152**, 70P (1960).
358. H. NEWEY and D. H. SMYTH, *J. Physiol.*, **164**, 527 (1962).
359. I. L. CRAFT and D. M. MATTHEWS, *Brit. J. Surg.*, **55** (2), 158 (1968).
360. I. L. CRAFT, D. GEDDES, C. W. HYDE, I. J. WISE and D. M. MATTHEWS, *Gut*, **9**, 425 (1968a).
361. I. L. CRAFT, D. M. GEDDES and D. M. MATTHEWS, *J. Physiol.*, **196**, 31P (1968b).
362. I. L. CRAFT, R. F. CRAMPTON, M. T. LIS and D. M. MATTHEWS, *J. Physiol.*, **200**, 111P (1969).
363. D. M. MATTHEWS, I. L. CRAFT, D. M. GEDDES, I. J. WISE and C. W. HYDE, *Clin. Sci.*, **35**, 415 (1968b).
364. D. M. MATTHEWS, M. T. LIS, B. CHENG and R. F. CRAMPTON, *Clin. Sci.*, **37**, 751 (1969).
365. A. M. ASATOOR, J. K. BANDOH, A. F. LANT, M. D. MILNE and F. NAVAB, *Gut*, **11**, 250 (1970).
366. B. CHENG and D. M. MATTHEWS, *J. Physiol.*, **210**, 37P (1970).
367. R. J. C. BARRY, M. J. JACKSON and D. H. SMYTH, *J. Physiol.*, **185**, 667 (1966).
368. D. M. MATTHEWS, D. M. GEDDES, I. L. CRAFT, I. J. WISE and C. W. HYDE, *Gut*, **9**, 365 (1968c).
369. G. SEMENZA, R. TOSI, M. C. VALLOTTON-DELACHAUX and E. MULHAUPT, *Biochim. Biophys. Acta*, **89**, 109 (1964).
370. R. K. CRANE, P. MALATHI, W. F. CASPARY et al., *Fed. Proc.*, **29**, 595 Abstr. (1970).

371. I. K. GOZITE, A. A. GRUZDKOV, V. V. EGOROVA, N. N. IEZUITOVA, G. I. LOGINOV, N. M. MITOSHOVA, Ya. Ya. NURKS, L. F. SMIRNOVA, N. M. TIMOFEEVA, E. Kh. TULJAGANOVA, A. M. UGOLEV, V. A. TSVETKOVA, M. U. CHERNYAKHOVSKAYA and G. G. SCHERBAKOV, 11th All-Union Congress of Physiol., 2, 280 Leningrad (1970).
372. E. BRESNICK and A. SCHWARTZ, Functional Dynamics of the Cell. Academic Press, New York and London (1968).
373. A. PARDEE, *Science*, 162, 632 (1968a).
374. A. B. PARDEE, *J. Gen. Physiol.*, 52, 279 (1968b).
375. I. K. GOZITE, N. M. MITUSHOVA and A. M. UGOLEV, Materials on Physiological Conference of Middle Asia and Kazakhstan, p. 76. Alma-Ata (1969).
376. N. M. MITUSHOVA, Materials on the All-Union Conference of Gastroenterology, p. 18. Riga (1970).
377. N. M. MITUSHOVA and A. M. UGOLEV, *Dokl. Akad. Nauk SSSR*, 195, 503 (1970).
378. A. M. UGOLEV, N. M. IEZUITOVA, R. I. KUSHAK and G. G. SCHERBAKOV, *Die Nahrung*, 14, 437 (1970a).
379. T. U. IZMAILOV, *Izv. Akad. Nauk Kazakh. SSSR*, N 5, 84 (1965).
380. A. Yu. UNUSOV, K. R. RAKHIMOV and Z. N. YAKUSCH, *Uzbeksk. Biologitscheschij Zhurnal*, 5, 32 (1965).
381. Z. N. YAKUSH, in: Problems of Physiology of Man and Animals under Hot Climate Conditions, p. 168. Tashkent (1965).
382. Z. N. YAKUSH, Materials of the IIIrd Conference of Physiologists of Middle Asia and Kazakhstan, p. 427. Dushanbe (1966).
383. Z. N. YAKUSH, in: Biostimulators in Animal Practice. Tashkent (1967).
384. N. U. BAZANOVA and T. U. IZMAILOV, Data of IIIrd Physiological Conference of Middle Asia and Kazakhstan, p. 54. Dunshanbe (1966).
385. R. M. SEIFULLINA, *Izvest. AN Kazakh. SSSR*, 1, (1967).
386. R. M. SEIFULLINA, Materials on the All-Union Conference on Gastroenterology, p. 81. Riga (1970).
387. R. I. KUSHAK, A. Ya. OZOL and A. G. BUIKE, Theses of Reports of IXth Conference on Physiology of Nutrition, p. 172. Odessa (1967).
388. R. I. KUSHAK, Vth Scientific Conference on Evol. Physiol. Dedicated to the Memory of Acad. L. A. Orbeli, p. 144. Leningrad (1968).
389. G. G. SCHERBAKOV, Materials of the Conference on Preventing Disease in Agricultural Animals and Birds. Pskov (1968).
390. G. G. SCHERBAKOV, Materials of the IVth Scient. Confer. of Physiologists, Biochem. and Pharmacol., p. 515. Krasnoyarsk (1969a).
391. G. G. SCHERBAKOV, in: Collection of Works of the Leningrad Veterinary Inst., 30 (1969b).
392. A. T. STEPANOVA, Materials of the All-Union Conference on Gastroenterology, p. 89. Riga (1970).

393. Sh. A. BERMAN, XIth Scientific Conference for the Study of Internal Water Basins, p. 7. Petrozavosk (1964).
394. Sh. A. BERMAN, Theses of the Scientific Conference. Sevastopol (1965).
395. Sh. A. BERMAN and I. K. SALENIETSE, *Vopr. Ikhtiologiyi*, **6**, 720 (1966).
396. V. A. PEGEL and V. A. REMOROV, Materials on the IIIrd All-Union Conference on Ecology, Physiology, Biochemistry and Morphology, p. 83. Novosibirsk (1967).
397. V. A. PEGEL, V. A. REMOROV, A. S. ANTIPIN and V. A. NOVAK, Vth Scientific Conference on Evol. Physiol. Dedicated to the Memory of Acad. L. A. Orbeli, p. 198. Leningrad (1968).
398. R. I. KUSHAK, in: Physiologically Active Components of Nutrition in Animals, p. 361. Izd. 'Znaniye', Riga (1969).
399. A. M. UGOLEV, IVth Scientific Conference on Evol. Physiol. Dedicated to the Memory of Acad. L. A. Orbeli, p. 260. Leningrad (1965b).
400. N. N. IEZUITOVA and Z. F. NOVIKOVA, Materials on IInd All-Union Conference on Ecology, Physiology, Biochemistry and Morphology, p. 49. Novosibirsk (1967).
401. N. M. TIMOFEEVA and L. E. YAROSLAVSKAYA, Data of IIIrd Conference on Ecology, Physiology, Biochemistry and Morphology, p. 105. Novosibirsk (1967).
402. A. M. UGOLEV, N. N. IEZUITOVA, N. M. TIMOFEEVA, M. Yu. CHERNYAKHOVSKAYA, G. G. LAPTEVA, Z. F. NOVIKOVA, V. A. TSVETKOVA and L. E. YAROSLAVSKAYA, IIIrd All-Union Conference on Ecol., Physiol., Biochem. and Morphol., Abstr. of Reports, p. 75. Novosibirsk (1967b).
403. A. M. UGOLEV, N. N. IEZUITOVA, E. D. KESAREVA, N. M. TIMOFEEVA, E Kh. TULYAGNOVA, V. A. TSVEKOVA, M. Yu CHERNYAKHOVSKAYA and G. G. SCHRABAKOV, Abrstracts of Reports of the 5th Enlarged Confr. Dedicated to the Memory of Acad. L. A. Orbeli, p. 255. Izd. 'Nauka', Leningrad (1968a).
404. A. M. UGOLEV, N. M. TIMOFEEVA, N. N. IEZUITOVA, E. D. KESAREVA, M. Yu. CHERNYAKHOVSKAYA, V. A. TSVETKOVA, G. G. SCHERBAKOV, E. G. GURMAN, Z. F. NOVIKOVA and L. E. YAROSLAVSKAYA, in: Symposium: 'Membrane Processes in the Organs of the Digestive System', p. 77. Lvov (1968c).
405. M. Yu. CHERNYAKHOVSKAYA, IIIrd All-Union Conference on Ecology, Physiology, Biochemistry, Morphology, Abstracts of Reports, p. 112. Novosibirsk (1967).
406. L. B. MENDEL and C. S. LEAVENWORTH, *Am. J. Physiol.*, **21**, 95 (1908).
407. J. J. SAMPSON, *J. Biol. Chem.*, **38**, 345 (1919).
408. T. G. KLUMPP and A. NEALE, *Am. J. Dis. Children*, **40**, 1215 (1930).
409. A. P. KRYTCHKOVA, *Fiziologischeskij Zhurnal SSSR*, **27**, 437 (1939).

410. W. F. WINDLE, Physiology of the Fetus. Philadelphia and London (1940).
411. V. ALLEREY, H. STERN and A. E. MIRSKY, Nature, **169**, 128 (1952).
412. J. VERNE, S. HEBERT and O. CHARPAL, C. R. Soc. Biol., **146**, 176 (1952).
413. W. D. KITTS, C. B. BAILEY and A. J. WOOD, Canad. J. Agric. Sci., **36**, 45 (1956).
414. I. A. ARSHAVSKII, Pediatriya, **7**, 24 (1957).
415. H. M. CUNNINGHAM, J. Anim. Sci., **18**, 964 (1959).
416. C. A. SMITH, The Physiology of the Newborn Infant. Springfield (1959).
417. P. A. HARTMAN, V. W. HAYS, R. O. BAKER, L. H. NEAGLE and D. V. CATRON, J. Anim. Sci., **20**, 114 (1961).
418. J. T. HUBER, N. L. JACOBSON et. al., J. Dairy Sci., **44**, 1494 (1961).
419. J. ROKOS, P. HAHN, O. KOLDOVSKY and P. PROCHAZKA, Cesk. fysiol., **11**, 210 (1962a).
420. J. ROKOS, P. HAHN, O. KOLDOVSKY and P. PROCHAZKA, Cesk. fysiol., **11**, 472 (1962b).
421. J. HELLER, Physiol. Bohemoslov., **12**, 526 (1963).
422. F. HOWARD and J. YUDKIN, Brit. J. Nutr., **17**, 281 (1963).
423. L. T. KAPRALOVA, Dokl. Akad. Nauk SSSR, **148**, 985 (1963).
424. Z. V. KUZNETSOVA, Fiziologitscheskij Zhurnal SSSR, **49**, 242 (1963).
425. M. M. NACHLAS, B. MONIS, D. ROSENBLATT and A. M. SELIGMAN, J. Biophys. Biochem. Cytol., **7**, 261 (1960).
426. H. ZETTERQVIST, The Ultrastructural Organization of the Columnar Absorbing Cells of the Mouse Jejunum. Stockholm (1956).
427. T. Ya. NADIROVA, in: Physiology and Pathology of Digestion. Brief Account of Reports of the Scient. Confer., p. 192. Lvov (1965).
    T. Ya. NADIROVA, R. A. OVDEICHUK and A. M. UGOLEV, Symposium: Problems of Biochemical Adaptation, p. 48. Moscow (1965a).
428. T. Ya. NADIROVA, V. A. TIMOFEEVA and A. M. UGOLEV, Bull. Experim. Biologii i Med., **59**, 29 (1965b).
429. R. G. DOELL and N. KRETCHMER, Science, **143**, 42 (1964).
430. N. N. IEZUITOVA, T. Ya. NADIROVA, N. M. TIMOFEEVA, N. V. TOROPOVA, Yu. M. TOROPOV. G. G. SCHERBAKOV and A. M. UGOLEV, Ann. Tartu University, fasc. 215, Works on medicine, XVIII Gastroenterology, p. 230. Tartu (1968).
431. N. N. IEZUITOVA, G. G. SCHERBAKOV, R. I. KUSHAK and A. M. UGOLEV, Dokl. Akad. Nauk SSSR, **192**, 680 (1970).
432. I. H. HOLT and D. H. MILLER, Biochim. Biophys. Acta, **58**, 239 (1962).
433. J. LARNER and R. E. GILLESPIE, J. Biol. Chem., **223**, 709 (1956).

434. J. L. POPE, J. C. McPHERSON and H. C. TIDWELL, *J. Biol. Chem.*, **241**, 2306 (1966).
435. J. B. RHODES, A. EICHHOLZ and R. K. CRANE, *Biochim. Biophys. Acta*, **135**, 959 (1967).
436. R. G. DOELL, G. ROSEN and N. KRETCHMER, *Proc. Nat. Acad. Sci. Washington*, **54**, 1268 (1965).
437. G. BOURNE, *Quart. J. Exper. Physiol.*, **32**, 1 (1943).
438. R. K. MORTON, *Biochem. J.*, **57**, 595 (1954).
439. K. WATANABE and W. H. FISHMAN, *J. Histochem. Cytochem.*, **12**, 252 (1964).
440. A. ROTHSTEIN, R. C. MEIER and T. C. SCHARFF, *Am. J. Physiol.*, **173**, 41 (1953).
441. G. HUBSCHER and G. R. WEST, *Nature*, **205**, 799 (1965).
442. M. H. FLOCH, S. VAN NOORDEN and H. M. SPIRO, *Gastroenterology*, **52**, 230 (1967).
443. A. M. UGOLEV, N. M. MITUSHOVA and M. K. GOZITE, *Fiziologitcheskij Zhurnal SSSR*, **55**, 1513 (1969a).

CHAPTER 8

Absorption of Protein Digestion Products

GERALD WISEMAN

		Page
8.1	FORMS IN WHICH PROTEIN MAY BE ABSORBED	364
	8.1.1 *Nature of the substances taken up by the intestinal epithelium*	365
	8.1.2 *Nature of the substances appearing in the subepithelial space*	372
	8.1.3 *Summary*	377
8.2	EPITHELIAL CELL PEPTIDASES	377
8.3	ACTIVE ABSORPTION OF AMINO ACIDS IN VIVO	380
8.4	TRANSPORT OF AMINO ACIDS AGAINST A CONCENTRATION GRADIENT	384
	8.4.1 *L-amino acids*	384
	8.4.2 *D-amino acids*	391
8.5	EFFECT OF SODIUM	392
8.6	EFFECT OF POTASSIUM	404
8.7	COMPETITION FOR ABSORPTION BY L-AMINO ACIDS	405
8.8	EFFECT OF CONCENTRATION ON AMINO ACID ABSORPTION	411
8.9	KINETIC STUDIES	414
8.10	TRANSMURAL POTENTIAL DIFFERENCE	418
8.11	MEMBRANE ATPase	425
8.12	TRANSAMINATION OF GLUTAMIC AND ASPARTIC ACIDS	426
	8.12.1 *In vitro*	426
	8.12.2 *In vivo*	427
8.13	EFFECT OF DIETARY RESTRICTION	429
	8.13.1 *In vitro*	429
	8.13.2 *In vivo*	433
8.14	AMINO ACID–SUGAR INTER-RELATIONSHIPS	434

8.15	REQUIRED MOLECULAR CONFIGURATION FOR ACTIVE TRANSPORT	440
	8.15.1 *Stereochemical form*	441
	8.15.2 *Carboxyl group*	442
	8.15.3 *α-amino group*	442
	8.15.4 *α-hydrogen atom*	443
	8.15.5 *Side chain*	443
8.16	SITE OF ABSORPTION	444
	8.16.1 *Human small intestine*	444
	8.16.2 *Dog small intestine*	446
	8.16.3 *Rat small intestine*	446
	8.16.4 *Hamster small intestine*	448
	8.16.5 *Colon*	449
8.17	ROUTE OF ABSORPTION	450
8.18	EFFECT ON PLASMA AMINO ACID LEVELS	451
8.19	DEVELOPMENT OF AMINO ACID ACTIVE TRANSPORT	456
8.20	EFFECT OF HORMONES	457
	8.20.1 *Thyroid*	457
	8.20.2 *Parathyroid*	458
	8.20.3 *Insulin*	458
	8.20.4 *Adrenal*	459
	8.20.5 *Pituitary*	459
	8.20.6 *Lactation and pregnancy*	460
8.21	EFFECT OF VITAMIN B_6	460
8.22	EFFECT OF VITAMINS C, D AND E	463
8.23	EFFECT OF pH	463
8.24	EFFECT OF PANCREATIC SECRETION AND BILE	464
8.25	EFFECT OF ALCOHOL	465
8.26	EFFECT OF X-IRRADIATION AND INTESTINAL MOTILITY	466
	REFERENCES	466

8.1 FORMS IN WHICH PROTEIN MAY BE ABSORBED

When protein enters the intestinal lumen, it is possible that it may be taken up intact by the intestinal epithelium and passed to the subepithelial space in that form, or it may be broken down by the epithelium and transmitted in a degraded form, or it may be digested in the intestinal lumen to molecules of various sizes and then taken up by the epithelium and passed on as such, or the digestion products may be further broken down during their passage across the epithelium. Thus we are faced with a two-fold problem: in which form is protein or its digestion products taken up by the epithelium?; and in which form is it or its digestion products released into the subepithelial space (and so to the blood, lymph or fluid bathing the intestinal serosal surface)? A vast amount of work has been devoted to the answering of these two questions.

8.1.1 Nature of the substances taken up by the intestinal epithelium

That ingested protein might be digested entirely to free amino acids within the intestinal lumen by secreted proteolytic enzymes seemed possible from the observations of Kutscher and Seeman [1] and Abderhalden [2], who found that a protein meal caused a fairly rapid appearance of several free amino acids in the luminal fluid. In addition, Loewi [3] demonstrated that an animal could be kept alive and well on a diet of hydrolysed protein, so that there seemed to be no need for the uptake of undegraded protein. He was aware that enzymatic hydrolysates gave better results than did acid-produced ones, and suggested that the latter were probably deficient in some essential amino acids. However, as enzymatic hydrolysates contained peptides as well as free amino acids, it was argued by Falta [4] that normally peptides would enter the blood. He emphasized that after a protein meal it took about 72 hours for all the ingested nitrogen to be excreted, this time lag being required, it was said, for the slow breakdown and metabolism of the absorbed peptides by the tissues. Levene and Kober [5], who supported the view that intact peptides could be absorbed, also observed that nitrogen elimination was faster after the administration of free amino acids than after protein, but they believed that the time lag for protein was due to its slow luminal digestion. That luminal digestion of protein is normally slow and incomplete has more recently been stressed by Fisher [6, 7] as part of his argument in favour of the uptake by the epithelium of fragments larger than amino acids (which does, in fact, occur, although its nutritional significance in the normal adult is as yet uncertain, but it is of considerable importance in some abnormal states). Nevertheless, much free amino acid is quickly liberated in the intestinal lumen after protein ingestion, even if the process of digestion is incomplete. This has been clearly shown by Wright and Wynn [8], who found that the duodenum of the intact dog soon contained enough free amino acids to account for half the beef ingested. These workers came to the opinion that intracellular hydrolysis of protein was not essential for total protein digestion, provided that adequate supplies of pepsin, trypsin, chymotrypsin and pancreatic carboxypeptidase were available. Liberation of free amino acids in the jejunum of the normal fully-fed rat during the absorptive phase also produced a wide spectrum of these substances, as can be seen in Table 8.1, which gives the concentration range for each amino acid three-quarters of an hour after a meal of 5 g Purina chow [9]. The total concentration of free amino acid in the

TABLE 8.1. Concentrations of free amino acids in jejunum of rats 45 minutes after feeding 5 g Purina chow.

Amino acid	Concentration range in jejunum (mM)
Cysteine	0.0-0.5
Histidine	0.5-1.0
Methionine	1.0-1.5
Aspartic acid	1.5-2.0
Isoleucine	1.5-2.0
Serine	1.5-2.0
Threonine	1.5-2.0
Glycine	2.0-2.5
Tyrosine	2.0-2.5
Valine	2.0-2.5
Phenylalanine	2.5-3.0
Glutamic acid	3.0-3.5
Alanine	3.5-4.0
Lysine	4.0-4.5
Arginine	4.0-4.5
Leucine	4.5-5.0

From Steiner and Gray [9].

jejunum of these rats was, therefore, considerable, amounting to at least 40 mM. In the human, Borgström et al. [10] have likewise recorded marked, though incomplete, duodenal hydrolysis of dietary protein after giving 25 g ^{131}I-labelled human serum albumin, and Nixon and Mawer [11] have concluded that dietary protein was normally about 80% digested to free amino acids in the first 50-100 cm of the jejunum. The non-amino acid material was believed to be resistant-to-hydrolysis fragments of test protein and endogenous protein.

According to Nixon and Mawer [12], the rate of amino acid release from ingested protein was rapid enough to account for the absorption of all the arginine, lysine, tyrosine, valine, phenylalanine, methionine and leucine in the free form, but not sufficiently fast for glycine, threonine, serine, glutamic acid, aspartic acid and the imino acids. The estimated times required to release all the amino acids from 15 g casein in the upper jejunum of the young adult human are given in Table 8.2; amino acids in Group A could theoretically be absorbed as free amino acids, but those in Group B would have to be absorbed as peptide. No free hydroxyproline was present at any time in the intestinal lumen even when gelatin, very rich in hydroxyproline,

TABLE 8.2. Estimated times to release all the amino acids of casein in the upper jejunum of the young adult human (15 g casein fed).

Group A (minutes)		Group B (hours)	
Arginine	66	Alanine	11.6
Lysine	72	Histidine	11.9
Methionine	104	Isoleucine	13.0
Tyrosine	108	Serine	13.6
Phenylalanine	175	Threonine	23.6
Valine	175	Glutamic acid	45.5
Leucine	358	Aspartic acid	71.0
		Proline	74.5
		Glycine	98.5

From Nixon and Mawer [12].

was fed. Other observations suggesting considerable (if not complete) hydrolysis of protein to free amino acids before absorption by man have been published by Crane and Neuberger [13, 14] and Crane [15], and reviews of *in vivo* digestion of protein by Gitler [16], Crane [17], and Dawson *et al.* [18] have also presented evidence indicating much intraluminal hydrolysis of dietary protein to free amino acids.

It should be borne in mind that it is not always easy to interpret the results of experiments in which intraluminal nitrogenous compounds have been measured after a test meal, because there is an increase in luminal protein, peptide and free amino acid even after a protein-free meal (Table 8.3), the source of these endogenous substances being secretions and desquamated cells [11, 19-21]. The quantity of protein secreted varies very considerably with the type of meal given and the experimental method, although its relative importance becomes less when larger amounts of ingested protein are used [11, 22]. Curiously, the accumulation of nitrogen in the caecum of rats has been reported to be much greater in germ-free than in normal animals, so that even on a protein-free diet, the caecal nitrogen content was 60-90% of that for rats on a 10-20% casein diet [23]. Campbell et al. [24] state that plasma albumin may enter the sheep jejunal lumen in amounts sufficient to raise its concentration there to one-tenth that of the plasma.

TABLE 8.3. Total amino acids* in gastrointestinal contents of dogs fed three types of test meals.

Test meal	Stomach (mg)	Duodenum (mg)	Jejunum (mg)	Ileum (mg)
Nonprotein	29	19	39	197
Egg albumin (3.3-7.4 g N)	167	113	270	664
Zein (2.9-3.3 g N)	89	94	240	205

* The sum of microbiologically available lysine, leucine, methionine, threonine, histidine, valine, tryptophan and phenylalanine.
From Nasset et al. [20].

Until about 1960, it was fairly widely accepted that, during protein digestion and absorption, only free amino acids were taken up by the epithelium of the small intestine. This was based on the observation that, in general, not even dipeptides appeared in significant amounts in the blood *in vivo* or in the fluid bathing the serosal surface of isolated intestine. Thus, there was little or no transference from mucosal to serosal solution of intact glycylglycine, L-leucylglycine, glycyl-L-leucine, glycyl-L-tyrosine, DL-alanyl-DL-alanine, L-alanyl-L-phenylalanine, DL-alanyl-DL-phenylalanine, L-leucyl-L-tyrosine or β-alanyl-DL-phenylalanine [25-28]. On the other hand, these peptides were mostly rapidly hydrolysed and their constituent amino acids often readily transported to the serosal fluid by an active mechanism. The possibility of uptake of peptide by the epithelium, followed by intracellular or surface hydrolysis to free amino acids, was largely ignored, even though, many years previously, Cajori [29] and Rhoads *et al.* [30] had found that peptides were absorbed from *in vivo* loops of dog jejunum, ileum and colon at rates faster than could be accounted for by the enzymic activity of the intestinal juices. It seemed to these latter workers that if peptides were not absorbed intact, then some must be broken down to amino acids at the epithelial cells' surface or actually within the cells, which they found to be rich in peptidase. That some peptide must be taken up by the epithelium was also argued by Fisher [6, 7] on the grounds that intraluminal digestion of protein is incomplete. Consistent support for the views of Cajori [29] and Rhoads *et al.* [30] started in 1960, when Newey and Smyth [31] also recorded that peptide uptake was not explicable in terms of intraluminal hydrolysis to free amino acids. In addition, they claimed that the rates of uptake of glycine and glycylglycine from the intestinal lumen of the rat, *in vivo* and *in vitro*, were very similar [31-33]. They concluded that their glycine-containing dipeptides (glycylglycine, glycyltyrosine, glycylleucine, leucylglycine and glycyltryptophan) were mainly hydrolysed intracellularly, although some hydrolysis might take place at the cell membrane. Crane and Neuberger [34], studying ^{15}N-labelled yeast protein absorption in man, supported this conclusion, but Dawson and Holdsworth [35] contended that no dipeptides entered the rat epithelial cells under normal physiological conditions, basing their argument on the absence of labelled peptide in the intestinal mucosa (or blood) after feeding ^{14}C algal protein. It was possible, however, that rapid intracellular hydrolysis prevented peptide accumulation. Clear evidence that certain intact peptides could penetrate the intestinal epithelium came in 1962, when Prockop *et al.* [36] demonstrated the presence of hydroxyproline peptides in the plasma and urine after feeding 25 g gelatin to patients with no gastrointestinal disease. They calculated that up to 2 g of intact peptides were so absorbed. This was not due to resynthesis of the peptides from free amino acids as feeding the latter yielded no extra peptide in the blood. The urinary excretion of hydroxyproline-containing peptide by man given gelatin orally has been confirmed by Hueckel and Rogers [37, 38], who also showed that hamsters and rats, but not dogs or monkeys, excreted the peptide. Hamster small intestine *in vitro* allowed passive movement of the peptide from mucosal to serosal solution, the movement being proportional to the peptide concentration in the mucosal solution and was unchanged by anoxia, cyanide or large amounts of proline or hydroxyproline. The peptide was particularly resistant to hydrolysis by the intestine.

The uptake of glycine and its peptides has recently been the object of study in man by Craft *et al.* [39] and in rats by Matthews *et al.* [40]. In the human, feeding free glycine, diglycine or triglycine led to only free glycine appearing in the peripheral blood, but the number of glycine molecules absorbed was greater when the peptides were fed than when an equal number of molecules of free glycine were given. The uptake of intact glycine peptide was at least as rapid as the uptake of free glycine, and this would result in higher intracellular levels of free glycine when the intracellular hydrolysis of the peptide occurred. It is possible, of course, that the uptake and hydrolysis of the peptide both took place at or very close to the cell border, but the uptake step must have preceded the hydrolytic one. Similarly in the rat *in vivo,* the rate of disappearance of glycine molecules from the lumen of tied-off loops of jejunum was greater for diglycine, triglycine and tetraglycine than for free glycine. Very little free glycine was liberated into the lumen when the peptides were used. Tetraglycine appeared to be approaching the maximum size for uptake of intact glycine peptide. When Peters *et al.* [41] infused glycine peptides (up to tetraglycine) into the duodenum of rats, they, unlike Craft *et al.* [39] for man, found diglycine (but not triglycine or tetraglycine) in both the portal and systemic blood. In addition, they reported that glycine peptides of 3 to 6 glycine units could be hydrolysed by isolated epithelial cell brush borders, but that diglycine could not. The latter peptide was hydrolysed, apparently, only intracellularly. This would be in agreement with the observations of Fern *et al.* [42], that although no leucylglycine, glycylleucine or leucylleucine accumulated in rat jejunal wall *in vitro,* glycylglycine was present in the intestinal wall in large amounts, presumably before undergoing intracellular hydrolysis. Matthews *et al.* [43] have also noted that glycine uptake from glycyl-methionine or methionylglycine was the same as glycine uptake from an equivalent solution of free glycine in the absence of free methionine, indicating no competition between glycine and methionine for uptake when these peptides were used. In contrast, when free glycine and free methionine were present together in the intestinal lumen, the methionine caused a marked inhibition of glycine uptake. Hence glycylmethionine and methionylglycine were taken up before hydrolysis in these experiments. Similar experiments showed that dimethionine and trimethionine were taken up intact. Some debate has been introduced into this problem by Fern et al. [42]. They stated that whereas leucylleucine, leucylglycine and glycylleucine had a substantial inhibitory effect on the uptake of free leucine by rings of rat jejunum, glycylglycine had little effect on the uptake of free glycine, and from this they concluded that the leucine peptides were hydrolysed extracellularly, while the glycylglycine was hydrolysed after being taken up. They also claimed that at low concentrations, free leucine plus free glycine had the same action as did leucylglycine and glycylleucine; at high concentrations the peptides were less effective than the free amino acids because the hydrolytic system was presumably already saturated and hence would yield proportionately less free amino acid for competitive action. Once free leucine was liberated within the cell, or accumulated there by pre-incubation, it had only a small inhibitory influence compared with extracellular leucylleucine, suggesting that back diffusion of leucine after intracellular hydrolysis of leucylleucine was not the cause of the inhibitory effect found with leucylleucine in the mucosal solution. The unusual behaviour of glycylglycine might be explained by the fact that its hydrolase is an enzyme of low activity. Indeed, Fern et al. [42] were of the opinion that the intracellular breakdown of glycylglycine might not be relevant to normal protein absorption.

It must be emphasized that there is evidence in normal man that not all peptides are absorbed faster than an equivalent mixture of free amino acids. For example, there were higher plasma levels of free β-alanine and L-histidine after feeding a mixture of them than after feeding an equivalent dose of β-alanyl-L-histidine (carnosine) [44]. Likewise, for man and rats, there was faster absorption of free phenylalanine than of an equivalent amount of diphenylalanine, and tryptophan was absorbed faster from a mixture of tryptophan and glycine than from glycyltryptophan. Yet such subjects absorbed diglycine and triglycine faster than an equivalent load of free glycine. In the dog, unlike man, the liver synthesizes carnosine from the free β-alanine and L-histidine present in the portal blood following carnosine ingestion, and this reconstituted carnosine appears in the systemic blood. Thus confusion as to the nature of the absorbed substance could easily arise [45].

In Hartnup disease, where there is a genetically determined defect in the intestinal uptake of certain neutral amino acids, there is subnormal absorption of free β-alanine and L-histidine, but the absorption of carnosine is good [44, 46]. Likewise, phenylalanine absorption is poorer from the free amino acid than from diphenylalanine, and tryptophan absorption is poorer from a mixture of tryptophan plus glycine than from glycyltryptophan [47]. No intact diphenylalanine or glycyltryptophan is found in the blood of such patients. Thus, although Hartnup cases may have little or no absorption of some free amino acids, they can take up sufficient amino acid for daily needs in the form of small peptides, from which the constituent amino acids are liberated into the blood.

8.1.2 Nature of the substances appearing in the subepithelial space

Among the early investigators to show that dietary protein was degraded before entering the blood were Cathcart and Leathes [48], who found that administering a pancreatic digest of protein caused an increase in the circulating nitrogenous material which was not precipitated by tannic acid; and a little later Howell [49], who dialysed portal blood during the period of maximum absorption of a protein-containing meal, noted that, although the peptones and proteoses remained the same, there was a rise in the amino acids in the dialysate. Subsequently, Folin and Denis [50] recorded that portal blood was always richer in nonprotein nitrogen than was systemic blood following a meal, and although the difference was not great, the large blood flow through the intestine during digestion made it conceivable that all the ingested protein might be absorbed in a degraded form. They came to the view that dietary protein entered the mesenteric blood as free amino acids and that these were stored in the muscles and other tissues. They pointed out [51] that the ammonia in portal blood came from bacterial action in the intestinal lumen (largely in the colon) and not from deamination by the intestine.

The change in plasma amino acid concentration following a protein-containing meal was also investigated in dogs by Van Slyke and Meyer [52]. After a 24-hour fast, the amino acid level of 3-5 mg per 100 ml often doubled when protein was eaten, and sufficient amino acid escaped the liver to raise the femoral artery concentration to near that in the mesenteric venous blood. A similar series of events occurred when 10 g alanine was fed instead of the intact protein, the total amino acid in the blood of the mesenteric vein rising from 3.9 to 6.3 mg per 100 ml. The amino acid concentration in the blood did not increase more markedly because of rapid uptake of amino acid by the liver and other tissues; thus only 1.5 g alanine was present in the circulation at 5 minutes after an intravenous injection of 12 g alanine, and 0.4 g after 35 minutes, although as little as 1.5 g was passed into the urine. It was concluded that protein was absorbed as free amino acids which entered the blood unchanged. That cells could take up amino acids from the blood was demonstrated by Van Slyke and Meyer [53], who used Van Slyke's [54] nitrous acid method to estimate amino groups (not necessarily free amino acids) in biological material. The intracellular amino acid concentration was normally 5-10 times that in the plasma, and when the latter was loaded with amino acid by intravenous injection, the tissues soon restored the gradient. Some amino acid was removed from the plasma by urea formation and protein synthesis in the liver, which other organs performed only slowly, while excess amino acid was excreted via the urine (although this could not always prevent death from abnormally raised plasma levels brought about by intravenous injections) [55].

Other early studies showing a gain in free amino acids in the blood after a protein meal were those of Abel *et al.* [56, 57], who, by 'vividiffusion' (dialysis of circulating blood), demonstrated an increase in alanine, valine and histidine. Later, Kalmykoff [58] found that after administering protein, there was more amino acid in portal blood than in hepatic venous blood, whereas the reverse was true for peptides. It seemed that free amino acids were taken up by the intestine and converted by the liver to peptide. Thus analysing systemic blood for evidence of the form in which protein was absorbed could be very misleading. Bollman *et al.* [59] emphasized this aspect of the problem by showing that there was a considerable rise in blood amino acids in the dog (and urea lower) after hepatectomy, and Luck [60] recorded the rapid accumulation of fed amino acids by rat liver. In rabbits, the ingestion of glycine or alanine caused a high plasma amino nitrogen level which was followed by a rise in urea and, with very high amino nitrogen levels, an increase in circulating peptide [61]. The rate at which the plasma amino acid level rose was, according to Wilson and Lewis [62], partially dependent on the speed with which the amino acids (alanine, glycine, leucine and glutamic acid) left the intestinal lumen, and this view was supported by the results of Kratzer [63]. The latter author also claimed that hydrolysed protein produced higher plasma amino nitrogen

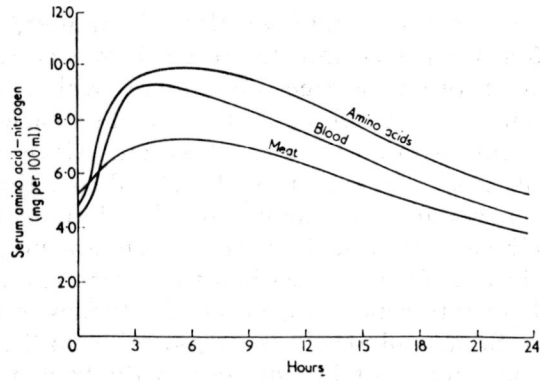

Fig. 8.1. Changes in serum amino acid nitrogen after the ingestion of 480 g protein, amino acids, or 2.25 litres whole blood by normal adult man. (From Free and Leonards [64].)

levels than did intact protein. This was in agreement with the work of Free and Leonards [64], who fed to normal men either 480 g meat or free amino acids and observed that the serum amino nitrogen rose from about 5 mg per 100 ml (starting value) to only about 7 mg per 100 ml after feeding the meat but to about 10 mg per 100 ml 6 hours after feeding the amino acids (Fig. 8.1). Free and Leonards [64] concluded that the limiting factor in protein utilization by normal subjects was the rate of uptake of amino acids. More recent investigations into this facet of protein utilization have been carried out by Levenson et al. [65] in dogs in vivo. They noted marked increases in free amino acids in superior mesenteric venous blood up to 6 hours after feeding liver or homologous plasma protein, both of which gave essentially the same results, and in both cases the plasma amino acid pattern was unlike that of the dietary protein (see also the section discussing the effect of protein ingestion on plasma amino acid levels). Apart from an amino acid conjugate containing much glycine, which appeared to be unrelated to the administered protein, no extra peptide was discovered and it seemed as if all the fed protein had been degraded to free amino acids by the time the blood was reached. Similar experiments in conscious dogs by Elwyn *et al.* [45] have also shown high concentrations of free amino acids in portal blood after a large meat meal, with lower levels in hepatic venous and arterial blood. The greatest increases in the portal blood were seen with valine, leucine and isoleucine, while threonine, proline, glycine and lysine gave intermediate values. This division fits quite well with that published by Nixon and Mawer [12] for probable amino acid release from protein during *in vivo* intraluminal digestion (Table 8.2). As expected (see section on transamination), glutamic and aspartic acids gave very low levels. There was more extra free amino acid in the portal blood than could be accounted for by that fed, the source being endogenous protein (see below). In addition to this latter complication of interpretation of changing amino acid levels in blood after a meal, there are the problems of transamination of glutamic and aspartic acids by the intestinal wall [66-70], rapid metabolism of ornithine, arginine and glutamine [71] by the intestinal wall, conversion of asparagine to aspartic acid [72] (but probably not glutamine to glutamic acid [73]), conversion of cystine to cysteine [74] and the general release of cellular amino acids when the plasma becomes loaded with one or more other amino acids [75]. The intestine can also turn appreciable amounts of methionine to methionine sulphoxide, which is then reconverted to methionine by the time it reaches the portal vein [76]. An account of the comparative enzyme apparatus of the intestinal mucosa has been given by Spencer and Knox [77].

As well as estimation of plasma amino acid after protein administration, a search has been made for alteration in peptide level, in order to determine whether or not these fragments enter the blood intact. Thus Christensen *et al.* [78] fed gelatin, which contains 25.5% glycine and 8.7% alanine, but found no increase in glycine or alanine peptides in human peripheral plasma, and after feeding 35 g casein or 1 quart milk [75] there was actually a decrease in peripheral peptide level. Similarly, in the dog, no convincing evidence was available for protein absorption causing a rise in plasma (portal) peptide [79, 80]. The conclusion was that fed protein was transferred to the blood as free amino acids and this opinion was supported by the subsequent results of Parshin and Rubel [81] for cats and dogs; Denton et al. [82], Denton and Elvehjem [83], Hartmann et al. [84] and Levenson et al. [65] for dogs; Stein and Moore [85] and Crane and Neuberger [13] for man; and Wheeler and Morgan [86] and Newey and Smyth [26] for rats.

In contrast to the view that all fed protein was degraded to free amino acids before entering the blood, Dent and Schilling [80] stated that homologous protein, unlike foreign protein, could be absorbed into the blood intact. However, subsequent work has not supported that claim. One would expect, for example, that if fed homologous protein entered the blood intact, its metabolic fate should be identical to that of homologous protein entering the blood via intravenous injection. But after feeding ^{14}C-labelled homologous protein to rats, the expired $^{14}CO_2$ in the first 3 hours was about 5 times that produced when the protein was given intravenously, suggesting that the fed protein entered the blood in a metabolically more available form [87]. Further, intravenously or intraperitoneally administered ^{14}C-labelled plasma protein rapidly entered the thoracic duct lymph, which did not occur when the protein was ingested, indicating once again that the ingested protein was altered during its passage across the intestinal epithelium [88]. Other work showing that ingested homologous protein was degraded during or before absorption was recorded by Yuile et al. [89]. They gave to dogs lysine-ϵ-^{14}C in either homologous plasma protein or in an amino acid digest and observed that the radioactivity of the blood was, in both cases, due mostly to nonprotein material. With time, the plasma nonprotein ^{14}C decreased and the protein ^{14}C increased, the former virtually disappearing by 7 hours and the latter just then reaching its peak (6-8% of the dose). After injection of the labelled intact protein, there was still one-fifth of the dose in the circulation after 7 days [90]. In addition to these results, Levenson et al. [65] have found that on feeding liver or homologous plasma protein to dogs, there were similar changes in the amino acid pattern of the blood. It was their opinion that both homologous and foreign protein were completely degraded to free amino acids before entering the mesenteric venous blood.

8.1.3 Summary

In summary, protein is digested in the intestinal lumen to yield free amino acids and small peptides, both groups of substances being taken up by the epithelial cells. Amino acids and peptides may exhibit markedly different rates of uptake. The free amino acids are mostly transferred unchanged to the blood, although some may be metabolized (ornithine, arginine, glutamic acid, aspartic acid, asparagine). During or after uptake, the peptides are almost all hydrolysed to their constituent amino acids, probably mainly at the cell surface or border, but some seem to be hydrolysed actually within the cell. A very small amount of a few dipeptides appear in the blood, but virtually all the original protein reaches the blood as free amino acids. (Immunological amounts of whole protein may enter the blood, and the newborn of some species can, for a short while, absorb colostral protein intact.)

8.2 EPITHELIAL CELL PEPTIDASES

With the discovery that small peptides could be taken up intact by the intestinal epithelial cell and released to the blood as free amino acids, interest was aroused as to the site of the peptidases. Observations described above suggested that some peptides were hydrolysed at the cell surface, whereas others seemed to be hydrolysed at a deeper location. Furthermore, the nature of the activity and specificity of epithelial cell hydrolases became of some relevance.

In 1941, Gailey and Johnson [91] reported that hog intestinal mucosa contained peptidases capable of splitting alanylglycine, glycylglycine and prolylglycine, and in dried preparations most of the activity was due to leucylpeptidase. Subsequently, Nachlas et al. [92, 93] demonstrated histochemically that leucine aminopeptidase was present in the epithelial cells themselves, its site being the brush border. An analysis of the enzymic behaviour of isolated brush borders has since shown that of the cell's peptidase activity, there may be sufficient in the brush border to make this an important locus of hydrolysis of alanylglycine, leucylglycine, leucylleucine, leucine-amide, leucylglycylglycine, glycylleucyltyrosine and trileucine [94-96]. Optimum pH was at or near 7. Glycylglycine was split at only a very low rate, making it possible that some would penetrate the cell more deeply and be hydrolysed in the cell sap, as claimed by Newey and Smyth [33] and by Robinson [97]. In addition, Peters et al. [41, 98] have noted that isolated brush borders hydrolysed glycine peptides up to hexaglycine with the exception of diglycine, showing that the latter was most probably split in the cell sap. Fern et al. [42] have also recorded that diglycine penetrates into the cell. Of the guinea-pig mucosal cell's total di- and tripeptidase content, Peters [99] found that only 5-10% of the dipeptidase attacking glycine-containing peptides was in the brush border, but up to 60% of the tripeptidase (also for glycine-containing peptides) was located there. It must be emphasized that nearly all workers in this field have employed mostly glycine-containing peptides, which makes hazardous broad generalization.

To account for the association of certain peptidases with discrete particulate fractions of microvillous membranes, it has been speculated [96] that the membranes have a mosaic substructure, each part of the mosaic representing perhaps only one peptidase, although several might co-exist.

According to Ugolev and Kooshuck [100], at least some peptidases lie on the external surface of the brush border. This was thought to be so because on incubating lengths of rat small intestine in solutions containing glycyl-L-leucine, glycyl-L-tyrosine or glycyl-DL-alanine, in the presence of fluoride to stop amino acid active transport, they found the cells to become loaded with more glycine than leucine or tyrosine. They argued that if the hydrolysis had been intracellular, the small glycine molecule would have diffused out of the cell faster than the slower moving leucine or tyrosine, leaving the cell relatively rich in leucine and tyrosine. But surface hydrolysis of the peptides would, they said, allow the small glycine molecule to enter by diffusion quicker than the larger molecules of leucine and tyrosine. Fern et al. [42] have supported this view that leucine peptides are split at the cell's surface. Other evidence in favour of some peptidase activity at the external surface of the cell has come from Cheng et al. [101] using dimethionine. They observed that moderate concentrations of L-amino acid oxidase in the mucosal solution somewhat reduced the uptake by rat mid-ileal rings of methionine from dimethionine while completely preventing uptake from free methionine. It seemed, in this case, that some of the free methionine produced by surface hydrolysis of dimethionine was accessible to the oxidase in the mucosal solution. Lower concentrations of the oxidase, although still stopping free methionine uptake, had no effect in dimethionine experiments.

Disagreement as to the functional site, regardless of the site of apparent production or storage, of intestinal epithelial dipeptidases has come from Dawson and Holdsworth [35], who have been supported by Josefsson and Sjöström [102]. The former group of workers claimed that intact dipeptides probably never entered normal epithelial cells under physiological conditions as they were unable to recover any radioactive peptide from the mucosa after feeding ^{14}C-labelled algal protein to rats. It was possible, of course, that dipeptide was hydrolysed by the cells too efficiently to allow accumulation of enough for detection. The second group of investigators observed that pig brush border preparations released their dipeptidases within 10-20 seconds when bathed by an aqueous solution of dipeptides (as would occur during digestion), resulting in less than 10% dipeptidase activity remaining in the brush border. Thus the site of dipeptide hydrolysis would be extracellular. Whether the normal brush border released its dipeptidases as rapidly as brush border preparations was unknown.

The distribution of mucosal peptidases along the rat small intestine has been studied by Robinson and Shaw [103], using leucylglycine, glycylglycine, glycylleucine and glycylalanine. The greatest activity for all these peptides was in the ileum. In the adult human, intestinal mucosal hydrolase activity for L-glutaminyl-L-proline and glycyl-L-proline was most evident in the jejunum, but also plentiful in the ileum [104]; a similar distribution was seen in the human foetus, which by 14-16 weeks had amounts of hydrolase comparable to that of the adult [105]. Peptidases attacking L-alanyl-L-glutamic acid, glycylglycine, glycyl-L-leucine and glycyl-L-valine have been found throughout the small intestine of the pig, apart from the proximal duodenum, with a wide zone of maximum activity which showed great variation from animal to animal, as did the relative reactivity rates of these peptidases [106].

The accumulated results show that, in addition to extracellular hydrolysis of peptides, there is peptide hydrolysis at the mucosal border of the intestinal epithelial cell, either within the brush border or at its external surface. Some peptides may be partly or only hydrolysed in the cell sap. Further aspects of this topic are reviewed by Ugolev [107].

8.3 ACTIVE ABSORPTION OF AMINO ACIDS *IN VIVO*

Until 1950 most published work on amino acid absorption supported the view that passage of amino acids across the intestinal epithelium was a passive event, the driving force being the difference in luminal to plasma amino acid concentration. A method often used between 1925 and 1950 was that of Cori [108]. It required the feeding of strong solutions of amino acid by stomach tube, washing out the entire gastrointestinal tract at the end of the experimental period, and analysing the collected material for unabsorbed amino acid. By comparison with animals given blank solutions, the rate of absorption per 100 g body weight per hour (referred to as the 'absorption coefficient') was derived. Employing this technique in unanaesthetized chicks, Kratzer [63] seemed to show that the rates of uptake of amino acids were inversely proportional to the molal volumes of the amino acids, and this finding fitted with that of Levene and Meyer [109], who noted that leucine was absorbed less rapidly than was alanine. Seth and Luck [110] had also recorded that the relatively small molecules glycine and alanine were absorbed more rapidly than were aspartic acid, glutamic acid, leucine or cystine; and Wilson [111, 112] had claimed that cystine was poorly taken up compared with glycine or alanine, although the rate of absorption of glycine was slower than that of the larger molecule alanine. Others to agree that amino acid absorption was dependent on amino acid molal volume were Chase and Lewis [113], Andrews *et al.* [114], Doty and Eaton [115], Lathe [116] and Hier and Bergeim [117].

As well as experiments with different amino acids, several early investigators studied the rates of absorption of the L-, D- and

DL-forms of a single amino acid. Their work also often gave rise to the conclusion that only simple diffusion accounted for amino acid absorption because no appreciable differences were found between the rates of uptake of the isomers. Thus, Johnston and Lewis [61] stated that L-alanine and DL-alanine were absorbed at the same rates in the rabbit, which Wilson and Lewis [62] confirmed for the rat. Similar results were obtained with the isomers of cystine [118-120], methionine [119, 121], phenylalanine [122], valine [123], tryptophan [124], leucine and isoleucine [125] and histidine [126]. Although Schofield and Lewis [127] reported that rats utilized L-alanine and DL-alanine at rates faster than D-alanine, the observation made no significant impression at the time.

In contrast to the then general opinion that amino acid absorption was a simple passive event, Höber and Höber [128, 129] pointed out that a special accelerating mechanism probably existed, because they observed that, by comparison with other absorbed substances, amino acids were taken up faster than could be expected from calculated molecular volumes, even when ignoring the shell of water dipoles, which as zwitterions, these acids would possess. Their deduction was mostly disregarded.

The first convincing demonstration that a special mechanism existed for the absorption of L-amino acids came from Gibson and Wiseman in 1951 [130] (a preliminary account had been given in 1950 [131]). They measured the rates of disappearance, from the lumen of the small intestine of the anaesthetized rat, of the L- and D-forms of racemic alanine, phenylalanine, methionine, histidine, valine, leucine, isoleucine, norleucine, aspartic acid, glutamic acid, α-aminoadipic acid, α-aminopimelic acid and lysine, and found that, in every case, the L-form was absorbed faster than its D-form. These experiments were of especial interest because it had recently become possible to estimate accurately, by enzymic techniques, one isomer of an amino acid in the presence of its other form. This enabled the absorption of the L- and D-forms of an amino acid to be estimated at the same time in the same experimental loop of intestine in the same animal. For histidine, the L:D absorption rate ratio was 6.0, while for L-methionine it was 1.6, with the other amino acids giving intermediate values (Table 8.4). Although absorption into the circulation had not been demon-

TABLE 8.4. Absorption of L- and D-amino acids from DL mixtures introduced into the lower ileum of anaesthetized rats.

DL-amino acid	Amount introduced μ mole	L-form absorbed μ mole	D-form absorbed μ mole	L/D ratio
Histidine	232	84	14	6.0
Phenylalanine	156	54	10	5.4
Glutamic acid	192	49	10	4.9
Lysine	168	38	10	3.8
Valine	222	69	22	3.1
Aspartic acid	206	46	17	2.7
α-Aminopimelic acid	244	46	19	2.4
Leucine	200	48	21	2.3
Alanine	256	48	22	2.2
Isoleucine	184	54	25	2.2
α-Aminoadipic acid	146	31	16	1.9
Norleucine	282	55	32	1.7
Methionine	272	104	64	1.6

From Gibson and Wiseman [130].

strated directly, this was assumed on the grounds that it was unlikely that the small length of intestine (about 15 cm) used could, in the time available, have metabolized as much amino acid as had disappeared. It was concluded that L-amino acids

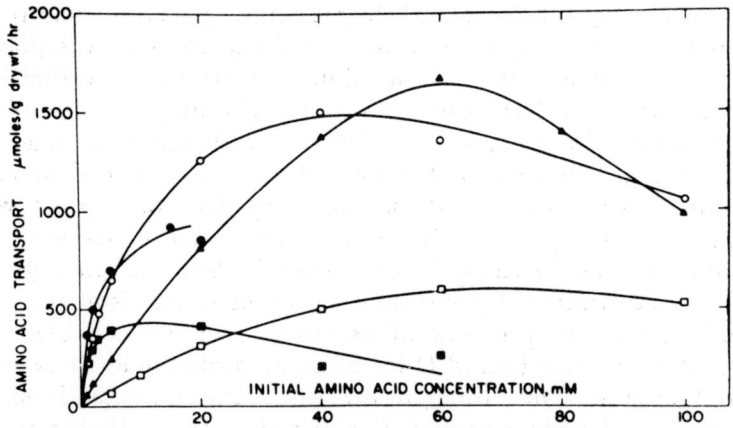

Fig. 8.2. Relation between initial amino acid concentration in mucosal and serosal fluids and the transport rate into the serosal fluid of sacs of everted small intestine of the hamster. ▲glycine; ○L-alanine; ●L-valine; ■L-leucine; □ α-aminoisobutyric acid. (From Matthews and Laster [226].)

were absorbed by an active process in addition to entry by simple diffusion. In confirmation, Clarke *et al.* [132] recorded that the L-forms of alanine and histidine were absorbed faster than their D-forms when a DL mixture was introduced into a Thiry-Vella loop of small intestine in the conscious dog. Following these reports, preferential absorption of L-amino acids has been observed in every intestine examined, including man [133-135]. That absorbed L-amino acids (alanine, phenylalanine, leucine) actually enter the blood faster than their D-forms has since been shown [136].

An interesting claim has been made by Aroska and Berg [137], to the effect that, while L-methionine was absorbed by the conscious rat (Cori technique) faster than D-methionine when both were present together, the D-form was absorbed faster than the L-form when each was used separately. This did not happen for valine, histidine, phenylalanine or tryptophan. Active transport of D-methionine is, of course, well-known, and it is probably the best transported of any of the D-amino acids. It is not, however, transported against its concentration gradient as well as is L-methionine.

While the results of the above experiments (and many others) indicate that absorption of L-amino acids can be via an active mechanism, passive absorption from a high to a low concentration may account for the entry of considerable quantities of certain amino acids. When the entry is passive, the amount absorbed should display a linear relationship with luminal amino acid concentration, and this will become more obvious when the luminal concentration needed to saturate the active mechanism has been greatly exceeded. It should be borne in mind that very high levels of some amino acids may, by damaging the intestinal epithelium, actually reduce absorption. This has been seen with 10-40 mM L-tryptophan, but not D-tryptophan [138, 139], over 20 mM L-phenylalanine [140] and over 10 mM L-valine [141]; and Fig 8.2 shows that the rates of absorption of L-leucine, L-alanine, glycine and α-aminoisobutyric acid were below their peak values when very high concentrations of the amino acids were used. The gastric and colonic epithelia appear to be resistant to the toxic action of L-tryptophan, which, at 40 mM, produces lesions in the small intestine within 5 minutes.

8.4 TRANSPORT OF AMINO ACIDS AGAINST A CONCENTRATION GRADIENT

8.4.1 L-amino acids

Shortly after the proof that a special mechanism existed for the uptake of L-amino acids, Wiseman [142, 143] designed experiments which demonstrated that the L-forms of alanine, phenylalanine, valine, isoleucine, histidine and methionine could be absorbed against their concentration gradients. An *in vitro* method was used, in which the mucosal and serosal surfaces of 15 cm lengths of rat small intestine were bathed with an oxygenated saline solution containing initially the same concentration of a DL-amino acid. At the end of 1 hour, the concentrations of L- and D-amino acid in the mucosal and serosal solutions were measured so that the movement of each isomer during the incubation could be assessed. All six of these L-amino acids were transported from mucosal to serosal solution so that there was a final serosal-to-mucosal L-amino acid concentration gradient greater than 1.0 (the initial value). Under those conditions, L-alanine gave the highest final concentration ratio of 2.8,

TABLE 8.5. Absorption of L-amino acids against their concentration gradients by sacs of everted small intestine of the hamster.

Amino acid	Final concentration ratio* (serosal/mucosal)
Proline	2.08
Threonine	1.90
Alanine	1.82
Glycine	1.65
Serine	1.61
Valine	1.42
Histidine	1.42
Hydroxyproline	1.36
Phenylalanine	1.24
Isoleucine	1.19
Leucine	1.18
Methionine	1.18

* Initial serosal and mucosal fluid contained 20 mM amino acid. Experimental period 1 hour. 37°C.
From Wiseman [147].

while the lowest value was 1.5 for isoleucine. The D-forms of the amino acids (in the presence of their L-forms) were not transported against their concentration gradients. Nor were the L- or D-forms of aspartic acid or glutamic acid. The active transport needed oxygen.

In an extension to this work, using sacs of everted small intestine of the hamster [144-146], glycine and the L-forms of proline, threonine, alanine, serine, valine, histidine, hydroxyproline, phenylalanine, isoleucine, leucine and methionine were also transported against their concentration gradients, and the values are given in Table 8.5 [147]. At the 20 mM concentration employed, there was no active transport of L-tryptophan. The reason for this negative result was not apparent; it seemed to be due to only small amounts of L-tryptophan being transported in comparison with the amount of water entering the serosal fluid. It had been expected that L-tryptophan would be actively transported because it had already been recorded by Pinsky and Geiger [148] that this amino acid could partially inhibit absorption of L-histidine by rats *in vivo*. In addition, its rate of movement down its concentration gradient was greater than that of the smaller molecules lysine and ornithine. Part of the explanation may have been the toxic effect of L-tryptophan at concentrations over about 10 mM [139]. Subsequently, L-tryptophan was shown to be actively transported by small intestine, but low initial concentrations had to be used [138], and this was so for 4-CH_3-, 5-CH_3-, 6-CH_3-DL-tryptophan and N-chloroacetyl-L-tryptophan [149]. An identical series of events occurred with L-lysine and L-ornithine; Wiseman [150] detected no active transport of these amino acids from an initial concentration of 20 mM, but they were actively transported (as was L-arginine) when their initial concentration was low [151, 152]. For the L-amino acids actively transported at 20 mM, there was no simple relation between molecular size and active transport, although the relatively small molecules glycine, alanine and serine were among those concentrated best. In contrast to the ranking order in Table 8.5 for amino acids at an initial concentration of 20 mM, the order found when the initial concentration in the serosal and mucosal solutions was 50 μM was quite different, as can be seen in Table 8.6. Ranking orders for some amino acids used at up to 100 mM can also be obtained from Fig. 8.3.

TABLE 8.6. Relation between active transport of amino acids and their apparent K_t and V_{max} values in sacs of everted small intestine of the rat.

	Final concentration ratio* (serosal/mucosal)	K_t (mM)	V_{max} (μmole/100 mg wet wt/1.5 hr)
Neutral acids			
DL-tryptophan	13	0.25	0.3
L-phenylalanine	9	1.4	2.1
L-isoleucine	15	1.6	0.9
L-leucine	10	2.2	3.4
L-valine	14	3.3	3.0
L-tyrosine	14	4.0	3.6
L-methionine	15	5.3	6.5
L-histidine	15	6.0	2.0
DL-proline	11	6.2	4.1
L-alanine	4	6.3	5.0
Glycine	5	10.0	4.7
L-threonine	6	13.0	6.7
Basic acids			
DL-ornithine	9	0.7	0.3
L-lysine	6	0.7	0.7
N-methylsubstituted acids			
Betaine	3	1.5	4.0
Sarcosine	2	1.9	0.9

* Initial serosal and mucosal fluid contained 50 μM amino acid.
From Larsen *et al.* [152].

The rate of change in amino acid concentration in the serosal fluid, the mucosal fluid and the wall of sacs actively absorbing ^{131}I-mono-iodo-tyrosine can be seen in Fig. 8.4 [153]. The initial concentration of amino acid in the mucosal and serosal fluids was 10^{-7} M; the sacs were made of everted mid-small intestine of the rat. Under those experimental conditions, the tissue amino acid concentration reached its peak in about 30 minutes, the serosal fluid in about 150 minutes, while the mucosal concentration, its volume being large in relation to the rest of the system, showed little change throughout. Thus the greater concentration of amino acid in the sac wall apparently caused movement (presumably by diffusion) of amino acid into the serosal fluid for about 2 hours after the tissue peak concentration had been reached. Similar results have been published for L-phenylalanine transport by sacs of hamster everted small intestine, and for L-valine transport by rat

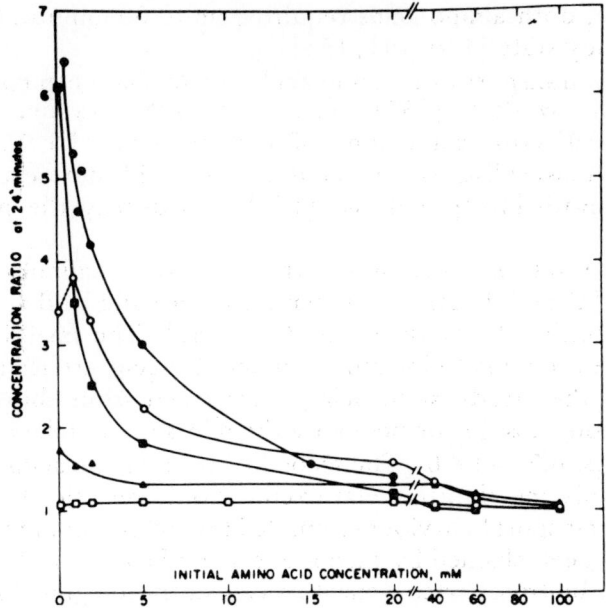

Fig. 8.3. Relation between the initial amino acid concentration in mucosal and serosal fluids and the concentration ratio (serosal/mucosal) developed at 24 minutes by sacs of everted small intestine of the hamster. ▲ glycine; ○ L-alanine; ● L-valine; ■ L-leucine; □ α-aminoisobutyric acid. (From Matthews and Laster [226].)

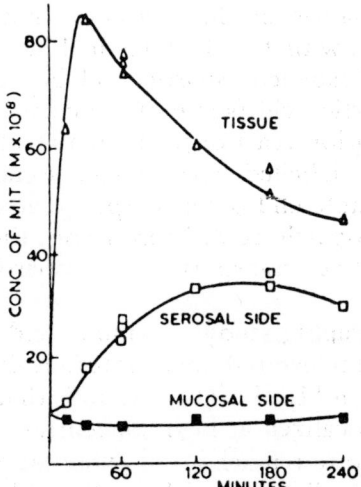

Fig. 8.4. Changes in mono-iodo-tyrosine concentration in mucosal fluid, serosal fluid and intestinal wall during active transport of the amino acid by sacs of everted mid-small intestine of the rat. The initial amino acid concentration inside and outside the sacs was 10^{-7} M. (From Nathans et al. [153].)

intestine, both amino acids requiring up to 60 minutes to come to a steady state [140, 141, 154].

Active transport of L-amino acids *in vitro* has been confirmed by many workers [155], who have added cystine, cysteine [156] and tyrosine and some of its derivatives [153, 157, 158] to an extensive list, which includes amino acids not occurring in nature. Birds [159] and fish [160] also display the phenomenon.

In the rat *in vivo*, active transport of α-aminoisobutyric acid-1-^{14}C and 1-aminocyclopentanecarboxylic acid-1-^{14}C has been described by Christensen et al. [161]. The small intestine developed plasma-to-intestinal lumen concentration ratios of 12-100, the steady state being approached from above (after intravenous dosage) or below by the release or uptake of amino acid (Figs. 8.5 and 8.6). The second of these amino acids was not appreciably metabolized, was excreted only very slowly, and its general transport behaviour resembled that of methionine.

It has been claimed by Jacobs and Lang [162] that L-aspartic acid can be transported against its concentration gradient by rat jejunum *in vivo*. This conclusion was based on the observation that radioactive aspartic acid left the intestinal lumen during a period when there was a net aspartic acid efflux from the epithelium. However, the disappearance of labelled aspartic acid via the transamination mechanism could keep its intracellular concentration below that in the intestinal lumen, accounting for passive influx of labelled aspartic acid in exchange for stable endogenous aspartic acid (some of which would also take part in the transamination reaction). The net result would be loss of some intraluminal labelled aspartic acid, some passive exchange of labelled for stable endogenous aspartic acid, and overall gain in total luminal aspartic acid. There seems, therefore, no reason to postulate active transport of L-aspartic acid in these experiments.

An autoradiographic study of amino acid active absorption by sacs of hamster everted small intestine has been made by Kinter and Wilson [163]. They found that the radioactive amino acid was localized at high concentration in the region of the microvilli, and to a much lesser extent at the serosal borders of the epithelial cells. The cells of the villus tips were most active, amino acid concentrating capacity becoming less as one passed down the villus, the crypt cells having virtually no concentrating ability at all.

ABSORPTION OF PROTEIN DIGESTION PRODUCTS

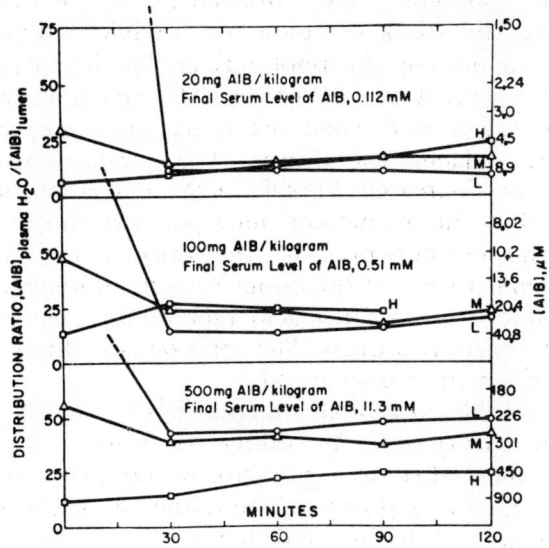

Fig. 8.5. Approach to steady-state distribution ratios for α-aminoisobutyric acid (AIB) in ileal loops with respect to the serum level, at three different concentrations in the organism. Scale at right shows the concentration of α-aminoisobutyric acid in the ileal lumen. H, upper ileum; M, mid-ileum; L, lower ileum. (From Christensen *et al.* [161].)

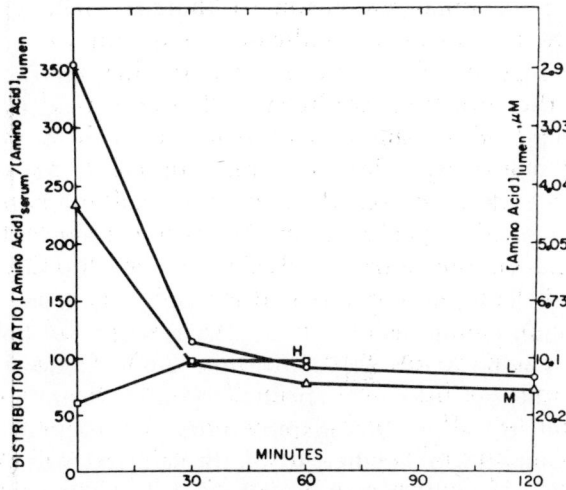

Fig. 8.6. Approach from above or below to steady-state distribution for l-aminocyclopentane carboxylic acid between serum and ileal loop contents. Scale at right shows the concentration of the amino acid in the ileal lumen, the serum concentration being 1 010 μM. H, upper ileum; M, mid-ileum; L, lower ileum. (From Christensen *et al.* [161].)

Recently, McLeod and Bressler [164] have reported increased incorporation of radioactive linoleic acid into lecithin (the major membrane phospholipid) at the site of most active amino acid transport during uptake of L-leucine or cycloleucine by sacs of rat everted small intestine. Inhibiting amino acid transport by replacing the sodium of the medium by potassium prevented the increased linoleic acid incorporation. It was speculated that the augmented incorporation might have been related to the transport process either through alterations in the physico-chemical state of the lipoprotein membrane, or through energy production, via fatty acid metabolism, needed for the amino acid transport process. The total phospholipid content of the intestine remained unchanged.

Heptane-soluble intestinal mucosal lipids may play a part in amino acid absorption. According to Reiser and Christiansen [165], such substances (possibly phosphoglycerides) could bind valine, glycine and lysine. The valine binding was inhibited by isoleucine, methionine and leucine, but not by anoxia, cyanide, dinitrophenol, glycine, lysine or glutamic acid. The leucine action on valine binding appeared to be competitive. Binding was complete in about 2.5 minutes, producing a complex (3.63 μmole valine per 100 mg lipid) which very rapidly broke down in aqueous solution to liberate unchanged valine.

Although the evidence indicates that amino acid active uptake is brought about by a system lying in the mucosal border of the intestinal epithelial cell, there may be a mechanism for extruding amino acid from the cell at the serosal border [33, 166-169]. Part of this argument rests on the observation that, with sacs of rat everted small intestine, serosal fluid proline had a greater inhibitory action than had mucosal fluid proline on the entry of glycine and methionine into the serosal fluid. Uptake of amino acid seemed, therefore, to be less affected than amino acid release. With strips of rabbit ileal mucosa, L-leucine at low concentration (2 mM) had a stimulating action on the net flux of L-lysine (10 mM) from mucosal to serosal fluid, the effect being apparently due to increase in the unidirectional flux of lysine out of the cell towards the serosa. At high mucosal concentration (10 mM), L-leucine reduced the net flux of lysine (10 mM or less) across the epithelium. The overall conclusion was that, for some amino acids at least, active transport was a 2-stage system, one for uptake at the cell's mucosal border, and one for release at the serosal border. Active transport of amino acid at the epithelial cell's serosal border has also been proposed for goldfish intestine [160].

When using the final serosal/mucosal fluid amino acid concentration ratio as an assessment of active transport of an amino acid, it is essential to demonstrate that the final serosal fluid amino acid concentration has risen. In addition, the serosal fluid volume should remain the same or increase (except under special circumstances). Obtaining a final ratio of more than 1.0 by virtue of a fall in mucosal amino acid concentration may merely reflect the destruction of the amino acid concerned either intraluminally or by the intestinal tissue. Metabolism of glutamic and aspartic acids occurs readily, and so does breakdown of arginine (by mucosal arginase). There is also some metabolism of ornithine, especially in the hamster proximal small intestine [170]. Most other amino acids, however, at least when used singly, are resistant to metabolic change during absorption.

8.4.2 D-amino acids

In the report of Gibson and Wiseman [130], in which they measured the rates of disappearance of DL-amino acids from the lumen of the ileum of anaesthetized rats, no opinion could be expressed as to whether or not the D-forms were absorbed passively or actively. All that could be said was that D-amino acids were absorbed much less rapidly than their L-forms, indicating a special system for uptake of the latter. Later, when Wiseman [142, 143] found, during the absorption of racemic amino acids (valine, phenylalanine, alanine, isoleucine and methionine) by rat small intestine *in vitro*, that only the L-forms were transported against a concentration gradient, it was concluded that D-amino acids in general were not actively transported by the intestinal epithelium. That conclusion was supported by Agar *et al.* [25, 171, 172] for D-histidine and D-phenylalanine, Hird and Sidhu [173] for D-methionine, Neil [156] for D-cystine and D-cysteine, and Korelitz and Frank [174] for D-alanine.

However, some D-amino acids can be actively absorbed, provided that they are used in the absence of their L-forms or other actively transported amino acids with a high affinity for the system. Thus, D-methionine can be actively transported by rat [175] and by hamster small intestine [176], D-norvaline and D-serine by rat small intestine [177], D-histidine and

D-methionine by semistarved rat small intestine [178], D-tyrosine by flounder fish intestine [179], and D-alanine is accumulated against its concentration gradient by rabbit ileal mucosa [180]. In the normal rat [181] and chicken [159, 182], D-methionine partially inhibits L-histidine and L-methionine absorption, while D-histidine and D-methionine (but not D-valine, D-alanine or D-glutamic acid) partially inhibit D-glucose active transport by the rat [183]. In addition, D-leucine and D-tryptophan appear to have an effect on L-histidine active uptake by the rat [184]. The affinities of D-amino acids (except D-methionine) for the L-amino acid systems are very low. In the fully-fed rat [178, 181] and in the fully-fed and semistarved mouse [185], active transport of D-histidine cannot be demonstrated, nor is there any apparent active transport of D-alanine, D-tryptophan or D-tyrosine by hamster small intestine [176], or D-alanine or D-norleucine by rat small intestine [177], or D-tryptophan by flounder fish [179] or killifish [186].

Despite the negative results enumerated above, it would seem likely that the small intestine possesses one or more active transport systems capable of transporting D-amino acids (except probably D-glutamic and D-aspartic acids) against their concentration gradients, but that the affinities of most D-amino acids for such systems are so low that active transport is often easily reduced to negligible proportions by competing exogenous or endogenous amino acids.

8.5 EFFECT OF SODIUM

Although Hewitt [187], as long ago as 1924, noted that glucose absorption by anaesthetized cats was increased by sodium ions, it was not until their obligatory role in glucose active transport by guinea-pig intestine was emphasized by Riklis and Quastel in 1958 [188] that interest was stimulated in the relation between sodium ions and amino acid transport. Since then, much work has shown that sodium ions play an essential part in amino acid absorption by the small intestine. Thus, Nathans et al. [153] found that active transport of L-mono-iodo-tyrosine by rat and hamster small intestine *in vitro* was completely abolished when sodium in the incubating medium was replaced by lithium, potassium, calcium, magnesium or sucrose; and requirement for sodium for maximal amino acid transport has been confirmed for L-tyrosine and L-phenylalanine in the frog [189], for L-tyrosine (Table 8.7) and L-valine in the rat [190, 191], for L-tryptophan (Table 8.8), L-lysine and L-methionine in the hamster [149, 192], for L-lysine and L-methionine in the guinea-pig and mouse [192], for glycine, L-lysine, L-methionine, L-alanine and D-alanine in the rabbit [180, 193, 194], for glycine and L-methionine in the chicken [195], and for

TABLE 8.7. Effect of alteration of sodium and potassium concentrations on the active transport of L-tyrosine by sacs of everted small intestine of the rat.

Sodium (m-equiv/l)	Potassium (m-equiv/l)	L-tyrosine final concentration ratio* (serosal/mucosal)	L-tyrosine transport into serosal fluid (μmole)
125	25	5.61 ± 0.41	1.35 ± 0.06 (22)
36	25	3.23 ± 0.27	0.97 ± 0.09 (22)
18	6	2.84 ± 0.24	0.68 ± 0.08 (24)
18	25	2.05 ± 0.12	0.20 ± 0.08 (22)

* Initial concentration ratio was 1.0. Values are Mean ± S.E.M.(n).
From Harrison and Harrison [190].

TABLE 8.8. Effect of replacing mucosal and serosal fluid sodium by potassium, lithium of choline on the active transport of L-tryptophan by sacs of everted small intestine of the hamster.

Cation replacement*	Concentration in medium (m-equiv/l)	L-tryptophan transport (% of normal)
Lithium	47.4	66.5
	94.7	36.2
	118.4	1.2
	127.0	0
Potassium	10	123
	30	74
	50	73
	95	70
	127	0
Choline	47.4	95.0
	94.7	74.5
	118.4	12.6

* Control serosal and mucosal fluid contained 128 m-equiv sodium per litre.
From Cohen and Huang [149].

L-tyrosine in the flounder fish [179]. Some degree of species variation may exist, however, as amino acid active transport was usually completely abolished in the absence of added sodium ions in the rat, rabbit and guinea-pig but was only partially inhibited in the mouse [192], flounder fish [179] and chicken [195]. Further, different amino acids may behave differently in a single species when sodium is absent, as in the hamster transport of L-mono-iodo-tyrosine [153], L-proline [196], L-lysine and L-methionine [192] was completely suppressed but for L-phenylalanine only 90% suppression has been reported [196]. Difficulty arises in deciding on the part played by species variation on the one hand and individual amino acid requirement for sodium on the other because amino acids have often been used under dissimilar conditions.

In addition to their role in active transport, sodium ions seem to be involved in the passive or carrier-mediated (but not active) passage of amino acids across the intestinal epithelium. This conclusion is based on the observation that in the presence of the metabolic inhibitors cyanide and fluoride, which abolish active transport, lowering the sodium concentration in the incubating medium produced a further fall in amino acid penetration into rabbit jejunum, indicating the need for sodium by some energy-independent carrier-mediated process in addition to the part played by sodium in the energy-dependent mechanism [193]. Chez *et al.* [197] have also found that sodium-dependent L-alanine influx in rabbit ileum occurred in the presence of cyanide, iodoacetate and 2,4-dinitrophenol. Hence, Csáky's suggestion [189, 198] that sodium was required intracellularly only as a non-specific link joining a specific carrier system to an energy-yielding process is, at best, incomplete.

Analysis of the relationship between amino acid transport and sodium has led to the hypothesis [199-202], worth considering in some detail, that the as yet unidentified specific carrier (or site) for amino acid is also capable of combining with free sodium (carrier-amino acid-sodium), yielding a complex (ternary complex) which is more stable than the carrier-amino acid complex without sodium. The fact that the intestinal epithelial cell during amino acid absorption becomes loaded with amino acid at a concentration higher than that in the mucosal or serosal fluid bathing the intestine [193, 203] places the carrier at the mucosal border of the epithelial cell, very likely in the brush border region. The carrier-amino acid and carrier-amino acid-sodium complexes are believed to 'move' (translocate) in an unspecified way across the cell 'membrane barrier', to a locus 'within' the cell where amino acid and sodium may then be liberated. These terms are not rigidly defined. Experiments show that the entry (influx) of amino acid into the cell, at constant extracellular sodium concentration, increases with increasing concentration of extracellular amino acid but shows a tendency towards saturation, as does active transport of amino acid from mucosal to serosal surface. In addition, absence of sodium from the mucosal fluid causes a

TABLE 8.9. Amino acid apparent K_t and V_{max} values at different concentrations of sodium. Absorption by rabbit ileum *in vitro*.

Amino acid	Extracellular sodium concentration (mM)	K_t (mM)	V_{max} (μmole/hr/cm^2)
L-alanine	140	9.1	6.1
	70	16.3	13.7
	22	31.2	7.8
	0	70.0	6.8
L-valine	140	5.0	4.3
	0	31.5	6.5
L-leucine	140	4.2	6.3
	0	29.0	4.5

From Curran et al. [200].

substantial depression of amino acid influx at any given concentration of extracellular amino acid, but the maximal rate of amino acid influx (V_{max}) can still be achieved by raising the extracellular amino acid to a concentration higher than that required when sodium is present. What happens is that amino acid affinity for the carrier is reduced, but not abolished, when the sodium concentration is lowered, and in Table 8.9 the concentrations (K_t) of amino acid required for half maximal rates of influx at different sodium concentrations in the extracellular fluid are shown. For L-alanine, L-valine and L-leucine absorption by rabbit ileum, therefore, deficiency of sodium causes an increase in the K_t but no change in the V_{max} [200].

(In contrast, sodium affects the maximal rate of uptake (V_{max}), but not the K_t, during 3-0-methyl-D-glucose absorption by the rabbit ileum [204].) The influx of L-alanine into rabbit jejunal epithelium shows characteristics similar to those for rabbit ileum, the K_t rapidly increasing with fall in mucosal sodium concentration, but the V_{max} remaining constant (Table 8.10) [205]. The rabbit jejunal influx rates are about twice those of the rabbit ileum under comparable conditions, with both displaying Michaelis-Menten type kinetics.

The amino acid influx in rabbit ileum appears to be independent of the intracellular concentration of sodium or amino acid, as can be seen from Tables 8.11 and 8.12.

TABLE 8.10. Effect of mucosal sodium concentration on apparent K_t value of L-alanine during its active uptake by rabbit jejunum *in vitro*.

Mucosal sodium concentration (mM)	L-alanine apparent K_t (mM)	V_{max} (μmole/hr/cm^2)
140	14.6 ± 1.6	12.8 ± 1.5 (5)
70	26.1 ± 2.3	10.1 ± 1.1 (9)
20	45.0 ± 3.9	11.3 ± 1.6 (6)
10	51.9 ± 5.5	8.7 ± 3.8 (3)
0	83.0 ± 20.0	7.4 ± 2.3 (4)

Values are Mean ± S.E.M.(n).
From Alvarez *et al.* [205].

Pre-incubation of intestine (Table 8.11) in a medium in which sodium was replaced by choline, to deplete the epithelial cells of their sodium, was followed by the same amino acid influx as normal provided that the extracellular sodium concentration was high in the test experiments (series A). Reduction of extracellular sodium during the test period (series B and C) reduced the amino acid influx, as expected. From these experiments one would conclude that it is extracellular sodium, therefore, and not intracellular sodium, which governs amino acid influx. (But see below for the opposing views of Newey *et al.* [206], working with L-methionine uptake by rat mid-small intestine.) Similar results were obtained when the sodium chloride in the mucosal solution was replaced by Tris chloride, lithium chloride, potassium chloride or mannitol instead of choline, showing that the stimulatory effect of sodium under those conditions was specific [207]. Sodium in the extracellular fluid was not entirely essential for L-alanine entry into the epithelial cells, however, some influx still occurring when choline was used for both the pre-incubation and the test

TABLE 8.11. Effect of intracellular and extracellular sodium on L-alanine influx in rabbit ileum *in vitro*.

Experimental series	Principal cation		L-alanine influx (μmole/hr/cm^2)
	Pre-incubation solution	Test solution	
A	Sodium	Sodium	2.2 ± 0.1 (10)
	Choline	Sodium	2.2 ± 0.2 (11)
B	Choline	Sodium	1.9 ± 0.2 (12)
	Choline	Choline	0.6 ± 0.1 (12)*
C	Sodium	Sodium	2.8 ± 0.2 (25)
	Sodium	Choline	1.9 ± 0.2 (22)*

* $P = < 0.01$.
Values are Mean ± S.E.M.(n).
From Schultz *et al.* [207].

TABLE 8.12. Effect of intracellular L-alanine on L-alanine influx in rabbit ileum *in vitro*.

Test series	L-alanine in		L-alanine influx (μmole/hr/cm^2)
	Pre-incubation solution (mM)	Test solution (mM)	
A	5	5	2.2 ± 0.2 (8)
B	0	5	2.6 ± 0.4 (8)
C	15	15	2.9 ± 0.7 (4)
D	0	15	3.1 ± 0.4 (4)

Values are Mean ± S.E.M.(n).
From Schultz *et al.* [207].

periods (series B). This sodium-independent amino acid influx was quite small; it should not be confused with active transmural amino acid flux which is completely abolished by sodium lack. (Work by Bihler and Adamic [208] has demonstrated that lithium may partially substitute for sodium as a stimulatory cation during entry of sugar into hamster small intestine. It enables sugar equilibration to occur faster than does choline, mannitol or potassium; but lithium will not support sugar active transport against a concentration gradient. Lithium has no effect on α-aminoisobutyric acid entry into hamster small intestine. Its action is thought to be via stimulation of the sugar carrier mechanism and not by interaction with the sodium pump.) Table 8.12 shows that the rate of L-alanine entry into rabbit ileal epithelium was unaffected by previous loading of the epithelial cells with the amino acid, even though pre-incubation with 5 mM L-alanine resulted in an intracellular L-alanine concentration of as much as 40 mM and the amino acid in the cells seemed to be in a free and osmotically active form [203, 209]. The build-up of amino acid within a cell via this system based on a flux of sodium into the cell is energy-independent, but the flux of sodium itself into the cell depends upon a low intracellular sodium concentration produced by the energy-dependent sodium pump. Hence the sodium asymmetry across the cell's mucosal border acts as an energy store or link between an energy-supplying system (the sodium pump) and amino acid movement.

The close association between sodium and amino acid movement across the rabbit ileal epithelium was also seen in the experiments of Curran et al. [210], who used mucosal strips treated with ouabain (to inhibit the sodium pump extrusion mechanism at the serosal pole of the epithelial cell). They found that sodium movement against its concentration gradient could be produced by a concentration difference of L-alanine across the epithelial mucosal border, sodium being expelled from the epithelium when the cellular L-alanine concentration was greater than that in the external fluid, whereas sodium was taken up against its concentration gradient by the epithelium when the external L-alanine concentration was greater than the cellular L-alanine concentration. The sodium-amino acid transport system was, therefore, reversible and symmetrical.

There is argument about a species variation in ability of intestine to recover from pre-incubation (30 minutes) in a sodium-free medium. Although rabbit intestine more or less regained its power to concentrate amino acid when sodium ions were returned to the mucosal fluid after a period of incubation in a sodium-free solution, rat and mouse small intestine were claimed by Robinson [211] (but not Newey et al. [206]) never to recover amino acid active transport capacity after such pre-incubation. Robinson [211] said it was as if the sodium pump was destroyed by the pre-incubation in these species, for the epithelial cells became loaded with sodium (instead of remaining sodium-low) when sodium ions were added to the mucosal fluid. Toad [212] and hamster [213] small intestine behaved as did rabbit small intestine. However, Newey et al. [206] found that rat mid-small intestine, after being in sodium-free saline for 30 minutes, recovered (in 2 minutes) up to about three-quarters of its capacity to take up L-methionine, the actual degree of recovery being to some extent dependent on potassium concentrations. They agreed, though, that recovery might be incomplete.

According to Curran [201], the carrier site for L-alanine in rabbit ileum must first combine with amino acid before sodium can enter into combination with the carrier. Such a sequence would, of course, allow transport of amino acid in the absence of sodium activation. Further, the translocation steps of the complexes must be slow relative to the association-dissociation reactions. The ratio of the increment of sodium influx to the increment of L-alanine influx when L-alanine was added to the mucosal solution bathing rabbit ileum has been investigated by Curran et al. [200], who recorded values between zero and unity. It was zero when the amino acid entry took place in the absence of sodium activation because of lack of sodium ions at the cell's mucosal surface, and rose to approach unity at an extracellular sodium concentration of about 140 mM, when most of the amino acid influx was sodium activated but a minimal influx might still have been independent of sodium. The net transmural flux rate for L-alanine when its initial concentration on each side of the rabbit ileum was 5 mM and when the solutions contained 140 mM sodium was 1.2 μmole/hour/cm^2, the mucosal-to-serosal flux being about 10 times the serosal-to-mucosal flux [180].

That the influx of amino acid into intestinal epithelium was closely related to movement of sodium was also shown by measuring the change in transmural potential difference due to the influx of sodium during amino acid uptake [214]. As soon as L-alanine was added to the mucosal solution bathing rabbit ileum, there was a rapid increase in sodium transport which was maintained more or less constant during the period of amino acid transport. This increased influx of sodium was reflected by a rise in the potential difference across the intestinal wall, the serosal surface becoming more positive. The phenomenon occurred within 10 seconds of addition of L-alanine, and reached maximum value by 1-2 minutes. In addition to L-alanine, it was produced by glycine, L-methionine, α-aminoisobutyric acid, L-lysine, L-glutamic acid, L-glutamine, D-methionine, D-alanine and D-glutamic acid. The effect could be inhibited by ouabain added to the serosal solution. The change in potential difference was not seen when sodium-free media were employed, nor when the amino acid was added only to the serosal solution. As well as in the mammal, the change in potential difference accompanying the extra sodium transport has been demonstrated in the goldfish, in which it could be produced by the L-forms of alanine, histidine, leucine, methionine, phenylalanine, serine, threonine and valine [160]. Further consideration of potential difference and amino acid transport is presented in the section dealing with potential difference.

After the carrier complex has arrived at the inner locus, it may return, associated or dissociated, to the outer position. Normally, the ternary complex returns to the outer position without sodium, because the intracellular sodium concentration is kept well below that at the mucosal surface of the epithelial cell (sodium inside is about 50 mM when sodium outside is about 140 mM). This loss of sodium from the ternary complex at the inner locus supposedly results in loss of stability of the carrier-amino acid complex, which in turn dissociates to give free amino acid and carrier. The latter then returns alone to the outer border of the cell. The release of amino acid in this fashion builds up the concentration of intracellular amino acid, causing a gradient from cell interior to serosal surface to be established, because of which diffusion of amino acid from cell to serosal surface occurs. The overall result is a flow of amino acid against its concentration gradient. The low intracellular sodium concentration is achieved by an energy-dependent mechanism which actively pumps sodium out of the cell, most probably at the serosal pole of the cell but possibly also laterally. In the rabbit ileum *in vitro* there is a sufficient supply of endogenous metabolizable substrate, but for isolated rat jejunum metabolizable substrate must be added to the incubation media for the intracellular sodium to be kept low [215]. If this sodium pump is prevented from working (by ouabain, which inhibits the sodium-potassium-activated ATPase, or by cyanide, 2,4-dinitrophenol or iodoacetate), however, the intracellular sodium concentration rises, which progressively inhibits liberation of sodium from the ternary complex at its inner locus, and this in turn, by keeping the complex stable, prevents amino acid liberation. Under such conditions, when the complex returns to the outer position it takes with it sodium and amino acid. Hence, according to this explanation, which presupposes that intracellular sodium is largely free (which it may not be) and is as effective as extracellular sodium in its action on the carrier, the rise in intracellular sodium results in greater flux (efflux) of sodium and amino acid from within the cell to the mucosal surface. When the sodium concentrations inside and outside the cell are equal, sodium and amino acid influx and efflux become equal, leading to no net sodium or amino acid accumulation by the cell. The results for L-alanine entry into the mucosal epithelium and net transmural flux in short-circuited ileum of the rabbit in the presence of ouabain (which must be in the serosal solution) and cyanide indicated that L-alanine influx was not affected by these inhibitors, but net transmural flux was very nearly abolished. Thus, L-alanine efflux must have been increased until it very nearly equalled L-alanine influx, which, if the hypothesis is correct, was brought about by rise in intracellular sodium due to inactivation of the sodium pump mechanism (there was a concomitant abolition of net transmural sodium flux in these experiments). It must be emphasized that whereas there is direct evidence of the action of extracellular sodium on amino acid influx, the action of intracellular sodium on amino acid efflux is only indirect.

A diagrammatic representation of amino acid and sodium fluxes under various conditions described above is presented in Fig. 8.7.

The stimulatory action of sodium on amino acid transport is pH sensitive, Frizzell and Schultz [216] having observed that at pH 2.5 mucosal fluid sodium ions had little or no effect on L-alanine influx into rabbit ileum, even though at this pH L-alanine influx was subject to competitive inhibition, showing that carrier-mediated transport was still involved. These workers concluded that the interaction between sodium and the

L-alanine-carrier complex at the brush border involved reactive groups whose pK was about 3.

Disagreement with the above general conclusion that it is chiefly extracellular sodium, rather than intracellular sodium (as originally proposed by Csáky [189, 198]), which controls amino acid active transport has come from Newey et al. [206]. They pre-incubated (30 minutes) sacs of everted mid-small intestine of the rat in saline containing no sodium (to deplete the epithelial cells of sodium) and then measured the rate of

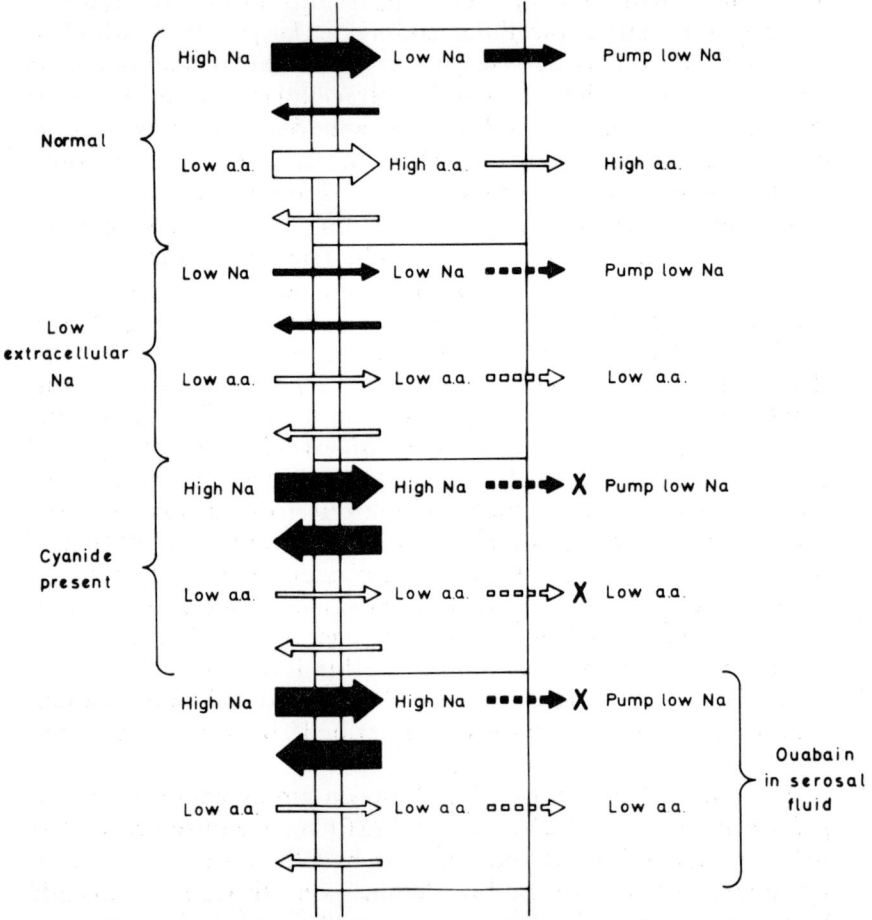

Fig. 8.7. Fluxes of amino acid and sodium across the intestinal epithelial cell under various conditions. The upper half of the diagram shows flux rates for high (normal) and low concentrations of sodium in the mucosal fluid. The lower half shows flux rates when the cell is affected by cyanide or ouabain. The thicker arrows indicate greater flux; dotted arrows indicate little or no flux; X indicates blocked pumping mechanism. Amino acid, open arrows; sodium, black arrows.

L-methionine uptake when these sacs were placed for a test period (5-120 seconds) in a saline containing sodium (143 m-equiv/l). Under such test period conditions there was presumably a downhill (mucosal fluid to cell interior) sodium gradient across the mucosal border of the epithelial cells. These sacs took up less L-methionine (Table 8.13) than did sacs which, having been pre-incubated for 30 minutes in a sodium (143 m-equiv/l) saline (to give a normal cellular sodium content), were then allowed during the test period to absorb L-methionine from a sodium-free saline (there now being an uphill sodium gradient from mucosal fluid to cell interior). It seemed, therefore, that having a normal intracellular content of sodium was more important than having extracellular sodium bathing the cell's mucosal border. The greatest L-methionine uptake occurred, however, when sacs were pre-incubated in a sodium saline and then tested for amino acid uptake in a sodium saline, indicating that as well as the need for normal sodium concentration within the cell, sodium outside the cell was also required for optimal L-methionine uptake. The rapidity with which the sodium-depleted rat intestine took up L-methionine (though less than normal) when placed in a sodium saline gave rise to the

TABLE 8.13. Effect of intracellular and extracellular sodium concentration on the uptake of 1 mM L-methionine by rat small intestine *in vitro*. Potassium concentration was 6 mM in all solutions.

Extracellular sodium	Intracellular sodium	L-methionine uptake period (sec)	Sodium-dependent* L-methionine uptake (%)
Normal	Normal	5	100
Low	Normal	5	94
Normal	Low†	5	60
Normal	Normal	30	100
Low	Normal	30	101
Normal	Low†	30	71
Normal	Normal	60	100
Low	Normal	60	63
Normal	Low†	60	68

* Obtained by the sum: L-methionine uptake observed minus L-methionine uptake under low intracellular and low extracellular sodium conditions.

† Intracellular sodium was reduced by 30 minutes pre-incubation in a sodium-free solution.

From Newey et al. [206].

speculation that the sodium-sensitive mechanism for amino acid uptake might be close to the cell's luminal border and might be the brush border ATPases.

The absorption of amino acids *in vivo* usually shows little or no dependence on exogenous sodium in the mucosal (luminal) fluid, because sufficient endogenous sodium rapidly enters such fluid even if it is entirely sodium-free initially. Thus Matthews *et al.* [40, 43] found added sodium ions unnecessary for maximal absorption of glycine, methionine and their peptides by the rat; and in the dog, L-alanine absorption from Thiry-Vella loops was reduced by only a small amount when the luminal solutions were at first free of sodium, which they soon gained [217]. Glycine disappearance from dog Thiry-Vella loops was also unchanged in the absence of sodium [218]. In man, jejunal absorption of glycine and L-alanine was not impaired when the perfusate was initially sodium-free [219].

8.6 EFFECT OF POTASSIUM

Although omission of potassium ions from the incubating media was noticed by Nathans *et al.* [153] to have no effect on the active transport of mono-iodo-tyrosine by sacs of hamster and rat small intestine, increasing the potassium to 20 mM or above caused depression of the active transport. An analysis by these workers by means of Lineweaver-Burk plots [220] showed potassium to behave as a competitive inhibitor. Harrison and Harrison [190] have also observed, at low sodium concentration, that potassium inhibited tyrosine active transport (Table 8.7), but unlike Nathans *et al.* [153] stated that some potassium was required for maximal transport. In support of Nathans *et al.* [153], however, was the report of Reiser and Christiansen [191], that potassium was not required for optimal L-valine transport by rat everted intestine. According to Read [221], extra potassium had no effect on the active uptake of cycloleucine by rat small intestine *in vitro*. Newey *et al.* [206] believe that potassium may be involved in amino acid entry into intestinal epithelial cells, possibly via a sodium-potassium-ATPase.

During the uptake of L-phenylalanine by rat jejunum, the intracellular potassium concentration fell (accompanied by a rise in sodium) [209]. This was due to water entry into the cells, the net potassium flux remaining apparently unchanged.

8.7 COMPETITION FOR ABSORPTION BY L-AMINO ACIDS

During his experiments on amino acid absorption by the rat *in vivo*, Cori [222] noted that the rate of uptake of glycine and DL-alanine when present together was less than the sum of the rates when each was present alone. It seemed that these amino acids competed for a single mechanism, thereby causing mutual partial inhibition. Similar results were obtained, by the Cori technique, by Kamin and Handler [223], who stated that the presence of an excess of an amino acid (glutamic acid, aspartic acid, leucine, isoleucine, arginine, histidine, threonine) almost invariably inhibited the absorption of another amino acid, and they were of the opinion that the nature of the competing amino acid appeared to be of little consequence. Pinsky and Geiger [148] likewise found, in the rat, that L-tryptophan decreased the uptake of L-histidine, although the latter appeared to have no effect on the former.

With the introduction of the sac of everted intestine technique [144-146, 224], detailed study of amino acid competition for absorption was greatly simplified and a vast literature on the topic has since accumulated. The first description came from Wiseman [150]. He measured the ability of hamster small intestine to transport against a concentration gradient single L-amino acids and pairs of amino acids (in equimolecular amounts). As can be seen from Tables 8.14 and 8.15, of the amino acids tested, only those that were actively transported when present alone had an observable effect on the active transport of others. In addition, those amino acids relatively poorly concentrated when present alone had a greater inhibitory action on those amino acids which, when present singly, were relatively well concentrated. Thus, when the initial amino acid concentration in the mucosal and serosal fluid was 20 mM, L-methionine completely inhibited the active transport of L-proline, glycine and L-histidine, whereas these three latter amino acids had little or no effect on L-methionine. Further, L-glutamic acid had no measurable effect on the active transport of L-proline, L-histidine or L-methionine, and L-lysine and L-ornithine caused no inhibition of L-histidine or L-methionine absorption. It seemed, therefore, that the small intestine had a special transporting mechanism common to the L-forms of the

TABLE 8.14. Competition for active transport among L-amino acids.

Amino acid*	Final concentration ratio† when present alone	Final concentration ratio† in the presence of an equimolar amount of					
		Proline	Glycine	Histidine	Methionine	Lysine	Ornithine	Glutamic acid
Proline	2.08		1.74	1.41	0.86			2.02
Glycine	1.65	1.07		1.04	0.78			
Histidine	1.42	1.14	1.30		0.75	1.40	1.29	1.52
Methionine	1.18	1.11	1.20	1.06		1.17	1.08	1.31
Lysine	0.82			0.81	0.95			
Ornithine	0.84			0.76	0.93			

* Initial concentration in mucosal and serosal fluid was 20 mM. Experimental period 1 hour. 37°C.
† Expressed as final serosal concentration/final mucosal concentration. Initial ratio was 1.0.
From Wiseman [150].

TABLE 8.15. Rates of absorption from mixtures of L-amino acids.

Amino acid*	Abs. rate† when present alone	Absorption rate† in presence of equimolar amount of					
		Proline	Glycine	Histidine	Methionine	Lysine	Ornithine	Glutamic acid
Proline	62.5		50.0	26.8	11.2			64.8
Glycine	45.1	12.5		8.1	0			
Histidine	23.7	13.4	15.2		0	25.5	18.3	30.8
Methionine	14.8	21.9	14.8	9.8		13.9	10.7	21.9

* Initial amino acid concentration in mucosal and serosal fluid was 20 mM. Experimental period 1 hour. 37°C.
† In μmole/100 mg dry weight of sac per hour.
From Wiseman [150].

neutral amino acids and that they competed for this mechanism, while the basic and acidic amino acids did not appear to be involved in active transport nor to interfere with the neutral amino acid mechanism. Since then, active transport of the basic amino acids has been shown to occur, but only when their initial concentration was comparatively low. Active transport of glutamic and aspartic acids has not yet been demonstrated, even at low initial concentrations. It may be that the transamination reaction completely overshadows a very low active transport capacity for these acidic amino acids (if such capacity exists at all).

The behaviour of L-proline in the presence of various concentrations of L-methionine, in the experiments described above, indicated that the former amino acid had an affinity for the active transport mechanism which amounted to only about one-twentieth that of L-methionine. As a working hypothesis, it was postulated [147] that L-methionine and other slowly, but actively, absorbed neutral L-amino acids formed more stable complexes with the transporting system and thereby acted as good inhibitors of the more rapidly absorbed amino acids, which formed relatively unstable complexes. Hence, for example, when each was present singly, L-methionine was absorbed slower than L-proline, but when the two amino acids were present together, L-methionine (a good inhibitor) was absorbed faster than the inhibited L-proline (provided that their initial concentrations were not too different). The more stable complexes would, of course, also account for the low rate of transport observed. This analysis was subsequently supported by the kinetic studies of Jervis and Smyth [225] and Matthews and Laster [226]. Graphs obtained by Nathans *et al.* [153] and Matthews and Laster [227], reproduced in Figs. 8.8 and 8.9, showed that when the reciprocal of the rate of uptake ($1/V$) of an amino acid was plotted against the reciprocal of the initial concentration ($1/S$) of that amino acid in the mucosal fluid, the line cut the ordinate at the same point whether or not the competing amino acid was present. When the competing amino acid was present, the slope of the line was steeper. Such results are typical of competitive inhibition, in which both amino acids form complexes with a transporting system and thereby reduce, to a variable extent, the active transport of each other.

In addition to the experiments described above, there have been many studies of amino acid competition for absorption in

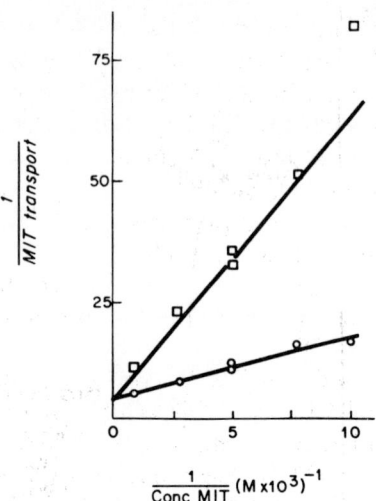

Fig. 8.8. Relation between initial amino acid concentration and rate of transport of mono-iodo-tyrosine (MIT) by sacs of everted small intestine of the rat in the presence (□) or absence (○) of L-phenylalanine, plotted by the Lineweaver-Burk method. MIT transport is given in μmole/100 mg wet wt/90 min. The results are indicative of competitive inhibition. (From Nathans et al. [153].)

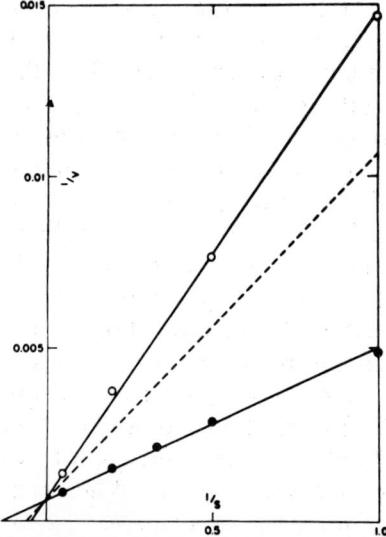

Fig. 8.9. Relation between initial amino acid concentration and rate of transport of L-alanine by sacs of everted small intestine of the hamster in the presence (○) or absence (●) of L-leucine. V, transport rate (μmole/g dry wt/hr); S, initial amino acid concentration (mM). The results are indicative of competitive inhibition. (From Matthews and Laster [227].)

man [135, 219, 228-230] and in experimental animals (mammals, birds, fish) *in vivo* [161, 231-233] and *in vitro* [138, 140, 141, 151, 153, 158, 167, 172, 179, 181, 182, 184, 234-242]. In man, for example, Adibi and Gray [229], who intubated the upper jejunum, found that L-threonine (20 mM) was absorbed at about 220 μmole per minute when present alone, but at only about 75 μmole per minute from an amino acid mixture. The outcome of this enormous endeavour has been a broad classification of amino acid pathways on the basis of mutual antagonism during absorption, although once one delves deeper than a broad classification, the field is confused by numerous claims and counter-claims. The following pathways for L-amino acids have been described as being discrete functionally:

1. A mechanism for the neutral amino acids (including histidine). This is the most efficient system and operates well at high concentrations (20 mM and above). It is hardly, if at all, affected by basic or acidic amino acids, or by betaine or N-dimethyl-glycine.

2. A mechanism for the basic amino acids (lysine, ornithine, arginine) which is also used by cystine (in addition to mechanism 1). This pathway is less efficient than mechanism 1, operates only at lower concentrations (below about 3 mM), and may be inhibited by some neutral amino acids (e.g. methionine). It has been claimed that leucine and tryptophan may also use this system [243, 244]. Cysteine may have a separate pathway in man [74].

3. A mechanism for proline, hydroxyproline, sarcosine, N-dimethyl-glycine and betaine [176, 238]. Proline and hydroxyproline also use mechanism 1 efficiently. Other neutral amino acids make little or no use of this pathway and have little or no inhibitory action on it, although Munck [238] has stated that this imino pathway in the rat can also be used by alanine and leucine. Proline has a high affinity for this mechanism, whereas its affinity for mechanism 1 is comparatively low. That betaine and N-dimethyl-glycine have available a transport pathway other than those for neutral and basic amino acids has been supported by the work of Butt and Wilson [245], who found that L-valine, glycine, L-proline and L-lysine were actively accumulated by foetal guinea-pig intestine at 35 days of gestation, yet it took 60 days of gestation (full gestation is 65-70 days) before betaine and n-dimethyl-glycine were actively taken up.

4. A mechanism in the rat for glycine and proline, distinct from mechanisms 1 and 3, has been claimed by Akedo and Christensen [246], Christensen [247] and Newey and Smyth [166, 248]. Proline was said to have a greater affinity than glycine for this pathway, while methionine had little affinity (in contrast to its high affinity for mechanism 1). According to Matthews and Laster [227], only mechanism 1 was available for glycine in the hamster, although Nelson and Lerner [195] thought that in the chicken there existed a pathway for glycine which was distinct from that preferentially used by methionine and nonpolar neutral amino acids. This glycine site was said not to be available to acidic or basic amino acids, β-alanine, α-aminoisobutyric acid, sarcosine or even imino acids. Spencer *et al.* [235] have tested a wide range of glycine analogues for their effect on glycine accumulation by hamster intestine and have recorded that, despite the close structural similarities of the substances examined, only alanine, allylglycine and N-phenyl-glycine inhibited glycine uptake. A pathway special to proline, hydroxyproline and glycine seems to exist in the human, for subjects with a genetic defect leading to prolinuria (in which proline, hydroxyproline and glycine are in the urine in excessive amounts) also displayed reduced efficiency in intestinal absorption of proline, and sometimes of hydroxyproline and glycine too [249, 250].

5. A mechanism, distinct from 1 above, for the active transport of valine, leucine and isoleucine has been described by Hagihira *et al.* [234].

It has been suggested by Alvarado [239] that instead of multiple separate pathways for the active transport of amino acids of different classes, with some functional overlap, there may exist a polyfunctional carrier capable of binding at different sites any actively transported amino acid (and also actively transported sugars), the characteristics of the carrier being such as to give rise to an allosteric effect when an amino acid of one or other class is bound.

During the absorption of amino acid mixtures *in vivo,* there is a fairly distinct pattern of absorption rates. Thus, Orten [135] found that, in the human ileum, amino acids absorbed from a mixture of 18 amino acids (each at the same initial concentration) fell into three main groups, as shown in Table 8.16; the pattern was discernible within 5 minutes and remained steady over 3 hours. A very similar pattern was later published by

Adibi and Gray [229], who perfused the human jejunum, about 30 cm beyond the ligament of Treitz, by means of a double-lumen tube. They observed that, at all concentrations from 5 to 20 mM, methionine was absorbed fastest, followed in descending order by isoleucine, leucine, valine, lysine, phenylalanine, tryptophan and threonine. Adibi *et al.* [230] confirmed these results, reporting that methionine, isoleucine, leucine and valine were absorbed (from an equimolar mixture of amino acids at 1-8 mM concentration for each) faster than proline, arginine, alanine and phenylalanine, while aspartic and glutamic acids were the slowest. In rats, the pattern of absorption of amino

TABLE 8.16. Pattern of absorption of L-amino acids from 50 ml of an equimolar mixture (each amino acid about 9 mM) introduced into the human ileum (Thiry loop). Absorption period 3 hours.

Fast	Intermediate	Slow
Arginine	Valine	Threonine
Isoleucine	Lysine	Histidine
Methionine	Phenylalanine	Glycine
Leucine	Cystine	Glutamic acid
	Tyrosine	
	Tryptophan	
	Proline	
	Aspartic acid	
	Alanine	
	Serine	

From Orten [135].

acids from an equimolar mixture (each at 2 mM), introduced into tied-off loops of small intestine *in vivo*, was very like that for the human, as can be seen from a comparison of Tables 8.16 and 8.17 [231].

8.8 EFFECT OF CONCENTRATION ON AMINO ACID ABSORPTION

Maximum absorption of casein hydrolysate, introduced directly into the human upper jejunum by intubation, was reported by Zetzel and Banks [251] to take place when the concentration of the administered solution was 10 g amino acid per 100 ml. For weaker solutions, the absolute amount absorbed decreased, although percentage absorption rose. Whatever the concentration initially used, the intestine brought the luminal concentration to about 2 mg nitrogen per ml within 30 minutes. The intubated region had to remain in functional continuity with the rest of the intestinal lumen for optimum amino acid uptake. In these experiments, the diluting effect of the stomach, which can be considerable, was largely lost. These results were in general agreement with those of Höber and Höber [128, 129], Lathe [116] and Hetenyi and Winter [252], who also observed that amino acid absorption *in vivo* (when absorbed amino acid

TABLE 8.17. Pattern of absorption of L-amino acids from 1 ml of an equimolar mixture (each amino acid 2 mM) introduced into the rat small intestine *in vivo*. Absorption period 15 minutes.

Fast	Intermediate	Slow
Cysteine	Phenylalanine	Aspartic acid
Methionine	Proline	Glutamic acid
Arginine	Tyrosine	
Isoleucine	Lysine	
Tryptophan	Alanine	
Leucine	Histidine	
Valine	Serine	
	Glycine	
	Threonine	

From Delhumeau *et al.* [231].

would be quickly removed by the blood) displayed saturation phenomena, which suggested a mode of uptake other than simple diffusion, and this view was supported by the finding by Free and Leonards [64] that, under ordinary conditions in man, the liberation of free amino acids from ingested protein was fast enough to lead to appreciable amino acid accumulation within the lumen because of saturation of the amino acid absorption mechanisms. Absorption of amino acids from an amino acid mixture by human ileum has been investigated by Orten [135] in a subject with an ileostomy. Uptake of total nitrogen was maximal at about isotonicity (which confirmed the report of Zetzel and Banks [251]), the increase in nitrogen absorption at higher concentrations being only very small. According to Adibi [253], threonine had such a low affinity for the uptake mechanism in human jejunum that, at concentrations of 5 to 150 mM, it appeared to be absorbed at rates proportional to its concentration. Because of this apparent lack of saturation of uptake, threonine was absorbed faster than leucine or valine at 100 mM, when the mechanism for the latter two amino acids had reached saturation point. The human ileum, however, unlike the jejunum, did show saturable kinetics for threonine (Table 8.18).

For *in vitro* work, when an amino acid is being transported against its concentration gradient, a high serosal/mucosal concentration ratio is developed when the initial concentration of the amino acid in the incubating media is low, although the absolute amount of amino acid transported may be small.

TABLE 8.18. Apparent K_t and V_{max} values for some L-amino acids during absorption by the human *in vivo*.

Amino acid	Site	K_t (mM)	V_{max} (μmole/min/cm)	Reference
Alanine	Jejunum	36	53-85	219
Glycine	,,	76	58-75	,,
Methionine	,,	11-28	17-37	257
Methionine	Ileum	2-6	4-8	,,
Leucine	Jejunum	21	26	253
Leucine	Ileum	12	8	,,
Threonine	,,	22	19	,,
Valine	Jejunum	40	50	,,

Conversely, for high initial concentrations of amino acid in the incubating media, a smaller serosal/mucosal concentration ratio will be developed, but the absolute amount of amino acid moved will be increased (Table 8.19; Fig. 8.3). Under *in vitro* conditions, when absorbed amino acid is not being removed but instead accumulates in the intestinal wall and serosal fluid, the rates of absorption of different amino acids do not keep the same ranking order for all initial concentrations. Thus, with an initial concentration of amino acid in the incubation media of 20 mM, the absorption rate (μmole per g dry weight per hour) has been found to be greatest for L-alanine, followed by glycine, L-leucine and α-aminoisobutyric acid, whereas for an initial concentration of 60 mM the descending order was glycine, L-alanine, α-aminoisobutyric acid and L-leucine (Fig 8.2) [226].

Other aspects of this topic are dealt with under the section on kinetic studies.

8.9 KINETIC STUDIES

The observation that the uptake of certain amino acids by the intestinal mucosa could be progressively saturated by increasing the concentration of the amino acids in the fluid bathing the mucosal surface has led to numerous reports on the effect of concentration on amino acid uptake (see also the section on the effect of concentration on amino acid absorption). Such experiments have been performed on a wide variety of animals, employing both *in vivo* and *in vitro* techniques, and by and large the kinetic results have much in common despite the considerable variations in experimental methods.

TABLE 8.19. Effect of initial concentration on L-proline active transport by sacs of everted small intestine of the hamster.

Initial concentration in mucosal and serosal fluid (mM)	Rate of transport into serosal fluid (μmole/100 mg dry wt/hr)	Final concentration ratio* (serosal/mucosal)
5	32.8 ± 1.6	3.27 ± 0.41 (7)
10	46.3 ± 1.5	2.51 ± 0.25 (7)
20	62.6 ± 3.2	2.08 ± 0.18 (9)

* Initial concentration ratio was 1.0. Experimental period 1 hour. 37°C.
Values are Mean ± S.E.M.(n).
From Wiseman [150].

Typical experiments, using glycine, L-alanine, L-valine, L-leucine and α-aminoisobutyric acid with segments of everted small intestine of the hamster, have been published by Matthews and Laster [226, 227, 254]; and by Larsen *et al.* [152] for sixteen amino acids and sacs of everted small intestine of the rat. The rate of uptake of each amino acid was found to reach a maximum, as can be seen in Figs. 8.2 and 8.10, the value depending on the test amino acid. Some amino acids achieved maximum transport at quite low initial mucosal fluid concentrations (about 10 mM for L-leucine), whereas for others much higher concentrations had to be used (about 60 mM for glycine). It should be noted that the ranking order for uptake

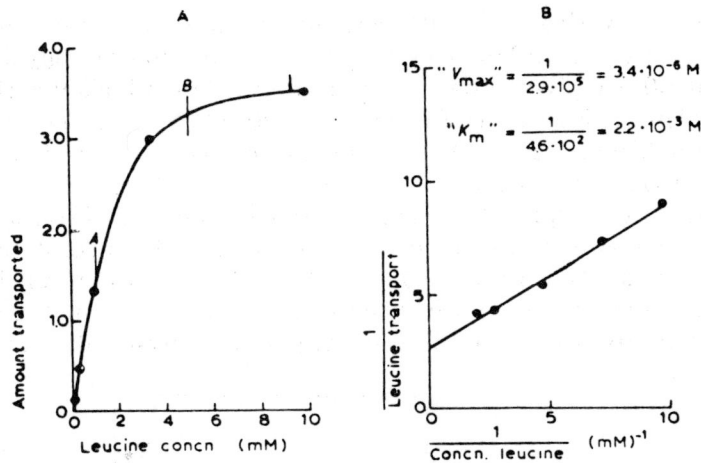

Fig. 8.10. A: Relation between the initial amino acid concentration (mM) and the rate of transport (μmole/100 mg wet wt/90 min) of L-leucine by sacs of everted small intestine of the rat. B: Lineweaver-Burk plot to determine the apparent K_t (mM) and V_{max} (μmole/100 mg wet wt/90 min). This analysis was made with results falling between points A and B on the left-hand graph. (From Larsen et al. [152].)

TABLE 8.20. Active transport of L-amino acids by sacs of everted small intestine of the hamster.

Amino acid	Final concentration ratio* (serosal/mucosal)	Amino acid transport into serosal fluid (μmole/100 mg dry wt/hr)
Proline	2.08 ± 0.06	62.5 ± 1.1 (9)
Threonine	1.90 ± 0.08	53.5 ± 0.7 (14)
Alanine	1.82 ± 0.09	51.4 ± 1.0 (12)
Glycine	1.65 ± 0.06	45.1 ± 0.8 (9)
Serine	1.61 ± 0.05	39.3 ± 0.6 (15)
Valine	1.42 ± 0.05	36.6 ± 0.7 (14)
Histidine	1.42 ± 0.06	23.7 ± 0.8 (12)
Hydroxyproline	1.36 ± 0.04	28.1 ± 0.5 (12)
Phenylalanine	1.24 ± 0.03	24.1 ± 0.7 (11)
Isoleucine	1.19 ± 0.03	17.9 ± 0.3 (13)
Leucine	1.18 ± 0.03	17.9 ± 0.5 (12)
Methionine	1.18 ± 0.03	14.7 ± 0.6 (11)
Tryptophan	0.90 ± 0.02	4.9 ± 0.3 (7)

* Initial amino acid concentration in mucosal and serosal fluid was 20 mM (each amino acid present separately). Initial concentration ratio was 1.0. Experimental period 1 hour. 37°C.
Values are Mean ± S.E.M.(n).
From Wiseman [147].

sometimes varied with different initial amino acid concentrations. Thus at 20 mM L-alanine was taken up faster than glycine, but at 60 mM glycine was transported faster than L-alanine (Fig 8.2). The ranking order for uptake at 20 mM by hamster sacs is given in Table 8.20, and the order for uptake at 5×10^{-5} M by rat sacs is given in Table 8.6.

When the reciprocal of the initial rate of amino acid transport is plotted against the reciprocal of the initial mucosal fluid amino acid concentration (Figs. 8.10 and 8.11), the result is a straight line (ignoring the highest concentrations) such as is obtained by the Lineweaver-Burk [220] analysis of enzyme

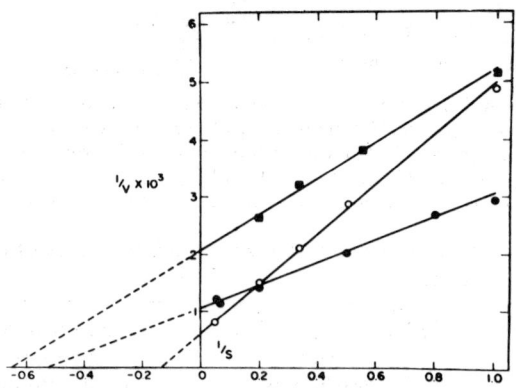

Fig. 8.11. Relation between initial amino acid concentration and rate of transport of L-alanine (○), L-valine (●), and L-leucine (■), by sacs of everted small intestine of the hamster, plotted by the Lineweaver-Burk method. The intersects on the abscissa give the apparent K_t and those on the ordinate the V_{max} for each amino acid. V, transport rate (μmole/g dry wt/hr); S, initial amino acid concentration (mM). (From Matthews and Laster [226].)

kinetics. The intersect on the abscissa gives the reciprocal of the initial concentration of the amino acid (equivalent to $1/K_m$ in enzyme studies) at which the active transport progresses at half the maximum velocity. This half-saturation value is referred to as the K_t, and is a concentration in molar terms. The intersect on the ordinate represents the reciprocal of the apparent maximum transport velocity ($1/V$) for the amino acid used. This maximum velocity is referred to as V_{max}, and is usually given in μmole of amino acid per unit time per unit of intestine. In such experiments the rate of the reaction is estimated by measuring the disappearance of amino acid, over as short a period as is practicable, from the intestinal lumen or mucosal fluid. A low value for K_t is obtained for an amino acid with a high affinity for its transporting mechanism, which means that it will, at any given concentration, occupy more active sites than will an amino acid with a higher K_t. Thus amino acids with a low K_t are better inhibitors of amino acid active transport than are amino acids with a high K_t. A good affinity (low K_t) is usually accompanied by a low V_{max} (which is determined by the rate of breakdown of the amino acid-transporting mechanism complex), as can be seen in Table 8.6. The maximum serosal/mucosal concentration ratios achieved by the amino acids in Table 8.6 bore no obvious relationship to the values for K_t.

The results of Matthews and Laster [226] for glycine, L-alanine, L-valine and L-leucine active transport by hamster small intestine, and those of Larsen *et al.* [152] for amino acid active transport by rat small intestine, showed that the rates of release (reflected by V_{max}) of amino acids from combination with their transporting systems were in reverse order to their affinities for their systems. For example, L-leucine joined rapidly with its system but was only slowly released and hence transported relatively poorly, while glycine joined less well with its system but was rapidly liberated and therefore well transported. An exception seemed to be α-aminoisobutyric acid, which had a relatively high K_t and a low V_{max}. Wiseman [150], studying competition for transport among amino acids, had earlier pointed out that those amino acids that were actively transported at a slow rate were good inhibitors of amino acids actively transported at a fast rate. When Lineweaver-Burk type analysis was made of the rate of active transport of mono-iodotyrosine in the presence or absence of L-phenylalanine (Fig. 8.8), and of L-alanine in the presence or absence of L-leucine (Fig. 8.9), the plots obtained were those expected for competitive inhibition, as were those for the uptake of L-valine in the presence or absence of L-leucine or L-histidine [141]. This problem has also been discussed by Gitler and Martinez-Rojas [255].

The active transport of amino acids into cells is often described in terms of a membrane carrier system. If a carrier does exist, the dissociation of the amino acid-carrier complex on the cell side of the membrane may or may not be reversible.

If there is reversibility, previously taken up amino acid A may compete with amino acid B taken up later for the carrier on the cell side of the membrane and thereby diminish the rate of efflux of B, causing an increase in the net transport of B. Finch [256], testing this hypothesis by incubating segments of rat small intestine first in ^{12}C-isoleucine (A) and then in medium containing ^{14}C-isoleucine (B), found that the net uptake of B was not enhanced by the pre-incubation. Hence, either the dissociation was not appreciably reversible, or the released amino acid in the cell was rapidly made unavailable to the hypothetical carrier at its internal position.

Variation in the K_t of an amino acid at different sites along the human small intestine has been reported by Schedl et al. [257] for L-methionine and by Adibi [253] for L-leucine. It was found, by intubation perfusion studies, that the K_t for L-methionine was 11-28 mM (V_{max} 15-33 mmole per hour per 15 cm) in the upper small intestine but only 2-6 mM (V_{max} 4-7 mmole per hour per 15 cm) in the distal region; similarly, L-leucine had a higher K_t (lower affinity) in the upper jejunum than in the upper ileum, with the expected higher V_{max} in the upper jejunum, so that absorption was faster in the jejunum than in the ileum. For threonine, there appeared to be no saturation of its mode of transport in the upper jejunum and hence no measurable K_t, but a K_t could be estimated for it in the upper ileum. The values are given in Table 8.18, together with some others for human small intestine. According to Spencer and Samiy [140], the K_t for L-phenylalanine was the same (1.8 mM) in the mid-small intestine and in the lower ileum of the hamster, but capacity for transport was much greater in the middle region, possibly because there are more active sites per cell in this part.

These and other observations suggest that active transport of amino acids by the small intestine is mediated by one or more enzymes, or that the transporting mechanism mimics an enzyme in its reaction characteristics. However, no enzyme or carrier has as yet been identified for amino acid transport by intestine.

8.10 TRANSMURAL POTENTIAL DIFFERENCE

In 1961 Schachter and Britten [258] reported changes in transmural potential difference during the active transport of amino acid by segments of rat small intestine. They found a potential difference across the duodenum of only 1.5-2.4 mV (serosal surface positive) when no substrate was added to the incubation media, but a potential difference of about 6 mV when 20 mM L-alanine was present. In contrast, 20 mM D-alanine gave a potential difference of about 3 mV, which was considered an insignificant rise. These workers concluded that active transport of certain non-electrolytes resulted in change in ion transport, reflected by change in potential difference. Since then, numerous experiments have confirmed that the active transport of amino acids (as well as other substances) is accompanied by a rise in transmural potential difference and an increase in the short-circuit current (I_{sc}). For example, Baillien and Schoffeniels [259] observed that addition of L-alanine or glycine to the fluid bathing the mucosal surface of isolated small intestine of the Greek tortoise increased the transmural potential difference, whereas dicarboxylic amino acids (not actively transported [143]) had no such effect, although they did increase the I_{sc}. Basic amino acids seemed to have no effect on either the potential difference or I_{sc}. For rabbit ileum *in vitro*, Zalusky and Schultz [260] recorded a rise in potential difference when L-alanine was present in the mucosal fluid, and, in contrast to the results of Baillien and Schoffeniels [259] and Gilles-Baillien and Schoffeniels [261], claimed that L-lysine and L-glutamic acid also caused a potential difference rise. All three amino acids apparently increased the I_{sc}. The conclusion was that active transport of amino acids was coupled directly or indirectly with sodium transport. It was also noted that actively transported amino acids gave potential difference rises independent of those produced by actively transported sugars, which suggested that different transporting mechanisms existed.

Fish, too, display a potential difference change during amino acid absorption, the potential difference across sacs of goldfish anterior intestine increasing during the active transport of the L-forms of alanine, threonine, serine, histidine, valine, methionine, phenylalanine and leucine [160]. There was a positive correlation between the transfer of amino acid to the serosal fluid and the potential difference increase. The actual potential difference developed for each amino acid depended both on the temperature to which the fish had been acclimatized and on the temperature at which the sacs were incubated [262]. When sacs were incubated at temperatures at or above the acclimatization temperature (8-25°C), there was an extra 1 μmole L-valine or L-threonine (initial mucosal concentration 10 mM) transferred to the serosal fluid per 100 mg wet weight of intestine per 2 hours for each rise of 3 mV (steady potential difference) above the 'basic' potential difference of 2 mV (this latter value being due to activity other than transport of added amino acid). This ratio of 1 to 3 applied also to the L-forms of alanine, serine, histidine, phenylalanine and leucine.

With samples of adult human ileum *in vitro*, Grady et al. [263] found that the transmural potential difference of about 4 mV was increased to about 6.6 mV by 20 mM L-alanine at the mucosal surface, but there was no change when D-alanine was used (in keeping with the above observation of Schachter and Britten [258]). Furthermore, when the ileum had achieved a maximum potential difference during glucose (12 mM) absorption, addition of L-alanine to the mucosal fluid produced a further rise in potential difference, although this did not occur when another actively transported sugar was added. The same phenomenon took place when 20 mM L-alanine (for maximum potential difference) was added first and glucose second. These results, similar to those of Zalusky and Schultz [260], supported the view that amino acids and sugars were actively transported by separate mechanisms (even though there is clear evidence of interaction between sugars and amino acids during their active absorption). These actively transported substrates gave a rise in ileal potential difference whether or not they were metabolized. The potential difference of about 10 mV (serosal surface positive) across the human colonic mucosal layer (not whole wall) was unaffected by amino acids, presumably because the colon does not actively absorb these substances. The potential difference across the ileum and colon was abolished within 15 minutes by 10^{-4} M 2,4-dinitrophenol.

As with adult human small intestine, foetal (14-21 weeks) human small intestine exhibited a rise in transmural potential difference during amino acid and sugar active absorption [264]. The effect with L-alanine (7 mM) was somewhat greater with sacs of everted ileum than jejunum, the potential difference rise for the former being about 2.6 mV (the value found for adult human ileum) and for the latter about 1.2 mV. The endogenous potential difference for foetal ileum was about 0.5 mV at 14 weeks gestation and about 1.0 mV at 20 weeks gestation; for children and adult humans, the ileal endogenous potential difference was about 4 mV. In confirmation of earlier work, D-alanine (7 mM) caused no rise in transmural potential difference, and the effects of amino acid and sugar were additive.

It is worth emphasizing that in direct contrast to the rise in potential difference brought about by actively transported amino acids, passively absorbed amino acids, when added to the fluid bathing the intestinal mucosal surface, may induce a reduction in transmural potential difference. This reduction is related to the osmotic load of the added substance and is associated with reduction in movement of sodium from mucosal to serosal surface [265, 266]. Indeed, a large enough osmotic load in the mucosal fluid may actually reverse the polarity across the intestinal wall so that the serosal surface becomes negative instead of the normal positive.

A detailed account of the effect (at 15 mM) of 21 L-amino acids, D-alanine, D-serine, D-leucine, glycylglycine, triglycine and glycyl-L-alanine on the potential difference across isolated rat small intestine has been published by Kohn *et al.* [266]. Apart from L-lysine and L-arginine, all the amino acids (which included L-glutamic and L-aspartic acids) and the three peptides stimulated a potential difference rise when present in the mucosal fluid, although the characteristics of the potential difference changes were not similar for all compounds. The time taken for the potential difference to achieve its new maximum value (P_{max}) when amino acid was added varied considerably, and was sometimes as much as several minutes, in contrast to a few seconds for actively transported hexoses. The P_{max} values for glycine, L-alanine and L-histidine were the same, but lower than that for L-methionine and asparagine (which were equal). For the peptides, the increase in potential difference was greater than that observed for the respective free amino acids. The amino acid-stimulated potential difference was most marked in the distal ileum, whereas it was greatest in the mid-small intestine for hexoses [268]. This variation in response along the rat small intestine has been confirmed *in vitro* and *in vivo* with 40 mM L-histidine by Cotterell and Kohn [269], who found that it was not related to the rate of amino acid uptake (*in vitro*) but was possibly attributable to differences in tissue resistance. For example, the mid-small intestine had a resistance of approximately 15 Ω per cm, while for the ileum it was about 28 Ω per cm. Such an explanation would allow the same stoichiometry between amino acid and sodium transport in both these regions. Munck [270], using 10 mM L-lysine, has also noted that the amino acid-stimulated potential difference was greater in rat ileum than in the jejunum, but was able to bring the jejunal response up to the ileal by substituting sulphate for chloride in the incubating media. The phenomenon appeared to be related to an increase in mucosal-to-serosal flux of chloride in the jejunum during active lysine absorption, which tended to neutralize the effect of the concomitant sodium flux. No such change in chloride flux was evident in the ileum. The lack of effect of L-lysine and the positive action of D-alanine in the experiments of Kohn et al. [267] were in disagreement with some reports of earlier workers [258, 260], for which no definite explanation was available apart from possible species or regional variation (Gilles-Baillien and Schoffeniels [261] found L-lysine to have no effect in the tortoise). However, active transport by the rat of L-lysine and L-arginine at high concentration (20 mM) is not demonstrable [150], although it may occur when the initial concentration of the amino acids is low (1 mM or less) [151, 152]. The observed increased potential difference with 15 mM D-alanine is, on the other hand, surprising, because experiments with this amino acid have not shown transmural active transport even at 5 mM concentration [176], although some concentrative accumulation of it by rabbit ileal mucosa has been described [180]. Furthermore, Hindmarsh et al. [183] found D-alanine, unlike D-methionine and D-histidine, to have no clear inhibitory effect on the active absorption of sugars. The increased potential difference with D-serine might be expected, as Randall and Evered [177] have recorded that this amino acid (5 mM) was actively transported by rat small intestine. The rise in potential difference seen when L-glutamic and L-aspartic acids were added to the mucosal fluid were possibly due to the active transport of L-alanine produced by transamination.

The increased transmural potential difference and short-circuit current during L-alanine absorption by rabbit ileum has been stated by Schultz and Zalusky [271] to be reflected probably entirely by the simultaneous mucosal-to-serosal sodium flux. However, some absorption of sodium by the rat jejunum (stimulated by mannose in the serosal fluid) may occur without change in transmural potential difference [272]. In addition, Taylor et al. [273] have suggested that secretion of chloride into the mucosal fluid may take place during active transport of D-galactose, and increased mucosal-to-serosal flux of chloride during L-lysine absorption has been claimed by Munck [270] (see above). Considerable care must, therefore, be exercised in the interpretation of transmural potential differences.

According to Thompson et al. [274], a low pH (6.3) increased the amino acid-stimulated potential difference across isolated rat jejunum when glycine, L-proline, L-leucine, L-valine and β-alanine were being absorbed. As there was also improved amino acid transport, the higher potential difference was presumably due to the accompanying enhanced sodium flux.

The potential differences across the intestinal mucosal epithelium have been measured in the hibernating Greek tortoise by Gilles-Baillien and Schoffeniels [261, 275] by the use of microelectrodes introduced into the epithelial cells. With only ordinary saline as the incubation medium, the cell interior was markedly negative (−20 mV) with respect to the mucosal border, while the serosal border was to a smaller extent positive to the mucosal border (+3 mV) (Fig. 8.12). On the addition of L-alanine to the mucosal solution, the cell interior showed no change (−20 mV), but the serosal surface became much more positive (+13 mV). Hence the amino acid-stimulated transepithelial potential difference (from +3 mV to +13 mV) was due to increased activity of the cell's serosal border. Similar, but not identical, values were obtained with active, rather than hibernating, tortoise intestine and L-alanine, but there was no alteration in any potential difference with either L-lysine or L-arginine in the mucosal fluid. With L-glutamic acid or L-aspartic acid in the mucosal fluid, there was no variation in transepithelial potential difference, but the cell interior became more negative with respect to the mucosal border (Fig. 8.13). It was thought that this was brought about by changes in membrane permeability to potassium and chloride.

For general considerations of intestinal membrane potentials readers should consult references 276 and 277.

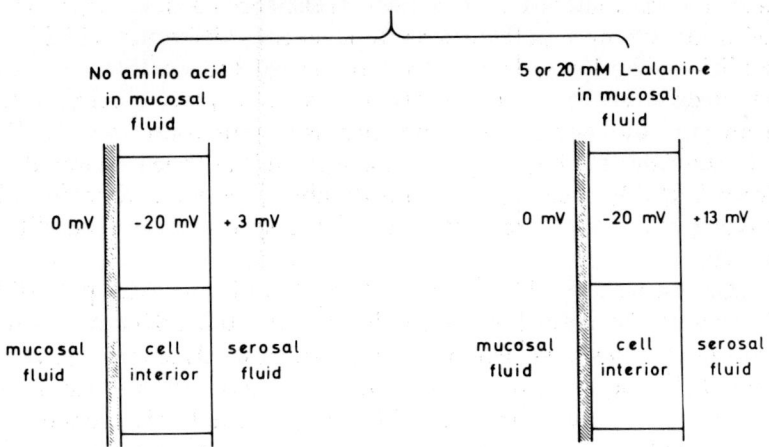

Fig. 8.12. Electrical potentials of the small intestine of the hibernating Greek tortoise in the absence and presence of L-alanine in the solution bathing the mucosal surface. (From Gilles-Baillien and Schoffeniels [275].)

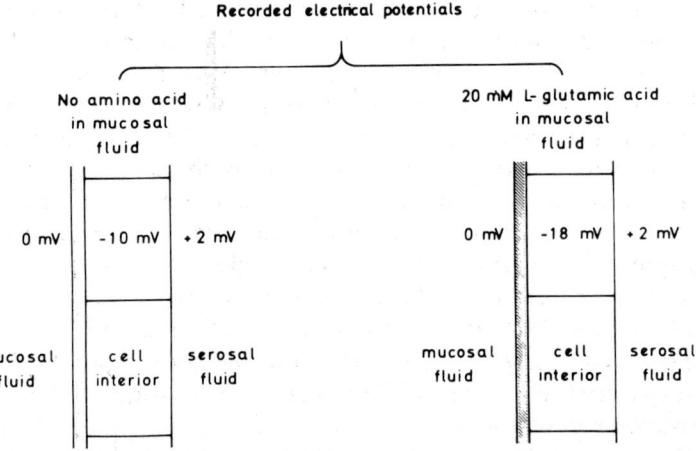

Fig. 8.13. Electrical potentials of the small intestine of the active Greek tortoise in the absence and presence of L-glutamic acid in the solution bathing the mucosal surface. (From Gilles-Baillien and Schoffeniels [261].)

8.11 MEMBRANE ATPase

It has become evident that the ouabain-sensitive sodium-potassium-activated ATPase which exists in cell membranes [278-281] may not be the linkage between the intestinal epithelial cell's energy supply and the sodium extrusion pump at the cell's serosal and/or lateral borders. If only this particular enzyme were involved, then sodium-dependent active transport of amino acid should be completely suppressed by ouabain in the fluid bathing the intestinal serosal surface. Although such inhibition is said to occur *in vitro* with rabbit ileum [197], the small intestine of the guinea-pig [192] and the jejunum of the rabbit [193] retain some active transport capacity in the presence of ouabain, and the small intestine of the hamster [192], mouse [192, 282] and rat [211, 282, 283] are more or less insensitive to the action of serosal ouabain. At a serosal fluid ouabain concentration of 5×10^{-5} M, rabbit jejunum had reduced, but still demonstrable, active transport of L-lysine and L-methionine [193], while with rat mid-small intestine, L-methionine was absorbed (about one-third normal rate) even in the presence of 10^{-2} M ouabain in the serosal fluid [283].

Supporting the view that serosal membrane ATPase may not be involved in sodium-dependent amino acid transport was the finding of Hollands and Smith [284], that histochemically there appeared to be very little (if any) ATP-specific sodium-potassium-ATPase in the basal membrane of rat intestinal epithelial cells. This was in striking contrast to the goldfish, where there was much such enzyme in the basal membrane, and in this species sodium-dependent sugar active transport was very sensitive to ouabain in the serosal fluid [285]. According to Berg and Chapman [286], the greatest concentrations of sodium-potassium-ATPase in the epithelium of rat small intestine were in the mitochondria and in the brush border (where the enzyme seemed to encrust the outer surfaces of the microvilli [278]), with little in the serosal or lateral membranes. Furthermore, analysis of the sensitivity of rat intestinal sodium-potassium-ATPase to various cardio-active glycosides (ouabain, scillaren A, scilliroside) showed that concentrations of these drugs which almost completely inhibited the ATPase had no effect on L-phenylalanine uptake by rat intestine [282]. Even when intestinal sodium-dependent active transport was ouabain-sensitive, Robinson [282] stated that there was no correlation between the sensitivity of the ATPase and of the transport systems to the cardio-active glycosides. Robinson [282] suggested that the function of the rat ATPase might be regulation of the cellular ionic content but not bulk sodium transit.

These accumulated results indicate that a link other than serosal membrane ATPase is required, at least in some species and possibly in all, to couple energy supply with intestinal sodium-dependent amino acid active transport. This presently undefined link may perhaps be the cytochrome system [282].

8.12 TRANSAMINATION OF GLUTAMIC AND ASPARTIC ACIDS

8.12.1 In vitro

During a study of the active absorption of amino acids by isolated segments of rat small intestine, Wiseman [143] discovered that glutamic and aspartic acids were neither transported against their concentration gradients nor accumulated within the intestinal wall, although they disappeared in appreciable amounts from the incubation fluids. The most likely explanation was transformation as a result of their participation in transamination reactions within the mucosa and this was subsequently shown to be the case by Matthews and Wiseman [66]. These workers found, by paper chromatography, that when the solutions bathing the mucosal and serosal surfaces contained originally equal concentrations of glutamic or aspartic acid, the serosal solution became rich in alanine. The mucosal solution contained only trivial amounts of alanine, which would be expected because this amino acid is well concentrated by rat intestine. When aspartic acid was used, small quantities of glutamic acid were also produced. Production of detectable alanine did not occur under these experimental conditions when histidine or phenylalanine were employed, nor when the initial solutions were amino acid-free.

Transamination of L-aspartic acid to alanine has also been reported for sacs of goldfish anterior intestine [160].

8.12.2 In vivo

Conversion of dicarboxylic amino acid to alanine during absorption occurs *in vivo*, as well as *in vitro*, as was demonstrated by Neame and Wiseman [67-70], using anaesthetized dogs, cats and rabbits. After the introduction of a dilute (0.15%) solution of L-glutamic acid into a washed-out loop of small intestine, the blood draining the loop contained only extra (i.e. above arterial level) alanine but no extra glutamic acid, although much glutamic acid disappeared from the intestinal lumen (Fig. 8.14). In fact, the glutamic acid in the venous blood from the experimental loop of intestine was below that in the arterial blood supplying the loop, so that endogenous as well as absorbed glutamic acid was being converted.

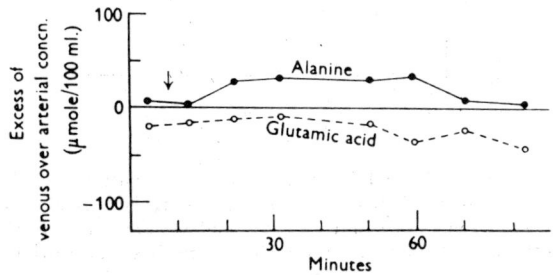

Fig. 8.14. At time indicated by arrow, 10 ml 0.15% glutamic acid solution was introduced into the lumen of the experimental portion of dog small intestine *in vivo*. There was an increase in alanine, but not glutamic acid, in the blood as it passed through the wall of the loop. (From Neame and Wiseman [68].)

When stronger solutions of L-glutamic acid were employed (0.5, 2 and 10%), there was extra alanine and glutamic acid in the blood draining the loop (Fig. 8.15), the amount of extra glutamic acid being greater with the more concentrated test solutions.

The production of alanine appeared to be limited to a loop venous level of about 200 μmole per 100 ml blood. (This agreed with the observation of Klingmüller *et al.* [287], who noted that the maximum concentration of alanine in the blood of rabbits did not depend on the quantity of glutamate given intravenously.) During the absorption of 0.5% glutamic acid, the venous extra alanine accounted for 40% and the extra glutamic acid for 28% of the glutamic acid disappearing from the intestinal lumen.

For L-aspartic acid (Fig. 8.16), the changes were in general similar to those seen with L-glutamic acid, except that some extra aspartic acid was present in the loop venous blood even when 0.15% aspartic acid was used as the test solution. During

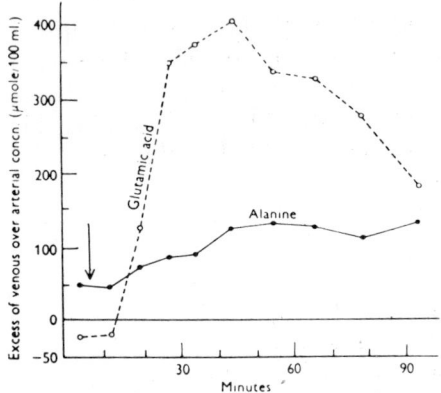

Fig. 8.15. At time indicated by arrow, 10 ml 2% glutamic acid solution was introduced into the lumen of the experimental portion of dog small intestine *in vivo*. There was a marked increase in both alanine and glutamic acid in the blood as it passed through the wall of the loop. (From Neame and Wiseman [68].)

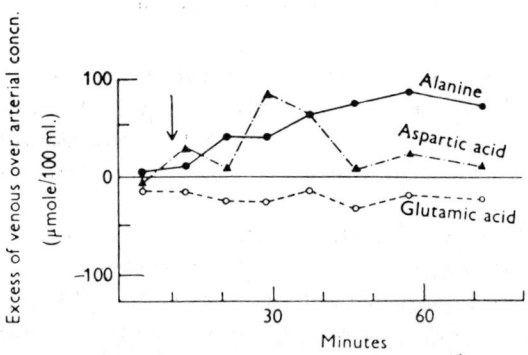

Fig. 8.16. At time indicated by arrow, 10 ml 0.5% aspartic acid solution was introduced into the lumen of the experimental portion of dog small intestine *in vivo*. There was an increase in both alanine and aspartic acid in the blood as it passed through the wall of the loop. (From Neame and Wiseman [68].)

uptake of the stronger solutions of aspartic acid, some extra glutamic acid appeared in the loop venous blood, which confirmed the *in vitro* finding of Matthews and Wiseman [66].

The phenomenon of transamination of glutamic acid to alanine during absorption has since been recorded by Peraino and Harper [288] for the rat and by Elwyn *et al.* [45] for the conscious dog.

During the transamination of absorbed glutamic acid, the concentrations of pyruvate and of α-oxoglutarate in the blood passing through the absorbing intestine remained unchanged [70]. Had all the α-oxoglutarate produced by the intestine entered the blood, its concentration would have increased tenfold. The extra alanine in the loop venous blood was usually greater than the total pyruvate supplied by the arterial inflow.

It seems very likely that after a protein meal, most of the glutamic and aspartic acid of the protein enters the blood as alanine.

8.13 EFFECT OF DIETARY RESTRICTION

8.13.1 *In vitro*

Reducing the daily dietary intake of young adult rats from the usual *ad libitum* amount of about 25 g to an intake of 5-10 g for up to 9 days caused enhancement of the normally occurring intestinal active transport of L-histidine and D-methionine, and, in addition, allowed active transport of D-histidine (which did not occur on the *ad libitum* diet) [178, 289-291]. The results for 9 days dietary restriction are given in Table 8.21, which shows the final serosal/mucosal concentration ratios and the amounts of amino acid transferred to the serosal fluid during 1 hour incubation of the sacs. It can be seen that with fully-fed intestine the serosal/mucosal concentration ratio for L-histidine (inital ratio 1.0) after 1 hour incubation was 1.81, whereas with dietary restricted intestine the ratio rose to 3.17. For D-histidine the control ratio was only 1.04 (indicating no active transport), which became 1.43 after dietary restriction; for D-methionine the respective ratios were 2.28 and 4.45. This was the first direct demonstration of intestinal active transport of D-histidine. The effect of dietary restriction was reversible, refeeding the semistarved animals bringing the enhanced ratios down to the control values in about 1 week. Feeding a restricted diet for less than 9 days produced intermediate final concentration ratios. The enhanced final concentration ratios were not due to alteration in net water movement; extra-distended sacs, which lost serosal volume, produced the same

TABLE 8.21. Effect of dietary restriction on the active transport of 2 mM L-histidine, 2 mM D-histidine and 5 mM D-methionine by sacs of everted mid-small intestine of the rat.

	Final concentration ratio* (serosal/mucosal)	Amino acid transport into serosal fluid (μmole/100 mg dry wt/hr)
L-histidine: fully fed	1.81 ± 0.05	2.81 ± 0.25 (35)
L-histidine: semistarved	3.17 ± 0.13	8.30 ± 0.50 (32)
D-histidine: fully fed	1.04 ± 0.03	0.78 ± 0.07 (14)
D-histidine: semistarved	1.43 ± 0.06	1.74 ± 0.13 (17)
D-methionine: fully fed	2.28 ± 0.09	7.78 ± 0.44 (12)
D-methionine: semistarved	4.45 ± 0.16	22.40 ± 1.20 (12)

* Initial concentration ratio was 1.0. Experimental period 1 hour. 37°C. Semistarved rats were fed 5-10 g food per day for 9 days.
Values are Mean ± S.E.M.(n).
From Neale and Wiseman [178] and Kershaw et al. [290].

ratios as standard sacs, which gained serosal volume during incubation. The phenomenon was observable throughout the length of the rat small intestine, but was rather greater in the middle region. Unlike semistarved rat intestine, the small intestine of semistarved hamsters showed no change in L-histidine active transport, even though the hamsters lost as much body weight and intestinal weight as did the underfed rats. No explanation for this species variation was available.

For a number of reasons, the augmentation of active transport of amino acid by semistarvation does not seem to be due to the thinning which the intestine undergoes during the period of dietary restriction. First, Dowling and Booth [292] have described enhanced glucose absorption by rat small intestine which had actually hypertrophied following extensive intestinal resection. Second, Dowling et al. [293] induced increased glucose and water absorption by rat jejunum by feeding a high-bulk low-calorie diet (by using kaolin) which had no effect on jejunal thickness. Third, Hindmarsh *et al.* [291] could find no change in amino acid active transport by semistarved hamster small intestine despite considerable thinning of the intestine. Fourth, the mid-small intestine of the normal rat is thicker than the lower region [224], yet the mid-region absorbs L-histidine better [291].

Enhanced active transport of amino acid by isolated small intestine of underfed rats has been recorded by Wiseman *et al.* [294], in whose animals a rapidly growing sarcoma was the cause of the undernutrition. Their results with L-histidine were very similar to those noted above. In both underfed normal and in sarcomatous rats the small intestine displayed a smaller diameter, was more translucent and had fewer villi; microscopically there was generalized atrophy of all coats and the epithelial columnar cells were small. There was no ulceration, inflammatory change or necrosis.

Other investigators have also found improved active transport of amino acid *in vitro* using intestine of underfed animals. Thus we have reports by Suda and Shimomura [295] for enhanced L-histidine transport in rats fed a low protein diet for 15 days; by Riecken *et al.* [296] for L-tryptophan in rats fed a high-bulk low-calorie diet for 9 months; by Wright and Barber [297] for L-histidine in rats completely deprived of food (but not water) for 5 days; and by Madge [185, 298] for L-histidine in guinea-pigs starved for 48 hours and in young mice (35 g) given a reduced diet for 7 days. In experiments by Kirsch *et al.* [299], sacs of mid-small intestine of rats fed a protein-free diet took up amino acid better than did sacs of normal intestine, but apparently the extra amino acid was not passed on to the serosal fluid. It was speculated that the increased uptake of amino acid might have been related to greater protein synthesis rather than to augmented transport. Workers not observing enhanced amino acid uptake by underfed rat intestine *in vitro* were Levin *et al.* [300] and Newey *et al.* [301]. The former group were of the view that fasting rats for 3 days reduced glycine uptake, although a recalculation by Hindmarsh *et al.* [291] indicated that the final concentration of glycine in the sac wall plus serosal fluid was higher for fasted intestine than for fully-fed intestine. The findings of Newey *et al.* [301] were somewhat more complicated. They agreed that, in general, in the absence of added glucose, upper and mid-small intestine of in the total α-amino-nitrogen and in individual free amino acids in portal vein blood after an oral test dose of 1 ml bovine plasma protein hydrolysate. In support of these observations, Steiner and Gray [9], who introduced 10 mM L-valine or a mixture of 10 essential amino acids including L-valine (each at 5 mM) into 6 cm loops of mid jejunum of anaesthetized rats which had been deprived of food (but not water) for 4 days, found that starved rats absorbed L-valine at a faster rate and in greater amount than did fully-fed rats. Absorption was measured per unit weight of intestinal mucosa, estimated by

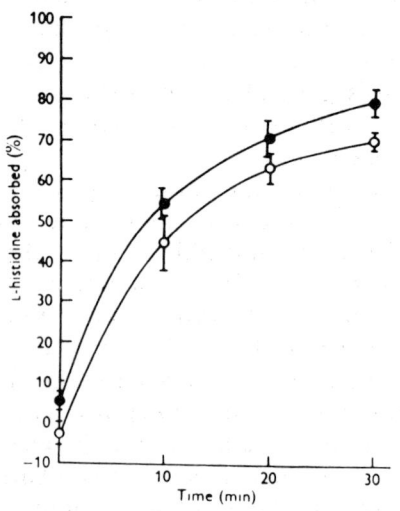

Fig. 8.17. Effect of dietary restriction on the rate of absorption of L-histidine from the whole of the small intestine of rats *in vivo*; 5 ml of 1 mM L-histidine with 0.3% D-glucose in bicarbonate saline was introduced into each small intestine. Values are means and 95% confidence intervals. ●fed 5 g food per day for 9 days; ○fed *ad libitum.* (From Kershaw *et al.* [290].)

scraping the intestine after the experiment. For 2 mM L-valine, and for casein hydrolysate containing 2 mM L-valine, starved and fully-fed rats gave the same absorption values. Enhanced absorption by loops of jejunum *in vivo* has also been recorded for underfed rats by Crampton *et al.* [306] for L-methionine, glycine, glycylglycine, triglycine and methionylmethionine. Their rats were fed 45-70% of normal dietary intake for 9-13 days (which prevented gain in body weight) and the augmented absorption rates were obtained when the results were expressed on a loop length basis or on a loop weight (wet or dry) basis.

rats deprived of food for 3 days showed improved transport of L-methionine and L-proline, but when D-glucose was added to the system, the transport of the amino acids was below normal. For the lower small intestine, the inhibitory effect of added D-glucose was said to be less marked. As absorption of amino acid normally takes place in the upper small intestine and as glucose is always available *in vivo*, Newey et al. [301] concluded that fasting would have an inhibitory effect on amino acid uptake in the intact rat. This conclusion, however, is not in agreement with the results of some of the *in vivo* experiments for rats described below.

It is of interest to note that the mid-small intestine (whole wall) of rats fed 5-10 g food per day for 9 days has only about 11 mg endogenous D-glucose per 100 g wet weight of tissue, whereas fully-fed intestine has about 38 mg D-glucose per 100 g [302].

Human small intestine seems to behave as does hamster small intestine during undernutrition; that is, there is no augmentation of amino acid active transport. Woodd-Walker *et al.* [303], using peroral jejunal biopsy material from children with kwashiorkor, found no abnormality in lysine or alanine uptake, and Steiner *et al.* [304] saw no alteration in L-valine uptake by peroral jejunal biopsy samples from three normal-weight adults on a protein-free diet. The latter investigators recorded reduced L-valine uptake by jejunal mucosa of obese subjects after 2 weeks complete starvation.

8.13.2 In vivo

As well as showing enhanced capacity for active transport of amino acid *in vitro*, Kershaw *et al.* [290] demonstrated that semistarved rats absorbed L-histidine *in vivo* faster than did fully-fed rats, when measured by the rate of disappearance of L-histidine from the whole of the washed-out small intestine (Fig. 8.17). Under their *in vivo* conditions, it was unlikely that thinning of the intestinal wall due to undernutrition could have significantly affected the results, as the blood supply removing absorbed material lies in close proximity to the epithelium. The luminal fluid contained 16.7 mM D-glucose in addition to the 1 mM L-histidine under test. Ziemlanski *et al.* [305] have also published work describing improved absorption of amino acid by undernourished rats. They gave their animals a protein-free diet for 2-3 weeks and obtained a greater than normal increase

Levin [307], too, has described experiments indicating better, and certainly no worse, absorption *in vivo* by the small intestine of underfed (3-day starved) rats. He noted that although the absolute amounts of amino acid (methionine, leucine, histidine, glycine, proline and alanine) taken up per whole loop (regardless of length or weight) were the same in starved and fully-fed rats, expressing the results on a loop weight basis gave greater absorption in starved rats for all the amino acids when used at 5 mM concentration. Increasing the concentration of an amino acid to 30 mM was claimed to make the uptake of some nonessential amino acids poorer in the starved rats. In contrast to the above observations, Kujalová and Fabry [308] were unable to find any change in amino acid absorption in rats fed intermittently, while Pérès *et al.* [309], who employed a modified version of Cori's [108] *in vivo* technique in rats, stated that a 3-day fast had no influence on the absorption of glycine.

There have been few satisfactory direct investigations of absorption of protein digestion products by the underfed human. According to Hardy and Schultz [310], who measured (by intubation) the rate of disappearance of casein hydrolysate from the jejunum of six normal and five hypoproteinaemic subjects, there was an average of 37% of the dose absorbed by the normals and 28% absorbed by the abnormals. It was speculated that tissue oedema might have been the cause of the reduced uptake, as the one non-oedematous hypoproteinaemic subject had normal absorption. These results were net figures and could have been complicated by differences in secretion of protein in the two groups. Studies of absorption in patients who have undergone extensive intestinal resection have shown quite clearly that the intestinal remnant, over a period of months, increased its absorptive capacity [311-313]. Whether this was due, even in part, to increased active absorption by the epithelium remains to be determined, although *in vitro* work with human material suggests that it may not be so.

8.14 AMINO ACID–SUGAR INTER-RELATIONSHIPS

Since about 1960, there has been much effort expended in the unravelling of the relationship between active transport of amino acid and sugar. It was recorded by Nathans *et al.* in

1960 [153] that the active transport of mono-iodo-tyrosine by isolated small intestine of the rat was stimulated by glucose and to a smaller extent by fructose, although this did not occur with metabolic intermediaries such as lactate, pyruvate, α-oxoglutarate, acetate, propionate, butyrate, valerate, phosphoglycerate and fructose diphosphate. They believed that glucose reduced amino acid efflux from the intestinal cell's mucosal surface, thereby improving net uptake. (Read [221], in contrast, has claimed that glucose diminished net amino acid uptake by increasing amino acid efflux.) A similar enhancement by glucose of amino acid (glycine) uptake from the mucosal solution was later reported by Newey and Smyth [314], who found that passively transported sugars (sorbose, fructose, mannose and 2-deoxyglucose) had no such effect. Phlorrhizin by itself also had no effect but it could abolish the action of glucose and, in addition, prevent the reduction of glycine uptake induced by galactose. Glucose could overcome the action of galactose, too. It was suggested that amino acids and sugars competed for a common requirement which was possibly energy provided by ATP; thus glucose, by generating more ATP than was required for its own transport, would act as a stimulant, while galactose, which does not generate ATP in the intestine, behaved only as a competitor for energy. Further work [315] *in vitro* supported these observations, showing that glucose stimulated uptake of proline (but, for some unknown reason, not methionine), while both proline and methionine were inhibited by galactose, 3-0-methyl-D-glucose and α-methylglucoside in the mucosal fluid. Once again, it was found that sorbose, 2-deoxyglucose and fructose had no effect, and nor did rhamnose, lactose or mannitol. For *in vivo* amino acid uptake by the rat [316], however, galactose caused no inhibition, presumably because the blood supplied enough glucose to overcome any galactose inhibition. Saunders and Isselbacher [317], who failed to obtain any effect of 25 mM glucose, xylose or ribose on the active transport of 5 mM alanine, valine, glycine or hydroxyproline by sacs of rat everted intestine, have speculated that the inhibition exerted by galactose might possibly be due to a toxic action of galactose-1-phosphate. As other actively transported sugars act as inhibitors *in vitro* and as galactose did not appear to have this effect *in vivo* [316] the proposition seems very doubtful. It is of interest in this regard that Casey and Felber [318] have reported that 30 mM galactose severely damaged the intestinal mucosa.

Reiser and Christiansen [319] have also concluded that sugars and amino acids compete for a common energy source, although they claimed that fructose, as well as glucose, could supply enough energy to overcome the inhibitory effect of sugars such as galactose. They recorded that the inhibitory action of a sugar was greater in the rat jejunum than in the ileum. When they pre-loaded intestine with actively transported sugars they could detect no effect on subsequent L-valine uptake, whereas pre-loading with L-leucine had a marked effect. Galactose, though, did cause counter-transport of accumulated valine and had, they said, a rapid action as a partially non-competitive inhibitor. On the other hand, Alvarado [239] concluded that galactose behaved as a partially competitive inhibitor of the amino acid cycloleucine.

According to Munck [320], although glucose increased the amount of amino acid (proline and valine) removed from the mucosal fluid bathing sacs of everted small intestine of the rat, it had no influence at all on the serosal fluid amino acid concentration at 1 hour, nor on the degree of tissue amino acid accumulation. The glucose caused more water to be taken up by the intestine and hence greater total quantities of amino acid (but at the usual concentration) were moved. The sugar achieved this, it was believed, by acting as a source of energy. When small sacs comparatively well distended were used, the effect of the glucose on amino acid transport was abolished because only insignificant amounts of extra water were absorbed against the higher intra-sac pressure. Glucose, therefore, had no direct action on the hypothetical amino acid carrier. This conclusion was supported by a further series of experiments by Munck [321]. In these, the intestinal epithelium was first loaded with radioactive sugar or amino acid by pre-incubation for 90 minutes, and then the effect of mucosal fluid sugar or amino acid (20 mM) on the efflux of the accumulated material was tested over a 10 minute period. If the mucosal fluid sugar or amino acid in the second period could block influx, then net efflux might be expected. It was found that the efflux of accumulated ^{14}C-leucine was unaffected by mucosal fluid galactose, 3-0-methyl-D-glucose or α-methylglucoside, but was increased by mucosal fluid L-leucine, L-alanine and L-arginine. In addition, the efflux of accumulated ^{14}C-galactose was unaltered by L-alanine, L-arginine, L-histidine, L-leucine or L-lysine, but was increased by D-glucose and α-methylglucoside. There was, therefore, no evidence in favour of a membrane carrier common to sugars and amino acids; and Munck [321] was of the opinion that sugar—amino acid interactions were not explicable in terms of energy limitation. Others to disagree with the view that sugars and amino acids were transported by a common carrier have been Chez *et al.* [322] and Burns and Faust [242]. The former group claimed that glucose and galactose both inhibited net alanine transport by rabbit ileum (*in vitro*) but that the sugars only behaved in this way after they had entered the epithelial cells. The sugars were said to increase the efflux of amino acid from the cells to the mucosal fluid, while having no effect on amino acid influx. (Casey and Felber [318] have also noted that galactose caused amino acid (phenylalanine) efflux from intestinal epithelium (rat), but their interpretation of their results was that sugars and amino acids did share a common transport system.) Burns and Faust [242] based their dissension on the findings of experiments on amino acid and sugar binding by isolated brush borders of hamster jejunal mucosa.

Contrary to the conclusions that might be drawn from the above publications of Munck [321], work by Alvarado [239] has led him to the view that a common carrier does in fact exist for the intestinal transport of amino acids and sugars. He observed that when rings of hamster everted small intestine were allowed to absorb D-galactose and cycloleucine, the D-galactose behaved as a partially competitive inhibitor, there being an increase in the K_t of the cycloleucine but no change in its V_{max}. An allosteric effect, due to binding of the sugar sufficiently close to the amino acid-binding site, was suggested. (This explanation seems unlikely as Bingham *et al.* [316] obtained no inhibition of amino acid uptake by galactose in the rat *in vivo* but only *in vitro*. An allosteric effect would, presumably, be demonstrable in both types of experiments.) Alvarado [239] also stated that arginine, galactose, glucose and α-methylglucoside seemed to elicit counter-transport of tyrosine; and that basic amino acids (arginine, lysine, ornithine) exerted a similar effect as did actively transported sugars (partially competitive inhibitors) on the uptake of neutral amino acids, whereas neutral amino acids (histidine, proline, methionine) behaved as fully competitive inhibitors when tested with other neutral amino acids (cycloleucine, tyrosine). The explanation offered in an attempt to combine all these results was that each group of substances (basic amino acids, neutral amino acids, actively transported sugars) had a separate binding site on a polyfunctional carrier (which also bound sodium ions), and the presence of one class on the carrier exerted an allosteric effect on the other classes. This hypothetical polyfunctional carrier was believed to exist on the external surface of the cell membrane [323]. This conclusion by Alvarado that amino acids and sugars competed for a common (polyfunctional) carrier was not in accord with an earlier contention of Read [221]. The latter, who used dogfish intestine *in vitro,* claimed that all actively transported sugars, including glucose, inhibited cycloleucine uptake, there being an appreciable time lag before the sugars exerted their effect. This was interpreted as showing that there was no direct competition for entry at the external surface of the cell; instead, it was proposed that the inhibitory behaviour was related to increased intracellular sodium brought about by sugar entry, the extra sodium then being available to increase the efflux of previously absorbed amino acid, thereby reducing net uptake of amino acid.

Attempts to obtain evidence about the nature of sugar—amino acid interaction during active absorption have been made by study of the change in transmural potential difference during absorption [214, 260, 263, 264]. All workers agree that when the maximal increase in transmural potential difference during the transport of an amino acid has been induced, addition of a second amino acid to the mucosal fluid has no effect, yet addition of an actively transported sugar gives a further rise in potential difference equal to the rise observed when that sugar is used alone. The same series of events takes place when a sugar is present first and an actively transported amino acid added second. Thus the transport mechanisms for amino acid and sugar, at least as reflected by transmural potential difference change, seem to be quite separate. Support came also from Butt and Wilson [245], on the grounds that amino acid active uptake in the foetal guinea-pig develops well before sugar active uptake.

In curious contrast to the many reports of the stimulatory

action of glucose on amino acid absorption *in vitro* (though Read [221] and McConnell and Cho [324] noted reduced amino acid transport *in vitro* in the presence of glucose), Orten [325] has found that in conscious man, glucose decreased the total absorption of amino acids from an equimolar mixture of 18 amino acids. Galactose had no discernible effect, but fructose may have caused a little acceleration. The region used was the ileum, and the amino acid absorption pattern remained unaltered in the presence of the glucose. Similarly, Annegers [326], using Thiry-Vella loops of jejunum in conscious dogs, recorded that increasing the luminal concentration of glucose resulted in lowered absorption rates of glycine, methionine, histidine and glutamic acid, and galactose reduced the absorption rates of alanine, methionine, histidine, lysine and glutamic acid. Fructose had no effect. The changes were not due to alteration in water absorption.

In addition to the inhibitory action of actively transported sugars on the active transport of amino acids, the latter reduce the active uptake of the sugars themselves [183, 327, 328]. Thus the active transport (by sacs of hamster everted small intestine) of D-glucose was found to be reduced by L-methionine, L-ornithine, L-leucine, L-valine, L-proline, L-histidine, L-alanine, D-methionine and D-histidine (all actively transported), but not by L-glutamic acid, L-aspartic acid, D-valine, D-alanine or D-glutamic acid (not actively transported). Other sugars to be affected were D-galactose, 3-0-methyl-D-glucose, D-fucose and D-xylose (all actively transported), but not D-mannose, L-sorbose, L-fucose, L-xylose or α-glucoheptose (not actively transported). Further, the active transport of L-glucose by the dietary-restricted rat was completely abolished by L-histidine [302]. Hence only actively transported amino acids and sugars are involved in this phenomenon. Although the small intestine readily produces L-alanine (a good inhibitor of D-glucose active transport) from L-glutamic or L-aspartic acid, the latter two amino acids had no effect on D-glucose active transport, showing that the site of action of amino acids on sugar uptake is not accessible to intracellularly produced L-alanine, and that the transamination step occurs beyond the point of sugar—amino acid interplay. The changes observed were not due to alteration in net water movement, and they were demonstrable along the whole length of the small intestine. The influence of L-amino acids on D-glucose active transport seems to be in some way related to the efficiency with which the amino acids are themselves concentrated (Table 8.22).

8.15 REQUIRED MOLECULAR CONFIGURATION FOR ACTIVE TRANSPORT

Apart from glycine and β-alanine, which have no asymmetric carbon atom, amino acids may exist in either the L- or D-form. Those derived from proteins (or otherwise found as intermediates in mammalian tissues) are referred to as the natural

TABLE 8.22. Effect of 20 mM L-amino acids on the active transport of D-glucose by sacs of everted mid-small intestine of the hamster.

Amino acid in mucosal and serosal fluid (20 mM)	D-glucose final concentration ratio* (serosal/mucosal)	D-glucose transport into serosal fluid (μmole/100 mg dry wt/hr)
None	3.23 ± 0.12	95.0 ± 5.0 (19)
Glutamic acid	3.24 ± 0.06	91.2 ± 2.6 (12)
Aspartic acid	2.90 ± 0.05	88.0 ± 3.7 (12)
Methionine	2.54 ± 0.07	66.8 ± 4.5 (12)
Ornithine	2.49 ± 0.07	54.7 ± 1.9 (12)
Leucine	2.41 ± 0.07	60.2 ± 3.9 (12)
Valine	2.27 ± 0.09	64.2 ± 5.4 (12)
Proline	2.12 ± 0.06	55.1 ± 2.1 (18)
Histidine	2.04 ± 0.06	44.0 ± 3.0 (17)
Alanine	2.03 ± 0.03	44.6 ± 2.6 (12)

* Initial concentration ratio was 1.0. Experimental period 1 hour. 37°C.
Values are Mean ± S.E.M.(n).
From Hindmarsh et al. [183].

amino acids, signified by the L. They all have the amino group attached to the α-carbon atom. Although D-amino acids are to be found in some plants and in microorganisms, there is no convincing evidence that they occur in proteins. When an amino acid rotates light to the right it is prefixed with +; when the light is rotated to the left it is prefixed with —. There may be asymmetry around other centres apart from the α-carbon, giving rise to additional stereoisomers. For details of the stereochemical configurations of amino acids see Neuberger [329]. The basic composition of an α-amino acid is

$$\begin{array}{c} H \\ | \\ R-C-COOH \\ | \\ NH_2 \end{array}$$

(with α indicating the α-carbon)

in which R may be a hydrogen atom or a radical of some complexity. In β-alanine, the amino group at the α-carbon is replaced by a hydrogen atom, leading to loss of the asymmetrical centre. In glycine, the R is a hydrogen atom, so that once again there is no asymmetric centre.

8.15.1 Stereochemical form

(i) L-amino acids. The neutral L-amino acids are usually readily transported against their concentration gradient by the small intestine, even when the initial concentration of the amino acid in the mucosal fluid is comparatively high (20 mM or more). An exception is L-tryptophan, for which active transport is only demonstrable when its initial concentration is low (a few mM). Active absorption was first shown for this group of amino acids and for a time it was believed that only the neutral L-amino acids could be actively absorbed. Histidine is included in this group even though it possesses two nitrogen atoms with basic properties.

The basic L-amino acids, unlike the neutral ones, are only actively transported to any degree when their initial concentration in the mucosal fluid is low (about 3 mM or less).

The acidic amino acids have never been shown to be actively transported by the small intestine. This may be because they are in fact not actively transported, or because such active transport as occurs is completely overshadowed by the transamination reaction in which these amino acids participate.

(See also section on transport of amino acids against a concentration gradient.)

(ii) D-amino acids. Some D-amino acids (methionine, histidine, tyrosine, norvaline, serine) can be absorbed against their concentration gradient to a small extent, but for the majority even this has not been demonstrated, despite numerous attempts. It is possible that inhibition of active transport by endogenous L-amino acids is the cause of lack of active transport for most D-amino acids. Analysis of the effects of L- and D-amino acids upon each other during absorption suggests that many D-amino acids have some affinity for the transporting system of their class (neutral, basic). Under ordinary conditions *in vivo*, it is unlikely that active absorption of D-amino acids occurs to any appreciable extent.

(See also section on transport of amino acids against a concentration gradient.)

8.15.2   Carboxyl group
For active transport, a free carboxyl group is essential at the α-carbon atom, neither the methyl ester of L-histidine [176] nor L-tyramine [153] appearing to have any affinity for a transporting mechanism. Replacing the carboxyl group by $-CH_2OH$, $-CO\ NHOH$ or by $-CO.C_6H_5$ also prevents active transport [330]. When a $-CO.NH_2$ group replaces the carboxyl group, active transport of the corresponding free amino acid occurs, deamination of the parent molecule presumably preceding the active transport step. According to Spencer *et al.* [330], analogues of the type $R-CHNH_2.PO_3H_2$ (phosphonic) have no inhibitory action on amino acid transport, although Nathans *et al.* [153] found that α-amino sulphonic acids can inhibit the uptake of mono-iodo-tyrosine, though less well than do α-amino carboxylic acids.

8.15.3 α-amino group
Replacement of the whole amino group by a hydrogen atom leaves a molecule (a fatty acid) which may still be actively transported, as are acetic, propionic and butyric acids [331]. Experiments by Nathans *et al.* [153] have shown that fatty acids and amino acids have separate systems, as acetate, propionate and butyrate had no effect on mono-iodo-tyrosine active transport. Transposing the amino group from the α- to the β-position, although it reduces active transport by the intestine, does not abolish it; β-alanine is actively absorbed, though less well than L-α-alanine [177]. Thus the α-amino group, as such, is not essential for intestinal active transport of a molecule. However, replacing it by an hydroxyl group, as in lactic acid (hydroxypropionic acid), prevents active transport of the new compound [332].

Modifying the α-amino group does not necessarily cause loss of active transport, as L-proline (an imino acid) is well transported against its concentration gradient, as are hydroxy-L-proline and N-methyl-glycine. The latter seems to be about the bulkiest N-substituted molecule to be actively transported by the neutral amino acid system, as neither N-ethyl-glycine [196] nor N-phenyl-glycine [333] is transported against its concentration gradient, although the latter compound can markedly inhibit the uptake of glycine [235]. It should be noted that N-dimethyl-glycine can be actively transported, but no longer by the neutral amino acid system, for which it has little or no affinity [334]. The acetylated forms of histidine [176], glycine, valine and methionine [153], which have one hydrogen atom of the amino group replaced by an acetyl group, have no effect on glycine active transport. Hence, size of the group at the α-carbon atom may be of some importance, but it is not the only factor (lactic acid is not actively absorbed).

Cyclic imino acids of the form $(CH_2)_n NHCHCOOH$ can be actively transported by hamster small intestine when $n = 2$ (azetidine-2-carboxylic acid), $n=3$ (proline) and $n=4$ (pipecolic acid) but not $n = 1$ (aziridine-2-carboxylic acid) [196]. The cyclic and non-cyclic imino acids show cross-inhibition.

8.15.4   α-hydrogen atom
Exchange of the α-hydrogen atom for a methyl group may much reduce (α-methylmethionine) or almost entirely abolish (α-methyltyrosine) active absorption [176]. Similarly, cyclo-leucine (1-amino,1-carboxyl-cyclopentane) is less well transported than is L-leucine [247].

8.15.5   Side chain
Amino acids with greatly differing side chains (R = H for glycine but $C_6H_5CH_2$ for phenylalanine) may be transported against their concentration gradient by small intestine even when their initial concentration in the mucosal fluid is high (20 mM or more). At low initial concentrations (1-3 mM), there is even active transport of L-tryptophan, in which R = C_6H_4NHCH:CCH_2; and of $4 - CH_3 -$, $5 - CH_3 -$, $6 - CH_3 -$, and N-chloroacetyltryptophan [149]. Generally, the side chain seems to be of comparatively little importance

provided that it does not bear a charge. When it does, active transport may be quite low (lysine, arginine, ornithine) or entirely absent (glutamic and aspartic acids). Thus Huang [335] observed that tyrosine derivatives with neutral side chains were actively absorbed, but not those with ionized groups; and Lin *et al.* [176] noted that the γ-methyl ester of L-glutamic acid, which has a neutral side chain, was well concentrated by hamster small intestine and was a good inhibitor of valine uptake. Agar *et al.* [172] and Nathans *et al.* [153] have also found that amino acids with large nonpolar side chains (tryptophan, methionine, leucine, isoleucine) were good inhibitors of amino acid active uptake, while those with polar side chains (glutamic acid, aspartic acid, lysine, arginine) were not.

Solubility of the side chain in a lipid part of the epithelial cell membrane is, according to Wilson [336], a decisive factor in allowing the groups attached to the α-carbon atom to gain access to the active site of the transporting system. He stressed the relationship between ready solubility in ethanol of amino acids with long alkyl side chains and their greater apparent affinity for active absorption. This hypothesis was amply supported by the work of Reiser and Christiansen [141], who showed that among the neutral L-amino acids inhibiting L-valine uptake by rat small intestine, those with side chains that were easily soluble in lipid possessed a lower K_t and were more effective competitive inhibitors.

8.16 SITE OF ABSORPTION

8.16.1 Human small intestine
In man, the site of absorption of protein digestion products has been determined in normal subjects by sampling luminal contents after the ingestion of a test meal or its introduction by intubation; or by measuring the rate of absorption or total absorption in persons subjected to various degrees of intestinal resection. In experiments of this sort, it is necessary to use an identifiable form of test protein so as to be able to distinguish it from endogenous protein. A detailed study performed in normal adults by Borgström *et al.* [10], who fed a mixture of ^{131}I-labelled human serum albumin, milk protein, corn oil, glucose and lactose, together with nonabsorbable polyethylene-

glycol to permit changes in dilution to be assessed, showed that 80-90% of the radioactive protein was absorbed (presumably as small peptides and free amino acids) in the proximal 50-100 cm of jejunum (130-180 cm from the nose) (Fig. 8.18). Similarly, Nixon and Mawer [11], who fed to normal persons 15 g casein, recorded that by the time the test meal had reached a point 230 cm from the nose about three-quarters of the fed protein had been absorbed when estimated on a net basis (which included endogenous secreted and desquamated protein). When they took into account endogenously derived protein (by feeding a

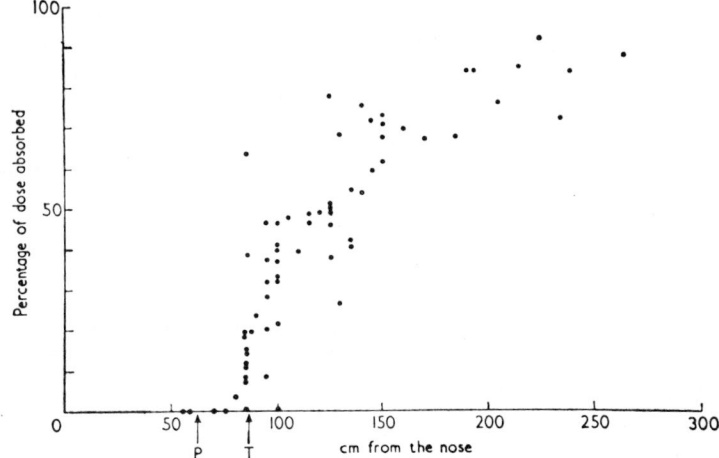

Fig. 8.18. Site of absorption of ^{131}I-labelled human serum albumin in the normal human small intestine. The test meal contained corn oil, glucose, lactose, protein and polyethyleneglycol. P, pylorus; T, ligament of Treitz. (From Borgström et al. [10].)

protein-free meal), it seemed most likely that the digestion products of all the fed protein could be absorbed by the time it had travelled 140 cm from the nose (i.e. in the upper half of the small intestine). Absorption started in the duodenum even though this area is relatively poor in mucosal peptidase. For a meal of 15 g gelatin, digestion and absorption were very much poorer than with casein, only a net 30% of the gelatin being absorbed by the time the meal reached 194 cm from the nose. The reason for the difference in the results with casein and gelatin might have been the latter substance's greater resistance to hydrolysis.

In general agreement with the above work, Booth and Mollin [337] found that protein digestion products were well absorbed in the region of the mid-small intestine. They used patients after intestinal resection. For free amino acids in normal man, Schedl et al. [257] and Adibi [253], using intraluminal perfusion techniques, reported that methionine, leucine and threonine were absorbed best in the jejunum.

With 6-100 mM L-methionine [257], rate-limiting kinetics were seen at all intestinal sites, the K_t being 11-28 mM in the jejunum and 2-6 mM in the ileum, while the jejunal V_{max} was 15-33 mmole per hour per 15 cm and the ileal V_{max} was 4-7 mmole per hour per 15 cm.

In man, therefore, it is quite clear that normally by far the greater part of ingested protein is absorbed in the upper half of the small intestine, although subjects with massive resection of the upper small intestine can still absorb adequate amounts of digested protein [338]. In cases of gastric resection in man, any inadequacy in protein absorption that occurs is not due to interference with intestinal amino acid uptake [339, 340].

8.16.2 Dog small intestine
In the dog, as in man, protein digestion products are normally absorbed in the proximal small intestine [341], although resection experiments [342] have shown that the distal half of the small intestine can also absorb an average amount of dietary protein. With loss of much of the ileum, however, there may be some decreased utilization of protein, possibly due to the associated steatorrhoea [343].

8.16.3 Rat small intestine
For the rat, the functional site of absorption of the digestion products of doses of 100 mg ^{131}I-labelled human serum albumin given by stomach tube seemed to be the lower jejunum and upper ileum, as these regions accumulated greatest amounts of radioactivity in the intestinal wall [344] (Fig. 8.19). Some radioactivity was found in the wall of the duodenum and upper jejunum soon after (5-10 minutes) the dose, but only insignificant amounts appeared in the wall of the lower ileum even after 60 minutes. That the region of greatest accumulation of amino acid in the intestinal wall of the rat is in general also the region of greatest transport across the wall has been claimed by Nathans et al. [153], Samiy and Spencer [154], Ramaswamy

and Radhakvishnan [345] and McConnell and Cho [324]. However, Schedl et al. [346], using α-aminoisobutyric acid as test material, have observed that the intestinal tissue concentration of the amino acid was least in the region where its absorption was greatest (in this case, the mid-jejunum), and even when the α-aminoisobutyric acid was given parenterally the wall of the mid-jejunum accumulated it less than did the wall of the rest of the intestine. Determination of absorption site by reference to tissue accumulation may, therefore, be sometimes misleading.

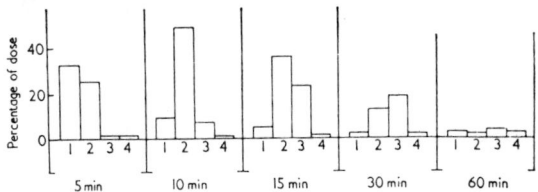

Fig. 8.19. Radioactivity in four equal segments of the small intestine of rats killed at various intervals after an oral dose of 0.8 μc ^{131}I-labelled human serum albumin. The protein load was 100 mg. Each value is the mean of three results. 1, duodenum and upper jejunum; 2, lower jejunum; 3, upper ileum; 4, lower ileum. (From Parkins et al. [344].)

Resection experiments by Nygaard [347] in the rat have also indicated that the mid-small intestine is probably the functional site of absorption of ingested protein, although he noted that nitrogen excretion was normal or only slightly increased 3 months after as much as 75% of the proximal small intestine had been removed. The last quarter of the small intestine must therefore be normally able to absorb adequate quantities of protein, or it can fairly soon become so. In contrast, nitrogen excretion rose progressively as more of the distal small intestine was resected, though this may have been due, at least partially, to the concomitant steatorrhoea. In general agreement with these conclusions, Ochoa-Solano and Gitler [348] stated that in the rat almost all exogenous protein was absorbed in the first half of the small intestine, but they thought that endogenous protein (chiefly pancreatic) was more resistant to digestion and that much of it would reach the lower half of the small intestine.

Experiments with free amino acids have suggested that the proximal and mid-small intestine best absorb L-cystine, L-cysteine [156], L-valine [141] and L-proline [349], whereas the lower ileum has been said by Nathans et al. [153] to be the best site for the transport of L-tyrosine and L-mono-iodo-tyrosine. According to Hird and Sidhu [173], L-histidine appeared to be equally well absorbed by the upper and lower small intestine, although Hindmarsh et al. [291] found the greatest transporting activity for this amino acid to be in the mid-small intestine (Fig. 8.20).

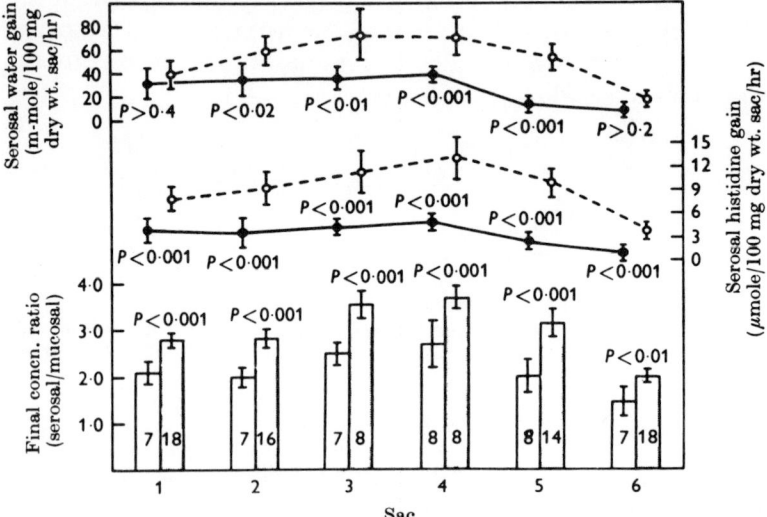

Fig. 8.20. Active transport of L-histidine by sacs of everted small intestine of fully-fed and semistarved rats. Control animals: open columns and continuous lines. Semistarved animals: stippled columns and interrupted lines. Histogram: ratio of L-histidine concentration in final serosal fluid to that in final mucosal fluid (initial serosal and mucosal fluid contained 2 mM L-histidine and 16.7 mM D-glucose). Lower two curves: L-histidine entry into serosal fluid. Upper two curved: water entry into serosal fluid. Semistarved rats were fed 4 g food per day for 1 week. Sac 1, upper jejunum; sac 6, lower ileum. Experimental period 1 hour. 37°C. Values are Mean ± 2S.E.M., with number of sacs in each column. (From Hindmarsh et al. [291].)

There can be no doubt that in the rat the functional site for absorption of digested protein is the more proximal part of the small intestine even though the middle and lower small intestine could probably replace it if necessary. It is of interest to note, in this regard, that the ileal mucosa is very rich in dipeptidases [103].

8.16.4 Hamster small intestine

With hamster small intestine *in vitro,* the site of greatest absorption of free amino acid has varied somewhat with the amino acid used. Thus the middle third of the small intestine has been found to be the region which best transports L-tyrosine [158] (although Nathans *et al.* [153] believe it to be the lower ileum), L-tryptophan [138, 149] and L-phenylalanine [140, 154]; for glycine, L-leucine, L-valine [226] (Fig. 8.21), L-arginine, L-ornithine, L-lysine [170] and L-selenomethionine [324] the region seems to be the distal jejunum and upper ileum (it should be noted that L-arginine and L-ornithine are metabolized to an appreciable extent by the small intestine); while for L-histidine all the small intestine appears equally good [291]. On the whole, the evidence suggests that in the hamster the middle and lower regions of the small intestine may have some role in the *in vivo* absorption of protein digestion products.

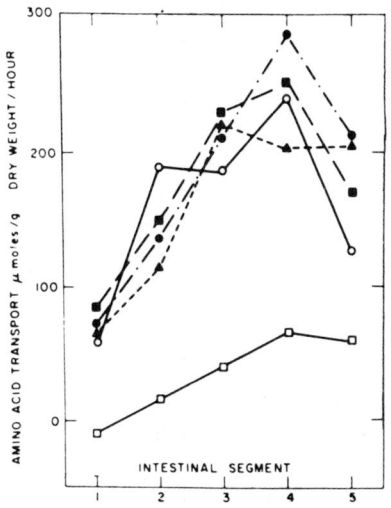

Fig. 8.21. Rate of amino acid transport at various points along the small intestine of the hamster. 1, duodenum; 5, ileum. ▲ glycine (initial concentration 10 mM); ○ L-alanine (initial concentration 5 mM); ● L-valine (initial concentration 5 mM); ■ L-leucine (initial concentration 5 mM); □ α-aminoisobutyric acid (initial concentration 10 mM). (From Matthews and Laster [226].)

8.16.5 Colon

Unlike the small intestine, the colon of man plays little or no part in the absorption of protein digestion products *in vivo* [350, 351]. In the dog, although Rhoads *et al.* [30] and Hauge and Krippaehne [352] have observed that amino acids and protein hydrolysate readily disappeared from loops of colon, it seems improbable that the colon plays an effective part in normal absorption of protein digestion products. According to Christensen *et al.* [161], rat colon *in vivo* probably does not actively transport amino acids (Fig. 8.22); nor does isolated colon of the adult rat [153, 349, 353], hamster [153, 354] or tortoise [355]. Active accumulation of L-proline by the colon of young rats (1 month) [349] and of L-valine by the colon of adult rats [353] is not accompanied by active transport of amino acid across the intestinal wall.

8.17 ROUTE OF ABSORPTION

The main route of absorption of amino acids is the portal blood, in which they are to be found in large quantities after the ingestion of protein, peptide or free amino acid. In addition, some amino acids may enter the mesenteric lymph, in the free form and also as newly synthesized protein [356]. According to Jacobs and Largis [357], the amino acid concentration in rat

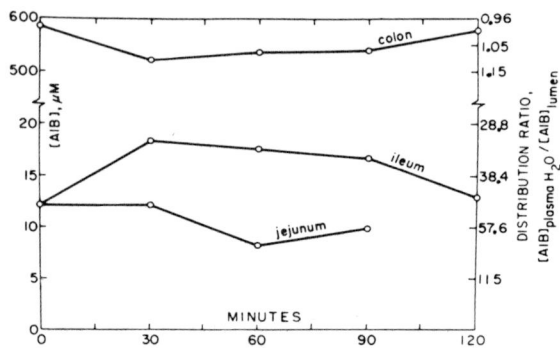

Fig. 8.22. Changes in α-aminoisobutyric acid concentration in intestinal loops of an anaesthetized rat. The right-hand scale shows the ratio of the amino acid concentration in the plasma to that in the intestinal lumen; the final plasma value is used. (From Christensen *et al.* [161].)

mesenteric lymph was equal to that in the plasma but the amino acid patterns were different in these two compartments (and also different from the pattern in the intestinal mucosa). The lymph protein in these experiments amounted to one-tenth to one-quarter of the protein in the plasma. The origin of mesenteric lymph protein need not necessarily be the intestinal mucosa, as protein from the blood readily enters the lymph via the capillaries [155, 358]. Inhibiting protein synthesis by cycloheximide or puromycin increased the lymph amino acid concentration and reduced that of the protein [359].

When protein is absorbed intact by the newborn, γ-globulin travels exclusively via the lymph [360], whereas appreciable amounts of intact albumin travel via the portal blood [361].

8.18 EFFECT ON PLASMA AMINO ACID LEVELS

It has been known for many years that ingestion of protein or its digestion products may lead to a rise in the free amino acid level of both portal and systemic blood, the response depending not only on the amount of protein taken, but also on its composition. This is because there is considerable variation in the rates of absorption of different amino acids, and some amino acids are very readily metabolized by the intestine. Thus, transamination of glutamic and aspartic acids may lead to little or none of these amino acids appearing in the mesenteric venous blood, but there is, in their place, the appearance of alanine. Arginine is quickly broken down by mucosal arginase, and ornithine and glutamine are also rapidly destroyed by the intestine. When systemic blood is examined, the portal blood amino acid pattern may have been markedly altered by the liver. A further complication is that the loading of cells with one amino acid (by raising the blood content) causes them to release some of the other free amino acids previously accumulated. In addition, amino acids of endogenous protein of secretions and desquamated cells are added to those of the exogenous protein. For all these reasons, interpretation of changing amino acid levels in the blood (especially systemic) may be difficult. It is usually desirable to use methods of analysis specific for each amino acid under consideration.

Ganapathy and Nasset [362] and Nasset et al. [363] have found that the molar ratios of the plasma free amino acids in the dog remained fairly constant after a meat meal, whether the total amino acid content rose or fell, which it did unpredictably. They were unrelated to the ratios of the free amino acids in the intestinal lumen. The accumulated results have been reviewed by Nasset [364]; some typical values are shown in

Table 8.23. Similarly, Nasset and Ju [365] observed that human peripheral plasma amino acid molar ratios were unlike those of a test meal. There was much variation from individual to individual in both the fasting and postprandial state. Others to find that the change in plasma amino acids did not fit the pattern of amino acids fed (either as intact protein or as hydrolysates) were Frame [366], Peraino and Harper [367], Yearick and Nadeau [368], Coulson and Hernandez [369], Anderson et al. [370] and Elwyn et al. [45]. In contrast to these various reports, Dent and Schilling [80], who examined dog portal and jugular vein blood by semi-quantitative paper chromatography, stated that the alteration in the plasma amino acid pattern in their animals after a casein meal could be accounted for (except for glutamic acid) by the amino acids in the administered protein; and Richmond and Girdwood [371] claimed that the increase in leucine, lysine, methionine and phenylalanine (but not glutamic acid) in plasma was proportional to their occurrence in casein when 25 g calcium caseinate was fed to fasting patients.

TABLE 8.23. Molar ratios of free amino acids in intestinal lumen, portal vein and mesenteric vein 2 hours after feeding zein to dogs.

	Intestinal lumen*	Portal vein*	Mesenteric vein*	Zein*
Isoleucine	1.00	1.00	1.00	1.00
Tryptophan	0.09	0.15	0.15	0.01
Methionine	0.31	0.59	0.57	0.25
Aspartic acid	0.69	0.20	0.21	0.45
Histidine	0.69	0.75	0.61	0.14
Threonine	1.03	3.42	3.72	0.41
Tyrosine	1.04	0.45	0.95	0.50
Phenylalanine	1.06	0.82	1.12	0.70
Proline	1.06	3.00	3.54	1.56
Arginine	1.13	0.67	0.65	0.17
Lysine	1.18	1.98	1.28	0.01
Glycine	1.27	2.37	2.68	0.10
Serine	1.41	2.28	2.53	1.32
Valine	1.44	2.40	2.83	0.40
Glutamic acid	1.49	0.97	1.43	3.47
Alanine	2.30	8.07	9.12	0.22
Leucine	2.45	3.80	3.84	3.30

* Isoleucine is taken as equal to 1.00.
From Nasset [364].

Despite the difficulties, change in plasma amino acid content and pattern after a test meal may be used to advantage. Thus the evidence that ingested protein enters the blood virtually entirely as free amino acid is based to a large extent on measurement of portal blood amino acid level. In addition, absorptive efficiency has been assessed by observation of the alteration in peripheral plasma amino acid after a test dose of amino acid. For example, feeding 25 g glycine to a normal person caused a rapid rise in the glycine and total amino acid concentration of peripheral plasma with a peak at about 1 hour (Figs. 8.23 and 8.24), whereas the increase was much delayed and less marked in cases of untreated malabsorption [372, 373]. Brisk responses were also produced by alanine, tyrosine and phenylalanine [155]. It should be emphasized that if the test amino acid is given with or incorporated in a meal, the expected rise (obtained when the amino acid is given alone) may be considerably reduced, as is shown for tryptophan in Table 8.24 [374]. This diminution is brought about by partial inhibition of tryptophan absorption by competing amino acids of the meal and digested secretions, and possibly by the rapid clearing of the blood of amino acids for digestive enzyme synthesis stimulated by the meal.

After the ingestion of appreciable quantities of intact protein, the increase in peripheral plasma free amino acids is not as rapid as after the ingestion of free amino acids. For the latter, the concentration peak is usually at about 1 hour after the test meal, but for intact protein the peak is usually at 1.5-3 hours. With free amino acids, the plasma level can be expected to return to the normal resting value by about 3 hours. If only very small amounts of intact labelled protein are fed, the maximum absorption rate, measured by peripheral plasma radioactivity, may occur as early as 40-50 minutes after the dose (Fig. 8.25) [13, 34, 375]. For human hepatic vein blood, Gay and Crane [376] found peak amino acid radioactivity 20-30 minutes after the introduction of intact labelled protein into the duodenum, with 25% of the peak value being reached within one minute in one experiment. When protein hydrolysate was put into the duodenum, the hepatic vein blood amino acids rose to their peak concentration within only a few minutes, followed by a rapid fall.

It is of interest to note, in regard to this topic, that David *et* al. [377] reported that free radioactive amino acid was taken up so rapidly by mouse intestine that within one minute there was amino acid in all regions of the epithelial cells and in all cell layers right up to the serosal lining.

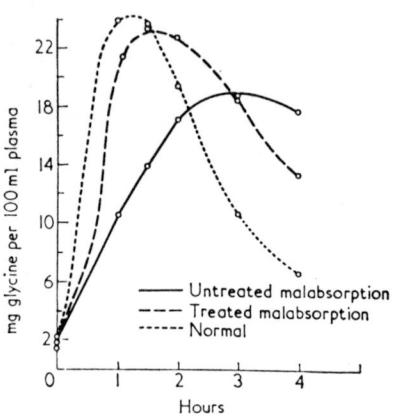

Fig. 8.23. Change in the plasma glycine concentration after ingestion of 25 g glycine by normal persons and by patients with malabsorption. (From Butterworth et al. [373].)

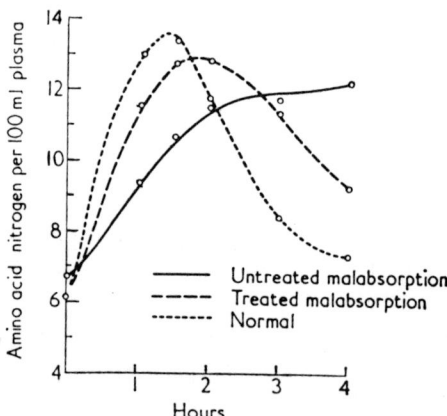

Fig. 8.24. Change in total amino acid nitrogen (mg per 100 ml) of plasma after ingestion of 25 g glycine by normal persons and by patients with malabsorption. (From Butterworth et al. [373].)

TABLE 8.24. Plasma tryptophan levels after ingestion of various amounts of the amino acid by normal persons.

Number of subjects	Tryptophan intake (g)	Plasma tryptophan (μg/ml) (minutes after dose)				
		0	30	60	90	240
3	0	10.3 ± 1.8	8.3 ± 1.5	8.4 ± 1.6	8.5 ± 0.9	8.5 ± 1.5
8	1	9.6 ± 0.6	43.2 ± 0.5	41.9 ± 0.8	29.6 ± 1.6	13.4 ± 1.2
1	2	10.0	50.0	57.6	61.6	23.7
2	3	8.1	85.5	96.6	86.3	35.8
2	1 plus 0.35 present in meal	10.0	33.0	30.0	26.0	16.0

Values are Mean ± S.E.M.
From French and Wertz [374].

8.19 DEVELOPMENT OF AMINO ACID ACTIVE TRANSPORT

Foetal, as well as postnatal, small intestine can absorb amino acids actively, the process developing at different gestational ages for different amino acids. Thus foetal rabbit small intestine can transport L-valine, L-lysine and L-methionine against small but definite concentration gradients at about the 22nd day of gestation (full gestation is 28-30 days), followed by L-proline and glycine a few days later, and then by betaine at about the 28th day [378]. Similarly, foetal guinea-pig small intestine of 35 days gestation (full gestation is 65-70 days) can actively

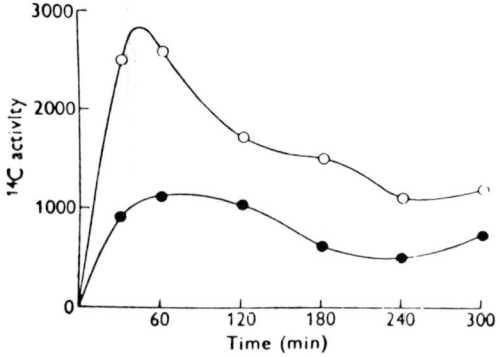

Fig. 8.25. Radioactivity (counts per minute per ml) of plasma at various intervals after feeding 35 mg ^{14}C-labelled *Chlorella* protein (containing 8 μc of ^{14}C activity) to rats. ○ portal plasma; ● systemic plasma. (From Dawson and Porter [375].)

accumulate L-valine, L-proline, L-lysine and glycine, with sarcosine following at 40 days and N-dimethyl-glycine and betaine at 60 days [245]. The process in the guinea-pig for all the amino acids is greatly enhanced at about the time of birth, and then declines to adult values in a few weeks. In the chicken, glycine active absorption is present at 4 days before hatching, the transport capacity at that time being the same as at 10 days after hatching (although for sugar it increases over that period) [379]. Active transport of glycine by the caecum is lost at hatching, probably because of the concomitant loss of the columnar epithelial cells. The human foetal jejunum and ileum (at 14 weeks gestation) can also actively transport amino acid (L-alanine) [264]. Although adult intestine requires a plentiful supply of oxygen for active absorption of amino acids, the small intestine of foetal and newborn rabbits can actively transport L-histidine and L-tryptophan under anaerobic conditions, the phenomenon being most marked during the first week of life and disappearing by about 35 days after birth [380].

In the rat, the ability of sacs of everted small intestine to actively transport L-valine was found by Ning *et al.* [381] to be greatest during the first 8 weeks of life, after which it fell fairly fast, so that at 40 weeks of age the concentrating ability, judged by the ratio of the final serosal valine concentration to the final mucosal valine concentration, was only about one-sixth that at 8 weeks. Tissue accumulation of L-valine at 40 weeks was about one-third that at 4 weeks. Fitzgerald *et al.* [382] have also observed that L-valine, L-lysine and glycine were transported better by the newborn (2-5 days) than by the adult rat; and Reiser *et al.* [383] have recorded that although the affinity of L-valine for the active transport mechanism was the same (K_t of 6-7 mM) in neonatal and adult animals, the rate of transport (V_{max}) was greater in the neonates. It has been claimed by Pénzes *et al.* [384] that 2-year old rats absorbed methionine (*in vivo*) faster than did 6-month old rats.

According to Batt and Schachter [349], rat colon actively accumulated L-proline during the first month or so of life.

It is clear, therefore, that in experiments on amino acid active absorption, the age of the animals may be of considerable importance.

8.20 EFFECT OF HORMONES

8.20.1 *Thyroid*

Excess thyroid hormone has been stated to have either no effect on, or to reduce, amino acid absorption. Thus, Matty and Seshadri [385] found that sacs made of everted intestine of thyroxine-treated rats took up less α-aminoisobutyric acid than normal in the proximal jejunum, while the mid-small intestine appeared to be unchanged. The dose of thyroxine was 100 μg per 100 g body weight. Likewise, Islam *et al.* [386], who gave rats 4 mg thyroxine parenterally on the first day and 8 mg daily for the next 3 days, obtained, with sacs of everted intestine, evidence of decreased absorption of L-valine when measured by net serosal fluid gain in valine, final serosal/mucosal fluid valine concentration ratio, tissue valine accumulation, or serosal-mucosal fluid concentration difference. Gelb and Gerson [387], too, have observed no alteration in amino acid (L-histidine) transport by sacs of rat intestine when thyroid hormone was used; while with rats *in vivo*, Althausen and Stockholm [388] recorded that absorption of 10% L-alanine fed by stomach tube was identical in normal and thyroid hormone-treated animals. In contrast, hypothyroidism may produce enhanced amino acid absorption by the rat. This has been shown by London and Segal [389] for L-histidine, L-valine and cycloleucine; by Islam *et al.* [386] for L-valine; and by Hall and Hershman [390] for L-thyroxine. The augmentation could be diminished by giving hypothyroid rats 15 μg triiodothyronine parenterally for 4-5 days [389], but merely adding thyroxine to the incubation media of sacs had no effect. According to Bronk and Parsons [391], surgical thyroidectomy seemed to make no difference to intestinal accumulation of an amino acid mixture, or to the incorporation of the amino acids into protein. Giving such animals triiodothyronine also produced no significant changes in amino acid accumulation or incorporation.

8.20.2 *Parathyroid*
Parathyroidectomy induced no alteration in the ability of rat small intestine *in vitro* to accumulate or transport L-valine [386].

8.20.3 *Insulin*
Isolated intestine of alloxan-diabetic rats showed increased active transport of histidine, tryptophan, methionine, tyrosine, lysine [392], phenylalanine [393] and α-aminoisobutyric acid [394], although proline was claimed to be unaffected [393]. The response persisted for at least 4 months [392]; it was not due to the alloxan itself, as it was seen in rats made diabetic by the use of streptozotocin [394]; and it could not be mimicked by causing hyperglycaemia by intravenous glucose infusion, or by pre-incubation of intestine for 45 minutes in glucose-rich media [394]. The augmented absorption disappeared when the rats were treated with insulin *in vivo*, whereas incubating intestine in insulin-containing media, even for 45 minutes, had no such effect [394]. As diabetic intestine

had a raised V_{max}, but normal K_t, Olsen and Rosenberg [394] suggested that lack of insulin *in vivo* in some way changed the permeability of the epithelial cell membrane. *In vitro*, however, insulin (1 unit per ml) added to the incubation media had no action on net L-alanine transport by normal rabbit ileum, nor on the unidirectional flux of the amino acid in either direction across the intestinal wall [395].

8.20.4 Adrenal

Following bilateral adrenalectomy, rats given tap water post-operatively had reduced absorption (*in vivo*) of glycine and glycylglycine, which could be restored to normal by replacing the tap water by 1% sodium chloride [300]. Similar results were found for glycine absorption *in vitro*. However, even in rats given tap water, some active transport of glycine remained, indicating that adrenocortical hormones were probably not essential for amino acid active absorption. According to Levin et al. [300], the reduction in absorptive capacity observed in these experiments was proportional to the loss in weight of the intestine, and rats starved for 3 days gave an identical absorption response to that seen in adrenalectomized rats drinking tap water. It should be noted that these subnormal values after adrenalectomy were obtained only when the results were expressed in terms of 'per sac' or 'per whole rat'; if the values are expressed as amount of amino acid absorbed per unit wet weight of intestine, the results are then not different from those found in sham-operated rats. This latter mode of expressing absorption capacity (per unit weight of intestine) after adrenalectomy has been used by Gelb and Gerson [387], who reported the operation to have no effect on histidine or leucine active absorption by rat intestine *in vitro*. It has been claimed by Shishova [396] that adrenalectomy in rats reduced amino acid uptake *in vivo*, and that cortisone considerably increased amino acid absorption by normal rats.

8.20.5 Pituitary

Finkelstein and Schachter [397] have recorded that active absorption of 10 mM L-proline by rat ileum *in vitro* was decreased 2 weeks after hypophysectomy. Curiously, no such change occurred in the jejunum. In these operated rats, there was severe atrophy of the intestinal mucosa. Giving growth hormone (0.5 mg daily for 8 days) did not alter proline absorption. After hypophysectomy, there may be greater flux of some amino acids across the intestinal epithelium [387].

8.20.6 Lactation and pregnancy
Absorption of glycine by the lactating rat appears to be unchanged when the results are expressed in terms of length of jejunal loop used *in vivo* [398]. However, on a unit weight of intestine basis, glycine absorption seems to be low in lactation because of the hypertrophy which the small intestine undergoes at this time [398-400]. Prior to lactation, before the intestinal hypertrophy occurs, glycine absorption by rat jejunum *in vivo* is normal on either a length of loop or a dry weight of intestine basis.

8.21 EFFECT OF VITAMIN B_6

There are a number of conflicting reports as to whether or not vitamin B_6 plays a specific role in the intestinal active transport of amino acids. For example, Lathe [401] found that glycine absorption by the jejunum and ileum of the dog *in vivo* was depressed when the animals had a diet deficient in vitamin B complex. Oral vitamin B complex brought absorption back to normal. Similarly, vitamin B_6 deficiency, produced in intact rats by the use of deoxypyridoxine (a pyridoxine antagonist) (Table 8.25) [402, 403], L-penicillamine [404, 405] or a vitamin B_6-inadequate diet [176, 406], was said to cause poor absorption of amino acids. In the rats with vitamin B_6-deficiency due to L-penicillamine, Akedo *et al.* [404] claimed that normal absorption of L-histidine, L-methionine and L-lysine could be restored within about 10 minutes by an intravenous dose of 0.5 mg vitamin B_6 (Fig. 8.26). Intravenous doses of vitamin B_1 or B_2 were of no value. When the intraperitoneal route was used for a dose of pyridoxine in deoxypyridoxine-treated rats, recovery of absorption of L-methionine to normal levels occurred within 1 hour (Table 8.25) [402]. Vitamin B_6 deficiency seemed to have no deleterious action on the absorption of the D-forms of histidine, methionine or lysine.

In addition to the above experiments *in vivo*, Fridhandler and Quastel [72] have stated that DL-alanine absorption by guinea-pig small intestine *in vitro* was reduced by 10 mM deoxypyridoxine. This effect was not overcome by the presence of 5 mM pyridoxal in the incubation media. Lack of beneficial action of pyridoxal on normal active transport of amino acid *in vitro* has been observed by Wiseman [407], who used sacs of hamster everted small intestine (Table 8.26). Indeed, active transport of 20 mM glycine was below normal when the pyridoxal concentration in the mucosal and serosal fluids was 1 mM, and increasing this concentration to 5 mM resulted in subnormal active transport of 20 mM L-histidine as well.

TABLE 8.25. Effect of pyridoxine and deoxypyridoxine on the absorption of 20 mM L-methionine circulated through loops of jejunum of anaesthetized rats.

Pre-absorption treatment of rats	L-methionine absorbed in 1 hour (% of amino acid load)
None (controls)	49.0 ± 2.0 (14)
Pyridoxine-treated*	61.0 ± 3.2 (9)
Deoxypyridoxine-treated*	31.9 ± 1.3 (9)
Deoxypyridoxine followed by pyridoxine	52.2 ± 2.3 (8)

* Intraperitoneal doses equivalent to 0.5 mg deoxypyridoxine hydrochloride, given 1 hour before start of perfusion.
Values are Mean ± S.E.M.(n).
From Jacobs *et al.* [402].

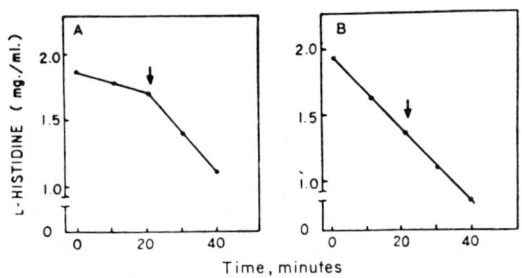

Fig. 8.26. Rate of absorption of L-histidine by normal and vitamin B_6-deficient rats *in vivo*. The graphs show the fall in amino acid concentration in the intestinal lumen with time. A, penicillamine-treated rat; B, vitamin B_6-treated rat. In both cases, 0.5 mg vitamin B_6 was injected into the caudal vein at the point indicated by the arrow. (From Akedo *et al.* [404].)

Munck [408] has also recorded that deoxypyridoxine inhibited amino acid (arginine, hydroxyproline, threonine) active transport by rat small intestine *in vitro,* but claimed that preincubation with the antagonist was necessary.

More recently, Chez *et al.* [409] have examined the influence of vitamin B_6 deficiency on L-alanine transport by rabbit ileum *in vitro.* The rabbits were fed a synthetic diet containing either

TABLE 8.26. Effect of 1 mM pyridoxal on the active transport of amino acids by sacs of everted small intestine of the hamster.

Amino acid	Final concentration ratio* (serosal/mucosal)		Transport of amino acid into serosal fluid (μmole/100 mg dry wt/hr)	
	Without pyridoxal	With pyridoxal	Without pyridoxal	With pyridoxal
L-proline	2.08 ± 0.06 (9)	1.95 ± 0.07 (9)	62.5 ± 4.7 (9)	54.2 ± 3.9 (9)
Glycine	1.65 ± 0.06 (9)	1.35 ± 0.03 (10)	45.3 ± 3.5 (9)	25.4 ± 3.0 (10)
L-histidine	1.42 ± 0.06 (12)	1.31 ± 0.04 (10)	23.8 ± 3.6 (12)	27.6 ± 1.1 (10)
L-methionine	1.18 ± 0.03 (11)	1.13 ± 0.03 (6)	14.6 ± 2.5 (11)	17.1 ± 2.6 (6)

* Initial serosal and mucosal fluid contained 20 mM amino acid.
Experimental period 1 hour. 37°C.
Values are Mean ± S.E.M.(n).
From Wiseman [407].

20 μg pyridoxine per g food or 0.05 μg pyridoxine per g food (former normal; latter grossly inadequate), were not allowed to consume their faeces, and were given a mixture of sulphathalidine-streptomycin-neomycin for 5 days prior to experimentation. Despite the moderate to severe vitamin B_6 depletion which ensued, there was no evidence of impairment of L-alanine absorption. The conclusion of earlier investigators that vitamin B_6 was specifically required for amino acid transport may not, therefore, be true for rabbit ileum, even though the accumulated results in other species seem to point to some involvement, if perhaps only of a general sort, of vitamin B_6 in amino acid absorption.

The mode of action, if any, of vitamin B_6, or its derivatives, in amino acid absorption, is unknown, but it has been suggested by Christensen [247] that it possibly decreased loss of amino acid from cells by altering membrane permeability. The formation of a Schiff base has also been proposed as an intermediate in amino acid uptake. However, ε-aminocaproic acid forms a Schiff base at pH 7.4 with pyridoxal-5-phosphate but it is not actively transported by rat intestine *in vitro* [177].

8.22 EFFECT OF VITAMINS C, D AND E

The jejunum of guinea-pigs made scorbutic by a vitamin C-deficient diet for 3-4 weeks displayed no disability in absorbing glycine, alanine or lysine, either *in vivo* or *in vitro* [410]. It is interesting to note that scorbutic guinea-pigs have a pronounced glycinuria.

Vitamin D medication has been said to have a beneficial effect on the transport of L-histidine by sacs of rabbit ileum [411], although no defect in alanine or lysine accumulation by human jejunal biopsy material has been observed in cases of nutritional rickets [303]. For control rabbit ileum, L-histidine had a K_t of 20.4 mM and a V_{max} of 125 mmole per 90 minutes per g intestine, whereas 18 hours after vitamin D administration to normal rabbits, the K_t was 12.8 mM and the V_{max} 167 mmole per 90 minutes per g intestine. With rachitic rabbits, the K_t was 38.5 mM and the V_{max} 83 mmole per 90 minutes per g intestine.

In vitamin E deficiency in rats, Imami *et al.* [412] have observed subnormal L-valine active transport by sacs of everted small intestine. There was no derangement of α-methyl-D-glucoside absorption.

8.23 EFFECT OF pH

Active transport of amino acids by small intestine can be achieved over a wide pH range, with the optimum pH falling, at least for some amino acids, well outside that normally found in the intestinal lumen. Experiments by Nathans *et al.* [153] on the active uptake of mono-iodo-tyrosine by isolated small intestine of the rat and hamster showed that absorption was best at about pH 6.0; and Thompson *et al.* [274] have reported that reducing the pH from 7.4 to 6.3 caused increased uptake of glycine, proline, leucine, β-alanine and sarcosine (but not methionine) by isolated jejunum of the rat. The latter workers, though, observed no such response for rat ileum or for hamster mid-jejunum. The optimal range for valine active transport by rat small intestine has been given as pH 5-8, the transport apparently remaining maximal between these two points [191]. It is perhaps surprising that even at pH 2.5 carrier-mediated influx of alanine still occurs (rabbit ileum), though this low pH does not allow the usual stimulatory effect of mucosal sodium and the influx is less than at pH 7.4 [216]. The partial inhibition produced by lowering the mucosal pH to this degree is reversible, normal alanine uptake returning in about 30 minutes after the pH is raised to 7.4. The alanine influx into rabbit ileum seems, therefore, to be composed of a pH-sensitive sodium-dependent part not working at pH 2.5 plus a pH-insensitive sodium-independent part.

Studies of the behaviour of peptides during absorption require carefully controlled pH. This is because peptidases can show quite different pH activity peaks. Thus, raising the pH from 6.5 to 8.0 produces considerable increase in hydrolysis of glycylleucine but not of leucylglycine [42].

8.24 EFFECT OF PANCREATIC SECRETION AND BILE

Inadequacy of pancreatic secretion has no obvious effect on the absorption of amino acids from an amino acid mixture (casein hydrolysate), although amino acid absorption from intact protein may be delayed or even reduced [413-415].

When biliary secretion is inadequate, subnormal amino acid absorption from ingested protein may occur, possibly as a result of the concomitant increase in faecal mass [416, 417]. According to Dawson and Matthews [418], taurocholate had no appreciable influence on the amount of amino acid transported by sacs of everted small intestine, and Pope et al. [419], who confirmed this, also found highly purified glycocholate to have no effect on lysine or leucine absorption by rat jejunum *in vitro*. On the other hand, Parkinson and Olson [420] have claimed that glycocholic and taurocholic acids inhibited leucine absorption by rat jejunum *in vitro* and they suggested that this

ABSORPTION OF PROTEIN DIGESTION PRODUCTS   465

may actually be a normal regulating action. It is possible that the conjugated bile salts employed by Parkinson and Olson [420] were contaminated by small amounts of deoxycholate, which is a powerful inhibitor of amino acid absorption [419]. Unconjugated bile acids, by damaging the epithelium, depress amino acid uptake [421].

8.25   EFFECT OF ALCOHOL

Orten [135], studying the absorption of amino acid mixtures by a loop of ileum in conscious man, observed that ethanol (about 20%) reduced the rate of disappearance of the faster moving amino acids (Table 8.16), so that they were all taken up at a similar slow rate. The effect lasted for about 2 days, which corresponded with the ileal epithelial regeneration rate in this subject. Israel et al. [422, 423], too, have found ethanol to have a deleterious effect on amino acid absorption by man. They perfused 15 cm lengths of jejunum in 10 subjects, and measured the influence of 2% ethanol on 3 mM ^{3}H-methionine (perfused at 5 ml per minute), making allowance (by use of polyethyleneglycol) for luminal water movement. In the control experiments, methionine was absorbed at 6.10 ± 0.27 (SEM) μmole per minute per 15 cm jejunum, whereas it was only 2.73 ± 0.50 μmole per minute per 15 cm when the ethanol was present. In fasted subjects, 200 ml of 20% ethanol was brought to a concentration of 2-3% in jejunal aspirates, suggesting that in heavy drinkers there could be appreciable interference with amino acid absorption. In rats, feeding 250 mg ethanol per 100 g body weight by stomach tube caused L-phenylalanine absorption to be 50% of normal; D-phenylalanine was unaffected [424].

As well as the above inhibitory action on amino acid absorption in vivo, Israel et al. [424] showed that ethanol markedly depressed the active transport of amino acid by sacs of everted small intestine of the rat. With 0.5% ethanol, L-phenylalanine active transport was cut down by about 60%, while with 2% ethanol the reduction was about 84%. For L-methionine, 2% ethanol completely prevented active transport, even though oxygen usage was still about 80% normal. Chang et al. [425] have also reported that ethanol interfered with amino acid uptake by sacs of rat small intestine, 3% ethanol abolishing active transport of glycine and the L-forms of phenylalanine, leucine, alanine, methionine and valine. Washing the intestine free of alcohol restored its absorptive capacity. Similar results were recorded for 3% methanol, 2% n-propanol and iso-propanol, 1% n-butanol and iso-butanol, and 0.5% cyclohexanol and phenethyl alcohol. Passive diffusion of D-phenylalanine across the intestinal wall was increased by ethanol.

8.26 EFFECT OF X-IRRADIATION AND INTESTINAL MOTILITY

There is no evidence that amino acid absorption as such is materially affected even by large doses of ionizing radiations, nor is there any induction of absorption of intact protein, provided that the epithelial lining remains intact. Shishova [396] has found that rats given 400-600 r (whole body dose) absorbed amino acids (*in vivo*) normally at the 3rd and 14th day, although a small degree of inhibition was seen after 500-600 r. Rabbits are also very resistant, absorbing (*in vivo*) 10 mM L- and D-methionine at normal rates 24 hours after being subjected to 800 r (whole body dose at 4.1 r per minute) [426]. When changes in amino acid uptake are seen, they probably result from alterations in gastric emptying time and intestinal motility [427]. Increasing the latter, according to Cummins and Almy [428], may sometimes improve the absorption of a test dose of methionine. Increased intestinal motor activity after the administration of physostigmine and urecholine was also accompanied by improved absorption, but this may have been due to increased blood flow.

REFERENCES

1. F. KUTSCHER and J. SEEMAN, *Z. Physiol. Chem.*, **34**, 528 (1901-2).
2. E. ABDERHALDEN, K. V. KOROSY and E. S. LONDON, *Z. Physiol. Chem.*, **53**, 148 (1907).
3. O. LOEWI, *Arch. Exp. Path. Pharmak.*, **48**, 303 (1902).
4. W. FALTA, *Dt. Arch. Klin. Med.*, **86**, 517 (1906).

5. P. A. LEVENE and P. A. KOBER, *Am. J. Physiol.*, **23**, 324 (1908-9).
6. R. B. FISHER, Protein Metabolism. Methuen, London (1954).
7. R. B. FISHER, *Proc. Nutr. Soc.*, **26**, 23 (1967).
8. R. D. WRIGHT and V. WYNN, *Aust. J. Exp. Biol. Med. Sci.*, **29**, 281 (1951).
9. M. STEINER and S. J. GRAY, *Am. J. Physiol.*, **217**, 747 (1969).
10. B. BORGSTROM, A. DAHLQVIST, G. LUNDH and J. SJOVALL, *J. Clin. Invest.*, **36**, 1521 (1957).
11. S. E. NIXON and G. E. MAWER, *Br. J. Nutr.*, **24**, 227 (1970).
12. S. E. NIXON and G. E. MAWER, *Br. J. Nutr.*, **24**, 241 (1970).
13. C. W. CRANE and A. NEUBERGER, *Biochem. J.*, **74**, 313 (1960).
14. C. W. CRANE and A. NEUBERGER, *Br. Med. J.*, **2**, 888 (1960).
15. C. W. CRANE, *Postgrad. Med. J.*, **37**, 745 (1961).
16. C. GITLER, Mammalian Protein Metabolism (H. N. Munro and J. B. Allison, eds.), p. 35. Academic Press, New York (1964).
17. C. W. CRANE, The Role of the Gastrointestinal Tract in Protein Metabolism (H. N. Munro, ed.), p. 333. Blackwell, Oxford (1964).
18. R. DAWSON, E. S. HOLDSWORTH and J. W. G. PORTER, The Role of the Gastrointestinal Tract in Protein Metabolism (H. N. Munro, ed.), p. 293. Blackwell, Oxford (1964).
19. L. DREISBACH and E. S. NASSET, *J. Nutr.*, **53**, 523 (1954).
20. E. S. NASSET, P. SCHWARTZ and H. V. WEISS, *J. Nutr.*, **56**, 83 (1955).
21. E. S. NASSET and J. S. JU, *J. Nutr.*, **74**, 461 (1961).
22. C. PERAINO, Q. R. ROGERS, M. YOSHIDA, M.-L. CHEN and A. E. HARPER, *Can. J. Biochem. Physiol.*, **37**, 1475 (1959).
23. W. J. LOESCHE, *Proc. Soc. Exp. Biol. Med.*, **129**, 380 (1968).
24. R. M. CAMPBELL, D. P. CUTHBERTSON, W. MACKIE, A. S. McFARLANE, A. T. PHILLIPSON and S. SUDSANEH, *J. Physiol., Lond.*, **158**, 113 (1961).
25. W. T. AGAR, F. J. R. HIRD and G. S. SIDHU, *J. Physiol., Lond.*, **121**, 255 (1953).
26. H. NEWEY and D. H. SMYTH, *J. Physiol., Lond.*, **145**, 48 (1959).
27. D. S. WIGGANS and J. M. JOHNSTON, *Fed. Proc. Fed. Am. Socs. Exp. Biol.*, **17**, 335 (1958).
28. D. S. WIGGANS and J. M. JOHNSTON, *Biochim. Biophys. Acta*, **32**, 69 (1959).
29. F. A. CAJORI, *Am. J. Physiol.*, **104**, 659 (1933).
30. J. E. RHOADS, A. STENGEL, Jr., C. RIEGEL, F. A. CAJORI and W. D. FRAZIER, *Am. J. Physiol.*, **125**, 707 (1939).
31. H. NEWEY and D. H. SMYTH, *J. Physiol., Lond.*, **152**, 367 (1960).
32. H. NEWEY and D. H. SMYTH, *J. Physiol., Lond.*, **157**, 15P (1961).
33. H. NEWEY and D. H. SMYTH, *J. Physiol., Lond.*, **164**, 527 (1962).
34. C. W. CRANE and A. NEUBERGER, *Br. Med. J.*, **2**, 815 (1960).
35. R. DAWSON and E. S. HOLDSWORTH, *Br. J. Nutr.*, **16**, 13 (1962).
36. D. J. PROCKOP, H. R. KEISER and A. SJOERDSMA, *Lancet*, **2**, 527 (1962).

37. H. J. HUECKEL and Q. R. ROGERS, *Fed. Proc. Fed. Am. Socs. Exp. Biol.*, **28**, 301 (1969).
38. H. J. HUECKEL and Q. R. ROGERS, *Comp. Biochem. Physiol.*, **32**, 7 (1970).
39. I. L. CRAFT, D. GEDDES, C. W. HYDE, I. J. WISE and D. M. MATTHEWS, *Gut*, **9**, 425 (1968).
40. D. M. MATTHEWS, I. L. CRAFT, D. M. GEDDES, I. J. WISE and C. W. HYDE, *Clin. Sci.*, **35**, 415 (1968).
41. T. J. PETERS, K. MODHA and M. T. MacMAHON, *Gut*, **10**, 1055 (1969).
42. E. B. FERN, R. C. HIDER and D. R. LONDON, *Biochem. J.*, **114**, 855 (1969).
43. D. M. MATTHEWS, M. T. LIS, B. CHENG and R. F. CRAMPTON, *Clin. Sci.*, **37**, 751 (1969).
44. A. M. ASATOOR, J. K. BANDOH, A. F. LANT, M. D. MILNE and F. NAVAB, *Gut*, **11**, 250 (1970).
45. D. H. ELWYN, H. C. PARIKH and W. C. SHOEMAKER, *Am. J. Physiol.*, **215**, 1260 (1968).
46. F. NAVAB and A. M. ASATOOR, *Gut*, **11**, 373 (1970).
47. A. M. ASATOOR, B. CHENG, K. D. G. EDWARDS, A. F. LANT, D. M. MATTHEWS, M. D. MILNE, F. NAVAB and A. J. RICHARDS, *Gut*, **11**, 380 (1970).
48. E. P. CATHCART and J. B. LEATHES, *J. Physiol., Lond.*, **33**, 462 (1905-6).
49. W. H. HOWELL, *Am. J. Physiol.*, **17**, 273 (1906-7).
50. O. FOLIN and W. DENIS, *J. Biol. Chem.*, **11**, 87 (1912).
51. O. FOLIN and W. DENIS, *J. Biol. Chem.*, **11**, 161 (1912).
52. D. D. VAN SLYKE and G. M. MEYER, *J. Biol. Chem.*, **12**, 399 (1912).
53. D. D. VAN SLYKE and G. M. MEYER, *J. Biol. Chem.*, **16**, 197 (1913-14).
54. D. D. VAN SLYKE, *J. Biol. Chem.*, **16**, 187 (1913-14).
55. D. D. VAN SLYKE and G. M. MEYER, *J. Biol. Chem.*, **16**, 213 (1913-14).
56. J. J. ABEL, L. G. ROWNTREE and B. B. TURNER, *J. Pharmac. Exp. Ther.*, **5**, 275 (1913-14).
57. J. J. ABEL, L. G. ROWNTREE and B. B. TURNER, *J. Pharmac. Exp. Ther.*, **5**, 611 (1913-14).
58. M. P. KALMYKOFF, *Arch. Ges. Physiol.*, **205**, 493 (1924).
59. J. L. BOLLMAN, F. C. MANN and T. B. MAGATH, *Am. J. Physiol.*, **72**, 629 (1925).
60. J. M. LUCK, *J. Biol. Chem.*, **77**, 13 (1928).
61. M. W. JOHNSTON and H. B. LEWIS, *J. Biol. Chem.*, **78**, 67 (1928).
62. R. H. WILSON and H. B. LEWIS, *J. Biol. Chem.*, **84**, 511 (1929).
63. F. H. KRATZER, *J. Biol. Chem.*, **153**, 237 (1944).
64. A. H. FREE and J. R. LEONARDS, *J. Lab. Clin. Med.*, **29**, 963 (1944).
65. S. M. LEVENSON, H. ROSEN and H. L. UPJOHN, *Proc. Soc. Exp. Biol. Med.*, **101**, 178 (1959).

66. D. M. MATTHEWS and G. WISEMAN, *J. Physiol., Lond.*, **120**, 55P (1953).
67. K. D. NEAME and G. WISEMAN, *J. Physiol., Lond.*, **133**, 39P (1956).
68. K. D. NEAME and G. WISEMAN, *J. Physiol., Lond.*, **135**, 442 (1957).
69. K. D. NEAME and G. WISEMAN, *J. Physiol., Lond.*, **138**, 41P (1957).
70. K. D. NEAME and G. WISEMAN, *J. Physiol., Lond.*, **140**, 148 (1958).
71. L. R. FINCH and F. J. R. HIRD, *Biochim. Biophys. Acta*, **43**, 268 (1960).
72. L. FRIDHANDLER and J. H. QUASTEL, *Archs. Biochem. Biophys.*, **56**, 424 (1955).
73. S. P. BESSMAN, J. MAGNES, P. SCHWERIN and H. WAELSCH, *J. Biol. Chem.*, **175**, 817 (1948).
74. L. E. ROSENBERG, J. C. CRAWHALL and S. SEGAL, *J. Clin. Invest.*, **46**, 30 (1967).
75. H. N. CHRISTENSEN, D. G. DECKER, E. L. LYNCH, T. M. MACKENZIE and J. H. POWERS, *J. Clin. Invest.*, **26**, 853 (1947).
76. T. SUGAWA, H. AKEDO and M. SUDA, *J. Biochem., Tokyo*, **47**, 131 (1960).
77. R. P. SPENCER and W. E. KNOX, *Fed. Proc. Fed. Am. Socs. Exp. Biol.*, **19**, 886 (1960).
78. H. N. CHRISTENSEN, P. F. COOPER, Jr., R. D. JOHNSON and E. L. LYNCH, *J. Biol. Chem.*, **168**, 191 (1947).
79. H. N. CHRISTENSEN, *Biochem. J.*, **44**, 333 (1949).
80. C. E. DENT and J. A. SCHILLING, *Biochem. J.*, **44**, 318 (1949).
81. A. N. PARSHIN and L. N. RUBEL, *Chem. Abstr.*, **45**, 7206 (1951).
82. A. E. DENTON, S. N. GERSHOFF and C. A. ELVEHJEM, *J. Biol. Chem.*, **204**, 731 (1953).
83. A. E. DENTON and C. A. ELVEHJEM, *J. Biol. Chem.*, **206**, 449 (1954).
84. F. HARTMANN, H. LENZ, C. LOPEZ-CALLEJA and H. MUNZENBERG, *Arch. Exp. Path. Pharmak.*, **228**, 403 (1956).
85. W. H. STEIN and S. MOORE, *J. Biol. Chem.*, **211**, 915 (1954).
86. P. WHEELER and A. F. MORGAN, *J. Nutr.*, **64**, 137 (1958).
87. I. A. ABDOU and H. TARVER, *J. Biol. Chem.*, **190**, 769 (1951).
88. I. A. ABDOU, W. O. REINHARDT and H. TARVER, *J. Biol. Chem.*, **194**, 15 (1952).
89. C. L. YUILE, A. E. O'DEA, F. V. LUCAS and G. H. WHIPPLE, *J. Exp. Med.*, **96**, 247 (1952).
90. C. L. YUILE, B. G. LAMSON, L. L. MILLER and G. H. WHIPPLE, *J. Exp. Med.*, **93**, 539 (1951).
91. F. B. GAILEY and M. J. JOHNSON, *J. Biol. Chem.*, **141**, 921 (1941).
92. M. M. NACHLAS, D. T. CRAWFORD and A. M. SELIGMAN, *J. Histochem. Cytochem.*, **5**, 264 (1957).

93. M. M. NACHLAS, B. MONIS, D. ROSENBLATT and A. M. SELIGMAN, *J. Biophys. Biochem. Cytol.*, **7**, 261 (1960).
94. J. H. HOLT and D. MILLER, *Biochim. Biophys. Acta*, **58**, 239 (1962).
95. J. B. RHODES, A. EICHHOLZ and R. K. CRANE, *Biochim. Biophys. Acta*, **135**, 959 (1967).
96. A. EICHHOLZ, *Biochim. Biophys. Acta*, **163**, 101 (1968).
97. G. B. ROBINSON, *Biochem. J.*, **88**, 162 (1963).
98. T. J. PETERS, K. MODHA and C. N. C. DREY, *Biochem. J.*, **119**, 20P (1970).
99. T. J. PETERS, *Biochem. J.*, **120**, 195 (1970).
100. A. M. UGOLEV and R. I. KOOSHUCK, *Nature, Lond.*, **212**, 859 (1966).
101. B. CHENG, F. NAVAB and D. M. MATTHEWS, *Clin. Sci.*, **37**, 874 (1969).
102. L. JOSEFSSON and H. SJOSTROM, *Acta Physiol. Scand.*, **67**, 27 (1966).
103. G. B. ROBINSON and B. SHAW, *Biochem. J.*, **77**, 351 (1960).
104. A. RUBINO, M. PIERRO, M. VETRELLA, L. PROVENZALE and S. AURICCHIO, *Biochim. Biophys. Acta*, **191**, 663 (1969).
105. A. RUBINO, M. PIERRO, G. LA TORRETTA, M. VETRELLA, D. DI MARTINO and S. AURICCHIO, *Pediat. Res.*, **3**, 313 (1969).
106. L. JOSEFSSON and T. LINDBERG, *Biochim. Biophys. Acta*, **105**, 162 (1965).
107. A. M. UGOLEV, *Physiol. Rev.*, **45**, 555 (1965).
108. C. F. CORI, *J. Biol. Chem.*, **66**, 691 (1925).
109. P. A. LEVENE and G. M. MEYER, *Am. J. Physiol.*, **25**, 214 (1909-10).
110. T. N. SETH and J. M. LUCK, *Biochem. J.*, **19**, 366 (1925).
111. R. H. WILSON, *J. Biol. Chem.*, **87**, 175 (1930).
112. R. H. WILSON, *J. Biol. Chem.*, **97**, 497 (1932).
113. B. W. CHASE and H. B. LEWIS, *J. Biol. Chem.*, **101**, 735 (1933).
114. J. C. ANDREWS, C. G. JOHNSTON and K. C. ANDREWS, *Am. J. Physiol.*, **115**, 188 (1936).
115. J. R. DOTY and A. G. EATON, *J. Biol. Chem.*, **122**, 139 (1937).
116. G. H. LATHE, *Revue Can. Biol.*, **2**, 134 (1943).
117. S. W. HIER and O. BERGEIM, *Fed. Proc. Fed. Am. Socs. Exp. Biol.*, **6**, 261 (1947).
118. J. C. ANDREWS and C. G. JOHNSTON, *J. Biol. Chem.*, **101**, 635 (1933).
119. W. C. HESS, *J. Biol. Chem.*, **181**, 23 (1949).
120. N. R. LAWRIE, *Biochem. J.*, **26**, 435 (1932).
121. H. A. HARPER and K. UYEYAMA, *Proc. Soc. Exp. Biol. Med.*, **68**, 296 (1948).
122. N. F. SHAMBAUGH, H. B. LEWIS and D. TOURTELLOTTE, *J. Biol. Chem.*, **92**, 499 (1931).
123. B. W. CHASE, *J. Biol. Chem.*, **100**, xxvii-xxviii (1933).
124. C. P. BERG and L. C. BAUGUESS, *J. Biol. Chem.*, **98**, 171 (1932).
125. B. W. CHASE and H. B. LEWIS, *J. Biol. Chem.*, **106**, 315 (1934).

126. R. M. FEATHERSTONE and C. P. BERG, *J. Biol. Chem.*, **146**, 131 (1942).
127. F. A. SCHOFIELD and H. B. LEWIS, *J. Biol. Chem.*, **168**, 439 (1947).
128. R. HOBER and J. HOBER, *Am. J. Med. Sci.*, **191**, 873 (1936).
129. R. HOBER and J. HOBER, *J. Cell. Comp. Physiol.*, **10**, 401 (1937).
130. Q. H. GIBSON and G. WISEMAN, *Biochem. J.*, **48**, 426 (1951).
131. S. R. ELSDEN, Q. H. GIBSON and G. WISEMAN, *J. Physiol., Lond.*, **111**, 56P (1950).
132. E. W. CLARKE, Q. H. GIBSON, D. H. SMYTH and G. WISEMAN, *J. Physiol., Lond.*, **112**, 46P (1951).
133. Y. KURODA and N. S. GIMBEL, *J. Appl. Physiol.*, **7**, 148 (1954-5).
134. H. L. WANG and H. A. WAISMAN, *J. Lab. Clin. Med.*, **57**, 73 (1961).
135. A. U. ORTEN, *Fed. Proc. Fed. Am. Socs. Exp. Biol.*, **22**, 1103 (1963).
136. D. M. MATTHEWS and D. H. SMYTH, *J. Physiol., Lond.*, **126**, 96 (1954).
137. J. P. AROSKAR and C. P. BERG, *Archs. Biochem. Biophys.*, **98**, 286 (1962).
138. R. P SPENCER and A. H. SAMIY, *Am. J. Physiol.*, **199**, 1033 (1960).
139. L. LASTER, P. T. WERTLAKE and M. WOODSON, *Gastroenterology*, **46**, 755 (1964).
140. R. P. SPENCER and A. H. SAMIY, *Am. J. Physiol.*, **200**, 501 (1961).
141. S. REISER and P. A. CHRISTIANSEN, *Am. J. Physiol.*, **208**, 914 (1965).
142. G. WISEMAN, *J. Physiol., Lond.*, **114**, 7P (1951).
143. G. WISEMAN, *J. Physiol., Lond.*, **120**, 63 (1953).
144. T. H. WILSON and G. WISEMAN, *J. Physiol., Lond.*, **121**, 45P (1953).
145. T. H. WILSON and G. WISEMAN, *J. Physiol., Lond.*, **123**, 116 (1954).
146. G. WISEMAN, *Methods in Medical Research*, **9**, 287 (1961).
147. G. WISEMAN, *J. Physiol., Lond.*, **133**, 626 (1956).
148. J. PINSKY and E. GEIGER, *Proc. Soc. Exp. Biol. Med.*, **81**, 55 (1952).
149. L. L. COHEN and K. C. HUANG, *Am. J. Physiol.*, **206**, 647 (1964).
150. G. WISEMAN, *J. Physiol., Lond.*, **127**, 414 (1955).
151. H. HAGIHIRA, E. C. C. LIN, A. H. SAMIY and T. H. WILSON, *Biochem. Biophys. Res. Commun.*, **4**, 478 (1961).
152. P. R. LARSEN, J. E. ROSS and D. F. TAPLEY, *Biochim. Biophys. Acta*, **88**, 570 (1964).
153. D. NATHANS, D. F. TAPLEY and J. E. ROSS, *Biochim. Biophys. Acta*, **41**, 271 (1960).
154. A. H. SAMIY and R. P. SPENCER, *Am. J. Physiol.*, **200**, 505 (1961).

155. G. WISEMAN, Absorption from the Intestine. Academic Press, London and New York (1964).
156. M. W. NEIL, *Biochem. J.*, **71**, 118 (1959).
157. K. C. HUANG, *J. Pharmac. Exp. Ther.*, **136**, 361 (1962).
158. E. C. C. LIN and T. H. WILSON, *Am. J. Physiol.*, **199**, 127 (1960).
159. C. M. PAINE, H. J. NEWMAN and M. W. TAYLOR, *Am. J. Physiol.*, **197**, 9 (1959).
160. T. B. MEPHAM and M. W. SMITH, *J. Physiol., Lond.*, **184**, 673 (1966).
161. H. N. CHRISTENSEN, B. H. FELDMAN and A. B. HASTINGS, *Am. J. Physiol.*, **205**, 255 (1963).
162. F. A. JACOBS and A. H. LANG, *Proc. Soc. Exp. Biol. Med.*, **118**, 772 (1965).
163. W. B. KINTER and T. H. WILSON, *J. Cell Biol.*, **25**, 19 (1965).
164. M. E. McLEOD and R. BRESSLER, *Proc. Soc. Exp. Biol. Med.*, **130**, 268 (1969).
165. S. REISER and P. A. CHRISTIANSEN, *J. Lipid Res.*, **9**, 606 (1968).
166. H. NEWEY and D. H. SMYTH, *J. Physiol., Lond.*, **170**, 328 (1964).
167. B. G. MUNCK, *Biochim. Biophys. Acta*, **109**, 142 (1965).
168. B. G. MUNCK and S. G. SCHULTZ, *Biochim. Biophys. Acta*, **183**, 182 (1969).
169. B. G. MUNCK and S. G. SCHULTZ, *J. Gen. Physiol.*, **53**, 157 (1969).
170. M. E. McLEOD and M. P. TYOR, *Am. J. Physiol.*, **213**, 163 (1967).
171. W. T. AGAR, F. J. R. HIRD and G. S. SIDHU, *Biochim. Biophys. Acta*, **14**, 80 (1954).
172. W. T. AGAR, F. J. R. HIRD and G. S. SIDHU, *Biochim. Biophys. Acta*, **22**, 21 (1956).
173. F. J. R. HIRD and G. S. SIDHU, *Biochim. Biophys. Acta*, **25**, 388 (1957).
174. B. I. KORELITZ and E. D. FRANK, *Gastroenterology*, **36**, 94 (1959).
175. E. L. JERVIS and D. H. SMYTH, *J. Physiol., Lond.*, **151**, 51 (1960).
176. E. C. C. LIN, H. HAGIHIRA and T. H. WILSON, *Am. J. Physiol.*, **202**, 919 (1962).
177. H. G. RANDALL and D. F. EVERED, *Biochim. Biophys. Acta*, **93**, 98 (1964).
178. R. J. NEALE and G. WISEMAN, *J. Physiol., Lond.*, **205**, 159 (1969).
179. W. R. ROUT, D. S. T. LIN and K. C. HUANG, *Proc. Soc. Exp. Biol. Med.*, **118**, 933 (1965).
180. M. FIELD, S. G. SCHULTZ and P. F. CURRAN, *Biochim. Biophys. Acta*, **135**, 236 (1967).
181. E. L. JERVIS and D. H. SMYTH, *J. Physiol., Lond.*, **145**, 57 (1959).
182. J. LERNER and M. W. TAYLOR, *Biochim. Biophys. Acta*, **135**, 991 (1967).
183. J. T. HINDMARSH, D. KILBY and G. WISEMAN, *J. Physiol., Lond.*, **186**, 166 (1966).

184. L. R. FINCH and F. J. R. HIRD, *Biochim. Biophys. Acta*, **43**, 278 (1960).
185. D. S. MADGE, *Comp. Biochem. Physiol.*, **32**, 1 (1970).
186. K. C. HUANG and W. R. ROUT, *Am. J. Physiol.*, **212**, 799 (1967).
187. J. A. HEWITT, *Biochem. J.*, **18**, 161 (1924).
188. E. RIKLIS and J. H. QUASTEL, *Can. J. Biochem. Physiol.*, **36**, 347 (1958).
189. T. Z. CSAKY, *Am. J. Physiol.*, **201**, 999 (1961).
190. H. E. HARRISON and H. C. HARRISON, *Am. J. Physiol.*, **205**, 107 (1963).
191. S. REISER and P. A. CHRISTIANSEN, *Am. J. Physiol.*, **212**, 1297 (1967).
192. H. J. BINDER, M. BOYER, H. M. SPIRO and R. P. SPENCER, *Comp. Biochem. Physiol.*, **18**, 83 (1966).
193. I. H. ROSENBERG, A. L. COLEMAN and L. E. ROSENBERG, *Biochim. Biophys. Acta*, **102**, 161 (1965).
194. R. E. FUISZ, S. G. SCHULTZ and P. F. CURRAN, *Biochim. Biophys. Acta*, **112**, 593 (1966).
195. K. M. NELSON and J. LERNER, *Biochim. Biophys. Acta*, **203**, 434 (1970).
196. R. P. SPENCER and K. R. BRODY, *Biochim. Biophys. Acta*, **88**, 400 (1964).
197. R. A. CHEZ, R. R. PALMER, S. G. SCHULTZ and P. F. CURRAN, *J. Gen. Physiol.*, **50**, 2357 (1967).
198. T. Z. CSAKY, *Fed. Proc. Fed. Am. Socs. Exp. Biol.*, **22**, 3 (1963).
199. R. K. CRANE, *Fed. Proc. Fed. Am. Socs. Exp. Biol.*, **24**, 1000 (1965).
200. P. F. CURRAN, S. G. SCHULTZ, R. A. CHEZ and R. E. FUISZ, *J. Gen. Physiol.*, **50**, 1261 (1967).
201. P. F. CURRAN, *Physiologist*, **11**, 3 (1968).
202. S. G. SCHULTZ and P. F. CURRAN, *Physiologist*, **12**, 437 (1969).
203. S. G. SCHULTZ, R. E. FUISZ and P. F. CURRAN, *J. Gen. Physiol.*, **49**, 849 (1966).
204. A. M. GOLDNER, S. G. SCHULTZ and P. F. CURRAN, *J. Gen. Physiol.*, **53**, 362 (1969).
205. O. ALVAREZ, A. M. GOLDNER and P. F. CURRAN, *Am. J. Physiol.*, **217**, 946 (1969).
206. H. NEWEY, A. J. RAMPONE and D. H. SMYTH, *J. Physiol., Lond.*, **211**, 539 (1970).
207. S. G. SCHULTZ, P. F. CURRAN, R. A. CHEZ and R. E. FUISZ, *J. Gen. Physiol.*, **50**, 1241 (1967).
208. I. BIHLER and S. ADAMIC, *Biochim. Biophys. Acta*, **135**, 466 (1967).
209. A. FAELLI, G. ESPOSITO and V. CAPRARO, *Med. Pharmac. Exp. (Basel)*, **17**, 483 (1967).
210. P. F. CURRAN, J. J. HAJJAR and I. M. GLYNN, *J. Gen. Physiol.*, **55**, 297 (1970).
211. J. W. L. ROBINSON, *Pflugers Arch. Ges. Physiol.*, **294**, 182 (1967).
212. T. Z. CSAKY and M. THALE, *J. Physiol., Lond.*, **151**, 59 (1960).

213. I. BIHLER and R. K. CRANE, *Biochim. Biophys. Acta*, **59**, 78 (1962).
214. S. G. SCHULTZ and R. ZALUSKY, *Nature, Lond.*, **204**, 292 (1965).
215. W. KOOPMAN and S. G. SCHULTZ, *Biochim. Biophys. Acta*, **173**, 338 (1969).
216. R. A. FRIZZELL and S. G. SCHULTZ, *Fed. Proc. Fed. Am. Socs. Exp. Biol.*, **28**, 651 (1969).
217. B. FLESHLER and R. A. NELSON, *Gut*, **11**, 240 (1970).
218. J. H. ANNEGERS, *Proc. Soc. Exp. Biol. Med.*, **116**, 933 (1964).
219. B. FLESHLER, J. H. BUTT and J. D. WISMAR, *J. Clin. Invest.*, **45**, 1433 (1966).
220. H. LINEWEAVER and D. BURK, *J. Am. Chem. Soc.*, **56**, 658 (1934).
221. C. P. READ, *Biol. Bull. Mar. Biol. Lab.*, *Woods Hole*, **133**, 630 (1967).
222. C. F. CORI, *Proc. Soc. Exp. Biol. Med.*, **24**, 125 (1926-7).
223. H. KAMIN and P. HANDLER, *Am. J. Physiol.*, **169**, 305 (1952).
224. T. H. WILSON and G. WISEMAN, *J. Physiol., Lond.*, **123**, 126 (1954).
225. E. L. JERVIS and D. H. SMYTH, *J. Physiol., Lond.*, **149**, 433 (1959).
226. D. M. MATTHEWS and L. LASTER, *Am. J. Physiol.*, **208**, 593 (1965).
227. D. M. MATTHEWS and L. LASTER, *Am. J. Physiol.*, **208**, 601 (1965).
228. B. FLESHLER and E. Y. COLIGADO, *J. Lab. Clin. Med.*, **70**, 883 (1967).
229. S. A. ADIBI and S. J. GRAY, *Gastroenterology*, **52**, 837 (1967).
230. S. A. ADIBI, S. J. GRAY and E. MENDEN, *Am. J. Clin. Nutr.*, **20**, 24 (1967).
231. G. DELHUMEAU, G. V. PRATT and C. GITLER, *J. Nutr.*, **77**, 52 (1962).
232. J. A. YOUNG and K. D. G. EDWARDS, *Am. J. Physiol.*, **210**, 1130 (1966).
233. J. H. ANNEGERS, *Am. J. Physiol.*, **216**, 1 (1969).
234. H. HAGIHIRA, M. OGATA, N. TAKEDATSU and M. SUDA, *J. Biochem., Tokyo*, **47**, 139 (1960).
235. R. P. SPENCER, J. WEINSTEIN, A. SUSSMAN, T. M. BOW and M. A. MARKULIS, *Am. J. Physiol.*, **203**, 634 (1962).
236. S. REISER and P. A. CHRISTIANSEN, *Biochim. Biophys. Acta*, **183**, 611 (1969).
237. K. MOCHIDA, K. SAKURAI and M. SUDA, *J. Biochem., Tokyo*, **57**, 497 (1965).
238. B. G. MUNCK, *Biochim. Biophys. Acta*, **120**, 97 (1966).
239. F. ALVARADO, *Science*, **151**, 1010 (1966).
240. J. W. L. ROBINSON, *European J. Biochem.*, **7**, 78 (1968).
241. W. G. BERGEN, *Proc. Soc. Exp. Biol. Med.*, **132**, 348 (1969).
242. M. J. BURNS and R. G. FAUST, *Biochim. Biophys. Acta*, **183**, 642 (1969).

243. B. G. MUNCK, *Biochim. Biophys. Acta*, 120, 282 (1966).
244. B. G. MUNCK, *Biochim. Biophys. Acta*, 126, 299 (1966).
245. J. H. BUTT and T. H. WILSON, *Am. J. Physiol.*, 215, 1468 (1968).
246. H. AKEDO and H. N. CHRISTENSEN, *J. Biol. Chem.*, 237, 113 (1962).
247. H. N. CHRISTENSEN, *Fed. Proc. Fed. Am. Socs. Exp. Biol.*, 21, 37 (1962).
248. H. NEWEY and D. H. SMYTH, *J. Physiol., Lond.*, 165, 74P (1963).
249. T. MORIKAWA, K. TADA, T. ANDO, T. YOSHIDA, Y. YOKOYAMA and T. ARAKAWA, *Tohoku J. Exp. Med.*, 90, 105 (1966).
250. S. I. GOODMAN, C. A. McINTYRE, Jr. and D. O'BRIEN, *J. Pediat.*, 71, 246 (1967).
251. L. ZETZEL and B. M. BANKS, *Am. J. Dig. Dis.*, 8, 21 (1941).
252. G. HETENYI, Jr. and M. WINTER, *Acta Physiol. Hung.*, 3, 49 (1952).
253. S. A. ADIBI, *Gastroenterology*, 56, 903 (1969).
254. D. M. MATTHEWS and L. LASTER, *Gut*, 6, 411 (1965).
255. C. GITLER and D. MARTINEZ-ROJAS, The Role of the Gastrointestinal Tract in Protein Metabolism (H. N. Munro, ed.), p. 269. Blackwell, Oxford (1964).
256. L. R. FINCH, *Biochim. Biophys. Acta*, 64, 556 (1962).
257. H. P. SCHEDL, C. E. PIERCE, A. RIDER and J. A. CLIFTON, *J. Clin. Invest.*, 47, 417 (1968).
258. D. SCHACHTER and J. S. BRITTEN, *Fed. Proc. Fed. Am. Socs. Exp. Biol.*, 20, 137 (1961).
259. M. BAILLIEN and E. SCHOFFENIELS, *Archs. Int. Physiol.*, 70, 140 (1962).
260. R. ZALUSKY and S. G. SCHULTZ, *Clin. Res.*, 11, 189 (1963).
261. M. GILLES-BAILLIEN and E. SCHOFFENIELS, *Life Sci.*, 7, 53 (1968).
262. T. B. MEPHAM and M. W. SMITH, *J. Physiol., Lond.*, 186, 619 (1966).
263. G. GRADY, K. MacGAFFEY, E. W. MOORE and T. C. CHALMERS, *Gastroenterology*, 50, 883 (1966).
264. R. J. LEVIN, O. KOLDOVSKY, J. HOSKOVA, V. JIRSOVA and J. UHER, *Gut*, 9, 206 (1968).
265. D. H. SMYTH and E. M. WRIGHT, *J. Physiol., Lond.*, 172, 61P (1964).
266. E. M. WRIGHT, *Nature, Lond.*, 212, 189 (1966).
267. P. G. KOHN, D. H. SMYTH and E. M. WRIGHT, *J. Physiol., Lond.*, 196, 723 (1968).
268. R. J. C. BARRY, S. DIKSTEIN, J. MATTHEWS, D. H. SMYTH and E. M. WRIGHT, *J. Physiol., Lond.*, 171, 316 (1964).
269. D. COTTERELL and P. G. KOHN, *Biochim. Biophys. Acta*, 203, 179 (1970).
270. B. G. MUNCK, *Biochim. Biophys. Acta*, 203, 424 (1970).
271. S. G. SCHULTZ and R. ZALUSKY, *J. Gen. Physiol.*, 47, 567 (1964).

272. R. J. C. BARRY, J. EGGENTON and D. H. SMYTH, *J. Physiol., Lond.*, **191**, 72P (1967).
273. A. E. TAYLOR, E. M. WRIGHT, S. G. SCHULTZ and P. F. CURRAN, *Physiologist*, **10**, 321 (1967).
274. E. THOMPSON, R. J. LEVIN and M. J. JACKSON, *Biochim. Biophys. Acta*, **196**, 120 (1970).
275. M. GILLES-BAILLIEN and E. SCHOFFENIELS, *Archs. Int. Physiol.*, **73**, 355 (1965).
276. P. F. CURRAN and S. G. SCHULTZ, Handbook of Physiology, Section 6, Vol. 3 (C. F. Code, ed.), p. 1217. Am. Physiol. Soc., Washington, D.C. (1968).
277. S. G. SCHULTZ and P. F. CURRAN, Handbook of Physiology, Section 6, Vol. 3 (C. F. Code, ed.), p. 1245. Am. Physiol. Soc., Washington, D.C. (1968).
278. C. T. ASHWORTH, F. J. LUIBEL and S. C. STEWART, *J. Cell Biol.*, **17**, 1 (1963).
279. J. C. SKOU, *Physiol. Rev.*, **45**, 596 (1965).
280. R. WHITTAM, Cellular Functions of Membrane Transport (J. F. Hoffman, ed.), p. 139. Prentice-Hall, Englewood Cliffs (1964).
281. R. WHITTAM and K. P. WHEELER, *A. Rev. Physiol.*, **32**, 21 (1970).
282. J. W. L. ROBINSON, *J. Physiol., Lond.*, **206**, 41 (1970).
283. H. NEWEY, P. A. SANFORD and D. H. SMYTH, *J. Physiol., Lond.*, **194**, 237 (1968).
284. B. C. S. HOLLANDS and M. W. SMITH, *J. Physiol., Lond.*, **175**, 31 (1964).
285. M. W. SMITH, *J. Physiol., Lond.*, **175**, 38 (1964).
286. G. G. BERG and B. CHAPMAN, *J. Cell. Comp. Physiol.*, **65**, 361 (1965).
287. V. KLINGMULLER, J. GAYER and F. BARMSTEDT, *Z. Physiol. Chem.*, **300**, 107 (1955).
288. C. PERAINO and A. E. HARPER, *Arch. Biochem. Biophys.*, **97**, 442 (1962).
289. K. D. NEAME and G. WISEMAN, *J. Physiol., Lond.*, **146**, 10P (1959).
290. T. G. KERSHAW, K. D. NEAME and G. WISEMAN, *J. Physiol., Lond.*, **152**, 182 (1960).
291. J. T. HINDMARSH, D. KILBY, B. ROSS and G. WISEMAN, *J. Physiol., Lond.*, **188**, 207 (1967).
292. R. H. DOWLING and C. C. BOOTH, *Clin. Sci.*, **32**, 139 (1967).
293. R. H. DOWLING, E. O. RIECKEN, J. W. LAWS and C. C. BOOTH, *Clin. Sci.*, **32**, 1 (1967).
294. G. WISEMAN, K. D. NEAME and F. N. GHADIALLY, *Br. J. Cancer*, **13**, 282 (1959).
295. M. SUDA and A. SHIMOMURA, *Osaka Univ. Med. J.*, **16**, 11 (1964).
296. E. O. RIECKEN, R. H. DOWLING, C. C. BOOTH and A. G. E. PEARSE, *Enzymol. Biol. Clin.*, **5**, 231 (1965).
297. C. L. WRIGHT and H. E. BARBER, *Biochem. J.*, **115**, 1075 (1969).
298. D. S. MADGE, *Comp. Biochem. Physiol.*, **30**, 295 (1969).

299. R. E. KIRSCH, S. J. SAUNDERS and J. F. BROCK, *Am. J. Clin. Nutr.*, **21**, 1302 (1968).
300. R. J. LEVIN, H. NEWEY and D. H. SMYTH, *J. Physiol., Lond.*, **177**, 58 (1965).
301. H. NEWEY, P. A. SANFORD and D. H. SMYTH, *J. Physiol., Lond.*, **208**, 705 (1970).
302. R. J. NEALE and G. WISEMAN, *J. Physiol., Lond.*, **198**, 601 (1968).
303. R. B. WOODD-WALKER, J. D. L. HANSEN and S. J. SAUNDERS, *Lancet*, **2**, 1428 (1969).
304. M. STEINER, G. C. M. FARRISH and S. J. GRAY, *Am. J. Clin. Nutr.*, **22**, 871 (1969).
305. S. ZIEMLANSKI, D. CIESLAK, B. PLISZKA and A. SZCZYGIEL, *Die Nahrung*, **11**, 559 (1967).
306. R. F. CRAMPTON, M. T. LIS and D. M. MATTHEWS, *J. Physiol., Lond.*, **206**, 66P (1970).
307. R. J. LEVIN, *Life Sci.*, **9**, 61 (1970).
308. V. KUJALOVA and P. FABRY, *Physiologia Bohemislovenica*, **9**, 35 (1960).
309. G. PERES, M. BUCLON and D. CARRUGE, *C. r. Seanc. Soc. Biol.*, **156**, 2080 (1962).
310. J. D. HARDY and J. SCHULTZ, *J. Appl. Physiol.*, **4**, 789 (1952).
311. T. L. ALTHAUSEN, R. K. DOIG, K. UYEYAMA and S. WEIDEN, *Gastroenterology*, **16**, 126 (1950).
312. R. H. DOWLING and C. C. BOOTH, *Lancet*, **2**, 146 (1966).
313. L. D. WEINSTEIN, C. P. SHOEMAKER, T. HERSH and H. K. WRIGHT, *Archs. Surg., Chicago*, **99**, 560 (1969).
314. H. NEWEY and D. H. SMYTH, *Nature, Lond.*, **202**, 400 (1964).
315. J. K. BINGHAM, H. NEWEY and D. H. SMYTH, *Biochim. Biophys. Acta*, **120**, 314 (1966).
316. J. K. BINGHAM, H. NEWEY and D. H. SMYTH, *Biochim. Biophys. Acta*, **130**, 281 (1966).
317. S. J. SAUNDERS and K. J. ISSELBACHER, *Biochim. Biophys. Acta*, **102**, 397 (1965).
318. M. G. CASEY and J. P. FELBER, *Gastroenterologia*, **107**, 209 (1967).
319. S. REISER and P. A. CHRISTIANSEN, *Am. J. Physiol.*, **216**, 915 (1969).
320. B. G. MUNCK, *Biochim. Biophys. Acta*, **150**, 82 (1968).
321. B. G. MUNCK, *Biochim. Biophys. Acta*, **156**, 192 (1968).
322. R. A. CHEZ, S. G. SCHULTZ and P. F. CURRAN, *Science*, **153**, 1012 (1966).
323. F. ALVARADO, *Nature, Lond.*, **219**, 276 (1968).
324. K. P. McCONNELL and G. J. CHO, *Am. J. Physiol.*, **213**, 150 (1967).
325. A. U. ORTEN, *Fed. Proc. Fed. Am. Socs. Exp. Biol.*, **20**, 2 (1961).
326. J. H. ANNEGERS, *Am. J. Physiol.*, **210**, 701 (1966).
327. J. T. HINDMARSH, D. KILBY and G. WISEMAN, *J. Physiol., Lond.*, **182**, 52 P (1966).

328. H. L. DUTHIE and J. T. HINDMARSH, *J. Physiol., Lond.*, **187**, 195 (1966).
329. A. NEUBERGER, *Adv. Protein Chem.*, **4**, 297 (1948).
330. R. P. SPENCER, K. R. BRODY and F. E. VISHNO, *Biochim. Biophys. Acta*, **117**, 410 (1966).
331. D. H. SMYTH and C. B. TAYLOR, *J. Physiol., Lond.*, **141**, 73 (1958).
332. T. H. WILSON, *Biochem. J.*, **56**, 521 (1954).
333. R. P. SPENCER, T. M. BOW and M. A. MARKULIS, *Am. J. Physiol.*, **202**, 171 (1962).
334. H. HAGIHIRA, T. H. WILSON and E. C. C. LIN, *Am. J. Physiol.*, **203**, 637 (1962).
335. K. C. HUANG, *Fed. Proc. Fed. Am. Socs. Exp. Biol.*, **20**, 246 (1961).
336. T. H. WILSON, Intestinal Absorption. Saunders, Philadelphia (1962).
337. C. C. BOOTH and D. L. MOLLIN, *Proc. Int. Congr. Gastroenterology*, 249 (1960).
338. K. A. ELSOM, F. W. CHORNOCK and F. G. DICKEY, *J. Clin. Invest.*, **21**, 795 (1942).
339. H. J. McCORKLE and H. A. HARPER, *Ann. Surg.*, **140**, 467 (1954).
340. A. H. JOHNSON, H. J. McCORKLE and H. A. HARPER, *Gastroenterology*, **28**, 360 (1955).
341. K. NAKAYAMA, T. NAKAMURA, K. YAMAMOTO and T. TAMIYA, *Gastroenterology*, **38**, 946 (1960).
342. E. C. WECKESSER, J. L. ANKENEY, A. F. PORTMANN, J. W. PRICE and F. A. CEBUL, *Surgery*, **30**, 465 (1951).
343. A. J. KREMEN, J. H. LINNER and C. H. NELSON, *Ann. Surg.*, **140**, 439 (1954).
344. R. A. PARKINS, A. DIMITRIADOU and C. C. BOOTH, *Clin. Sci.*, **19**, 595 (1960).
345. K. RAMASWAMY and A. N. RADHAKVISHNAN, *Indian J. Biochem.*, **3**, 138 (1966).
346. H. P. SCHEDL, D. L. MILLER, H. D. WILSON and P. FLORES, *Am. J. Physiol.*, **216**, 1131 (1969).
347. K. NYGAARD, *Acta Chir. Scand.*, **132**, 731 (1966).
348. A. OCHOA-SOLANO and C. GITLER, *J. Nutr.*, **94**, 249 (1968).
349. E. R. BATT and D. SCHACHTER, *Am. J. Physiol.*, **216**, 1064 (1969).
350. C. S. WELCH, E. G. WAKEFIELD and M. ADAMS, *Archs. Intern. Med.*, **58**, 1095 (1936).
351. H. G. SAMMONS, *Biochem. J.*, **80**, 30P (1961).
352. C. W. HAUGE and W. W. KRIPPAEHNE, *Am. J. Surg.*, **119**, 67 (1970).
353. D. F. EVERED and P. B. NUNN, *European J. Biochem.*, **4**, 301 (1968).
354. N. CORDERO and T. H. WILSON, *Gastroenterology*, **41**, 500 (1961).
355. M. BAILLIEN and E. SCHOFFENIELS, *Biochim. Biophys. Acta*, **53**, 521 (1961).

356. F. A. JACOBS, E. E. LARGIS and E. J. HANSON, *Fed. Proc. Fed. Am. Socs. Exp. Biol.*, **26**, 302 (1967).
357. F. A. JACOBS and E. E. LARGIS, *Proc. Soc. Exp. Biol. Med.*, **130**, 692 (1969).
358. K. BERGSTROM and B. WERNER, *Acta Chir. Scand.*, **131**, 413 (1966).
359. F. A. JACOBS and E. E. LARGIS, *Proc. Soc. Exp. Biol. Med.*, **130**, 697 (1969).
360. R. S. COMLINE, H. E. ROBERTS and D. A. TITCHEN, *Nature, Lond.*, **167**, 561 (1951).
361. W. E. BALFOUR and R. S. COMLINE, *J. Physiol., Lond.*, **148**, 77P (1959).
362. S. N. GANAPATHY and E. S. NASSET, *J. Nutr.*, **78**, 241 (1962).
363. E. S. NASSET, S. N. GANAPATHY and D. P. J. GOLDSMITH, *J. Nutr.*, **81**, 343 (1963).
364. E. S. NASSET, *Am. J. Dig. Dis.*, **9**, 175 (1964).
365. E. S. NASSET and J. S. JU, *Proc. Soc. Exp. Biol. Med.*, **132**, 1077 (1969).
366. E. G. FRAME, *J. Clin. Invest.*, **37**, 1710 (1958).
367. C. PERAINO and A. E. HARPER, *J. Nutr.*, **80**, 270 (1963).
368. E. S. YEARICK and R. G. NADEAU, *Am. J. Clin. Nutr.*, **20**, 338 (1967).
369. R. A. COULSON and T. HERNANDEZ, *Am. J. Physiol.*, **212**, 1308 (1967).
370. H. L. ANDERSON, N. J. BENEVENGA and A. E. HARPER, *Am. J. Physiol.*, **214**, 1008 (1968).
371. J. RICHMOND and R. H. GIRDWOOD, *Clin. Sci.*, **22**, 301 (1962).
372. L. A. ERF and C. P. RHOADS, *J. Clin. Invest.*, **19**, 409 (1940).
373. C. E. BUTTERWORTH, Jr., R. SANTINI, Jr. and E. PEREZ-SANTIAGO, *J. Clin. Invest.*, **37**, 20 (1958).
374. G. P. FRENCH and A. W. WERTZ, *J. Nutr.*, **73**, 57 (1961).
375. R. DAWSON and J. W. G. PORTER, *Br. J. Nutr.*, **16**, 27 (1962).
376. M. GAY and C. W. CRANE, *Gut*, **7**, 711 (1966).
377. H. DAVID, I. MARX and D. KRANZ, *Expl. Path. (Jena)*, **3**, 34 (1969).
378. J. J. DEREN, E. W. STRAUSS and T. H. WILSON, *Devl. Biol.*, **12**, 467 (1965).
379. C. D. HOLDSWORTH and T. H. WILSON, *Am. J. Physiol.*, **212**, 233 (1967).
380. T. H. WILSON and E. C. C. LIN, *Am. J. Physiol.*, **199**, 1030 (1960).
381. M. NING, S. REISER and P. A. CHRISTIANSEN, *Proc. Soc. Exp. Biol. Med.*, **129**, 799 (1968).
382. J. F. FITZGERALD, S. REISER, C. F. JOHNSON and P. A. CHRISTIANSEN, *Clin. Res.*, **17**, 526 (1969).
383. S. REISER, J. F. FITZGERALD and P. A. CHRISTIANSEN, *Biochim. Biophys. Acta*, **203**, 351 (1970).
384. L. PENZES, G. SIMON and M. WINTER, *Expl. Geront.*, **3**, 257 (1968).
385. A. J. MATTY and B. SESHADRI, *Gut*, **6**, 200 (1965).

386. S. ISLAM, S. REISER and P. A. CHRISTIANSEN, *Am. J. Dig. Dis.*, **13**, 266 (1968).
387. A. M. GELB and C. D. GERSON, *Am. J. Clin. Nutr.*, **22**, 305 (1969).
388. T. L. ALTHAUSEN and M. STOCKHOLM, *Am. J. Physiol.*, **123**, 577 (1938).
389. D. R. LONDON and S. SEGAL, *Endocrinology*, **80**, 623 (1967).
390. S. W. HALL, Jr., and J. M. HERSHMAN, *Am. J. Physiol.*, **215**, 1049 (1968).
391. J. R. BRONK and D. S. PARSONS, *J. Physiol., Lond.*, **184**, 942 (1966).
392. P. MANDELSTAM, *Fed. Proc. Fed. Am. Socs. Exp. Biol.*, **25**, 695 (1966).
393. M. G. CASEY, J. P. FELBER and A. VANNOTTI, *Am. J. Proctol.*, **20**, 64 (1969).
394. W. A. OLSEN and I. H. ROSENBERG, *J. Clin. Invest.*, **49**, 96 (1970).
395. D. FROMM, M. FIELD and W. SILEN, *Surgery*, **66**, 145 (1969).
396. O. A. SHISHOVA, *Biokhimiya*, **24**, 812 (1959).
397. J. D. FINKELSTEIN and D. SCHACHTER, *Am. J. Physiol.*, **203**, 873 (1962).
398. I. L. CRAFT, *Clin. Sci.*, **38**, 287 (1970).
399. B. F. FELL, R. A. SMITH and R. M. CAMPBELL, *J. Path. Bact.*, **85**, 179 (1963).
400. R. BOYNE, B. F. FELL and I. ROBB, *J. Physiol., Lond.*, **183**, 570 (1966).
401. G. H. LATHE, *Revue Can. Biol.*, **2**, 143 (1943).
402. F. A. JACOBS, L. J. COEN and R. S. L. HILLMAN, *J. Biol. Chem.*, **235**, 1372 (1960).
403. F. A. JACOBS, R. C. FLAA and W. F. BELK, *Fed. Proc. Fed. Am. Socs. Exp. Biol.*, **19**, 183 (1960).
404. H. AKEDO, T. SUGAWA, S. YOSHIKAWA and M. SUDA, *J. Biochem., Tokyo*, **47**, 124 (1960).
405. K. UEDA, H. AKEDO and M. SUDA, *J. Biochem., Tokyo*, **48**, 584 (1960).
406. F. A. JACOBS and R. S. L. HILLMAN, *J. Biol. Chem.*, **232**, 445 (1958).
407. G. WISEMAN, *J. Physiol., Lond.*, **136**, 203 (1957).
408. B. G. MUNCK, *Biochim. Biophys. Acta*, **94**, 136 (1965).
409. R. A. CHEZ, E. O. HORGER, S. G. SCHULTZ and P. F. CURRAN, *Biochim. Biophys. Acta*, **183**, 244 (1969).
410. S. J. SAUNDERS and R. E. KIRSCH, *Metabolism*, **17**, 386 (1968).
411. M. SUGAI and I. MATSUDA, *Biochim. Biophys. Acta*, **170**, 474 (1968).
412. R. H. IMAMI, S. REISER and P. A. CHRISTIANSEN, *J. Nutr.*, **100**, 101 (1970).
413. C. D. WEST, J. L. WILSON and R. EYLES, *Am. J. Dis. Child.*, **72**, 251 (1946).
414. H. ANFANGER and R. M. HEAVENRICH, *Am. J. Dis. Child.*, **77**, 425 (1949).

415. B. J. COHEN and J. H. ANNEGERS, *Gastroenterology*, **25**, 67 (1953).
416. R. J. COFFEY, F. C. MANN and J. L. BOLLMAN, *Am. J. Dig. Dis.*, **7**, 143 (1940).
417. J. R. HEERSMA and J. H. ANNEGERS, *Am. J. Physiol.*, **153**, 143 (1948).
418. A. M. DAWSON and D. M. MATTHEWS, *J. Physiol., Lond.*, **159**, 57P (1961).
419. J. L. POPE, T. M. PARKINSON and J. A. OLSON, *Biochim. Biophys. Acta*, **130**, 218 (1966).
420. T. M. PARKINSON and J. A. OLSON, *Life Sci.*, **6**, 393 (1963).
421. A. M. DAWSON and K. J. ISSELBACHER, *J. Clin. Invest.*, **39**, 730 (1960).
422. Y. ISRAEL, J. E. VALENZUELA, I. SALAZAR and G. UGARTE, *J. Nutr.*, **98**, 222 (1969).
423. Y. ISRAEL, J. E. VALENZUELA and G. UGARTE, *Gastroenterology*, **56**, 1170 (1969).
424. Y. ISRAEL, I. SALAZAR and E. ROSENMANN, *J. Nutr.*, **96**, 499 (1968).
425. T. CHANG, J. LEWIS and A. J. GLAZKO, *Biochim. Biophys. Acta*, **135**, 1000 (1967).
426. A. R. MEHRAN and R. BLAIS, *Archs. Int. Physiol. Biochim.*, **75**, 27 (1967).
427. L. R. BENNETT, S. M. CHASTAIN, A. B. DECKER and J. F. MEAD, *Proc. Soc. Exp. Biol. Med.*, **77**, 715 (1951).
428. A. J. CUMMINS and T. P. ALMY, *Gastroenterology*, **23**, 179 (1953).

CHAPTER 9

Immunological Proteins

I. G. MORRIS

*Department of Zoology,
University College of North Wales,
Bangor*

		Page
9.1	IMMUNOGLOBULINS	486
	9.1.1 *Structure and properties*	486
	9.1.2 *Distribution in colostrum and milk*	490
9.2	ABSORPTION IN PRIMATES, RABBIT AND GUINEA PIG	491
9.3	ABSORPTION IN RAT AND MOUSE	494
	9.3.1 *Homologous immunoglobulins*	494
	9.3.2 *Selection*	498
	9.3.3 *Competitive absorption*	502
	9.3.4 *Histology and cytochemistry*	504
	9.3.5 *Absorption in hedgehog*	508
9.4	ABSORPTION IN CARNIVORES	509
9.5	ABSORPTION IN RUMINANTS	512
	9.5.1 *Homologous immunoglobulins*	512
	9.5.2 *Selection*	515
	9.5.3 *Cytology*	516
	9.5.4 *Absorption accelerators*	517
9.6	ABSORPTION IN THE PIG	519
	9.6.1 *Homologous immunoglobulins*	519
	9.6.2 *Selection*	522
	9.6.3 *Cytology*	525
9.7	CONCLUSION	529
	REFERENCES	532

Intact dietary proteins do not reach the absorptive cells of the small intestine of the adult mammal in appreciable amounts [1, 2]. In the lumen, proteins are rapidly digested and absorbed as amino acids and small peptides. The peptides are further broken down intracellularly and amino acids alone reach the portal circulation. However, there is indirect evidence that in man, and in infants especially, small amounts of dietary proteins escape complete digestion and reach the circulation in sufficient quantities to stimulate the body to produce circulating antibodies against them [3-11]. Also, mature animals can be effectively immunized against protein antigens given to them orally [12]. Passage of antigens into the circulation in such cases could occur at the extrusion zones of effete cells on the tips of the intestinal villi by a process for which the name persorption has been suggested [13], or by absorption through the remnants of the mechanism used in the foetus or newborn for the transmission of passive immunity (*vide infra*). Absorption through intercellular seepage is effectively prevented by the tight junctions between the epithelial cells [14]. The low grade intestinal absorption of intact protein in mature animals is enhanced when digestion is hampered by enzyme inhibitors [15, 16] or pH changes [17, 18], or when protein is administered directly into the small intestine [19, 20]. In marked contrast to mature animals, the intestine in the foetus, neonate or both in some species displays a transient but pronounced permeability to some proteins. This is an adaptation to allow the transmission of passive immunity from the mother to the offspring.

Protection against invasive and potentially pathogenic organisms depends on the body's ability to produce circulating antibodies against them. Proteins having antibody activity are termed immunoglobulins and are found mainly in the γ-globulin fraction of the serum, although several types migrate electrophoretically with the β-globulins (Fig. 9.1). Immunity is usually developed gradually as a result of repeated innocuous exposures to a potential pathogen, and an animal thus immunized is able to respond to a massive pathogenic invasion by producing specific antibodies rapidly and copiously. This ability is deficient in the newborn [21], which has therefore to depend on the antibodies it receives passively from the mother for protection.

The transmission of passive immunity from the mother to young, reviewed recently by Brambell [22], occurs at different times in different species (Table 9.1). Before birth, maternal antibodies may be transmitted across the placenta or may be

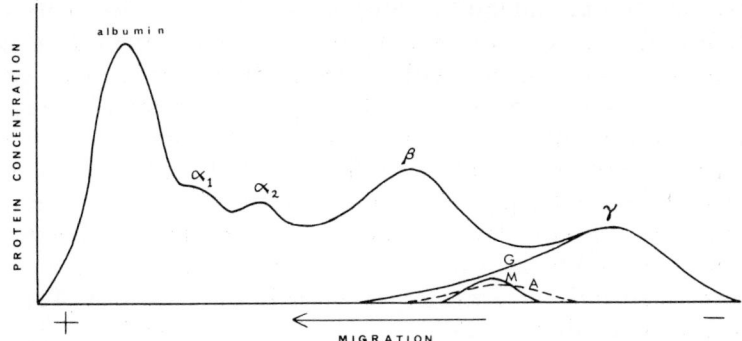

Fig. 9.1. An electropherogram of the serum proteins of a mammal showing the locations of IgG, IgA and IgM.

absorbed from the uterine fluid by the yolk sac. In the newborn, transmission is across the intestine from ingested colostrum or milk. The passive immunity thus received not only protects the newborn but also influences its ability to produce its own antibodies later [23-36]. Another requirement in the

TABLE 9.1. Transmission of passive immunity in animals.

Species	Route of transmission	Time of transmission	
		prenatal	postnatal
Primates	Placenta	+++	
	intestine	0	0
Rabbit	Yolk-sac	+++	
	intestine	0	0
Guinea pig	Yolk-sac	+++	
	intestine	?	?
Carnivores	Unknown	+	
	intestine		++ (1-2)
Hedgehog	Unknown	+	
	intestine		++ (40)
Rodents	Yolk-sac	+	
	intestine	±	++ (16-20)
Ungulates	Intestine	0	+++ (1½-2)

Figures in parentheses = transmission in days.
After Brambell [22].

newborn is an adequate supply of amino acids. Different adaptations are encountered in the gut in different species to cater for both needs either sequentially or concurrently. Newborn ungulates receive their passive immunity rapidly within a few hours of birth, the gut thereafter ceasing to transmit antibodies and assuming the full role of a digestive organ. In rodents, postnatal transmission of passive immunity persists for several days during which time the gut has to act both as antibody transmitter and amino acid producer. Some aspects of these remarkable processes in the newborn of several species will be discussed.

9.1 IMMUNOGLOBULINS

9.1.1 Structure and properties

The structure and properties of the immunoglobulins have recently been extensively reviewed [37-40]. In mammals they comprise three main classes, IgG, IgA and IgM. The IgG class predominates in normal sera. Small amounts of other classes, IgD and IgE, have also been identified in human serum. Nothing is known about the immune properties of IgD, but the IgE class comprises the reagins which are the antibodies associated with allergies. An analogous class may also occur in mouse, rat, rabbit and dog. The main physical properties of the human immunoglobulins are given in Table 9.2. Unlike other serum

TABLE 9.2. Properties of human immunoglobulins.

Class and subclasses	Concentration mg/100 ml	Molecular weight $\times 10^{-3}$	$S_{20.w}$	Carbohydrate % w/w
IgG1, 2, 3, 4	1240 ± 270	150-160	6.5-7.0	2.6-2.9
IgM1, 2	120 ± 35	900	18-20	10.0-11.8
IgA1, 2, 3	280 ± 70	150-160 (monomer)	7-17	7.0-7.5
IgD	0.3-40.0	160-180	6.2-6.8	12
IgE	0.01-0.07	200	8	11

Complied from Cohen and Milstein [37], Grey [39] and Ishizaka [40].

proteins the immunoglobulins display a wide range of heterogeneity in properties but all are characterized by a similar basic structure which was first elucidated for the IgG class.

The structure of the extended IgG molecule is illustrated in Figure 9.2. The shorter light (L) polypeptide chains and the

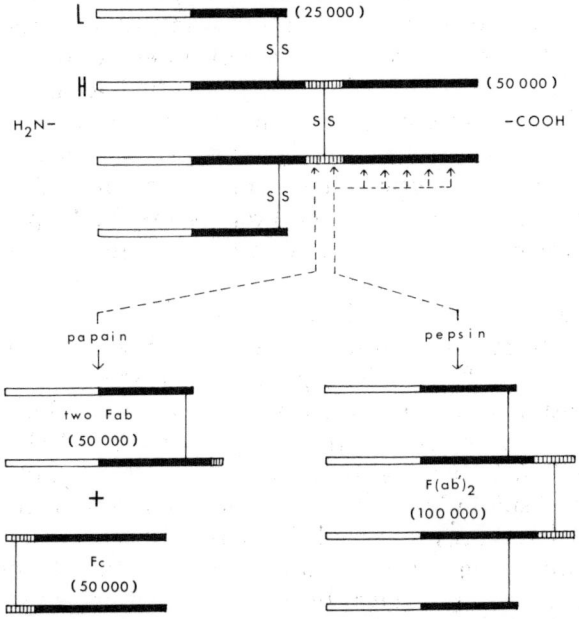

Fig. 9.2. Schematic structure of the IgG molecule. The open and solid parts of the polypeptide chains represent the variable and constant parts respectively. The cross-hatched parts represent the hinge regions. Figures in parentheses denote the molecular weights.

longer heavy (H) chains consist of about 230 and 460 amino acids respectively. The two L chains or the two H chains in a molecule are identical. The aminoterminal half of each L chain and an equivalent length of each H chain show variable amino acid sequences between molecules and give the molecules their diverse antibody specificities. The remaining parts of the chains are invariable and determine the properties characteristic of their class. Parts of the H chains in the region of the disulphide bond linking them together are very susceptible to proteolysis. Papain cleaves the H chains on the aminoterminal side of the bond, giving two identical antibody-reactive Fab fragments and one Fc fragment which carries most of the carbohydrate, antigenicity and biological activity of the intact molecule (Fig. 9.2). Pepsin cleaves the H chains on the carboxy-terminal side of the disulphide bond into successively smaller peptides leaving a major $F(ab')_2$ fragment which has two antibody-reactive sites. The enzyme-susceptible regions contain several proline residues which impart a degree of flexibility to the molecule, forming hinge regions. Intra-chain disulphide bonds help to maintain the tertiary structure of the rest of the molecule, which appears as a tripartite Y-shaped structure under the electron microscope. Each arm (65 Å long) represents an Fab portion and the stem (50 Å) represents the Fc portion. The angle between the arms is variable due to the flexibility of the hinge regions.

This basic 7S structure is common to all classes of immunoglobulins. Immunoglobulin-A and IgM have the capacity to polymerize, the former into polydisperse structures of 7-17S, and the latter into a pentamer of five 7S Y-shaped units, joined together by the carboxyterminal ends of their H chains into a star-shaped structure [41]. Reduction and alkylation of the interchain disulphide bonds allow the separation of the L and H chains. Heavy chains derived from the different classes in any species are antigenically distinct whilst the L chains are identical. Subclasses have been discovered in several species when the H chains within a class were found to have distinct as well as common antigenic determinants. In man there are four IgG, three IgA and two IgM subclasses. In other species the IgG comprises an electrophoretically slow IgG_2 and a fast IgG_1 subclass, each having distinct biological properties. Generally, but not invariably, the IgG_1 can induce passive cutaneous anaphylaxis (PCA) in the homologous species but cannot fix complement, whereas the IgG_2 can induce PCA in heterologous species and can fix complement. Exceptions to these rules obscure the homologies of the subclasses in different species. The elucidation of this problem awaits further work to determine the amino acid sequences. From some of the available results (Table 9.3), it is evident that the H chains of one class of immunoglobulins from different species resemble one another more closely than do the H chains of different classes in one species.

IMMUNOLOGICAL PROTEINS 489

TABLE 9.3. Amino acid sequences in the caboxyterminal part of immunoglobulin H chains.

H chain		Sequence*	Reference
Rabbit	IgG	Met His Glu Ala Leu His Asn His Tyr Thr Gln Lys Ser Ile Ser Arg Ser Pro Gly	252
Bovine	IgG$_1$	Thr Lys Ala	181
	IgG$_2$	Met Thr Lys Ala	181
Human	IgG1		252
	IgG3	Arg Leu Leu	252
	IgM	Met Ser Asx Thr Ala Gly Thr Cys Tyr Phe Thr Leu Leu	253
	IgA	Met Ala Glu Val Asp Gly Thr Cys Tyr	253

* All the sequences are as for rabbit except where indicated otherwise.

9.1.2 Distribution in colostrum and milk

Immunoglobulins also occur in external secretions; their distribution in colostrum is shown in Table 9.4. The concentration of IgA in colostrum greatly exceeds that in the serum due

TABLE 9.4. The colostrum:serum concentration ratios for the immunoglobulins.

Species	IgG		IgM	IgA	Reference
	IgG_2	IgG_1			
Man	very low	very low	low	high	42
Mouse	low	low	0	high	46
Rat				high	43, 254
Rabbit	low	low	low	high	42
Dog	1.5	1.1	0.4	14	255
	1.0	4.0	0.5	80	163
Pig	12.6		0.9	3	255
	4.2		1.2	6	255
	2.7		2.9	6	48
Cow	0.1	3.4	1.5	5	255
	very low	high			63
	4.4		2.0	15	45
Sheep	0.1	1.9	0.7	43	255
	<0.1	0.2	0.2	3	255
	0.1	1.1	2.1	17	255
	2.9		3.4	8	45
	7.0	1.9	4.5	7	255
Goat		2.6	2.4	5	45

to the mode of production and secretion. In man, it is largely produced locally in the mammary gland, and then transferred across the acinar cells to the colostrum where it appears mostly (80%) in the form of dimeric (11.6S) molecules bound to a secretory (S-) piece produced by the acinar cells [42]. This secretory IgA (SIgA) functions differently from the serum immunoglobulins since it reacts with potential pathogens outside the body. The S-piece changes the characteristics of the IgA, conferring on it a distinct antigenicity and a greater resistance to proteolysis. The latter acquisition must enhance the efficacy of the molecules, which may be antibodies against enteropathogens, in the gut of the suckling newborn. A similar secretory system for IgA occurs in the monkey, rabbit, dog, mouse, rat [43], cow [44], goat and sheep [45]. There is also evidence for the existence of an S-piece in the colostral IgA in these species except the rat where, nonetheless, the secretory IgA differs in many respects from its serum counterpart [43]. In the newborn rabbit and human, which have already obtained their full complement of passive immunity, the main immunoglobulin received through suckling is SIgA [42]. In rodents also, where the young continue to absorb antibodies from the milk for prolonged periods, the main colostral and milk immunoglobulin is SIgA [43, 46]. In the dog, pig and ruminants, where the newborn derive their passive immunity entirely within a few hours of birth, the SIgA forms a minor component of the colostrum [44, 45, 47-49], but becomes relatively more important later in lactation when the gut of the newborn has ceased to absorb antibodies.

There is a close serological relationship between the serum and colostral or milk proteins for the other immunoglobulin classes in the species tested [47, 50-55], the colostral and milk immunoglobulins being derived unaltered from their serum counterparts [54-60]. In ruminants and pig the serum immunoglobulins are concentrated by the mammary gland to attain higher than serum levels in the colostrum (Table 9.4). Active transfer occurs by pinocytosis and reverse pinocytosis at opposite poles of the acinar cells [61]. It is a selective process, IgG in the pig [44] and IgG_1 in ruminants [57, 62-65] being preferentially concentrated in the colostrum to form the predominant immunoglobulins.

9.2 ABSORPTION IN PRIMATES, RABBIT AND GUINEA PIG

The transmission of passive immunity from mother to young in primates occurs entirely before birth, the newborn having circulating levels of IgG equalling or slightly exceeding the maternal serum levels [22]. The route of transmission is across the chorio-allantoic placenta, and only IgG is transmitted. Under normal conditions only small amounts of maternal IgG and even less of IgA or IgM reach the amniotic fluid. When this is swallowed by the late foetus, insignificant amounts of the immunoglobulins, if any, are transferred intact across the gut. Also, when the immunoglobulin concentrations in the amniotic fluid are substantially increased as a result of the injection of antiserum into the amniotic cavity, the amounts transmitted to the foetal circulation are still very low. The opportunity for intestinal transmission of immunoglobulins is greater after birth when the newborn ingests colostrum and milk. The colostral levels of IgG and IgM are low, but the SIgA concentration greatly exceeds the maternal serum level, as evidenced by direct measurement [66] and by the occurrence in colostrum of blood group iso-agglutinins or agglutinins against enteric pathogens in higher titres than in the serum [67-72]. However, the gut of the newborn infant is impermeable to immunoelectrophoretically detectable amounts of the immunoglobulins [73-75]. It is also impermeable to serologically detectable amounts of a variety of antibodies, of human or bovine origin, administered orally in colostrum, milk or immune serum [68-71, 76-80]. The impermeability of the newborn gut to intact immunoglobulins does not preclude the possibility of the transmission of their smaller molecular weight degradation products in the same way as the low molecular weight (1.6S) milk-specific proteins are transmitted. Components closely resembling these proteins appear in appreciable amounts in the urine during the first 70 hours of life [58]. This type of transmission, however, cannot account for the results of Leissring and colleagues [81]. These authors found that newborn infants could absorb *Salmonella paratyphi* agglutinins or tetanus antitoxins efficiently from the gut for at least 5 days, and saline agglutinins were transmitted across the gut more readily than incomplete and possibly degraded antibodies. Despite the lack of significant absorption of colostral immunoglobulins in the newborn, the beneficial effects of SIgA antibodies against potential enteropathogens are considerable [42]. The survival of these antibodies in the gut must be prolonged by their greater resistance to proteolysis [42], by the low acidity of the stomach [82] and by the action of the trypsin inhibitors in the colostrum and milk [83].

Similar conclusions can be drawn about the intestinal transmission of immunoglobulins in the rabbit and guinea pig [22]. Before birth, maternal antibodies (7S but not 19S in rabbits) reach the amniotic fluid by molecular filtration through the foetal membranes, but they attain much lower concentrations in the fluid than in the foetal serum. When the late foetus drinks the amniotic fluid, the ingested immunoglobulins are concentrated in the stomach due to the rapid absorption of water, but they are not transferred into the circulation in serologically detectable amounts. Only the results of Leissring and Anderson [84] for the foetal guinea pig are at variance with this conclusion. They led the authors to believe that the foetal intestine replaced the yolk sac as the absorptive organ for maternal antibodies after the 50th day. Other results obtained by the same authors [85] imply that the permeability of the intestine persists at a very high level for at least 7 days into the neonatal period. During this time the newborn could absorb antibodies efficiently from immune sera administered to them by stomach tube, in spite of the apparent maturity of the digestive tract enzymes [86]. These results contradict the results of Schneider and Szathmary [87] and Jo-Keiichiro [88] showing a lack of intestinal permeability to a variety of antibodies in the newborn, but support the results of Brogi and Biagini [89] who induced haemolytic syndromes in newborn guinea pigs by feeding them with haemolytic antiserum. Also, Kellner and Hedal [90] showed that the red blood cells of newborn rabbits became strongly sensitized with antibody when the young were fostered by mothers having incompatible, incomplete isohaemagglutinins in their milk. In most of these studies the detection of transmitted antibodies relied on sensitive indirect techniques which would also measure incomplete and possibly degraded antibodies. The results probably reflect the ability of the intestine, irrespective of age, to absorb proteins in antigenic amounts.

Man, rabbit and guinea pig amongst other species pass through a stage in development when the absorptive cells of the small intestine display a pronounced capacity to pinocytose luminal contents. During this time the morphology of the cells closely resembles that of the immunoglobulin-transmitting ileal cells of the suckling rodent or pig (*vide infra*). From cytological and macromolecule ingestion studies, this stage has been shown to extend from the 25th gestational day to the 20th postnatal day in the rabbit [91-93], from the 50th gestational day to within a few hours after birth in the guinea pig [85, 93], and from the 14th gestational week until shortly before birth in man [94, 95]. The ileal cells can absorb trypan blue-protein complex (rabbit [93]), ferritin (rabbit [96]), fluorescien-labelled serum proteins (guinea pig [85]), antibodies (rabbit [97]) or radio-opaque substances (man [98]) into vacuoles, but the penetration of the marker molecules across the absorptive cells into the lamina propria was only seen in the guinea pig. Lysosomal enzymes are present within the cytoplasmic vesicles of the pinocytosing ileal cells in the human foetus [99]. There is a high degree of correlation between the distribution of lysosomal enzymes and of absorbed marker molecules inside the pinocytosing ileal cells of rabbits [93, 97] and guinea pigs [93]. Lysosomal enzymes and absorbed marker molecules coexist within the same cytoplasmic vesicles inside the pinocytosing ileal cells of the rabbit and guinea pig [93]. All of these observations support the possibility that most, if not all, of the substrates absorbed by the ileal cells are degraded intracellularly. Williams and Beck [93] suggested that this form of digestion may play an important role in the nutrition of the immature animal before efficient luminal digestion is established, and especially when this is accomplished some time after birth.

9.3 ABSORPTION IN RAT AND MOUSE

9.3.1 *Homologous immunoglobulins*

Although maternal antibodies are transmitted to the young rodent before birth, transfer of immunity occurs mainly after birth [22]. The most important prenatal route is from the uterine fluid and absorption by the yolk sac, but some transmission may also occur across the intestine from swallowed amniotic fluid. Antibodies of the IgG_1 and IgG_2 subclasses are transmitted readily [100]. Since the serum at birth lacks IgM [46] and IgA [43, 46], both prenatal routes discriminate against these proteins.

There are appreciable amounts of IgG_1 and IgG_2 in the rodent mammary secretions [46] and higher than serum levels of IgA [43, 46, 101], but no IgM is present [46, 59]. Immunoglobulin-G is derived from the serum [59, 60, 102], but the IgA is produced locally [101] and differs from the serum IgA in mobility [43] and in having an intermediate (11S) size [43, 101]. The postnatal transmission of maternal immunoglobulins to the young is also limited to the transmission of IgG$_1$ and IgG$_2$. The concentrations of these immunoglobulins in the serum rise rapidly with suckling to attain almost adult values in 7 to 12 days [46, 103] and then remain constant until the end of the second (mouse) or third (rat) week of lactation. Afterwards, upon the loss of intestinal permeability, the concentrations decline rapidly. Immunoglobulin-A and IgM first become detectable in the serum after 2 to 4 weeks due to autogenous production. Antibody absorption from immune milk also reflects the permeability of the intestine, and the serum titres in the sucklings attain the maternal serum levels during the first 3 days of life [104]. Absorption continues for several days more as reflected in the constancy of the circulating titres in the rapidly growing young. Antibodies from immune sera fed by stomach tube are also absorbed rapidly from the intestine, and they attain maximum circulating levels 2 to 3 hours after feeding [105, 106]. After 15 days in the mouse [107] and 18 days in the rat [105], the permeability of the intestine to antibodies is rapidly lost (Fig. 9.3).

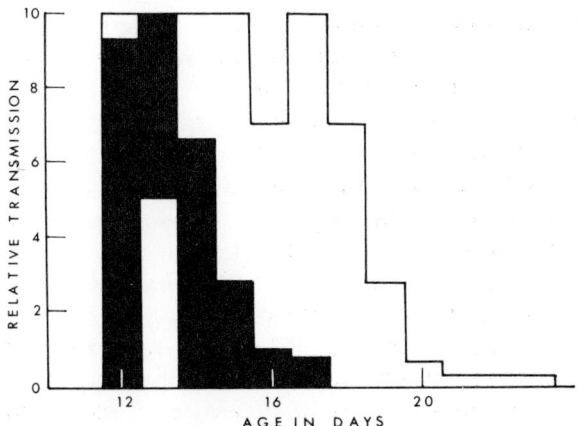

Fig. 9.3. The relative efficiency of transmission of agglutinins across the gut of rats ■ and mice □ fed with standard doses of homologous anti-*Salmonella pullorum* serum at various ages. (After Halliday [105, 107].)

The intestine of the 12-days-old rodent is impermeable to serum antibodies from primary immunized rats (mostly IgM antibodies) [108], and to milk SIgA (*vide supra*). This selectivity varies from a low to a high order from birth to weaning [109]. With increasing age the intestine becomes more and more impermeable to milk whey or serum proteins in inverse relation to their electrophoretic mobilities. On the first day, all the major electrophoretic fractions are transmitted equally, but by the 7th, 14th and 21st day respectively, albumin, β-globulins and γ-globulins cease to be transmitted. The final loss of intestinal permeability to proteins (Fig. 9.3), termed closure by Lecce and Morgan [110], is dependent on physiological age but independent of diet [111]. Preclosure rats transferred to foster mothers at markedly earlier or later lactational stages from their own mothers continue to absorb antibodies from suckled milk or orally administered immune sera until they are 18 to 20 days old. Young rats begin to ingest solid food around the time of closure, but if they are deprived of this diet and compelled to live on milk they still experience closure at the normal age. Conversely, young rats fed prematurely on solid food continue to absorb antibodies for the normal period provided they are adequately nurtured.

Until at least the 12th day of age in the rat, luminal digestion of ingested immunoglobulins is deficient due to the immaturity of the digestive system, but the intracellular degradation of intestinally absorbed protein is considerable (*vide infra*). Closure in the young rodent could result from a rapid onset of luminal digestion, a decrease in the efficiency of the intestinal absorptive cells to take up macromolecules, or the onset of complete intracellular degradation of absorbed protein. There are pronounced increases in the acidity of the stomach contents [86, 112], in gastric peptic activity [113, 114] and in pancreatic and intestinal tryptic activity [113] between the 17th and 25th days of age. These changes must drastically reduce the amount of ingested proteins reaching the absorptive terminal ileum intact. They may also alter conditions in the lumen that previously facilitated transmission. Thus it has been found (W. A. Hemmings, personal communication) that antibodies from immune sera administered orally to 12-days-old rats were absorbed more efficiently when milk whey was given simultaneously, suggesting that factors in the whey facilitated transmission. The ileum of the mature rat retains to some extent the ability to absorb and to transmit antibodies across the mucosal surface in the absence of the luminal digestive enzymes. Bamford [115] incubated sacs of everted ileum from preclosure and postclosure rats in physiological media containing immune sera, and found that antibody transmission into the sacs was considerable in all cases (Fig. 9.4). Morris and

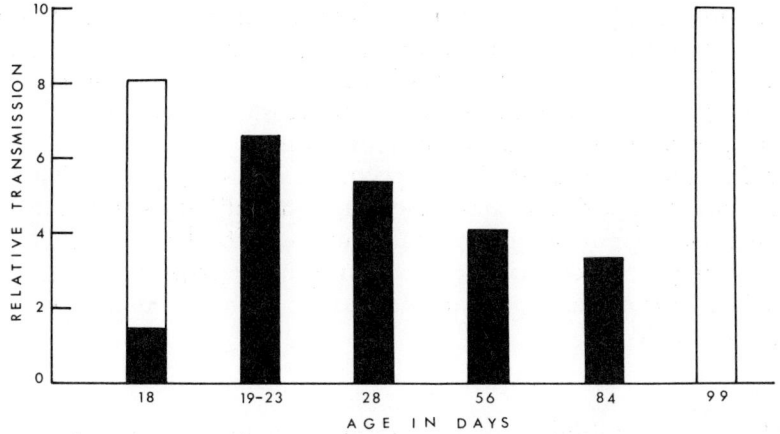

Fig. 9.4. The relative rates of transmission of rat □ or rabbit ■ anti-*Salmonella pullorum* agglutinins from the mucosal to the serosal sides of sacs of everted ileum prepared from rats of different ages. (After Bamford [115].)

Steel [20] mention briefly that, in 28-days-old rats, antibodies from immune sera were absorbed from the intestine provided they were injected directly into the duodenum after this organ had been washed out with saline and isolated from the stomach by ligature and from the liver and pancreas by severing the cystic and pancreatic ducts. Nevertheless, the intestine of the postclosure rat must become more selective towards the type of macromolecule it absorbs. In preclosure rats, indigestible polyvinyl pyrrolidone (PVP) of 160,000 molecular weight is readily absorbed into, but not transmitted across, the intestinal epithelium [116, 117]. After 18 days of age PVP ceases to be absorbed. The stimulus for closure in these experiments remains obscure, but it seems to affect the differentiating cells in the crypts rather than the villus cells directly [117]. During closure, absorbing cells on the villus are gradually displaced by nonabsorptive cells migrating from the crypts. Changes occurring between 18 and 21 days of age in the intracellular hydrolase activities of the epithelium of the ileum are such as to preclude the possibility of attributing closure to an increased efficiency in intracellular proteolysis. While the peptidase activity showed little change [118], the protease activity showed a marked decline [118, 119]. Closure can be induced prematurely in young rodents by the administration of adrenocortical steroids [107], or delayed by adrenalectomy [120]. Young rodents experiencing nutritional [107] or immunological (graft versus host reaction) stress [121] also show a premature closure. Cortisone-induced premature closure takes two days for completion [107]; this is also the time it takes for ileal epithelial cells to migrate from crypt to villus tip [117]. Thus a hormonally induced increase in luminal digestion and a hormonally induced decrease in mucosal cell absorptive efficiency may both contribute to closure in the rodent, though the primary stimulus for the hormone release remains to be determined.

9.3.2 Selection

The impermeability of the young rodent intestine towards homologous SIgA and IgM also applies to the IgM of other species. Halliday and Kekwick [108], using an aqueous-ethylether precipitation technique, separated the agglutinins in immune sera taken from several species undergoing primary or extensive secondary immunizations into several fractions. They also fed the immune sera to 12-days-old rats, and their results are shown in Table 9.5. Antibody transmission in rats fed with immune sera having most of their antibody activity associated with non-IgG immunoglobulins was markedly less than in rats fed with the predominantly IgG-active immune sera. More conclusively, it has been shown that detectable amounts of ^{131}I-labelled rabbit IgM failed to reach the circulation of 12-days-old rats fed with this preparation (W. A. Hemmings 1962, unpublished observation). Rabbit or bovine IgM agglutinins to *Brucella abortus* isolated from immune sera by gelfiltration or ion-exchange chromatography were not transmitted [122, 123]. Detectable amounts of IgM agglutinins to *B. abortus* from the cow failed to appear in the circulation of 15-days-old rats fed repeatedly over 5 days with large doses of immune serum [35].

The young rat intestine absorbed antibodies readily from rat

TABLE 9.5. Absorption of bacterio-agglutinins from the gut in 12-days-old rats.

Immune serum administered			% antibody activity in fractions				$-\log_2 (CQ)^d$ of fed rats
Specificity	Origin	Type	Alb.	$G2/1^a$	$G2/2^b$	$G3^c$	
Anti-S. *pullorum*	Guinea pig	hyper-immune	18	2	9	71	6.1 ± 0.18
	Rabbit	h.i.	6	1	1	93	5.4 ± 0.11
	Rat	h.i.	3	3	5	90	5.4 ± 0.13
	Rat	primary	0	33	50	16	< 7.0
Anti-B. *abortus*	Guinea pig	h.i.	44	0	11	44	8.3 ± 0.08
	Rabbit	h.i.	32	1	4	64	7.3 ± 0.18
	Rat	h.i.	0	50	25	25	7.9 ± 0.17

[a] α- and β-lipoproteins [b] α- and β-globulins (IgA, IgM) [c] γ-globulins (IgG)

[d] $CQ = \dfrac{\text{serum titre of fed rat}}{\text{titre of serum given}}$

After Halliday and Kekwick [108].

or mouse immune sera, less readily from rabbit immune serum and not at all from cow, sheep or fowl immune sera [105, 124]. Young rats fed with ^{131}I-labelled serum proteins of rat, rabbit or monkey absorbed the homologous IgG preferentially but only 7.5% of the administered amount was recoverable in the circulation 3 hours later [125]. In the mouse, the intestinal absorption of antibodies from immune sera decreases in the order mouse, guinea pig and rabbit [126]. These observations can be attributed to one or several factors other than the ability of the gut to discriminate between antibodies and IgG according to their species of origin [108, 127]. These factors include (a) the variable distribution of the antibody activity of the immune sera between different classes of immunoglobulins between which the gut discriminates, (b) the variable interference (*vide infra*) with antibody absorption by non-antibody immunoglobulin, (c) the variable digestion of ingested antibodies, (d) the variable absorption-facilitating activities of the immune sera, and (e) the variable amount of IgG in the doses administered. Information relating to factor (c) was provided by the results of experiments where variations (a b d e) were controlled (W. A. Hemmings, personal communication). Using standard oral doses of ^{131}I-labelled rat or bovine IgG, it was found that the disappearance of the doses from the gut of 12-days-old rats was almost complete within 4 hours of feeding, but the amounts reaching the circulation intact were only small fractions of the amounts administered. The fraction transmitted was greater for the homologous IgG. There was no evidence for IgG digestion in the gut lumen, and this is compatible with the immaturity of the digestive system [86, 113, 114]. Evidently, selection occurred during transmission of the proteins across the gut, after their absorption into the mucosal cells but before their entry into the circulation. Selection operates as between the proportions of the absorbed immunoglobulins which escape degradation during transmission rather than as between the proportions degraded. In the light of this work and of other studies where variations (a b d e) [106] or (a b c d e) [115, 128] were controlled, it has been established that the gut of the suckling rodent can discriminate between IgG immunoglobulins according to their species of origin (Fig. 9.5). The mice were fed with IgG solutions at constant volumes but varying concentrations and the immunoglobulins were

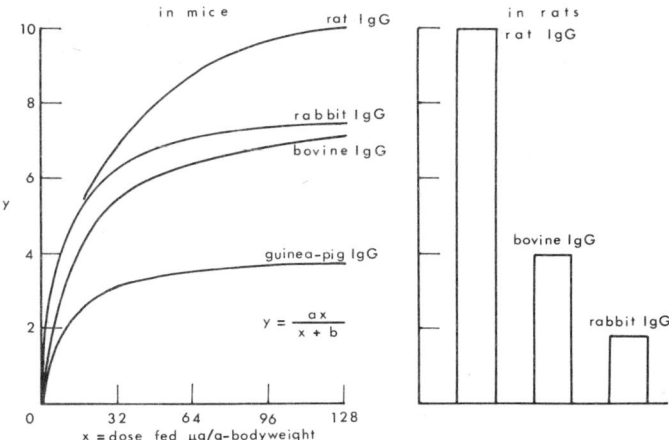

Fig. 9.5. The relative rates of transmission of immunoglobulins across the intestine in 11-days-old mice *in vivo*, and of ^{131}I-labelled immunoglobulins in 12-days-old rats *in vitro*. (After Morris [106] and Bamford [115].)

estimated as antigens in their sera 4 hours later. The results for the rats were determined *in vitro* using washed sacs of everted ileum, incubated for one hour in physiological media containing a standard concentration of the requisite ^{131}I-labelled IgG. Considerable amounts of trichloroacetic acid-soluble radioactivity appeared on both the mucosal and serosal sides of the sacs during incubation, indicating heavy intracellular proteolysis of absorbed protein; degradation of protein in physiological media in which ileal sacs had been previously incubated was negligible. In fed rats, the intracellular degradation of absorbed IgG molecules once initiated is complete since major fragments of the molecules (Fab or Fc) do not subsequently appear in the circulation [97].

There is also evidence that the young rat intestine favours the transmission of bovine IgG$_2$ relative to the IgG$_1$ [123, 129], but transmits the corresponding proteins of homologous origin equally (personal observations). The selectivity displayed towards the bovine IgG subclasses may account for the results of Brambell and colleagues [130] who fed young rats with a saline solution of ^{131}I-labelled IgG prepared from the serum of cows immune to *B. abortus*. The preparation comprised IgG$_1$ mainly and the agglutinins were almost exclusively of this subclass. Appreciable amounts of bovine protein reached the circulation in fed rats but none of it had antibody activity. Another explanation was favoured by the authors who suggested that the bovine protein was altered during transmission and lost its antibody activity but not its antigenic specificity. This seems possible in the light of later work on the transmission of anti-*B. abortus* agglutinins in young rodents. Young rats fed 4 hours previously with immune sera from rabbits undergoing primary or longstanding immunizations acquired serum titres which showed no correlation with the titres of the sera administered [122]. The latter contained 19S and 7S agglutinins of the complete type which agglutinated *B. abortus* organisms in saline suspension. Only the 7S agglutinins were transmitted in fed rats but the antibodies appearing in the circulation were incomplete types which failed to agglutinate the antigen in saline suspension and required special techniques for their detection. The change in the antibodies before and after feeding was effected during transmission across the gut rather than after transmission in the circulation. It was slight and reversible, since the incomplete antibodies recovered from the sera of fed rats partially reverted into the complete type on gel-filtration and osmotic concentration. Immunoglobulin-G_2 anti-*B. abortus* agglutinins of bovine origin may also be treated in the same way during transmission in the young rat [129]. The nature of the change imposed on the transmitted antibody remains obscure.

9.3.3 Competitive absorption
Another phenomenon encountered in the transmission of antibodies across the gut of suckling rodents has been termed interference. Normal sera from certain species reduce the transmission of antibodies derived from other species when mixed with them and administered orally [124, 131]. Interference could not be accounted for by simple dilution of the antibodies presented to the gut for absorption. In mice, interference with the absorption of rabbit antibodies decreased with the sera from rabbit, man, guinea pig and cow in that order, whilst sera from sheep, hamster, rat or mouse did not interfere significantly. In the rat, sera from rabbit, man, monkey, guinea pig and cow but not from sheep, hamster, mouse or rat interfered with the transmission of homologous antibodies. Interference was effected solely by the IgG of the interfering sera [132, 133], indicating

that it was probably a reflection of a competitive absorption of immunoglobulins from different species. Bovine IgG_1 is transmitted less readily and interferes less effectively than bovine IgG_2 [123, 129]. When tested separately for their interfering capacities in young mice, the Fab fragments of rabbit IgG were found to be ineffective, but the Fc fragment which determines the distinctive properties of the immunoglobulin class was at least 3 times as effective as the intact molecule [133] even though it was itself transmitted 11 times less readily [106].

TABLE 9.6. Dose-response relationships for feeding immunoglobulins to 11-days-old mice.

IgG fed	Regression constants	
	a(g/100 ml)	b
Rat	0.03	24.4
Bovine	0.02	15.2
Guinea pig	0.01	9.4
Rabbit	0.02	8.4

Each relationship is described by $y = ax(x + b)^{-1}$ where y = conc. of fed IgG (g/100 ml) attained in the serum of fed mice, x = μg of IgG fed per g bodyweight, and a and b are constants. (After Morris [106].)

That selection and interference are interrelated effects and probably different expressions of a competitive gut IgG-transfer was made feasible when it was seen that, in young mice at least (Fig. 9.5) [106], the rate of transmission y of IgG across the gut could be described by the relationship

$$y = ax(x + b)^{-1} \qquad (9.1)$$

where x is the oral dose of IgG administered and a and b are constants. The values of a and b for the immunoglobulins tested are given in Table 9.6. The constant a is a parameter of selection and represents the theoretical maximum transmission of a given IgG at high dosage concentration. The constant b is a parameter of interference since its value is inversely related to the capacity of an IgG to interfere with the transmission of other immunoglobulins, and is directly related to the susceptibility of an IgG to interference by others (Fig. 9.6). These conclusions were

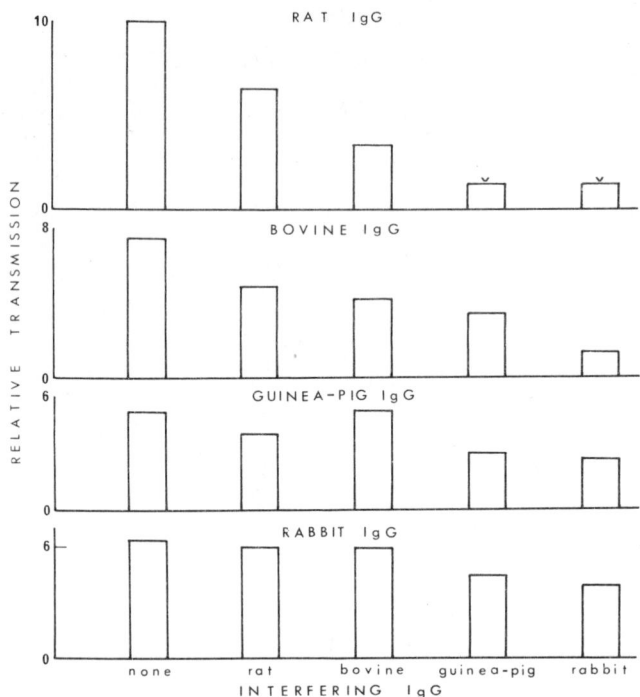

Fig. 9.6. Interference with the absorption of IgG in the gut of 11-days-old mice by other immunoglobulins. Mice were fed with 64 μg of test IgG per g body weight, alone or mixed with an equal amount of another IgG. The concentrations of the test IgG in their sera were determined 4 hours after feeding. (After Morris [106].)

substantiated by observations on the transmission of several other antibodies [106]. The expression in equation (9.1) is formally identical with the Langmuir adsorption isotherm or the Michaelis-Menten kinetic relationship.

9.3.4 Histology and cytochemistry

When macromolecular substances such as horseradish peroxidase [134, 135] or radioactive colloidal gold [136], PVP [117] or IgG (R. E. Jones, personal communication) are fed to suckling rats their absorption from the gut is limited to the lower jejunum and ileum. The mucosal absorptive cells in these regions are columnar in type, with their nuclei situated basally (Fig. 9.7a). They are characterized by the presence within them of supranuclear vesicles which become progressively larger in cells higher up the villi from the crypts. The larger vesicles are

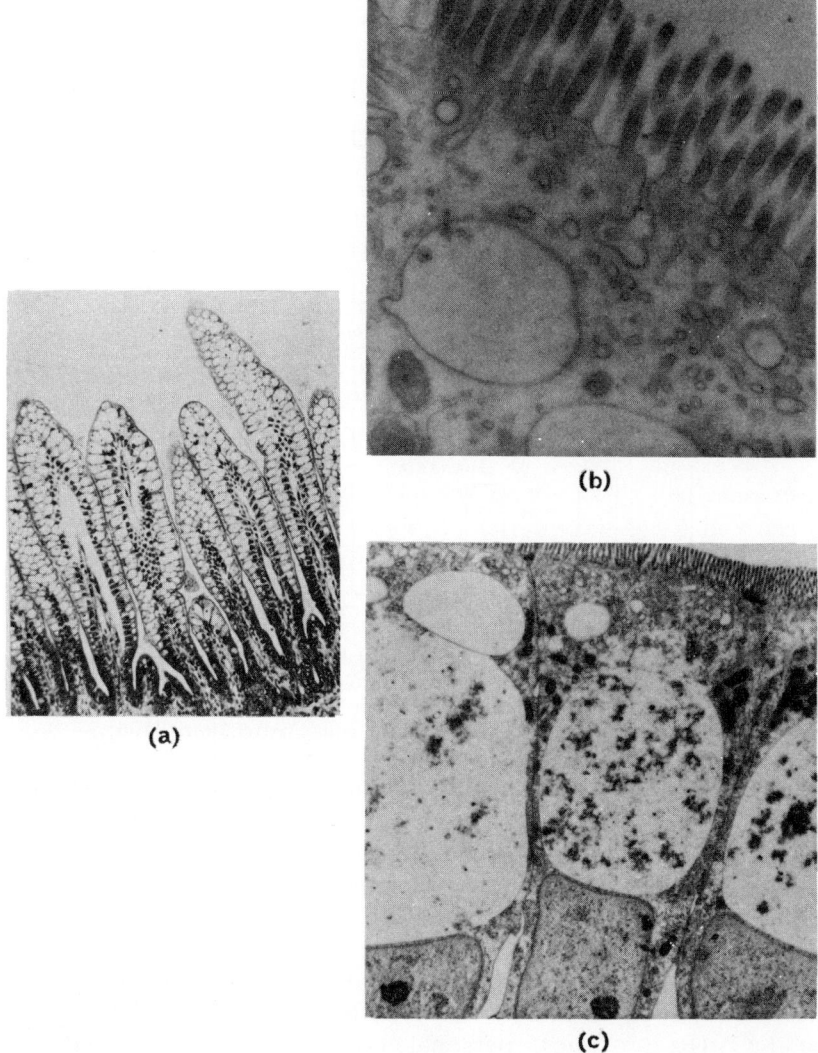

Fig. 9.7. (a) An intestinal villus from an 18-days-old rat showing the development of phagocytic vesicles within the absorptive cells (×65). (b) The apical tubular system of the ileal absorptive cells communicates with the intestinal lumen through caveoli between the bases of the microvilli (×20,000). (c) Luminal material sequestered into the tubular system is transported by vesicles into the giant supranuclear vesicle (×4000). (Photographs by courtesy of Dr. M. Taylor of the Charles Salt Research Centre, Oswestry, England.)

presumably formed by the fusion or merging of smaller ones as the cells continue to absorb luminal material on their migration up the villi. They impart to the intestine a distinctive yellowish-brown colour which disappears, with the vesicles, after closure [119].

An electron microscope study of the entry of ingested macromolecules into the intestinal absorptive cells in suckling rodents was first undertaken by Clark [137]. Subsequently, several other studies were carried out [92, 93, 96, 97, 134, 135, 138-144]. Adjacent cells are connected by desmosomes and have their lateral surfaces closely apposed above the level of the nucleus, but elsewhere the intercellular space is dilated (Fig. 9.7c). The free surfaces of the cells are raised into a well developed brush border of microvilli. An extensively anastomosing system of tubules and small vacuoles occupy the region of the terminal webb and communicates with the intestinal lumen by occasional caveoli at the bases of the microvilli (Fig. 9.7b). Between this system and the nucleus the cytoplasm is largely occupied by a giant vesicle, apically to which there are several smaller vesicles, mitochondriae and membrane-lined dense bodies considered to be lysosomes (Fig. 9.7c). The smaller vesicles and dense bodies sometimes appear to be discharging their contents into the giant supranuclear vesicle. The Golgi complex lies apico-lateral to the nucleus. After the ingestion of electron-dense materials such as Evans blue-protein complex, saccharated iron oxide, colloidal gold [137], trypan blue-protein complex [93, 137], ferritin [96, 139, 143] or peroxidase [134, 135, 142], dense particles appear between the microvilli and in the caveoli and apical tubular system at first, and later in the cytoplasmic vesicles and giant vesicle. Evidently, absorption into the cell is non-selective. Clark [137] believed that macromolecules and particles were absorbed into the cells pinocytotically by the invagination and pinching off of the cell surface membrane at the bases of the microvilli to form vesicles which then moved through the terminal webb into the cell. The occurrence in the vesicles of alkaline phosphatase [135, 138, 141, 145], which is normally associated with the surface membrane [145, 146], would support this contention. More recently, however, it has been suggested that the apical tubular system and caveoli are a relatively stable system and that entry of macromolecules into them occurs by diffusion [143]. Movement of sequestered macromolecules thence to the giant vesicle may occur through the pinching off of small vesicles from the apical tubular system or may involve a separate vacuolar system. There is morphological evidence to support the latter contention [144]. Vacek [138] suggested that the vacuolar system represented a certain form of the endoplasmic reticulum; conceivably they may represent swellings within the microtubular system extant also in the cells of mature animals [147, 148].

The giant vesicle and adjacent bodies contain acid phosphatase [93, 97, 134, 135, 138, 141, 142] and other lysosomal enzymes [93, 135], and can therefore be regarded as phagolysosomes. Jeal [134, 142] found that the acid phosphatase was concentrated most prominently in small granules forming a band around the apical surface of the giant vesicle in positions corresponding to the sites of dense bodies or lysosomes in electron micrographs. The enzyme also occurred within the giant vesicle (Fig. 9.8a). After feeding horseradish peroxidase to

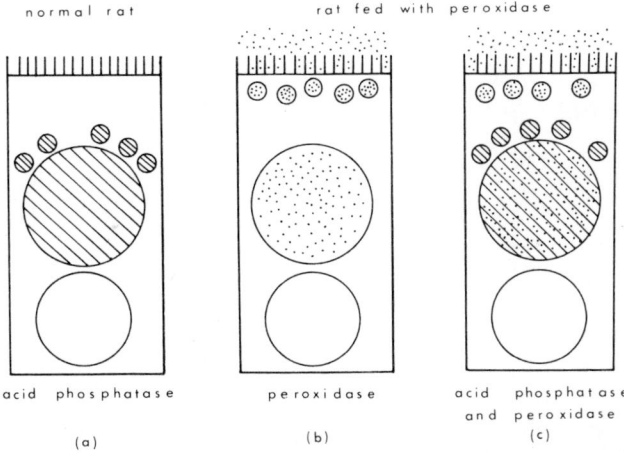

Fig. 9.8. The localization of acid phosphatase (cross hatching) and horseradish peroxidase (dotting) in the ileal mucosal cells of 15-days-old rats fed with peroxidase. The larger circles in each cell represent the basal nucleus and the giant vesicle above it. The smaller circles represent lysosomes near the giant vesicle or pinocytotic vesicles beneath the apical brush border. (After Jeal [134].)

suckling rats, this enzyme could be detected, by the benzidine-blue reaction within the ileal mucosal cells in a narrow band of deeply staining granules just beneath the brush border. The enzyme was also present at a much lower concentration within the giant vesicle (Fig. 9.8b). When sections of ileum from peroxidase fed rats were stained conjointly for acid-phosphatase and peroxidase, it was found that the giant vesicles in the mucosal cells stained clearly for both enzymes. The smaller vesicles towards the apex of the cells did not appear to contain acid phosphatase but this enzyme was concentrated, alone, in a few granules above the giant vesicle (Fig. 9.8c). This distribution of the two enzymes would suggest that phagosomes and lysosomes fuse independently with the giant vesicle. This work has been substantiated by similar studies [135] and other work [93] using trypan blue-protein complex as substrate.

Unfortunately, the investigations hitherto discussed employed marker substances which are not transmitted across the young rodent intestine. Although the results illustrate the uptake of macromolecules into and possibly their fate within the absorptive cells, they furnish no information concerning the exit route of transmitted protein. The giant vesicle may be in communication with the dilated intercellular space via the Golgi complex [92, 96]; a Golgi associated enzyme, thiamine pyrophosphatase, has also been detected in the giant vesicle [135]. The localization of absorbed IgG, anti-peroxidase antibodies or anti-ferritin antibodies within the ileal absorptive cells of fed young rats has been determined in tissue sections with fluorescein-labelled anti-IgG antibody [137] or with the electron-dense peroxidase [97] or ferritin [140]. These studies also failed to provide convincing information concerning the exit route of transmitted IgG but showed that, whereas the absorbed proteins were evenly distributed throughout the smaller apical vesicles, they were concentrated towards the periphery of the giant vesicle as though attracted to its wall. This distribution differs from that found in the suckling pig where absorbed fluorescein-labelled IgG entirely fills all of the vesicles (Fig. 9.13).

9.3.5 *Absorption in hedgehog*

The transfer of maternal antibodies to the young hedgehog was studied by immunizing females with killed *B. abortus* or *Salmonella pullorum* organisms before mating them [149-151]. Agglutinins against *B. abortus* were not transmitted in detectable amounts before birth and only at a very low level after birth, but agglutinins against the S. *pullorum* were transmitted before birth and more readily afterwards. The route of transmission before birth is not known. Transmission via the milk continued for up to 41 days, several days beyond the time when solid food is first ingested. The titres of the *S. pullorum* agglutinins in the colostrum were half of those in the maternal serum and this level was maintained in the milk throughout lactation. The titres of the *B. abortus* agglutinins in the colostrum equalled the titres in the maternal serum but fell to a quarter of this level in the milk after 4 days and then remained constant throughout lactation. These differences suggested that the agglutinins against the two antigens were associated with different immunoglobulin classes and that the mammary gland of the mothers and the intestine of the sucklings could discriminate between them. Later work [152] supported this contention, since the anti-*B. abortus* agglutinins in hyperimmune hedgehog sera were found to be predominantly (62%) β-globulins (IgG_1 or IgM), whilst the *S. pullorum* agglutinins in hyperimmune hedgehog sera were predominantly (80%) γ-globulins (IgG_2). The gut of the newborn hedgehog can discern more subtle differences than this in immunoglobulin structure, since it selectively transmits IgG_2 antibodies from orally administered antisera prepared in different species against the same antigen [153]. Homologous antibodies are transmitted preferentially, then guinea pig or rabbit antibodies and finally rat antibodies. A premature closure could not be induced in young hedgehogs by the administration of cortisone acetate [154]. However, closure in the hedgehog, as in rodents, has been attributed to the maturation of the luminal digestive function which occurs between the 4th and 5th week, when the stomach pH falls sufficiently low to activate pepsin [20]. Pepsin is present in the stomach wall in appreciable quantities at 9 days of age and attains adult levels by 28 days of age.

9.4 ABSORPTION IN CARNIVORES

Immunoglobulins could not be detected in the sera of full term kittens by electrophoresis [155]. The transmission, if any, of these proteins from the mother to the foetus must therefore be of a very low order. Agglutinins against *Salmonella montevideo* were transmitted to the foetuses from hyperimmune mother cats but attained only 0.07 of the maternal serum concentration by term [156]. The route of transmission remains unknown. Although the maternal colostral titres were only an eighth of the serum titres, the titres in the sera of the sucklings had increased fivefold in 24 hours; the milk titres had decreased by half in this time. These results reflect an efficient mechanism for absorbing antibodies in the newborn intestine. The milk titres fall by another 90% during the following week but the serum titres in the growing young remained constant, thus suggesting continued absorption for some time after the first day. Large vacuoles within the absorptive cells of the small intestine of late foetal and neonatal cats reflect the pinocytotic activity of these cells [157]. After suckling, these vesicles and the lacteals in the lamina propria contained colostral material.

The mink also lacks immunoglobulins in its serum at birth [158] but acquires them from the milk, attaining almost maternal serum levels by 8 days of age in spite of the lower concentrations of the immunoglobulins in the mammary secretions. Transmission across the intestine must be efficient during this time.

Young dogs receive their passive immunity from the mother before and after birth but the proportion received prenatally is very low [26, 159-161]. The route of prenatal transmission is not known, but it must be selective in that some antibodies [26, 159, 160] but not others [162] are transmitted. Immunoglobulin-G_1 and IgG_2 are concentrated in the colostrum to between 1 and 4 times the maternal serum level [49, 163], but their concentrations in the milk fall rapidly with suckling [49]. Immunoglobulin-A is a minor component in the colostrum though present at 14 to 80 times the maternal serum concentration. Later it becomes the major immunoglobulin in the milk at 5 times the maternal serum level. The concentrations of IgM in the colostrum and milk remain constant at half of the maternal serum concentrations. Antibodies against *Salmonella typhi,* diphtheria, *Escherichia coli* [159] and canine hepatitis virus [161] are concentrated in the colostrum to higher than serum levels but their concentrations fall rapidly below serum levels with suckling. Their serum to milk concentration ratios identify these antibodies as IgG_1 or

IgG_2. They are transmitted to the suckling puppy rapidly, and maximum circulating titres are attained after 3 days [160, 161], although most of these are transmitted during the first day [26]. It was not known whether this pattern of transmission reflected the changing levels of the antibodies in the milk or whether the gut of the newborn lost its ability to transmit antibodies soon after the first day of life. The results of Gillette and Filkins [164] support the latter possibility. Doses of homologous anti-*S. pullorum* sera were fed to newborn puppies at intervals after birth and the agglutinin titres attained in their sera estimated 5, 10 or 15 hours later. The efficiency of absorption increased from birth until 6 to 8 hours later and then declined and ceased by 24 hours. Absorption was a relatively slow process since the sera of puppies sampled 15 hours after feeding had higher titres than the sera sampled 10 hours after feeding. When fluorescein-labelled IgG was fed, it was present within the absorptive cells of the jejunum and ileum after 1 to 2 hours but did not appear in the lamina propria until 1 to 2 hours later [165]. Premature closure in suckling puppies could not be induced by injections of hydrocortisone presumably due to the delay (longer than 24 hours) before the expression of hormonal effect, but puppies born to mothers treated 24 hours prepartum with ACTH or hydrocortisone absorbed antibodies from the gut less efficiently than normal puppies [164]. The effect of diet on closure was also investigated [165]. Antibody absorption ceased in suckling puppies at 24 hours, but persisted for up to 12 hours more in puppies deprived of colostrum and starved or maintained on a sucrose or synthetic protein diet. However, the colostrum-deprived puppies had been surgically delivered near term and the extended intestinal permeability in them was attributed to their slight immaturity at delivery. The authors concluded that closure in the puppy was an age-dependent or a hormonally controlled phenomenon rather than a diet-dependent one.

Hitherto, nothing is known about the transmission of IgA or IgM in the puppy. Although the intestinal permeability to IgG is limited to the first day of life, the absorption of partly digested proteins or small molecular weight milk components may continue for some time further as reflected in the marked proteinuria which occurs for the first 4 to 6 days and which persists at lower levels for another 8 days [58].

9.5 ABSORPTION IN RUMINANTS

9.5.1 Homologous immunoglobulins

The transmission of passive immunity does not occur before birth in the ruminants [22]. They are hypogammaglobulinaemic at birth and derive their immunoglobulins upon the ingestion of colostrum. The immune-lactoglobulin constitutes 76-80% of the proteins in the whey of bovine colostrum [63, 166], and most of it is IgG_1 (Table 9.4). Low molecular weight (S < 3) components, including the α-lactoglobulins, β-lactoglobulins, and lactalbumin are also present in considerable amounts [167]. Ingested colostrum is clotted in the newborn calf's stomach by the gastric renin, and the whey proteins pass rapidly into the small intestine. Luminal digestion of colostrum is limited by the buffering action of the secretion itself, by the low peptic [168, 169] and tryptic activities [170] of the gut and by the activity of the trypsin inhibitor of the colostrum [83]. The slight digestion occurring in the abomasum (probably due to rennin rather than pepsin) and duodenum is insufficient to effect a difference between the absorption of orally administered ^{131}I-labelled IgG or indigestible PVP of 160,000 molecular weight (Fig. 9.9) [171]. This result also emphasizes the non-selective nature of transmission in the calf. Colostral immunoglobulins are absorbed in the terminal part of the small intestine [172] and are transmitted across the absorptive cells without degradation [171, 173]. They reach the systemic circulation via the lymphatics and thoracic ducts [173, 174]. Within a few hours, immunoglobulins appear in the circulation of the suckling calf to approach or equal the maternal serum concentrations [167, 175-180], and the serum protein profile approximates closely to that of the colostrum whey [36, 62]. The IgG and IgM appear in the circulation in the same proportions as those in the ingested colostrum [36, 180]. This lack of selectivity in absorption (Fig. 9.10) also applies to IgG_2 which is present in colostrum in minute quantities (Table 9.4); when serum is fed to suckling calves, components S2 and S3 (subsequently identified as IgG_1 and IgG_2 respectively [181]) are transmitted equally to the circulation [63]. In suckling calves, the smaller molecular weight proteins of the colostrum [182, 183], and immunologically reactive antibody fragments of intestinally degraded

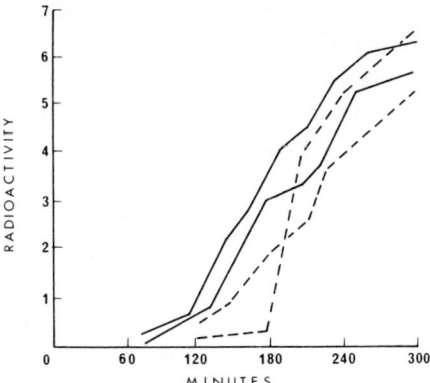

Fig. 9.9. The intestinal absorption in newborn calves, of ^{131}I-labelled PVP (M.Wt.=160,000) ——— or bovine serum IgG --- infused with cow colostrum into the duodenum. The blood radioactivity is expressed as a percentage of the radioactivity of the preparation administered. (From Hardy [171]. Courtesy *J. Physiology*, Editorial Board.)

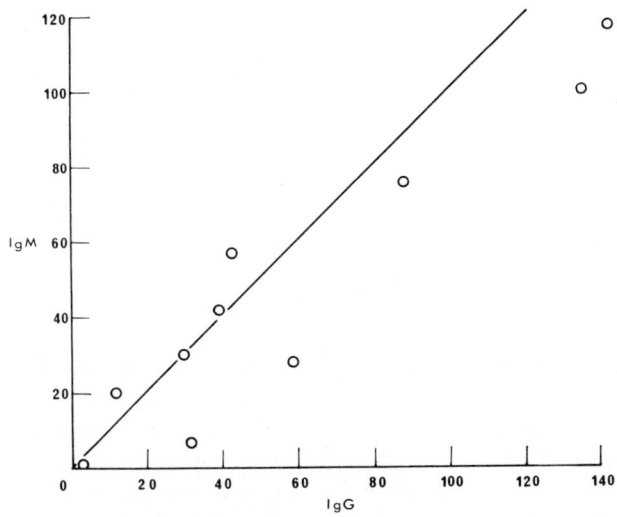

Fig. 9.10. The correlation between the amounts of colostral IgG and IgM absorbed in suckling calves during the first 24 hours of life. The serum concentrations attained are expressed as percentages of the concentration in the colostrum ingested. The line of unit slope represents the relationship for equal absorptive efficiency for IgM and IgG. (After Klaus, Bennett and Jones [180].)

colostral antibodies [183] are also absorbed and transmitted to the circulation, but these components are rapidly filtered by the kidney to appear in the urine. Proteinuria persists in the calf for as long as the intestine remains permeable to proteins [58].

Antibodies appear in detectable amounts in the sera of calves 1 to 2 hours after they begin to ingest immune colostrum [29, 36, 184-187]. This delay cannot be accounted for entirely by the time it takes for the ingested colostrum to reach the distal small intestine, nor by the insensitivity of the antibody titration techniques employed, since a delay of this order also occurs before the appearance in the thoracic duct lymph of IgG infused directly into the duodenum [173, 188]. The delay varies directly with the molecular weight of the transmitted protein [188]. The absorbed colostral antibodies in suckling calves attain maximum circulating titres after 5 to 6 hours according to some authors [29, 184] or 18 to 24 hours according to others [185, 187]. Stone [36] found a wide range of variation (4 to 16 hours) in this respect, and suggested that it could be due to the varying amounts of colostrum ingested. Transmission has virtually ceased within 24 hours of birth [174].

Many authors have studied the effects of diet on the absorption of antibodies in the newborn calf, with variable results. Some [29, 189], but not others [190, 191], found that previous feeding with proteins reduced absorption. Some authors failed to delay closure by depriving calves of colostrum and starving them [175] or maintaining them on sugar solutions given orally [179, 182, 192] or parenterally [182], whereas others succeeded [191]. Feeding immunoglobulins with antidigestants to postclosure calves [182], or infusing colostrum directly into their small intestines [190], failed to induce absorption. Thus the onset of an efficient luminal digestion is not solely responsible for closure.

The main colostral immunoglobulin presented to the suckling lamb and kid for absorption is also IgG_1, since this protein amounts to 90% of the colostral immunoglobulins, the IgM and IgA respectively amounting to only 6% and 4% [45]. Relative to the maternal serum levels, the colostral levels of IgG_1 and IgM are 3 to 4 times higher, and the IgA level 5 to 40 times higher (Table 9.4). Several authors have described the transmission of colostral antibodies and immunoglobulins to the newborn lamb [58, 193-199] and kid [58, 172, 178, 182, 194, 200]. Transmission resembles that seen in the calf in efficiency, duration and route and in the occurrence of a transient proteinuria. The antibodies investigated were IgG_1 or IgM types according to their relative concentrations in the maternal serum and colostrum. Both types are transmitted equally in the lamb [201]. In starved lambs the intestinal permeability to immunoglobulins can be maintained for at least 24 hours beyond the normal time [110], indicating that closure is normally not an age or birth dependent phenomenon as it appears to be in the dog, but a dietary dependent one. The lamb resembles the young pig in this respect. Dietary induced closure is an irreversible process. Absorption after closure cannot be restimulated after a period of starvation [110]. However, hormonally induced premature closure may have been induced in lambs born to mothers experiencing severe climatic stress, since the permeability of the intestine to colostral immunoglobulins was seriously impaired in these lambs [202].

9.5.2 Selection

The inability of the newborn calf and lamb intestine to discriminate between homologous immunoglobulins extends to a variety of other macromolecules. All the main electrophoretic fractions of bovine colostrum [182, 203] or serum [203], conalbumin and ovalbumin [182], bovine and human albumin [204], human and caprine IgG [182], gelatin [192], insulin [205], dextran [204] and PVP [188], are absorbed readily by the newborn calf. The lamb intestine is permeable to PVP, hen's egg proteins and bovine colostral proteins [110]. Intestinal transmission in the newborn kid is also qualitatively non-selective [172, 178]. Dextran was absorbed more readily in the calves when it was administered in colostrum, indicating that factors in the colostrum facilitated intestinal transmission. Variation in the concentrations of such factors in oral test doses given to young calves may account for the more rapid absorption of colostral IgG than gelatin [192], since the gelatin was given in boiled milk. Strictly quantitative comparisons of the absorptive efficiency of the intestine towards different proteins should be made using substrates in a standardized solution or by administering them together. The latter was done by Bangham and colleagues [203] using ^{131}I-labelled serum or colostrum. In fed calves, the relative proportions of the electrophoretic fractions transmitted were similar to those in the preparation administered (Fig. 9.11). This was most evident shortly after

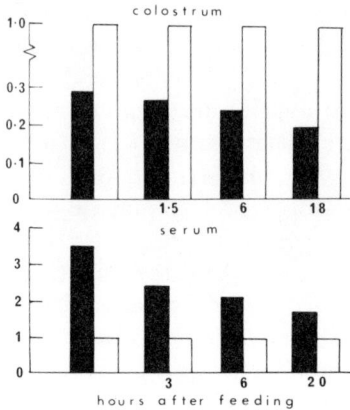

Fig. 9.11. Changes in the relative circulating concentrations of albumin ■ and globulins □ in calves fed with ^{131}I-labelled colostral or serum proteins. The ratios in the preparations administered are shown on the extreme left. (From Morris [127] after Bangham and others [203]. Courtesy American Physiological Society.)

feeding, but not thereafter due to the variation in the blood clearance rates of the different fractions. Despite the non-selective nature of transmission in the young ruminant, there is a selective route of entry of protein into the circulation after its passage through the absorptive cells. Low molecular weight proteins, dextran or PVP pass directly into the portal circulation as well as into the lacteals [188, 204, 205]. Factors in the colostrum affect transmission by these routes differently (*vide infra*) and in this way may provide a basis for selection according to molecular size of substrate.

9.5.3 *Cytology*
Detailed cytological studies on colostrum absorption in the young ruminant are lacking. Information is limited to light-microscopic observations for the calf [206], lamb and kid [86, 207]. During the transmission of passive immunity in these species, the absorptive cells of the jejunum and ileum contained cytoplasmic globules of varying diameters seemingly consisting of absorbed colostral proteins. Material with similar

staining characteristics sometimes occurred in the lacteals as well. From the size and distribution of the globules in the lamb and kid, it appeared that colostrum was absorbed in a finely divided form and larger globule formation occurred later within the cells. In cells containing several globules the nuclei became displaced towards the luminal borders. Absorption in the ruminants thus resembles the process in the pig (*vide infra*) more closely than that in the rodent.

9.5.4 Absorption accelerators

Dextran fed to newborn calves was absorbed more readily when it was given in colostrum [204]. The accelerator activity of colostrum was attributed to a low molecular weight, heat stable protein fraction which was effective only in the presence of inorganic phosphates and glucose-6-phosphate [173]. These components were isolated from colostrum and their effects on the absorption of ^{131}I-labelled IgG from a chloride solution infused into the duodenum investigated (Fig. 9.12). Although

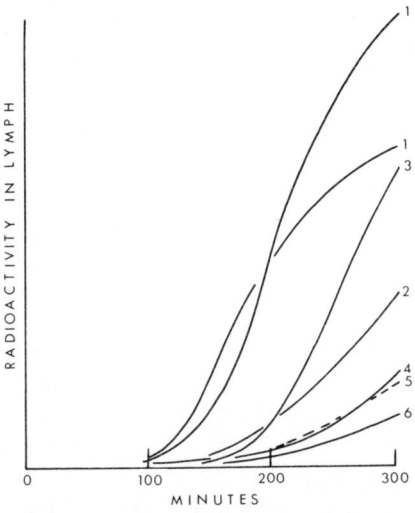

Fig. 9.12. The recovery of ^{131}I-labelled bovine IgG from the thoracic duct lymph of calves infused duodenally with a protein solution also containing 1, colostrum whey filtrate with small protein accelerator; 2, whey filtrate without protein accelerator; 3, small protein accelerator, inorganic phosphates and glucose-6-phosphate; 4, small protein accelerator and glucose-6-phosphate; 5, small protein accelerator and inorganic phosphate; 6, small protein accelerator. (From Morris [127], after Balfour and Comline [173]. Courtesy American Physiological Society.)

the combination of all these components simulated absorption from colostrum, the full accelerator effect of colostrum could not be achieved, thus indicating the existence of other unrecognized colostral accelerators. How the small protein accelerator worked was unknown but the authors suggested three possibilities: (a) It acted as a surface active agent to enhance pinocytotic uptake into the intestinal cells. (b) It combined, after absorption, with a mucopolysaccharide component of the intestinal cell vacuoles to facilitate their discharge on the lacteal side. (c) It possessed an intracellular function not directly associated with the absorption of immunoglobulins. Another possibility was considered by Hardy [188], who suggested that the small molecular weight proteins of absorbed colostrum acted osmotically to induce a higher rate of lymph flow through the lacteals for more efficient removal of absorbed proteins. Whatever the mechanism concerned, the small protein accelerator was not directly responsible for absorption and its effects were limited to the normal period of intestinal permeability [173]. The phosphates possibly operated through the intracellular metabolism, to provide energy for the active transport of the IgG in vesicles across the absorptive intestinal cells. Glucose and lactose were ineffective in this respect.

The effects of other accelerators in bovine boiled colostrum whey on the intestinal transmission of IgG, albumin or PVP, infused into the duodenum in newborn calves, were studied by Hardy [188]. The factors were equally effective in accelerating the transmission of IgG and PVP of 160,000 molecular weight. Their effects could be closely simulated by giving chloride solutions of the substrates with simple anions such as lactate or pyruvate. Acetate, formate, propionate or isovalerate were slightly less effective, but butyrate or isobutyrate were considerably more effective. Chloride, bicarbonate or citrate lacked accelerator activity. However, it was unlikely that any of these effective anions contributed significantly to the accelerator activity of colostrum whey, since none of them was present in sufficient quantity to be effective. A study of their action was considered profitable since they presumably function in a similar way to the unidentified colostral accelerators. In newborn calves, the absorption of 160,000 molecular weight PVP into the intestinal absorptive cells from an aqueous solution infused into the duodenum was considerable even in the absence of any accelerators, but the release of PVP into the lacteals was prevented. Later infusion of boiled colostrum whey into the duodenum of these animals, or of lactate intravenously, resulted in the rapid appearance of PVP in the thoracic duct lymph. In the former case, the appearance of PVP in the lymph preceded the arrival of the colostrum whey at the absorptive part of the small intestine. It was concluded that some of the accelerators of the colostrum whey reached the ileum via the circulation after their absorption from the proximal part of the small intestine. How the accelerators effect the release of absorbed macromolecules from the intestinal cells remains conjectural. Hardy [188] suggested, as did Balfour and Comline previously [173], that they contributed to the metabolic energy of the cells.

Hardy [188] also showed that, in the young calf, the colostral accelerators and the size of intestinally transmitted protein affected the flow of lymph in the lacteals. When boiled colostrum whey containing IgG or PVP of 160,000 molecular weight was infused into the duodenum, the lymph flow increased markedly although transiently during the first hour, and again later when IgG or PVP appeared in the lymph. Similar changes were observed when albumin in whey was administered, but the secondary increased flow, as albumin was transmitted, was markedly enhanced. When IgG was infused without whey or with butyrate, only a slight early increase in lymph flow was observed even though, with butyrate, IgG appeared copiously in the lymph. It was concluded that the early rise in lymph flow reflected the rapid absorption of water from the infusion. The secondary rise in flow was probably effected by osmotically active components of colostrum such as the small protein accelerator described by Balfour and Comline [173], and this effect was accentuated by the albumin when this was also transmitted.

9.6 ABSORPTION IN THE PIG

9.6.1 *Homologous immunoglobulins*
Prenatal transmission of maternal antibodies to the young pig, if any occurs, is clearly insignificant in comparison with postnatal

transmission [22]. Hardy [208] states that the total protein and immunoglobulin concentrations are higher in porcine than in bovine colostrum. Immunoglobulin-G accounts for 80% of the total immunoglobulins in porcine colostrum, whilst the IgA and IgM account for 14% and 6% respectively [44, 48]. A small proportion (2%) of the IgG exists as a 19S polymer [48, 209]. The IgA is produced locally in the mammary gland and ranges in size from 6S to 18S units. The IgG and IgM levels in the mammary secretions, but not the IgA level, decrease markedly during the first 3 days of lactation and the IgA becomes the predominant immunoglobulin [44]. Digestion of colostral proteins in the gastrointestinal tract in the suckling pig is considerably greater than in the newborn calf [208, 210], in spite of several inhibitory factors. The higher protein concentration of porcine colostrum and its buffering action on the gastric acidity, together with the relatively low peptic activity [211, 212], must inhibit excessive gastric proteolysis, whilst the copious trypsin inhibitor in the colostrum [213] also retards digestion [208, 214, 215] in the small intestine by the relatively high tryptic activity from the pancreas [211]. Colostral immunoglobulins are absorbed in the middle third of the small intestine [216] and transmission across the absorptive cells occurs without degradation [216] despite the presence of several lysosomes inside the cells [217]. During the first day of life, but not thereafter [218], the lysosomes either lack proteolytic enzymes or fail to intercept the absorbed proteins. The small molecular weight degradation products from ingested immunoglobulins broken down in the intestinal lumen are also transmitted in significant quantities but are later excreted in the urine [210, 214]. The serum proteins of the unsuckled piglet are mainly α-globulins, but upon suckling, colostral β-globulins and γ-globulins are rapidly absorbed until they eventually more than double the pre-existing serum protein concentrations [50, 194, 219-226]. These earlier studies investigated the transmission of the main electrophoretic fractions of the colostral proteins, amongst which the γ-globulin fraction mainly represented 7S IgG. Recently, Porter [48, 209] showed that the 19S IgG and the IgM are transmitted as readily as the 7S IgG. However, only the smaller molecular weight components of the IgA are transmitted, and Porter attributes this to the masking, by S-piece, of the structures on SIgA which correspond to cellular receptors for intestinal transmission. This explanation is difficult to reconcile with the observations that a variety of non-immunological proteins and non-protein macromolecules are also readily transmitted across the intestine when they are fed to the young pig (*vide infra*).

Absorption of colostral antibodies continues for 24-36 hours after birth though most of the transmission occurs during the first 12 hours [219, 223, 226, 227]. The absorptive efficiency of the intestine varies inversely with the amount of colostrum previously ingested and has a half-life of about 3 hours [215, 225, 228]. The falling efficiency of absorption is not wholly dependent on the diet and the gradually increasing digestive efficiency of the gut [211, 212, 229], since a similar though much less marked decline also occurs with age in fasted animals towards the absorption of indigestible PVP [229]. Nevertheless, the intestine of the starved pig will retain to some extent the ability to transmit immunoglobulins and other macromolecules for several days after birth [110, 215, 230-232]. This independence of complete closure from physiological age and birth may be of considerable survival value to the last born piglet in a litter which may appear several hours after the first born. A premature closure can be induced in the starved piglet by giving it cortisone acetate [232]. The absence of a birth dependent hormonally induced closure in piglets may possibly reflect the lower degree of stress experienced by the sow and piglets at parturition compared to that experienced by the carnivores and possibly the ruminants.

The intestinal absorption of immunoglobulins in the young pig is also dependent on accelerators in the colostrum. Payne and Marsh [233] found that IgG from an orally administered solution was not absorbed during the first day of life, and they suggested that some material within colostrum must alter the permeability of the gut wall in normally suckling pigs. Hardy [229] found that the absorption of PVP of 160,000 molecular weight (and presumably of IgG [214]) required factors present in sow colostrum. The absorption of PVP of 40,000 molecular weight (and presumably of albumin [214]) was accelerated to a lesser extent by the same or similar factors. Unlike the accelerators which are effective in the calf, the porcine accelerators are heat labile and their activity cannot be simulated by phosphate, lactate or pyruvate. Similar heat-labile accelerators effective in the piglet also occur at lower levels of activity in cow and goat colostrum and at a higher level of activity in ewe colostrum. As in the calf, the accelerators facilitate the release of absorbed protein from the intestinal absorptive cells into the lacteals.

9.6.2 Selection

Except for the SIgA, the intestine of the suckling pig cannot discriminate quantitatively between the homologous immunoglobulins it transmits into the circulation. Nor can it discriminate qualitatively between a variety of other macromolecules. The intestine is permeable to colostral immunoglobulins of bovine [231, 234, 235], ovine or human origin [231], serum immunoglobulins of porcine, ovine, equine [236] or bovine origin [237], all of the major electrophoretic fractions of porcine [230] or equine serum [238], bovine lactalbumin [235], hen's egg proteins [235], porcine lactic dehydrogenase isoenzymes [239], insulin [224] and PVP [235, 237]. Few of these studies yield information about the capacity of the intestine to discriminate quantitatively between the proteins transmitted. This capacity may be masked by (a) the varying susceptibility of unrelated proteins to intestinal digestion, (b) the variation in volume and constitution (accelerators) of the test doses, and (c) the variation in the plasma clearance rates of transmitted proteins according to their molecular size. In the work of Nordbring and Olsson [215, 230, 234, 238], variation (b) was closely controlled. The results are summarized in Table 9.7. Evidently the gut of the young pig cannot discriminate between proteins according to their species of origin since similar proteins from different species are transmitted equally. However, the recovery of IgG from the circulation of fed pigs always exceeds the recovery of other administered serum proteins. This result may be attributable to either or both of variations (a) and (c) as well as, or rather than, the ability of the intestine to discriminate between IgG and other proteins.

Pierce and Smith [210] found that when human serum albumin, porcine and bovine colostral IgG were fed separately, but under similar conditions, to newborn pigs, these proteins were transmitted equally into the circulation. When human serum albumin was fed with bovine colostral IgG it did not affect the absorption of the IgG, but when equimolecular

TABLE 9.7. The intestinal transmission of serum proteins in young pigs.

Serum fed	Volumes fed	Composition of dose fed	No. of pigs fed	Mean % of administered dose recovered in circulation*
Pig	75 ml serum and 75 ml milk replacer	1.5 g IgG 1.5 g β-globulin 2.5 g albumin	5	11.0 6.4 3.4
	75 ml serum and 75 ml milk replacer	1.0 g IgG 1.5 g β-globulin 2.0 g albumin	4	10.1 4.5 5.4
Horse	75 ml serum and 75 ml milk replacer	1.9 g IgG 1.6 g β-globulin 1.6 g albumin	16	11.3 5.7 3.5

*Determined 16 hours after feeding.
(After Morris [127] and compiled from data in Olsson [234] and Nordbring and Olsson [230].)

amounts of bovine and porcine colostral IgG were fed together the homologous IgG was transmitted preferentially, though each IgG inhibited the transfer of the other. Since marked degradation of the administered proteins occurred in the gut lumen the interpretation of the results was uncertain, but the authors concluded that the intestine could discriminate to some extent between different proteins. Segre and colleagues [236, 240] fed newborn pigs with saline solutions of porcine, ovine or equine immunoglobulins separated from the sera of animals immunized against tetanus and diphtheria. The porcine and equine antitoxins were transmitted preferentially, and the tetanus antitoxins of each species were transmitted more readily than the diphtheria antitoxins. These results were attributed to the ability of the intestine to discriminate between the antitoxins according to their immunoglobulin classes; 6.6S antitoxins were transmitted preferentially to 18S antitoxins. These results may also be attributed to factors (a) or (b) above. The test doses were followed by a non-colostral diet possibly deficient in accelerators and not conducive to the absorption of the larger molecular weight proteins. Perry and Watson [241] fed newborn pigs with albumin-diminished porcine serum labelled with ^{131}I, before allowing them to suckle their dam. Twelve hours later, analyses of their sera indicated a preferential absorption of 7S over 4.5S or 19S proteins and of 4.5S over 19S proteins. The authors suggested that the intestine discriminated between the serum proteins according to their surface configurations or the transmission mechanism at some stage involved molecular filtration. Whereas factors (a) and (c) above could account for selection between the 4.5S and the other proteins, factor (a) alone could account for the selection between the 7S and 19S proteins. From the work of Hardy [208] in which effects due to variations (a), (b) and (c) were eliminated or carefully assessed, it was concluded that selection between porcine and bovine IgG did not occur.

Most experimental variations can be eliminated in *in vitro* experiments using rings or sacs of everted intestine incubated in nutrient media containing a protein substrate. Lecce [242] found that the absorption of porcine IgG into the absorptive cells of everted rings required no other exogenous inducers. The cells absorbed protein until they became engorged. There was no evidence for the secretion of absorbed protein into the lamina propria, indicating the absence of accelerators from the incubation media. Uptake into the cells was an active, sodium-dependent process that could be reversibly inhibited by various metabolic inhibitors. In the studies of Pierce and Smith [216], everted intestinal sacs were incubated in dialysed bovine colostrum containing accelerators to facilitate protein transmission across the intestinal wall. Human serum albumin in the incubation medium markedly reduced the transfer of the bovine IgG. Given the equal opportunity on a weight basis of transferring albumin or IgG, the intestine transferred 20 molecules of albumin for each molecule of IgG. Since the transmission of bovine IgG was reduced in the presence of L-leucine (facilitated diffusion across mucosal cell membranes) or L-methionine (facilitated diffusion and active transport) but not L-alanine (active transport) or D-methionine, the authors [243] concluded that the rate of uptake of proteins into the absorptive cells depended on the specificities of the proteins for binding sites on the luminal border of the cells. Proteins bound preferentially (albumin) to the luminal membrane would be pinocytosed and transmitted the most readily and would inhibit the transfer of other proteins (IgG) the most markedly [216]. Support for this contention should preferably reflect selectivity between proteins of similar size, since the transmission of proteins across the additional barriers of the muscular layers and the peritoneum of the *in vitro* preparations could also involve intercellular seepage and pronounced molecular filtration, similar to that which is known to occur across the comparatively thin membrane of the rabbit chorion [244, 245]. However, there is evidence [246] that bovine IgG is transmitted *in vitro* more readily than porcine, equine, ovine or human IgG. Whether the intestine selected the proteins according to their species of origin or according to the porportions of IgG_1 and IgG_2 in the preparations used remains to be determined; the authors recalled the preferential transmission of IgG_1 over IgG_2 across the acinar cells of the ungulate mammary gland.

9.6.3 Cytology

Comline and colleagues [157] described the histological changes occurring in the intestinal mucosal cells of the suckling pig upon the absorption of colostral material. Within 6 to 8 hours of commencing to suckle, globules 0.5μ to 10μ in diameter, and with the same staining characteristics as the luminal contents, appear in the apical parts of the cells, with the nuclei occupying a basal position. Nine hours later the position of the globules relative to the nuclei are reversed, and material with similar staining characteristics as the globules is detectable within the lacteals. Studies on the absorption of radioactively-labelled or fluorescein-labelled proteins indicate that protein is absorbed in a finely divided form at the luminal border of the mucosal cells [231, 232, 236]. The resulting globules sink deeper into the cells and fuse or merge together to form larger globules (Fig. 9.13). Absorption into the cells is qualitatively non-selective. Porcine, equine, bovine and human IgG, porcine and bovine β-globulins and bovine α-globulin are readily absorbed [231, 232]. In these studies, Hardy [229] maintains that the occurrence of giant globules in the basal part of the absorptive cells several hours after the ingestion of labelled protein is probably due to the engorgement of the cells by retained absorbed protein in the absence of accelerators. The release of protein into the intercellular spaces probably occurs from smaller globules in the region of the Golgi complex, as

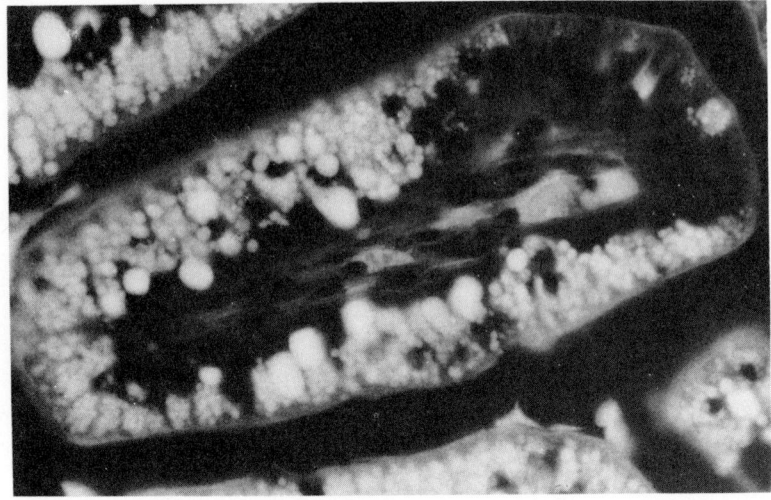

Fig. 9.13. Section through an intestinal villus taken from a newborn pig 5 hours after feeding with fluorescein isothiocyanate-labelled IgG. (Photograph by courtesy of Dr. L. C. Payne, University of Nebraska, U.S.A.)

suggested by more recent ultrastructural studies [247-249]. Unfortunately these studies relate to the prevalently lipid-absorbing duodenum or proximal jejunum and not to the mainly protein-absorbing distal small intestine. In unsuckled pigs or in suckled pigs after closure, the absorptive cells closely resemble those in adult pigs except that the Golgi complex changes from a pre-colostral infranuclear position to a post-colostral supranuclear one. Before suckling, the microvilli are long and slender and rootlets from them pass into an organelle-free terminal web zone in the apical cytoplasm with only a few pinocytotic vesicles and caveoli above it and in between the bases of the microvilli respectively. Between the terminal web and the central nucleus the cytoplasm is filled with large mitochondriae interposed with rough endoplasmic reticulum and a few dense bodies. The microvilli and caveoli have an extraneous [247] or hirsute [248, 249] layer which is poorly developed in comparison with post-colostral stages [248]. The intercellular spaces are undilated [247]. During colostrum absorption drastic changes occur in the cells. Initially an extensive anastomosing tubular system located within and beneath the terminal web originates by invagination of the cell membrane at the bases of the microvilli [249]. Bulbous, colostrum-containing end-pieces from this system are pinched off to form vacuoles which sink deeper into the cells. These fuse together into larger vacuoles which course deeper into the cell past the nucleus (Fig. 9.14). The extraneous, hirsute layer is also incorporated into the vesicles whose cytoplasmic surface appears to have bristles [247] or spinelike extensions [248] of

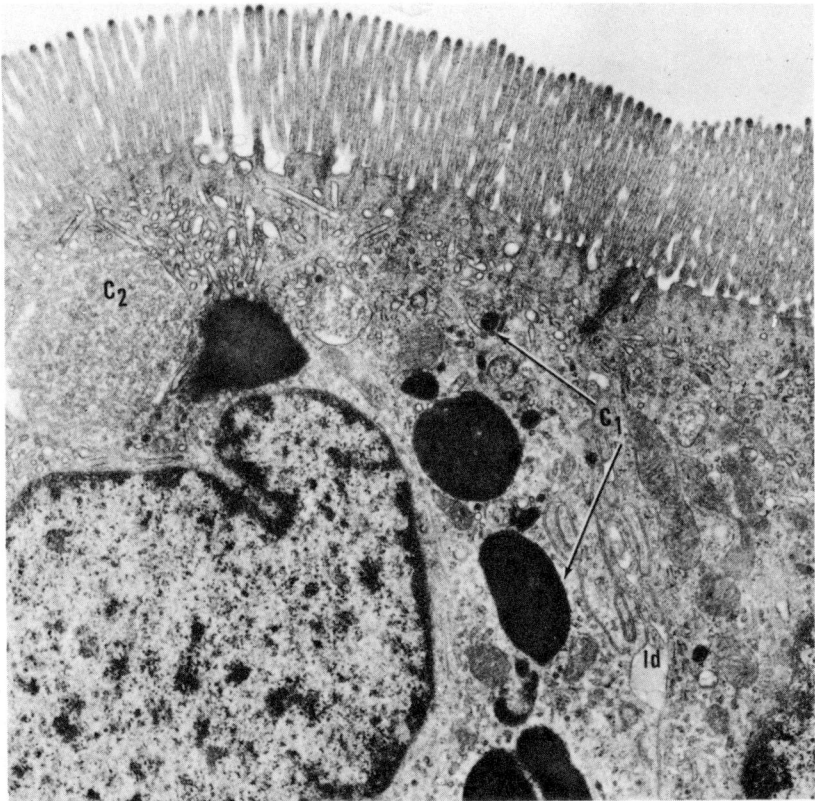

Fig. 9.14. Duodenal absorptive cell from a newborn pig one hour after receiving colostrum. An apical tubular system connects with the lumen by caveoli between the bases of the microvilli. Colostrum-filled bulbous end pieces of this system are pinched off to form vacuoles which sink deeper into the cells and course past the nucleus towards the basal part ($\times$ 16,500). (From Staley, Jones and Corley [249]. Courtesy American Veterinary Medical Association.)

unknown function projecting into their lumina. Protein and lipid appear to enter the cells in distinct vacuoles and only lipid vacuoles can definitely be seen discharging their contents by a reversal of pinocytosis at the lateral and basal parts of the cells [249]. Some of the larger colostrum-containing vesicles have small bud-like protrusions seemingly being pinched off into the surrounding cytoplasm, whilst at the base of the cells or lateral to the nucleus small colostrum containing vesicles appear to be in transit to the cell membrane. There is suggestive evidence that protein transfer from the larger vacuoles into the intercellular spaces occurs via the Golgi complex (Fig. 9.15).

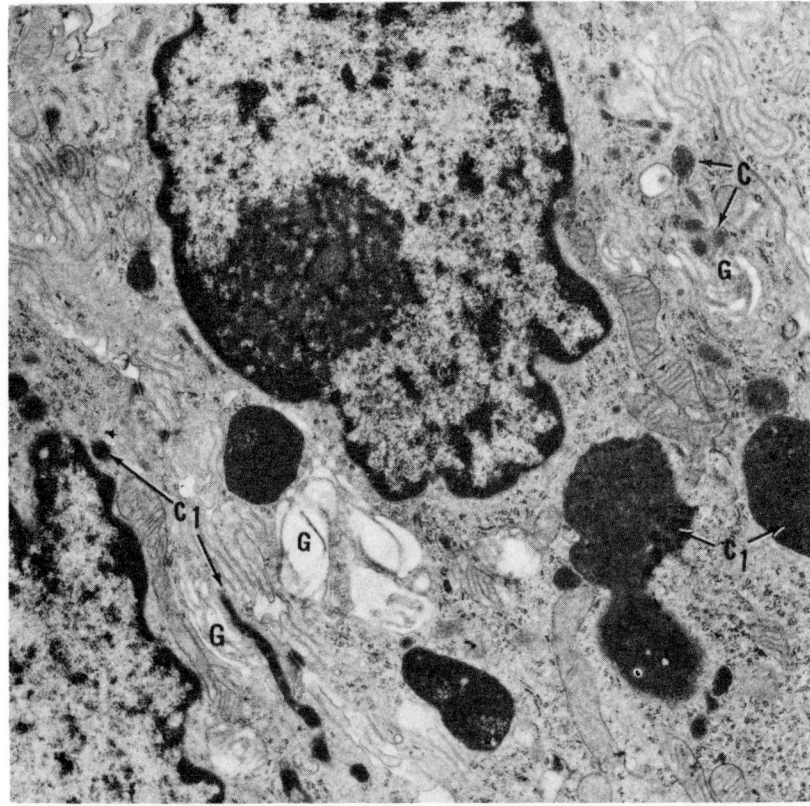

Fig. 9.15. Duodenal absorptive cells from a newborn pig 16 hours after it began to suckle. Colostral protein C1 occurs in the central (right) and lateral (left) cisternae of the subnuclear Golgi complex (X 13,000). (From Staley, Jones and Corley [249]. Courtesy American Veterinary Medical Association.)

The intercellular spaces become greatly distended and contain colostral protein [247]. In suckling pigs the sub-apical tubular system persists for at least 36 hours although the degree of pinocytosis decreases markedly by this time. After 48 hours the cells assume the adult state with the loss of the tubular system, a change in position of the Golgi complex [249] and the disappearance of the intercellular spaces [247]. The change in position of the Golgi complex may reflect a role in the control of absorption of large molecular weight proteins at the cellular level [248, 249]. Kraehenbuhl and Campiche [97] studied the absorption of rabbit IgG antibodies to horseradish peroxidase by the newborn pig jejunal cells, and the results agree closely with those already described for colostrum absorption. Acid phosphatase activity, indicative of lysosomal enzymes location, could not be detected in the Golgi areas nor in the antibody-containing vacuoles.

9.7 CONCLUSION

Animals which are known to receive some or all of their passive immunity from the mother after birth fall into two main groups—those in which transmission is intense and completed within a few hours of birth, and those in which transmission is a more protracted process which may take several days or even weeks for completion. The ungulates and carnivores belong to the first group. In the young of these species, considerable amounts of the ingested colostral proteins reach the absorptive parts of the small intestine intact despite the comparatively well developed state of the luminal digestive system. Proteins are then absorbed rapidly and non-selectively by the mucosal cells and are transmitted across them without degradation. The release of absorbed proteins from the cells is facilitated by circulating colostral factors previously absorbed in the upper small intestine. Closure of the intestine to the transmission of protein is a phenomenon dependent on birth or diet and ensues soon after the acquirement of the full complement of passive immunity. Rapid closure and the onset of efficient luminal digestion are essential to prevent the entry of excessive amounts of foreign proteins into the circulation and to provide immediately a supply of amino acids. In rodents and insectivores, which belong to the second group, the transmission of passive immunity to the newborn is most intense during the first few days of life when, in the rat at least, it may resemble transmission in the ungulates and carnivores in its lack of selectivity towards colostral and milk proteins [109]. Due to the extremely rapid growth of young rodents and insectivores, the transmission of immunity is necessarily prolonged in order to maintain in them useful levels of circulating antibodies. The luminal digestive efficiency remains at a low level to facilitate continued intestinal absorption of milk antibodies, and the absorptive mucosal cells assume the role of digestion due presumably to the rising levels of their lysosomal proteases [119] (also, R. E. Jones, 1970, personal communication). These cells also become more selective towards the type of proteins they transmit. Jeal [134] suggested that the lysosomal system of the ileal mucosal cells in the young rodent is designed primarily to provide the developing animal with adequate supplies of amino acids at a time when the gut luminal digestion is inefficient, and that the antibody transfer mechanism may be a secondarily developed modification of this intracellular digestive system. Williams and Beck [93] have also expressed a similar opinion, and their work goes some way to substantiate it.

A hypothesis to account for the selective intestinal transmission of antibodies in the young rodent and for the experimental phenomenon of interference has been postulated by Brambell and colleagues [22, 132]. This hypothesis, modified in the light of the foregoing discussion, is shown schematically in Fig. 9.16. Ingested proteins enter the absorptive mucosal cells by non-selective fluid endocytosis. The resulting endocytic vesicles move deeper into the cells and fuse with others to form larger ones which ultimately discharge their contents into the giant supranuclear vesicle. The latter also receives the contents of secondary lysosomes. Within the endocytic vesicles and the giant vesicles, absorbed IgG molecules display a reversible affinity for the walls of these structures by virtue of attachment sites on the Fc portions of the H chains, and whilst attached to the walls they are protected from digestion by lysosomal proteases. Under experimental conditions, IgG molecules from other species also show an affinity for the receptors on the walls of the vesicles since they resemble the homologous IgG

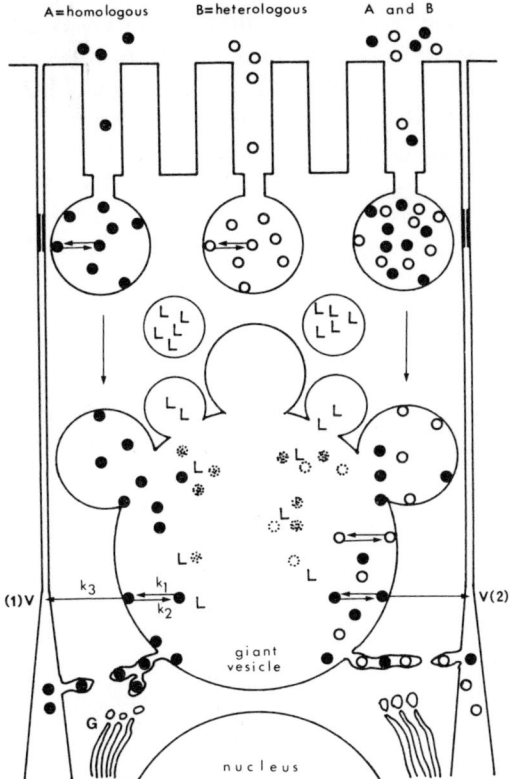

Fig. 9.16. A schematic theoretical representation of the transmission of IgG across the ileal mucosal cell of the suckling rodent. The transmission rates are represented by the relationships:

(1) $V_A = k_3 A Sn[A]([A] + K_A n)^{-1}$,
(2) $V_A = k_3 A Sn[A]([A] + K_A n \; [B] K_A/K_B)^{-1}$,

where S = total surface area bearing receptors, n = density of receptors on these surfaces, [] = concentration, and $K = (k_2 + k_3)/k_1$. The letter L denotes lysosomal protease.

molecules over extensive regions of their H chains (see Table 9.3). Hemmings [250] suggested that some of the tyrosine residues of the molecular surface of IgG molecules played a part in their selective transmission across the rodent mucosal cells, since the transmission of the rabbit IgG molecule was decreased progressively as successive groups were modified by iodination. A similar conclusion may also apply to several other predominating amino acid residues on the H chains, especially if attachment of the IgG molecules to mucosal cell receptors involves extensive areas of the molecules; modification in any of these residues could alter the spatial distribution of the remaining residues and their goodness of fit to the receptors. Competition between IgG molecules for receptors provides the basis of selection and the blocking of receptors forms the basis of interference. A proportion of the membrane bound IgG is transmitted into the intercellular spaces. How this is accomplished remains conjectural. Conceivably, it could occur by a process of secondary internal endocytosis. Such a mechanism is implied by the results of Kraehenbuhl and colleagues [92, 96], when the Golgi complex was believed to be involved. Although the kinetics of such a mechanism are in agreement with the quantitative results on selection and interference (equation 9.1, page 000), this model still requires more direct experimental support. The participation of the lysosomal system in the transmission of antibodies across the mucosal cells is strongly supported by the observation that transmission in the young rat is markedly enhanced when cortisone acetate is administered simultaneously by mouth or parenterally [107]. It is known that cortisone and its analogues stabilize the membranes of lysosomes [251].

REFERENCES

1. C. GITLER, Mammalian Protein Metabolism (H. N. Munro and J. B. Allison, eds.), p. 35. Academic Press, New York (1964).
2. G. WISEMAN, Handbook of Physiology, Alimentary Canal (C. F. Code, ed.), sect. 6, vol. 3, p. 1277. Williams and Wilkins, Baltimore (1968).
3. O. M. SCHLOSS, Harvey Lectures, Series XX, p. 156. Williams and Wilkins, Baltimore (1924).
4. J. P. STANFIELD, *Acta Paediat., Stockh.*, **48**, 85 (1959).
5. M. GUNTHER, R. ASCHAFFENBERG, R. R. MATTHEWS, W. E. PARISH and R. R. A. COOMBS, *Immunology*, 3, 296 (1960).
6. C. COLLINS-WILLIAMS, *Int. Archs Allergy Appl. Immun.*, **20**, 38 (1962).
7. A. BURGIN-WOLFF and E. BERGER, *Experientia*, **19**, 22 (1963).
8. R. D. A. PETERSON and R. A. GOOD, *Pediatrics, Springfield*, **31**, 209 (1963).
9. S. SAPERSTEIN, D. W. ANDERSON, Jr., A. S. GOLDMAN and W. T. KINKER, *Pediatrics, Springfield*, **32**, 580 (1963).

10. R. M. ROTHBERG and R. S. FARR, *Pediatrics, Springfield*, 33, 571 (1965).
11. I. D. BERNSTEIN and Z. OVARY, *Int. Archs Allergy Appl. Immun.*, 33, 521 (1968).
12. R. M. ROTHBERG, S. C. KRAFT and R. S. FARR, *J. Immun.*, 98, 386 (1967).
13. G. VOLKHEIMER, *Arztliche Paraxis*, 21, 3711 (1969).
14. J. S. TRIER, Handbook of Physiology, Alimentary Canal (C. F. Code, ed.), sect. 6, vol. 3, p. 1125. Williams and Wilkins, Baltimore (1968).
15. M. LASKOWSKI, Jr., H. A. HAESSLER, R. P. MIECH, R. J. PEANASKY and M. LASKOWSKI, *Science*, N.Y., 127, 1115 (1958).
16. E. DANFORTH, Jr. and R. O. MOORE, *Endocrinology*, 65, 118 (1959).
17. M. BRUNNER and M. WALZER, *Archs Intern. Med.*, 42, 172 (1928).
18. R. HECHT, M. M. MOSKO, J. LUBIN, M. B. SULZBERGER and R. L. BAER, *J. Allergy*, 15, 9 (1944).
19. B. L. KABACOFF, A. WOHLMAN, M. UMHEY and A. VAKIANS, *Nature, Lond.*, 199, 815 (1963).
20. B. MORRIS and E. D. STEEL, *J. Zool. Lond.*, 152, 257 (1967).
21. R. A. GOOD and B. W. PAPERMASTER, *Adv. Immun.*, 4, 1 (1964).
22. F. W. R. BRAMBELL, The Transmission of Passive Immunity from Mother to Young. North-Holland Publishing Co., Amsterdam (1970).
23. A. B. HOERLEIN, *J. Immun.*, 78, 112 (1957).
24. G. J. V. NOSSAL, *Aust. J. Exp. Biol. Med. Sci.*, 35, 549 (1957).
25. R. D. BROWN, *J. Hyg.*, Camb., 56, 435 (1958).
26. J. A. BAKER, D. S. ROBSON, J. H. GILLESPIE, J. A. BURGHER and M. F. DOUGHTY, *Cornell Vet.*, 49, 158 (1959).
27. J. STERZL, J. KOSTKA, L. MANDEL, I. RIHA and M. HOLUB, Mechanism of Antibody Formation (M. Holub and L. Jaroskova, eds.), p. 130. Czechoslovak Acad. Sci., Prague (1960).
28. D. SEGRE and M. L. KAEBERLE, *J. Immun.*, 89, 782 (1962).
29. J. H. GRAVES, *J. Immun.*, 91, 251 (1963).
30. J. M. AIKEN and I. C. BLORE, *Am. J. Vet. Res.*, 25, 1134 (1964).
31. D. SEGRE and W. L. MYERS, *Am. J. Vet. Res.*, 25, 413 (1964).
32. J. STERZL, L. MANDEL, I. MILLER and I. RIHA, Molecular and Cellular Basis of Antibody Formation (J. Sterzl and others, eds.), p. 351. Czechoslovak Acad. Sci., Prague (1965).
33. M. L. KAEBERLE, *Iowa St. Univ. Vet.*, 29, 119 (1967).
34. T. GOSCICKA, W. RUDNICKA and B. ZABLOCKI, *Experientia*, 24, 603 (1968).
35. R. HALLIDAY, *J. Path. Bact.*, 96, 137 (1968).
36. S. S. STONE, *Immunology*, 18, 369 (1970).
37. S. COHEN and C. MILSTEIN, *Adv. Immun.*, 7, 1 (1967).
38. N. M. GREEN, *Adv. Immun.*, 11, 1 (1969).
39. H. M. GREY, *Adv. Immun.*, 10, 51 (1969).
40. K. ISHIZAKA, Biology of the Immune Response (P. Abramoff and M. F. LaVia, eds.), p. 35. McGraw-Hill, New York (1970).

41. R. M. E. PARKHOUSE, B. A. ASKONAS and R. R. DOURMASHKIN, *Immunology*, **18**, 575 (1970).
42. T. B. TOMASI, Jr. and J. BIENENSTOCK, *Adv. Immun.*, **9**, 1 (1968).
43. T. S. BISTANY and T. B. TOMASI, Jr., *Immunochemistry*, **7**, 453 (1970).
44. P. PORTER, D. E. NOAKES and W. D. ALLEN, *Immunology*, **18**, 245 (1970).
45. J. J. PAHUD and J. P. MACH, *Immunochemistry*, **7**, 679 (1970).
46. J. L. FAHEY and W. F. BARTH, *Proc. Soc. Exp. Biol. Med.*, **118**, 596 (1965).
47. R. J. GENCO, L. YECIES and F. KARUSH, *J. Immun.*, **103**, 437 (1969).
48. P. PORTER, *Biochim. Biophys. Acta*, **181**, 381 (1969).
49. J. P. VAERMAN and J. F. HEREMANS, *Immunochemistry*, **6**, 779 (1969).
50. J. F. FOSTER, R. W. FRIEDELL, D. CATRON and M. R. DIECKMANN, *Archs Biochem. Biophys.*, **31**, 104 (1951).
51. L. A. HANSON, *Int. Archs Allergy Appl. Immun.*, **17**, 45 (1960).
52. L. A. HANSON and B. G. JOHANSSON, *Int. Archs Allergy Appl. Immun.*, **20**, 65 (1962).
53. D. O. MORGAN and J. G. LECCE, *Res. Vet. Sci.*, **5**, 332 (1964).
54. A. FEINSTEIN and M. J. HOBART, *Nature, Lond.*, **223**, 950 (1969).
55. A. L. SULLIVAN, R. A. PRENDERGAST, L. J. ANTUNES, A. M. SILVERSTEIN and T. B. TOMASI, *J. Immun.*, **103**, 334 (1969).
56. B. A. ASKONAS, P. N. CAMPBELL, J. H. HUMPHREY and T. S. WORK, *Biochem. J.*, **56**, 597 (1954).
57. F. J. DIXON, W. O. WEIGLE and J. J. VASQUEZ, *Lab. Invest.*, **10**, 216 (1961).
58. A. E. PIERCE, *Proc. R. Soc. Med.*, **54**, 996 (1961).
59. C. B. LAURELL and E. H. MORGAN, *Biochim. Biophys. Acta*, **100**, 128 (1965).
60. S. M. JORDAN and E. H. MORGAN, *Q. Jl Exp. Physiol.*, **52**, 422 (1967).
61. J. D. FELDMAN, *Lab. Invest.*, **10**, 238 (1961).
62. F. A. MURPHY, O. AALUND, J. W. OSEBOLD and E. J. CARROLL, *Archs Biochem. Biophys.*, **108**, 230 (1964).
63. A. E. PIERCE and A. FEINSTEIN, *Immunology*, **8**, 106 (1965).
64. A. L. SULLIVAN and T. B. TOMASI, *Clin. Res.*, **12**, 452 (1964).
65. D. D. S. MACKENZIE and A. K. LASCELLES, *Aust. J. Exp. Biol. Med. Sci.*, **46**, 285 (1968).
66. R. W. NEWCOMB, D. NORMANSELL and D. R. STANWORTH, *J. Immun.*, **101**, 905 (1968).
67. W. A. TIMMERMAN, *Z. ImmunForsch. Exp. Ther.*, **70**, 388 (1931).
68. L. SCHNEIDER and G. PAPP, *Archs Kinderheilk.*, **114**, 91 (1938).
69. J. SCHUBERT and A. GRUNBERG, *Schweiz. Med. Wschr.*, **79**, 1007 (1949).
70. F. NORDBRING, *Acta Paediat.*, Stockh., **46**, 481 (1957).

71. K. E. BOORMAN, B. E. DODD and M. GUNTHER, *Archs Dis. Childh.*, **33**, 24 (1958).
72. S. SUSSMAN, *Pediatrics*, Springfield, **27**, 308 (1961).
73. R. M. DuPAN, J. J. SCHEIDEGER, P. WENGER, B. KOECHLI and J. ROUX, *Blut*, **5**, 104 (1959).
74. E. GUGLER and G. von MURALT, *Schweiz. Med. Wschr.*, **89**, 925 (1959).
75. E. SCHNEEGANS, G. von MURALT and C. DIERHEIMER-VAUR, *Arch. Franc. Pediat.*, **19**, 663 (1962).
76. F. WEISS, *Ann. Paediat. Basel*, **152**, 193 (1939).
77. I. A. B. CATHIE, *Br. Med. J.*, **2**, 650 (1947).
78. B. VAHLQVIST and C. HOGSTEDT, *Pediatrics*, Springfield, **4**, 401 (1949).
79. F. J. DIXON, W. KUHNS, W. O. WEIGLE and P. TAYLOR, *J. Immun.*, **83**, 437 (1959).
80. B. CAMPBELL and W. E. PETERSEN, *Dairy Sci. Abstr.*, **25**, 345 (1963).
81. J. C. LEISSRING, J. W. ANDERSON and D. W. SMITH, *Am. J. Dis. Child.*, **103**, 160 (1962).
82. S. MASON, *Arch. Dis. Childh.*, **37**, 387 (1962).
83. M. LASKOWSKI, Jr. and M. LASKOWSKI, *J. Biol. Chem.*, **190**, 563 (1951).
84. J. C. LEISSRING and J. W. ANDERSON, *Am. J. Anat.*, **109**, 149 (1961).
85. J. C. LEISSRING and J. W. ANDERSON, *Am. J. Anat.*, **109**, 175 (1961).
86. K. J. HILL, *Q. Jl Exp. Physiol.*, **41**, 421 (1956).
87. L. SCHNEIDER and J. SZATHMARY, *Z. ImmunForsch. Exp. Ther.*, **95**, 465 (1940).
88. K. JO-KEIICHIRO, *Jap. J. Med. Sci. Biol.*, **6**, 299 (1953).
89. G. BROGI and R. BIAGINI, *Clin. Pediat.*, **32**, 585 (1950).
90. A. KELLNER and E. F. HEDAL, *J. Exp. Med.*, **97**, 51 (1953).
91. J. J. DEREN, E. W. STRAUSS and T. H. WILSON, *Devl Biol.*, **12**, 467 (1965).
92. J. P. KRAEHENBUHL, E. GLOOR and B. BLANC, *Z. Zellforsch. Mikrosk. Anat.*, **70**, 209 (1966).
93. R. M. WILLIAMS and F. BECK, *J. Anat.*, **105**, 487 (1969).
94. F. BIERRING, H. ANDERSEN, J. EGEBERG, F. BRO-RASMUSSEN and M. MATTHIESSEN, *Acta Path. Microbiol. Scand.*, **61**, 365 (1964).
95. J. E. JIRASEK, J. UHER and O. KOLDOVSKY, *Acta Histochem.*, **22**, 33 (1965).
96. J. P. KRAEHENBUHL, E. GLOOR and B. BLANC, *Z. Zellforsch. Mikrosk. Anat.*, **76**, 170 (1967).
97. J. P. KRAEHENBUHL and M. A. CAMPICHE, *J. Cell Biol.*, **42**, 345 (1969).
98. O. ROSA, *Gynec. Obstet.*, **50**, 463 (1951).
99. H. ANDERSEN, F. BIERRING, M. MATTHIESSEN, J. EGEBERG and F. BRO-RASMUSSEN, *Acta Path. Microbiol. Scand.*, **61**, 377 (1964).

100. N. CARRETTI and Z. OVARY, *Proc. Soc. Exp. Biol. Med.*, **130**, 509 (1969).
101. R. ASOFSKY and M. B. HYLTON, *Fedn Proc. Fedn Am. Socs Exp. Biol.*, **27**, 617 (1968).
102. E. H. MORGAN, *Biochim. Biophys. Acta*, **154**, 478 (1968).
103. R. HALLIDAY and R. A. KEKWICK, *Proc. R. Soc.*, **146**, 431 (1957).
104. R. HALLIDAY, *Proc. R. Soc.*, **144**, 427 (1955).
105. R. HALLIDAY, *Proc. R. Soc.*, **143**, 408 (1955).
106. I. G. MORRIS, *Proc. R. Soc.*, **160**, 276 (1964).
107. R. HALLIDAY, *J. Endocr.*, **18**, 56 (1959).
108. R. HALLIDAY and R. A. KEKWICK, *Proc. R. Soc.*, **153**, 279 (1960).
109. S. M. JORDAN and E. H. MORGAN, *Aust. J. Exp. Biol. Med. Sci.*, **46**, 465 (1968).
110. J. G. LECCE and D. O. MORGAN, *J. Nutr.*, **78**, 263 (1962).
111. R. HALLIDAY, *Proc. R. Soc.*, **145**, 179 (1956).
112. I. A. MANVILLE and R. W. LLOYD, *Am. J. Physiol.*, **100**, 394 (1932).
113. B. MOSINGER, Z. PLACER and O. KOLDOVSKY, *Nature, Lond.*, **184**, 1245 (1959).
114. A. BOASS and T. H. WILSON, *Am. J. Physiol.*, **204**, 101 (1963).
115. D. R. BAMFORD, Ph.D. Thesis, University of Wales (1965).
116. R. M. CLARKE and R. N. HARDY, *J. Physiol.*, Lond., **204**, 113 (1969).
117. R. M. CLARKE and R. N. HARDY, *J. Physiol.*, Lond., **204**, 127 (1969).
118. R. NOACK, O. KOLDOVSKY, M. FRIEDRICH, A. HERINGOVA, V. JIRSOVA and G. SCHENK, *Biochem. J.*, **100**, 775 (1966).
119. K. BAINTNER, Jr. and B. VERESS, *Experientia*, **26**, 54 (1970).
120. R. HALLIDAY, CIBA Foundation Symposium on 'Somatic Stability in the Newly Born' (G. E. W. Wolstenhome and M. O'Connor, eds.), p. 241. J. and A. Churchill, London (1961).
121. S. E. KYFFIN, *Immunology*, **13**, 319 (1967).
122. I. G. MORRIS, *Proc. R. Soc.*, **163**, 402 (1965).
123. I. G. MORRIS, *Immunology*, **17**, 139 (1969).
124. R. HALLIDAY, *Proc. R. Soc.*, **148**, 92 (1957).
125. D. R. BANGHAM and R. J. TERRY, *Biochem. J.*, **66**, 579 (1957).
126. W. A. HEMMINGS and I. G. MORRIS, *Proc. R. Soc.*, **150**, 403 (1959).
127. I. G. MORRIS, Handbook of Physiology, Alimentary Canal (C. F. Code, ed.), sect. 6, vol. 3, p. 1491. Williams and Wilkins, Baltimore (1968).
128. D. R. BAMFORD, *Proc. R. Soc.*, **166**, 30 (1966).
129. I. G. MORRIS, *Immunology*, **13**, 49 (1967).
130. F. W. R. BRAMBELL, R. HALLIDAY and W. A. HEMMINGS, *Proc. R. Soc.*, **153**, 477 (1961).
131. I. G. MORRIS, *Proc. R. Soc.*, **148**, 84 (1957).

132. F. W. R. BRAMBELL, R. HALLIDAY and I. G. MORRIS, *Proc. R. Soc.*, **149**, 1 (1958).
133. I. G. MORRIS, *Proc. R. Soc.*, **157**, 160 (1963).
134. F. JEAL, Ph.D. Thesis, University of Wales (1965).
135. R. CORNELL and H. A. PADYKULA, *Am. J. Anat.*, **125**, 291 (1969).
136. N. D. GALLAGHER, *Nature, Lond.*, **222**, 877 (1969).
137. S. L. CLARK, Jr., *J. Biophys. Biochem. Cytol.*, **5**, 41 (1959).
138. Z. VACEK, *Cslka Morfol.*, **12**, 292 (1964).
139. D. O. GRANEY, *Anat. Rec.*, **148**, 286 (1964).
140. J. W. ANDERSON, *J. Cell Biol.*, **23**, 4A (1964).
141. P. SHERVEY, *Anat. Rec.*, **154**, 422 (1966).
142. F. JEAL, *Irish J. Med. Sci.*, ser. 7, **1**, 327 (1968).
143. D. O. GRANEY, *Am. J. Anat.*, **123**, 227 (1968).
144. S. L. WISSIG and D. O. GRANEY, *J. Cell Biol.*, **39**, 564 (1968).
145. R. M. WILLIAMS and F. BECK, *Histochem. J.*, **1**, 531 (1969).
146. S. L. CLARK, Jr., *Anat. Rec.*, **139**, 216 (1961).
147. E. SANDBORNE, P. F. KOEN, J. D. McNABB and G. MOORE, *J. Ultrastruct. Res.*, **11**, 123 (1964).
148. O. BEHNKE, *J. Ultrastruct. Res.*, **11**, 139 (1964).
149. B. MORRIS, *Nature, Lond.*, **184**, 1151 (1959).
150. B. MORRIS, *Proc. R. Soc.*, **152**, 137 (1960).
151. B. MORRIS, *Proc. R. Soc.*, **154**, 369 (1961).
152. B. MORRIS and R. W. BALDWIN, *Proc. R. Soc.*, **155**, 551 (1962).
153. B. MORRIS, *Proc. R. Soc.*, **158**, 253 (1963).
154. B. MORRIS and E. D. STEEL, *J. Endocr.*, **30**, 195 (1964).
155. C. CAUSSE-VAILLS, A. VERAIN and A. VERAIN, *C.R. Seanc. Soc. Biol.*, **155**, 737 (1961).
156. S. K. HARDING, D. W. BRUNER and I. W. BRYANT, *Cornell Vet.*, **51**, 535 (1961).
157. R. S. COMLINE, R. W. POMEROY and D. A. TITCHEN, *J. Physiol., Lond.*, **122**, 6P (1953).
158. D. D. PORTER, *Proc. Soc. Exp. Biol. Med.*, **119**, 131 (1965).
159. L. SCHNEIDER and J. SZATHMARY, *Z. ImmunForsch. Exp. Ther.*, **95**, 177 (1939).
160. J. H. GILLESPIE, J. A. BAKER, J. BURGHER, D. ROBSON and B. GILMAN, *Cornell Vet.*, **48**, 103 (1958).
161. L. E. CARMICHAEL, D. S. ROBSON and F. D. BARNES, *Proc. Soc. Exp. Biol. Med.*, **109**, 677 (1962).
162. J. W. ANDERSON, *Anat. Rec.*, **145**, 200 (1963).
163. J. S. JOHNSON and J. H. VAUGHAN, *J. Immun.*, **98**, 923 (1967).
164. D. D. GILLETTE and M. FILKINS, *Am. J. Physiol.*, **210**, 419 (1966).
165. M. E. FILKINS and D. D. GILLETTE, *Proc. Soc. Exp. Biol. Med.*, **122**, 686 (1966).
166. V. V. MICUSAN, *Revue Roum. Biochim.*, **6**, 147 (1969).
167. P. JOHNSON and A. E. PIERCE, *J. Hyg.*, Camb., **57**, 309 (1959).
168. N. J. BERRIDGE, J. G. DAVIS, P. M. KON, S. K. KON and F. R. SPRATLING, *J. Dairy Res.*, **13**, 145 (1943).

169. M. J. HERSCHEL, W. B. HILL and J. W. G. PORTER, *Proc. Nutr. Soc.*, **20**, xl (1961).
170. J. T. HUBER, N. L. JACOBSEN, R. S. ALLEN and P. A. HARTMAN, *J. Dairy Sci.*, **44**, 1494 (1961).
171. R. N. HARDY, *J. Physiol.*, Lond., **205**, 453 (1969).
172. R. S. COMLINE, H. E. ROBERTS and D. A. TITCHEN, *Nature, Lond.*, **167**, 561 (1951).
173. W. E. BALFOUR and R. S. COMLINE, *J. Physiol.*, Lond., **160**, 234 (1962).
174. A. D. SHANNON and A. K. LASCELLES, *Q. Jl Exp. Physiol.*, **53**, 415 (1968).
175. R. G. HANSEN and P. H. PHILLIPS, *J. Biol. Chem.*, **171**, 223 (1947).
176. E. L. SMITH, *J. Dairy Sci.*, **31**, 127 (1948).
177. E. L. SMITH, and A. HOLM, *J. Biol. Chem.*, **175**, 349 (1948).
178. R. G. HANSEN and P. H. PHILLIPS, *J. Biol. Chem.*, **179**, 523 (1949).
179. A. E. PIERCE, *J. Hyg.*, Camb., **53**, 247 (1955).
180. G. G. B. KLAUS, A. BENNETT and E. W. JONES, *Immunology*, **16**, 293 (1969).
181. C. P. MILSTEIN and A. FEINSTEIN, *Biochem. J.*, **107**, 559 (1968).
182. H. F. DEUTSCH and V. R. SMITH, *Am. J. Physiol.*, **191**, 271 (1957).
183. A. E. PIERCE and P. JOHNSON, *J. Hyg.*, Camb., **58**, 247 (1960).
184. R. B. LITTLE and M. L. ORCUTT, *J. Exp. Med.*, **35**, 161 (1922).
185. M. L. ORCUTT and P. E. HOWE, *J. Exp. Med.*, **36**, 291 (1922).
186. J. G. McALPINE and L. F. RETTGER, *J. Immun.*, **10**, 811 (1925).
187. A. McDIARMID, *Vet. Rec.*, **58**, 146 (1946).
188. R. N. HARDY, *J. Physiol.*, Lond., **204**, 607 (1969).
189. V. R. SMITH, R. E. REED and E. S. ERWIN, *J. Dairy Sci.*, **47**, 923 (1964).
190. V. R. SMITH and E. S. ERWIN, *J. Dairy Sci.*, **42**, 364 (1959).
191. F. SCHOENAERS and A. KAECKENBEECK, *Annls Med. vet.*, **108**, 205 (1964).
192. A. E. PIERCE, *J. Physiol.*, Lond., **156**, 136 (1961).
193. J. H. MASON, T. DALLING and W. S. GORDON, *J. Path. Bact.*, **33**, 783 (1930).
194. I. P. EARLE, *J. Agric. Res.*, **51**, 479 (1935).
195. L. SCHNEIDER and J. SZATHMARY, *Z. ImmunForsch. Exp. Ther.*, **95**, 169 (1939).
196. M. BARR, A. T. GLENNY and J. W. HOWIE, *J. Path. Bact.*, **65**, 155 (1953).
197. E. F. McCARTHY and E. I. McDOUGALL, *Biochem. J.*, **55**, 177 (1953).
198. K. LINDQVIST, *Nord. VetMed.*, **1**, 283 (1962).
199. D. E. ULLREY, C. H. LONG, E. R. MILLER and B. H. VINCENT, *J. Anim. Sci.*, **21**, 1031 (1962).
200. G. C. REYMANN, *J. Immun.*, **5**, 227 (1920).
201. R. HALLIDAY, *J. Immun.*, **95**, 510 (1965).

202. R. HALLIDAY, *Nature, Lond.,* **205,** 614 (1965).
203. D. R. BANGHAM, P. L. INGRAM, J. H. B. ROY, K. W. G. SHILLAM and R. J. TERRY, *Proc. R. Soc.,* **149,** 184 (1958).
204. W. E. BALFOUR and R. S. COMLINE, *J. Physiol.,* Lond., **148,** 77P (1959).
205. A. E. PIERCE, P. C. RISDALL and B. SHAW, *J. Physiol.,* Lond., **171,** 203 (1964).
206. R. S. COMLINE, H. E. ROBERTS and D. A. TITCHEN, *Nature, Lond.,* **168,** 84 (1951).
207. K. J. HILL and W. S. HARDY, *Nature, Lond.,* **178,** 1353 (1956).
208. R. N. HARDY, *J. Physiol.,* Lond., **205,** 435 (1969).
209. P. PORTER, *Immunology,* **17,** 617 (1969).
210. A. E. PIERCE and M. W. SMITH, *J. Physiol.,* Lond., **190,** 1 (1967).
211. C. J. LEWIS, P. A. HARTMAN, C. H. LIU, R. O. BAKER and D. V. CATRON, *J. Agric. Fd Chem.,* **5,** 687 (1957).
212. P. A. HARTMAN, V. W. HARP, R. O. BAKER, L. H. NEAGLE and D. V. CATRON, *J. Anim. Sci.,* **20,** 114 (1961).
213. M. LASKOWSKI, B. KASSELL and G. HAGERTY, *Biochim. Biophys. Acta,* **24,** 300 (1957).
214. R. N. HARDY, *J. Physiol.,* Lond., **176,** 19P (1965).
215. F. NORDBRING and B. OLSSON, *Acta Soc. Med. Upsal.,* **63,** 25 (1958).
216. A. E. PIERCE and M. W. SMITH, *J. Physiol.,* Lond., **190,** 19 (1967).
217. N. BJORKMAN and M. SIBALIN, *Experientia,* **23,** 339 (1967).
218. P. BROWN, M. W. SMITH and R. WITTY, *J. Physiol.,* Lond., **198,** 365 (1968).
219. F. NORDBRING and B. OLSSON, *Acta Soc. Med. Upsal.,* **62,** 193 (1957).
220. L.RUTQVIST, *Am. J. Vet. Res.,* **19,** 25 (1958).
221. R. A. McCANCE and E. M. WIDDOWSON, *J. Physiol.,* Lond., **145,** 547 (1959).
222. J. G. LECCE and G. MATRONE, *J. Nutr.,* **70,** 13 (1960).
223. E. R. MILLER, D. E. ULLREY, I. ACKERMAN, D. A. SCHMIDT, J. A. HOEFER and R. W. LUECKE, *J. Anim. Sci.,* **20,** 31 (1961).
224. J. M. ASPLUND, R. H. GRUMMER and P. H. PHILLIPS, *J. Anim. Sci.,* **21,** 412 (1962).
225. E. R. MILLER, B. G. HARMON, D. E. ULLREY, D. A. SCHMIDT, R. W. LUECKE and J. A. HOEFER, *J. Anim. Sci.,* **21,** 309 (1962).
226. C. G. RAMIREZ, E. R. MILLER, D. E. ULLREY and J. A. HOEFER, *J. Anim. Sci.,* **22,** 1068 (1963).
227. G. A. YOUNG and N. R. UNDERDAHL, *Cornell Vet.,* **39,** 120 (1949).
228. V. C. SPEER, H. BROWN, L. QUINN and D. V. CATRON, *J. Immun.,* **83,** 632 (1959).
229. R. N. HARDY, *J. Physiol.,* Lond., **204,** 633 (1969).
230. F. NORDBRING and B. OLSSON, *Acta Soc. Med. Upsal.,* **63,** 41 (1958).
231. L. C. PAYNE and C. L. MARSH, *J. Nutr.,* **76,** 151 (1962).

232. L. C. PAYNE and C. L. MARSH, *Fedn Proc. Fedn Am. Socs Exp. Biol.*, **21**, 909 (1962).
233. L. C. PAYNE and C. L. MARSH, *Fedn Proc. Fedn Am. Socs Exp. Biol.*, **18**, 118 (1959).
234. B. OLSSON, *Nord. VetMed.*, **11**, 441 (1959).
235. J. G. LECCE, G. MATRONE and D. O. MORGAN, *J. Nutr.*, **73**, 158 (1961).
236. M. L. KAEBERLE and D. SEGRE, *Am. J. Vet. Res.*, **25**, 1096 (1964).
237. J. G. LECCE, D. O. MORGAN and G. MATRONE, *J. Nutr.*, **84**, 43 (1964).
238. B. OLSSON, *Nord. VetMed.*, **11**, 375 (1959).
239. I. R. BALCONI and J. G. LECCE, *J. Nutr.*, **88**, 233 (1966).
240. R. F. LOCKE, D. SEGRE and W. L. MYERS, *J. Immun.*, **93**, 576 (1964).
241. G. C. PERRY and J. H. WATSON, *Anim. Prod.*, **9**, 557 (1967).
242. J. G. LECCE, *J. Physiol.*, Lond., **184**, 594 (1966).
243. M. W. SMITH and A. E. PIERCE, *Nature, Lond.*, **213**, 1150 (1967).
244. A. E. WILD, *Proc. R. Soc.*, **163**, 90 (1965).
245. A. E. WILD, *J. Embryol. Exp. Morph.*, **24**, 313 (**1970**).
246. R. WITTY, P. BROWN and M. W. SMITH, *Experientia*, **25**, 310 (1969).
247. M. SIBALIN and N. BJORKMAN, *Expl Cell Res.*, **44**, 165 (1966).
248. T. E. STALEY, E. W. JONES and A. E. MARSHALL, *Anat. Rec.*, **161**, 497 (1968).
249. T. E. STALEY, E. W. JONES and L. D. CORLEY, *Am. J. Vet. Res.*, **30**, 567 (1969).
250. W. A. HEMMINGS, *Nature, Lond.*, **186**, 399 (1960).
251. G. WEISSMANN, Lysosomes in Biology and Pathology (J. T. Dingle and Honor B. Fell, eds.), p. 276. North-Holland Publishing Co., Amsterdam (1968).
252. R. C. WEIR, R. R. PORTER and D. GIVOL, *Nature, Lond.*, **212**, 205 (1966).
253. C. A. ABEL and H. M. GREY, *Science*, N.Y., **156**, 1609 (1967).
254. D. R. NASH, J. P. VAERMAN, H. BAZIN and J. F. HEREMANS, *J. Immun.*, **103**, 145 (1969).
255. J. P. VAERMAN, Thesis, Universite Catholique de Louvain, Louvain (1970), p. 211.

CHAPTER 10

Intestinal Absorption of Glucose

ROBERT K. CRANE

*Department of Physiology,
The College of Medicine and Dentistry of New Jersey,
Rutgers Medical School,
New Brunswick, New Jersey, U.S.A.*

		Page
10.1	INTRODUCTION	541
10.2	SPECIFICITY AND GLUCOSE-CARRIER INTERACTION	542
10.3	THE Na^+-DEPENDENCE OF GLUCOSE ABSORPTION	545
10.4	THE MEANING OF Na^+-DEPENDENCE FOR TRANSLOCATION	548
10.5	THE MEANING OF Na^+-SUGAR COTRANSPORT FOR ENERGY COUPLING	549
	REFERENCES	551

10.1 INTRODUCTION

The absorption of carbohydrates has been the subject of two extensive reviews by the author [1, 2]. The first covered the literature up to 1961, and the second the work between 1961 and 1968. It will be assumed that these are available to the reader, and hence no attempt will be made to recover this material. The present chapter is restricted to a discussion of some modern aspects of carbohydrate absorption in which there are still some unresolved differences of viewpoint.

Although it is generally conceded that absorption of glucose from the lumen of the gut into the blood stream ultimately depends on the activity of an energy-coupled specific translocation process located in the brush border pole of the epithelial cells of the gut mucosa, there is disagreement about the nature of the process. On the one hand, there are researchers who subscribe to some modification of the Na^+-gradient hypothesis of Crane, Miller and Bihler [3] as perhaps best exemplified by Schultz and Curran [4]. On the other hand, there are researchers as may be exemplified by Forster and Hoos [5] who argue that Na^+ has nothing to do with the process. This being the case, discussion of the data with the view to providing some guidelines toward reconciliation of these divergent views would seem to be in order. Accordingly, this brief chapter will be addressed to three fundamental questions having to do with the membrane transport step in the absorption of glucose: Firstly, there is the question of the interaction of glucose with its specific membrane carrier and the meaning of transport specificity for the nature of this interaction. Secondly, there is the question of Na^+ dependency and whether it is a constant of glucose transport by the gut. Thirdly, there is the question of the meaning of Na^+ dependency for the molecular nature and energy coupling of transport.

10.2 SPECIFICITY AND GLUCOSE-CARRIER INTERACTION

Glucose and galactose share the same transport system [1]. The system involves interaction of the sugar with a specific component of the brush border membrane preceding translocation of the sugar from the lumen of the gut into the epithelial cells and, as far as they have gone, studies of this system [4], are interpretable along familiar lines of membrane carrier theory [6]. It is within this framework that the discussion to follow is constructed.

The specificity of transport can be used from two points of view. It can be used to delineate those characteristics of the sugar molecule which are involved in the interaction; that is, binding, with the carrier. It can also be used to determine which, if any, functional groups of the sugar molecule may be involved in a reaction during translocation. Generally speaking, characteristics which when altered influence but do not prevent transport would be interpreted as being of the first kind; groups which are required for transport *may* be of the second kind.

Early studies established that of all the functional groups of glucose only the hydroxyl group at carbon-2 could be said to be specifically required for transport [1]. It could be shown that analogues lacking a free hydroxyl group at carbons 1, 3 or 6 were transported rather well whereas 2-deoxyglucose was substantially inert. Mannose, the epimer of glucose at carbon-2, was extremely poorly transported in contrast to galactose which is the epimer at carbon-4 and is, of course, a natural substrate of the same process. From the sum of these studies it was argued that if there were a reaction of glucose during translocation, that reaction had to be with the hydroxyl group at carbon-2. By use of O^{18} and of the analogue 2-C-hydroxymethyl glucose, it was then shown that reactions involving transfer and replacement of oxygen or oxidation and reduction were excluded. What remained untested were reactions involving an intermediate of the type $\equiv$C—OR where only the hydroxyl-hydrogen at carbon-2 was displaced. Recently, a reaction of this type has been proposed by Barnett *et. al.*, [7]. They envisaged the formation and disruption of a covalent ester bond between carbon-2 and the carrier as a necessary part of translocation. No direct evidence was adduced for this proposed step. However, it was felt [8] that some anomalies among weakly transported analogues could be better explained on the assumption of this kind of covalent glucose-carrier interaction. Swaminathan and Eichholz [9] have made a test of this proposal by using D-2-^{3}H glucose as a substrate for transport by hamster intestine *in vitro*. According to their data, tritium is not exchanged during transport. Consequently, it must be concluded either that there is no covalent bond of the type $\equiv$C—OR formed during transport or else that the hydroxyl-hydrogen atom displaced during formation of the glucose-carrier bond is quantitatively conserved and replaced during disruption of the bond. The kind of reaction envisaged by Barnett and his colleagues might qualify. The reaction cannot be an energy source for active transport but it could represent the second step in transport which Caspary *et al.* [10] identified by their studies of L-fucose inhibition of transport and postulated as a conformational shift of the sugar-carrier complex. There is currently no clear way to distinguish between the two possibilities, nor to decide whether both are correct or neither.

Crane [11] made it clear that among the reactions involving an intermediate of the type $\equiv$C–OR and not ruled out by the early experiments was phosphorylation-dephosphorylation, *per se*; only particular proposals for phosphorylation-dephosphorylation, e.g. those involving carbons 1 and 6, were ruled out. Nonetheless, based on the absence of clear evidence in support of any chemical mechanism involving the sugar [11], an early commitment was made to the concept that energy coupling for active transport was delivered to the carrier rather than to the substrate. The experiments of Swaminathan and Eichholz [9], though perhaps indecisive with respect to the occurrence of an intermediate covalent bond, do decisively eliminate the last remaining possibility for a reaction of the sugar to supply energy for active transport. Energy coupling for intestinal glucose transport must be with the carrier. The nature of the coupling will be considered later.

As reviewed by Barnett and Munday [8] recent studies have greatly improved understanding of the interaction of groups other than carbon-2 with the carrier and of the conformational requirements for interaction along the guidelines laid down by Crane [2]. The data favour the interpretation that hydrogen-bonds donated by the carrier to the oxygens at carbons 1, 3, 4 and 6 (in addition to the interaction at carbon 2, whatever its nature) are important though not individually critical for establishing the fit between substrate and carrier. The ring oxygen has been found to be uninvolved in the transport interactions inasmuch as substitution of sulfur has no significant effect [7, 12].

It has been known for some time that intestinal sugar transport will accommodate very large substituents at carbon-1 [13, 14] in contrast to the severe limitations of substituent size at carbons 3 and 6 [15]. An obvious conclusion from this difference is that large substituents at carbon-1 lie outside the confines of the "active site" of the carrier. This being so it is then probable that these same large substituents interact with the membrane matrix in the immediate vicinity of the carrier [16] and that, appropriately studied, they could serve as probes of the nature of this portion of the matrix. Preliminary work in this direction has been reported by Ramaswamy, *et al.* [17].

10.3 THE Na^+-DEPENDENCE OF GLUCOSE ABSORPTION

As indicated above and as is generally known there is disagreement about the involvement of Na^+ in the process of glucose absorption by the small intestine. In an effort to clarify the issues, the subject of Na^+ involvement will be taken up as not one but a series of questions. The question to be taken up first is whether Na^+ has an effect on the permeability of the gut mucosa to glucose.

As matters now stand, there seems to be no room for further doubt that the brush border membrane of the intestinal epithelial cell has the species-independent property of being more permeable to glucose (and its actively transported analogues) when Na^+ is present in the bathing fluid than when it is absent. Whatever else may be a matter of speculation or hypothesis requiring further proof, this property may be taken as established.

Bihler et al. [18] first demonstrated the effect of Na^+ on permeability independently of its effect on active transport by use of energy-depleted tissue preparations from the hamster. Incubation of tissue in Na^+-containing buffer as compared to Na^+-free buffer enhanced the anaerobic entry of compounds like 6-deoxyglucose but not of compounds like 2-deoxyglucose. In line with carrier theory, this result was taken to indicate the "existence of a specific, Na^+-dependent and energy-independent process mediating the rapid equilibration of certain sugars between the tissue and the medium". It would now appear that this observation, most fundamental to an understanding of the various hypotheses for the mechanism of Na^+-dependent active transport, has finally been confirmed in a clear way. Hopfer et al. [19] have succeeded in preparing isolated rat brush border membranes in a vesicular form. These vesicles, devoid of cytoplasm and out of contact with energy-yielding reactions, nonetheless displayed the characteristics anticipated from the deductions of Bihler, et al. [18]. D-glucose was taken up and released from the vesicles faster than L-glucose. Na^+, specifically among all ions tested, increased the initial rate and extent of D-glucose uptake. Countertransport and inhibition by phlorizin and D-galactose were also demonstrated. Numerous other *in vitro* studies [4] with intact tissue carried out by test methods even more discriminating than those used by Bihler *et al.* [18] have given results which are readily interpreted in the same way; that is, the presence of Na^+ enhances membrane glucose permeability, irrespective of differences in kinetics as between, say, rabbit ileum and hamster jejunum. There is, in fact, no remaining dispute about this particular aspect of Na^+-dependency of transport. Dispute arises because this property of the brush border membrane is not clearly expressed when sugar absorption is studied *in vivo*.

Early studies with *in vivo* loops of intestine seemed to confirm the anticipated Na^+-dependency of glucose absorption [20]. However, with the introduction of peroral perfusion methods the situation changed. Olsen and Ingelfinger [21], for example, found that sugar absorption *in vivo* is not as strictly dependent on luminal Na^+ as would be expected from *in vitro* studies. Saltzman *et al.* [22] similarly have studied glucose absorption *in vivo* with the triple-lumen tube technique and have extended observations on ileal absorption in man to the rat and the dog. They find that glucose absorption is minimally or not at all reduced when the Na^+ concentration in the solutions perfused through the chosen segment of gut is reduced from 140 mEq per litre to concentrations (e.g. 2.5 mEq per litre) which are known to be ineffective in supporting glucose permeability *in vitro*. The use of the rat and the dog by these researchers has been an important addition to the information base even though the effect of Na^+ on glucose uptake by human intestine *in vitro* is readily demonstrated [23]. Removed now from consideration is any possibility that the results, either way, are species dependent. The problem can be stated clearly: It would have been predicted from the properties of the brush border membrane that glucose uptake under the conditions used would be strongly influenced by the Na^+ concentration of the perfusate. It is not. The question is why.

An answer to this question is not at hand. It will probably not be a simple answer when it is. There are several possibilities which can be considered. For example, it may be entertained that the effect of Na^+ on glucose entry is strictly an *in vitro* phenomenon; that the brush border membrane in the living cell is unaffected by Na^+ and the appearance of Na^+-sensitivity is a phenomenon of functional degeneration following disruption of the normal supply of oxygen and nutrients from the circulation.

As matters currently stand, to take this view would be to retreat to vitalism. There are no studies on the properties of the brush border membrane of the absorbing cells of the gut, *in vivo*. Such studies as have been done are of the mucosal surface as a whole in which, for one example, secretory phenomena [24, 25] may have been overlooked as a contributing factor to overall function. This is substantially the conclusion reached by Fleshler and Nelson [26] from their studies of L-alanine absorption from Thirty-Vella loops in dogs. They believe that the necessary Na^+ ions for absorption are contributed by the mucosal wall, perhaps as discussed earlier by Crane [27].

Semenza [28] takes the view that the difference between *in vivo* and *in vitro* is not at the level of the membrane but at the level of a barrier to diffusion, an unstirred layer, adjacent and external to the membrane. An unstirred layer can be demonstrated, *in vitro* [29, 30]. However, the unstirred layer that Semenza considers is in addition to and, based on comparative kinetic data [31], much more substantial than the one observed *in vitro*. It is composed of the spaces between the tightly packed microvilli, the fuzzy coat and a layer of mucous. It is Semenza's view that the efflux of Na^+ noted by Fleshler and Nelson [26] goes first into the unstirred layer before appearing in the lumen and would be available at that location to activate the glucose permeability of the membrane.

Some attention to this possibility is being paid in Fordtran's laboratory [32] by using replacement substances for Na^+ in the perfusion solutions other than mannitol. Mg^{++} has been used in an effort to displace Na^+ from a presumed negatively charged boundary at the mucosal surface. With the results of these studies in hand, it is still found that 50%, at least, of active sugar transport by the ileum is independent of Na^+ added to the perfusion solution.

Clearly, there is as yet no mutual ground for accommodation of these divergent views. Semenza [28] notes that the numerous methods for study of *in vitro* intestinal transport have been devised because of their individual advantages for study of one or another portion of the overall process of intestinal absorption. One may suggest that the difficulty in resolving the Na^+ question arises because some important component of the process has not yet been measured *in vitro* and yet another new and discriminating method will be needed.

10.4 THE MEANING OF Na^+-DEPENDENCE FOR TRANSLOCATION

In the section just preceding Na^+ dependence of glucose translocation across the brush border membrane was considered in its simplest terms as merely an increase in the permeability of the membrane for glucose. The phenomena underlying the increase in permeability are now to be considered.

It was, early on, possible to imagine that Na^+ merely conditioned the membrane; that is, that it activated transport without participating in it. Nonetheless, Crane et al. [3] postulated cotransport of Na^+ and sugar as a basis for the perceived need to couple energy with the carrier. Supportive evidence for cotransport came along [33] in the form of the influence of a reversed Na^+ gradient on sugar flux and [34] in the form of an effect of sugar on the *in vitro* transmural potential and Na^+ flux. However, substantial proof of cotransport came from direct measurements of the simultaneous flux of Na^+ and sugar in rabbit ileum by Goldner et al. [35]. Carrier-mediated sugar transport was found to be inactive in the absence of Na^+. When Na^+ was added both sugar and Na^+ entered the cell in a molecular ratio of 1:1. The role of Na^+ appeared to be the formation of a ternary complex (Na^+-carrier-sugar) capable of translocation across the brush border membrane. Kinetically this is seen as an increase in maximal rate.

The situation in hamster small intestine is kinetically different [36]; that is, Na^+ has the effect of changing the affinity of the carrier for sugar, which is much the same as the effect of Na^+ on amino transport in rabbit ileum [37]. The difference with sugar transport in rabbit ileum may have to do with the fact that hamster gut is almost entirely jejunum [38]. However, comparative studies of rabbit jejunum are lacking. Studies by means of transmural potential [39] have indicated that the effect of Na^+ on sugars is the same in the rat as in the hamster and that sugar has the same effect on Na^+; that is, it influences its affinity for the carrier. In these cases, the ratio of Na^+ and substrate entering the cell will necessarily vary with the concentration [37].

No matter which is the case, however, an effect of the substrates on affinity or on maximal rate, what is important to recognize is their interdependence. In all cases studied there is, concomitant with an increase of sugar permeability of the brush border membrane by the presence of Na^+, an increase in Na^+ permeability by the presence of sugar. One must view sugar-dependent Na^+ transport in the same way and as an expression of the same membrane events as one views Na^+-dependent sugar transport. The consequences of doing so may be surprising.

The failure of the *in vitro* studies on Na^+-dependent sugar transport to predict the *in vivo* consequences has been discussed above. There is, however, no like failure of *in vitro* studies on sugar-dependent Na^+ transport to predict the *in vivo* consequences. Sugar-dependent Na^+ transport occurs *in vivo* and can be measured in the intact human intestine [40]. Moreover, it would appear that this sugar-dependent Na^+ transport forms the basis of a reliable treatment for the devastating salt and water loss of cholera [41, 42].

10.5 THE MEANING OF Na^+-SUGAR COTRANSPORT FOR ENERGY COUPLING

There is little room for doubt that Na^+ and sugar are contransported on the same membrane carrier. There can be as little doubt that *the potential exists* for energy to be coupled to the transport mediated by that carrier by means of an imposed gradient or flux of one of the partners [43]. Such a coupling for sugar has, in fact, been demonstrated in the gut cell by Crane [33]. It has been demonstrated for amino acid in the pigeon red cell by Vidaver [44] and in the Ehrlich cell by Eddy [45]. Consequently, the question is not whether this kind of energy coupling can or does occur. The question is whether this kind of energy coupling occurs naturally and whether, if it occurs, there is sufficient driving force available to account for the observed active transport. These questions have been approached at length and in various ways and there is no clear answer for them. Schultz and Curran [4] have discussed a number of approaches. Semenza [28] refers to others which may be more particularly related to the intestine.

Most of the recent work on these questions has been with cell types other than the gut and hence, of necessity, with amino acids. At the present time, both the view that cation gradients are not sufficient and the view that they are seem to be supported. On the one hand, calculations of energy input versus energy need indicate that only half the needed energy can be provided by cation gradients [46]. On the other hand, when the cation gradients are specifically and selectively abolished by gramicidin [47] so also is amino acid *active* transport.

It is clear that calculations of ion gradients are inherently in error to the extent that the ions may not be homogeneously distributed throughout the cell. This problem is currently receiving attention [48]. It is also clear, however, that calculations of amino acid accumulation, and hence of energy need, are inherently in error to the extent that intracellular membranes contribute to accumulation and to apparent active transport. Since amino acid accumulation by nuclei is a well-documented phenomenon [49] it would be, to say the least, helpful were someone, using perhaps the superb technique of Stirling [50], to measure the proportion of amino acid accumulation contributed by the prominent nuclei of the Ehrlich cell, for example, thereby enabling the calculations of energy requirement to be appropriately corrected.

Some work has been done with the intestine and as made quite clear by Smyth [51] the question of energy coupling is still an open one. Overall, it may be said, the intestinal epithelial cell may be a more favourable cell for coupling to occur than some others. The reason for this is that the necessary events seem to be clearly separated. Na^+ and sugar enter together at the brush border membrane. Na^+, it is fairly clear, is ejected from the cell at the lateral membrane [50, 52]. Thus the possibility is evident for the brush border →lateral membrane flux of Na^+ to drive sugar accumulation without the need to establish large, overall lumen-cell Na^+ gradients. The gut cell, in this respect, may be seen to resemble a cylinder in which closer to one end than the other there is a band of Na^+ pump activity. Assuming the closer end to be relatively more permeable to Na^+ than the other, it can be seen that a strong through flux of Na^+ could be maintained without substantial disturbance of the larger volume of the cylinder. Effects of this sort need to be considered when evaluating the possible contribution of cation gradients to sugar accumulation. To aid understanding of this point, it can be pointed out that the terms gradient-coupled and flux-coupled are, for the polarized gut cell, synonymous. It will be seen, then, that tests of the contribution of cation *gradient* coupling to substrate accumulation should be done with the cells in a state of energy depletion so as to abolish transcellular flux. Having paid no attention to this point, the tests made by Kimmich [53] can be seen to be indecisive.

REFERENCES

1. R. K. CRANE, *Physiol. Rev.*, **40**, 789 (1960).
2. R. K. CRANE, *Absorption of Sugars*, in: Handbook of Physiology (C. F. Code, ed.). Section 6, Alimentary Canal. Vol. III. Intestinal Absorption, p. 1323. Washington, D.C., Am. Physiol. Soc. (1968).
3. R. K. CRANE, D. MILLER and I. BIHLER, *The Restrictions on the possible mechanisms of intestinal active transport of sugars*, in: Membrane Transport and Metabolism (A. Kleinzeller and A. Kotyk, eds.), p. 439. Academic Press, New York (1961).
4. S. G. SCHULTZ and P. F. CURRAN, *Physiol. Rev.*, **50**, 637 (1970).
5. H. FORSTER and I. HOOS, *Hoppe-Seyler's Z. Physiol. Chem.*, **353**, 88-94 (1972).
6. P. F. CURRAN and S. G. SCHULTZ, *Transport across Membranes: General Principles*, in: Handbook of Physiology (C. F. Code, ed.). Section 6, Alimentary Canal. Vol. III. Intestinal Absorption, p. 1217. Washington, D.C., Am. Physiol. Soc. (1968).
7 J. E. G. BARNETT, A. RALPH and K. A. MUNDAY, *Biochem. J.*, **118**, 843 (1970).
8. J. E. G. BARNETT and K. A. MUNDAY, *Structural requirements for active intestinal transport in the hamster*, in: Transport across the intestine. A Glaxo Symposium (W. L. Burland and P. D. Samuel, eds.), p. 111. Churchill Livingstone, Edinburgh and London (1972).
9. N. SWAMINATHAN and A. EICHHOLZ, *Biochim, Biophys. Acta*, **298**, 724 (1973).
10. W. F. CASPARY, N. R. STEVENSON and R. K. CRANE, *Biochim. Biophys. Acta*, **193**, 168 (1969).
11. R. K. CRANE, *Fed. Proc.*, **21**, 891 (1962).
12. D. R. CRITCHLEY, A. EICHHOLZ and R. K. CRANE, *Biochim. Biophys. Acta*, **211**, 224 (1970).
13. B. R. LANDAU, L. BERNSTEIN and T. H. WILSON, *Am. J. Physiol.*, **203**, 237 (1962).
14. F. ALVARADO and R. K. CRANE, *Biochim. Biophys. Acta*, **93**, 116 (1964).
15. T. H. WILSON and B. R. LANDAU, *Am. J. Physiol.*, **198**, 99 (1960).

16. R. K. CRANE, Functional Organization contributing to Carbohydrate Economy. Comprehensive Biochemistry. Vol. 17, Carbohydrate Metabolism (M. Florkin and E. Stotz, eds.), p. 1. Elsevier, Amsterdam (1969).
17. K. RAMASWAMY, B. R. BHATTACHARYYA and R. K. CRANE, *Gastroenterology*, **62**, 858 (1972).
18. I. BIHLER, K. A. HAWKINS and R. K. CRANE, *Biochim. Biophys. Acta*, **59**, 94 (1962).
19. U. HOPFER, K. NELSON, J. L. PERROTTO and K. J. ISSELBACHER, *J. Biol. Chem.*, **248**, 25 (1973).
20. F. PONZ and M. LLUCH, *Rev. Espan. Fisiol.*, **20**, 179 (1964).
21. W. A. OLSEN, and F. J. INGELFINGER, *J. Clin. Invest.*, **47**, 1133 (1968).
22. D. A. SALTZMAN, F. C. RECTOR, Jr. and J. S. FORDTRAN, *J. Clin. Invest.*, **51**, 876 (1972).
23. L. J. ELSAS, R. E. HILLMAN, J. H. PATTERSON and L. E. ROSENBERG, *J. Clin. Invest.*, **49**, 576 (1970).
24. T. R. HENDRIX and T. M. BAYLESS, *Ann. Rev. Physiol.*, **32**, 139 (1970).
25. S. G. SCHULTZ and R. A. FRIZZELL, *Gastroenterology*, **63**, 161 (1972).
26. B. FLESHLER and R. A. NELSON, *Gut*, **11**, 240 (1970).
27. R. K. CRANE, *Fed. Proc.*, **24**, 1000 (1965).
28. G. SEMENZA, *Some aspects of intestinal sugar transport*, in: Transport across the intestine. A Glaxo Symposium (W. L. Burland and P. D. Samuel, eds.), p. 78. Churchill Livingstone, Edinburgh and London (1972).
29. J. DIETSCHY, V. L. SALLEE and F. A. WILSON, *Gastroenterology*, **61**, 932 (1971).
30. M. DUGAS, Ph.D. Thesis, Rutgers University (1973).
31. D. S. PARSONS and J. S. Prichard, *Biochim. Biophys. Acta*, **126**, 471 (1966).
32. F. A. BIEBERDORF, personal communication (1972).
33. R. K. CRANE, *Biochem. Biophys. Res. Comm.*, **17**, 481 (1964).
34. S. G. SCHULTZ and R. ZALUSKY, *J. Gen. Physiol.*, **47**, 1043 (1964).
35. A. M. GOLDNER, S. G. SCHULTZ and P. F. CURRAN, *J. Gen. Physiol.*, **53**, 362 (1969).
36. R. K. CRANE, G. FORSTNER and A. EICHHOLZ, *Biochim. Biophys. Acta*, **109**, 467 (1965).
37. P. F. CURRAN, S. G. SCHULTZ, R. A. CHEZ and R. E. FUISZ, *J. Gen. Physiol.*, **50**, 1261 (1967).
38. H. MAGALHAES, in: The Golden Hamster (R. A. Hoffman, P. F. Robinson and H. Magalhaes, eds.), p. 91. Iowa State Univ. Press, Ames (1968).
39. I. LYON and R. K. CRANE, *Biochim. Biophys. Acta*, **112**, 278 (1966).
40. D. B. SACHAR, J. O. TAYLOR, J. R. SAHA and R. A. PHILLIPS, *Gastroenterology*, **56**, 512 (1969).

41. N. HIRSCHORN, J. L. KINZIE, D. B. SASHAR, R. S. NORTHDROP, J. O. TAYLOR, S. Z. AHMAD and R. A. PHILLIPS, *New Eng. J. Med.*, **279**, 176 (1968).
42. D. R. NALIN, R. A. CASH, M. RAHMAN and M. D. YUMUS, *Gut*, **11**, 768 (1970).
43. R. K. CRANE, *Gradient Coupling and the Membrane Transport of Water Soluble Compounds: A general mechanism?* in: Vol. XV, Colloquia on the Protides of the biological fluids (H. Peeters, ed.), p. 227. Elsevier, Amsterdam (1967).
44. G. A. VIDAVER, *Biochemistry*, **3**, 795 (1964).
45. A. A. EDDY, *Biochem. J.*, **108**, 489 (1968).
46. J. A. SCHAFER and E. HEINZ, *Biochim. Biophys. Acta*, **249**, 15 (1971).
47. P. M. TERRY and G. A. VIDAVER, *Biochem. Biophys. Res. Comm.*, **47**, 539 (1972).
48. C. PIETRZYK and E. HEINZ, *Some observations on the nonhomogenous distribution inside the Ehrlich cell* in: Na^+-linked transport of organic solutes (E. Heinz, ed.), p. 84. Springer-Verlag, Berlin-Heidelberg-New York (1972).
49. V. G. ALLFREY, R. MEUDT, J. W. HOPKINS and A. E. MIRSKY, *Proc. Natl. Acad. Sci., U.S.*, **47**, 907 (1961).
50. C. E. STIRLING, *J. Cell. Biol.*, **53**, 704 (1972).
51. D. H. SMYTH (1971), *Phil. Trans. Roy. Soc. Lond. B*, **262**, 121 (1971).
52. J. T. TOMASINI and W. O. DOBBINS, *Am. J. Digest. Diseases*, **15**, 226 (1970).
53. G. A. KIMMICH, *Biochemistry*, **9**, 3669 (1970).

Subject Index

A

Absorption
  accelerators, 515, 517, 519, 521, 522
  coefficient, 255
  in primates, rabbit and guinea pig, 489
  in rat and mouse, 492
  meanings, 241
  of bacterio-agglutinins, 497
  of water, 324
  polarity of, 242
  stages in, 240
Absorptive efficiency, 519
Acclimatization temperature, 418
Acetate, 433
Acetyl-β-D-glucosaminidase, 66, 79, 82
Acid phosphatase, 79, 82, 129, 505, 506
ACTH, 509
Activation energy, 317
Active ion transport, 183
Active transport, 149
Activity, 161
Adaptation of enzyme activity to diet, 92
Adenosine triphosphatase, 21
Adrenal, 457
Adrenalectomy, 496
Adrenocortical hormones, 457, 496
Adsorbed enzymes, 289, 309
Agammaglobulinaemia, changes in, 106
Agglutinins, 490
Agranular reticulum, 129
Alanine, 364, 365, 371, 372, 381, 382, 383, 389, 409, 411, 413, 417, 418, 424, 425, 432, 438, 449, 450, 451, 461
β-alanine, 369, 408, 421, 439, 462, 440
D-alanine, 389, 390, 391, 398, 403, 418, 419, 420, 437
L-alanine, 381, 384, 393, 394, 395, 396, 397, 398, 399, 402, 411, 412, 414, 415, 417, 421, 435, 437, 456, 522, 545
β-alanyl-DL-phenylalanine, 366
DL-alanyl-DL-alanine, 366
DL-alanyl-DL-phenylalanine, 366
L-alanyl-L-phenylalanine, 366
Alanylglycine, 375, 376
Alcian blue, 21, 51
Alcohol dehydrogenase, 72
Alkaline phosphatase, 60, 61, 62, 87, 294, 301, 307, 504
  during development, 88
  effect of actinomycin, 88
  effect of cortisone, 88
  effect of cyclohexamide, 88
  effect of puromycin, 88
  effect of weaning, 88
  function, 64
  localization, 88
Alkaline phosphate in celiac sprue, 98
Allosteric effect, 408, 435, 436
Alloxan-diabetic rats, 456
Allyl-glycine, 408
Alteration of villous structures, 320
Amino acid active transport, 392
Amino acid composition of membrane proteins, 143
Amino acid enantiomorphs, rates of absorption, 379
Amino acid pathways, 407

SUBJECT INDEX

Amino acid sequences, 487
Amino acid-sugar inter-relationships, 432
D-amino acids, 439
L-amino acids, 382
Amino acids,
  active absorption, 378
  concentration in intestinal lumen, 364
  rate of release in digestion, 365
ϵ-aminocaproic acid, 461
L-aminocyclopentanecarboxylic acid, 386
α-aminoisobutyric acid, 381, 386, 396, 398, 408, 411, 412, 415, 445
Aminopeptidase, 64, 305, 307
Amylase, 290, 291, 295, 305
Anomalous osmosis, 210
Anomalous transport, 227
Antibiotics and membrane conductance, 153
Antibody absorption, 493
Antitoxins, 521
Apical region of cell, 23
Appearance in blood stream, 253
Appearance in urine, 254
APUD-cells, 80
Aqueous pores in membranes, 148
Arachidonic acid, 141
Argentaffin cells, 11, 35, 36, 80, 81
Arginase, 449
Arginine, 364, 365, 373, 403, 407, 409, 435, 442, 449, 450, 460
L-arginine, 420, 421, 447
Argyrophil cells, 80, 83
Artificial membranes, permeability of non-carrier substances, 153
Aspartic acid, 364, 365, 373, 389, 403, 405, 409, 424, 438, 442, 449, 450
D-aspartic acids, 390
L-aspartic acid, 386, 421, 426, 437
ATP-ase, 60, 61, 62, 424
Autoradiography, 6, 21, 29, 386

B

Bacterial flora, 321
Balance experiments, 245, 246

Basement membrane, 11, 35, 128
  permeability, 35
Basic amino acids, 405, 407
Basigranular cells, 80
Betaine, 384, 407, 454
Bilateral adrenalectomy, 457
Bilayer properties, 150, 152
Bilayer structure, 147
Bilayer system, 153
Bilayers and antibiotics, 154
Bi-ionic diffusion potentials, 184
Bi-ionic potentials, 180, 182, 189
Biopsy, 268
Bound water, 145, 146, 147
Boundary diffusion potential, 168
Breathalyser, 254
Brunner's glands, 13
Brush border, 14, 15, 17, 48, 285, 292
  binding properties, 22
  enzymes, 54
  pores, 325
  structure, 11
Butyrate, 433

C

Capillaries, 35
Capillary fenestration, 36, 37
Carboxypeptidase, 305
Carnosine, 369, 370
Carrier complex, 398
Carrier hypothesis, 148
Casein hydrolysate, 409
Caseum permeability, 185
Cation gradients, 548
Cational selectivity, 184, 186
Cations permeation, 189
Caveoli, 524
Cavital digestion, 284
Cavital hydrolysis, 326
Celiac sprue, 98, 99, 105
Celiac sprue, effect on biopsies, 105
L-cells, 83
Cell changes,
  adrenal cortex, 94
  antibiotics, 96
  celiac sprue, 96

SUBJECT INDEX

colchicine, 96
  effect of adrenalectomy, 94
  effect of alloxan, 94
  effect of chiquoine, 94
  effect of cortisone, 94
  effect of diet, 92
  effect of glucagon, 94
  effect of hormones, 93
  effect of insulin, 94
  effects of protein malnutrition, 93
  effect of thyroxine, 93, 94
  hypophysectomy, 94
  in fat absorption, 92
  in protein absorption, 92
  kwashiorkor, 93
  malabsorption, 96
  neomycin, 96
  penicillin, 96
  terramycin, 96
  tetracycline, 96
  triparanol, 96
  vitamin D effect, 93
  X rays, 94, 95
E-cell distribution, 84
Cell fractionation, 268
Cell turnover, 6, 8, 9
  effect of antimetabolic drugs, 8
  effect of flora, 8
  effect of folic acid, 8
  effect of irradiation, 8
  effect of lactation, 8
  effect of nutrition, 8
  effect of pregnancy, 8
  effect of protein malnutrition, 8
  effect of vitamin B12, 8
  infestation, 9
  in germ free animal, 8
  villous shape, 9
E-cells, 79, 84
EC cells, 81, 82, 83
L cells, cytochemistry, 83
L-cells, function, 83
Cetrimide, 314
Chain mobility, 144
Change in antibodies, 500
Chemosorption, 312
Chicken, 454
Chloride unidirectional flux, 184

N-chloroacetyltryptophan, 441
Chloroplast membranes, 131
Cholecystochinin, 84
Cholesterine-esterhydrolase, 302
Cholesterol, 53
Cholesterol and permeability of membranes, 153
Chylomicrons, 25, 39
Closure, 496, 524, 527
Colchicine on mitosis, 6
Collection, 259
Colostral accelerators, 516
Colostral antibodies, 519
Colostral levels of IgG and IgM, 490
Colostrum, 483, 490, 507, 508, 510, 512, 515, 518, 520
  serum concentration, 488
Common carrier, 435
Compensatory changes after resections, 9
Competition for absorption by L-amino acids, 403
Competitive absorption, 500, 501
Competitive gut IgG-transfer, 501
Competitive inhibition, 405, 415, 402, 436
Composite membranes, 212
Composition of membranes, 138
Concentration gradients, 270
Concentration in gut wall, 275
Concentrations of immunoglobins, 493
Conductance measurements, 190
Conjugate driving force, 203
Conjugate properties, 201, 202
Constant field equation, 178, 184, 185
Convective flow, 223, 323
Cori technique, 254
Cortisone, 457, 507
Cortisone-induced premature closure, 496
Cotransport, 546
Counter-transport, 435, 543
Coupling of metabolism and transport, 234
Covalent bonds, 174
Cross coefficients, 204
Crypt cells, 11

Crypt cells of Liebekuhn, 29
Crypt villous function, 6
Crypts, 6, 9
    number and arrangement, 9
Current voltage relation, 192
Cyanide, 392, 399
Cycle of cell turnover, 6
Cyclic imino acids, 441
Cycloleucine, 388, 435, 436, 441
Cyclopentene, 139
Cysteine, 364, 373, 386
D-cysteine, 389
L-cysteine, 445
Cystine, 373, 386, 407, 409
D-cystine, 389
L-cystine, 445
Cytochemistry, 45
Cytoplasmic membranes, 125

D

Deamination, 370
Degree of coupling, 231, 232
Degree of hydration of membrane, 189
Deoxy-pyridoxine, 459, 460
2-deoxyglucose, 433, 541, 543
6-deoxyglucose, 543
Na^+-dependence, 543, 540
Desmosomes, 15, 16, 17, 23, 125, 128, 504
Desorption, 310, 314, 315
Desorption kinetics, 309
Development of amino acid active transport, 454
Development of enterocytes, 86
Dextran, 514, 515
Dietary dependent, 513
Dietary restriction, 427
Differential scanning calorimetry, 146
Differential ultracentrifugation, 45
Diffusion, 160, 177
    coefficient, 161, 162, 164
Diffusional water permeability, 166
Digestion, effect of turbulence, 322
Digestive conveyor, 290
Digestive surface, 245

Digestive-transport conveyor, 329, 332, 334, 335
Digestive-transport ensembles, 335
Diglycine, 368
Dilution potentials, 180, 182, 183
Dimethionine, 369, 377
N-dimethyl-glycine, 407, 441, 454
2, 4-dinitrophenol, 392, 399
Dipeptidases, 22, 296, 305
Dipeptides, 367, 377
Diphenylalanine, 369, 370
Dipole interactions, 174
Disaccharidases, 22, 54, 294, 296
Disaccharide malabsorption, 107
Discrepancy in osmotic and hydrostatic measurements, 168
Discrimination between IgG immunoglobulins, 498
Dispersion forces, 145
Dissipation function, 201, 203
Distribution in colostrum and milk, 488
DNA, 6
Dogfish intestine, 436
Dose-response relationships for feeding immunoglobulins, 501
Double membrane hypothesis, 215
Double membrane model, 217, 218, 219
Driving forces, 202
Duration of migration of cells up villi, 9
During transmission, 500

E

Effect of alcohol, 463
Effect of concentration on amino acid absorption, 409
Effect of pancreatic secretion and bile, 462
Effect of pH, 461
Effect of potassium, 402
Effect of sodium, 390
Effect of vitamin B_6, 458
Effect of vitamins C, D and E, 461
Effect of X-irradiation and intestinal motility, 464

SUBJECT INDEX

Effective pore radius, 296
Efficiency of transport, 230
Eisenman's sequences, 186
Electrical changes, 268
Electrochemical potential, 161, 202
  gradients, 178
Electrokinetic streaming potentials, 168
Electrolyte permeation, 177
Electro-osmosis, 177, 224, 181
Endocrine cells, 79
Endocytosis, 528, 530
Endohydrolases, 287
Endoplasmic reticulum, 23, 25, 27, 29, 73, 74, 75, 83, 89, 125, 126, 129, 131, 505
  enzymes, 73, 75
    alkaline phosphatase, 74
    cholinesterase, 74
    functions, 75
    glucose-6-phosphatase, 73
    lipase, 75
    non-specific estarases, 75
    nucleoside diphosphatase, 74
    non-specific esterase, 74
    proteosynthesis, 75
  glucose-6-phosphatase, 89
  glyceride biosynthesis, 75
  in development, 89
  non-specific esterase, 89
Endothermic transition temperature, 144, 145
Energetics of transport, 229, 234
Energy conversion in transport, 229
Enteric enzymes, 289
Enterochromafin cells, 11, 80
Enterocyte changes
  in absorption of nutrients, 91
  in absorption of water, 91
  in lipid absorption, 91
  in various conditions, 90
Enterocyte, polarity, 242
Enterocytes, cytochemistry, 86
Enteroglucogen, 83
Enterosorption, 243
Entropy, 203
Enzyme localization, 297
Enzyme-membrane complexes, 317
Enzymes, effect of adsorption, 318

Eosinophils, 84
Equilibrium affinities, 189
Equivalent pore radius, 176
Ether-water partition coefficient and reflexion coefficient, 173
N-ethyl-glycine, 441
Everted sac,
  fluid transfer, 266
  metabolic experiments, 266
  solute transfer, 266
  technique, 262 et seq.
Exchange diffusion, 180, 184, 288
Exohydrolases, 287, 290
Exsorption, 243
Extensive properties, 201, 203
Extra situm analysis, 47
Extracellular digestion, 284
Extrusion zones, 11, 482

F

Facilitated diffusion, 180
Factors in whey, 494
F bodies, 27
Ferritin, 27, 492, 506
Fibroblasts, 11, 84
Fibrocrypts, 11
Fick equation, 161, 162
Field strength of membrane sites, 186
Final concentration
  difference, 274
  gradient, 274
  in erythrocyte, 275
Final loss, 494
Fistula, lymph, 257
Fistulas, 255
  Thiry Vella, 255
  use of, 255
  validity of, 256
  water tight junction, 256
Five routes of absorption, 244
Fixed neutral sites, 191
Fluid circuit theory, 220
Fluorescence antibody technique, 54
Fluorescein-labelled, 509

SUBJECT INDEX

Fluoride, 392
Flux equation, 162
Flux ratio equation, 179, 180, 183, 228
Foetal guinea-pig, 454
Foetal gut, 454, 482
Foetal ilium, 418
Foetal rabbit, 454
Foetus, 482
Folate conjugase, 69
Freeze etching, 135, 136, 137
Frictional coefficients, 209, 210
Frictional forces, 208
Frictional interaction, 167, 180, 208
Fructose, 433, 437
Fructose diphosphate, 433
D-fucose, 437
L-fucose, 437, 541
Functional groups, 540
Fuzz, 127
Fuzzy layer, *see* glycocalyx

G

Galactose, 433, 434, 435, 540, 541
D-galactose, 437, 543
β-D-galactosidase, 58, 68
β-galactosidases, 58
Gastrectomy, changes in, 106
Gastric peptic activity, 494
General flux equation, 161, 177
Germ-free animals, 85, 320, 366
Ghosts, 137
Giant supranuclear vesicle, 528
Giant vesicle, 504, 505, 506
Giardiasis, changes in, 106
Glucose, 435, 540
D-glucose, 390, 543
D-glucose binding, 22
L-glucose, 543
Glucosidases, 305
β-D-glucuronidase, 66
Glutamic acid, 364, 365, 372, 373, 388, 389, 403, 404, 405, 409, 424, 438, 442, 450
D-glutamic acid, 398, 437

L-glutamic acid, 398, 403, 417, 421, 425, 426, 437
Glutamine, 373, 449
L-glutamine, 398
Glutamyl transpeptidase, 64
Glutaraldehyde fixation, 20
Gluten free diet, 104
a-glycerophosphate dehydrogenase, 823
Glycine, 364, 365, 368, 372, 373, 382, 384, 388, 391, 398, 402, 403, 404, 407, 408, 409, 411, 412, 413, 414, 415, 417, 431, 432, 450, 451, 454, 455, 457, 460, 462
Glycocalyx, 17, 45, 48, 52, 55, 86, 245, 287, 311, 315, 323
  pores, 325
Glycolytic cycle enzymes, 72
Glycosidases, 87
Glycylalanine, 377
Glycylglycine, 366, 367, 375, 377, 431
Glycylleucine, 367, 377
Glycyl-L-leucine, 366
Glycyl-leucine-dipeptidase, 291
Glycyl-L-leucine-dipeptidase, 290, 304
Glycylleucyltyrosine, 376
Glycyltryptophan, 367
Glycyltyrosine, 367
Glycyl-L-tyrosine, 366
Goblet cells, 3, 6, 11, 14, 21, 29, 51, 76, 77, 129
  cytochemistry, 76
  electron microscopy, 76
  enzymes, 77
  functions, 78
  mucous, 78
  secretion control, 77
Goldfish, 424
Golgi afferents, 72, 73
Golgi apparatus, 23, 25, 73
  acid phosphatase, 73
  alkaline phosphatase, 73
  nucleoside diphosphatase, 73
  participation in fat absorption, 73
  synthesis of glycoproteins, 73

thiamine pyrophosphatase, 73
Golgi complex, 504, 506, 523, 524, 526, 527
Gradient-coupled, 548
Na$^+$-gradient hypothesis, 540
SH-groups, 51, 53
SS-groups, 51

Hydroxyproline, 368, 382, 383, 407, 408, 413, 460
17-β-hydroxysteroid dehydrogenase, 72
5-hydroxytryptamine, 35
Hypophysectomy, 457
Hypothyroidism, 456
Hypovolaemic shock, effect on gut, 23

H

Haemachromatosis, 27
Hale's reaction, 51
Hartnup disease, 370
Hedgehog, 506
Heterogeneous catalysis, 285, 321
Hexaglycine, 376
Histidine, 364, 365, 371, 381, 382, 383, 398, 403, 404, 409, 413, 417, 418, 424, 432, 436, 438, 439, 450, 456
D-histidine, 389, 390, 420, 427, 428, 437
L-histidine, 369, 370, 384, 390, 403, 415, 427, 428, 429, 437, 455, 456, 458, 459, 460, 461
Histology of mucous membrane, 11
Homogenous membranes, 212
Homologous IgG, 498
Homologous immunoglobulins, 492, 510, 517
Hormonally induced closure, 519
Horseradish peroxidase, 502, 507, 527
Human foetal jejunum and ileum, 454
Human immunoglobulins, 484
Human small intestine, 442
Hyaluronic acid, 85
Hyaluronidase, 46
Hydration energies, 186
Hydraulic conductivity, 165, 167, 207
Hydrocarbon chain, 134
Hydrocortisone, 509
Hydrogen bonds, 163, 173, 174
Hydrostatic pressure, 168, 215, 218
2-C-hydroxymethyl glucose, 541

I

'Ice' structure of water, 174
Immunocytes, 11, 85
Immunocytochemical localization, 56
Immunofluorescent techniques, 11
Immonoglobulin A (IgA), 11
IgA immunoglobulin, 85
IgG immunoglobulin, 85
IgM immunoglobulin, 85
Immunoglobulins, 484
Impermeability of the newborn gut, 490
Infant malnutrition, changes in, 106
Influx and efflux, 243
Inhibitory factors, 518
In situ localization, 54
Insorption, 243
Insulin, 456
Intensive properties, 201
Intercellular route, 244, 245
Intercellular seepage, 523
Intercellular spaces, 527
Interference, 500, 501
Intermediate junction, 128
Intermolecular forces, 173
Intestinal absorption of antibodies, 498
Intestinal fistulas, 255
Intestinal permeability to proteins, 494
Intestinal transmission of immunoglobulins, 490
Intracellular digestion, 285
Intracellular hydrolysis, 368
Intravenous tolerance, 254

Intrinsic factor and vitamin B_{12}, 22
Intubation methods, 250, 252
Invertase, 290, 291, 317
In vitro and *in vivo* methods, 241
In vitro,
    condition of cells, 271
    disadvantages of, 271
    methods of improving, 271
    preparations, 262
    preparations, in air, 262
    techniques, 261, 267
    techniques, historical, 261
    techniques, interpretation, 269
    techniques, route of absorption, 269
In vivo, methods, 250
Iodoacetate, 392, 399
Ion-exchange membranes, 189
Ionic conductances, 183
Ionic radius and selectivity, 186
Iron, 22
Iron absorption, 27
Irradiation, 308
Irreversible thermodynamics, 214
Isolated brush border, 303
Isolated fractions of gut wall, 267
Isoleucine, 364, 365, 373, 382, 383, 388, 389, 403, 408, 409, 413, 450
L-isoleucine, 384
Isotonic water transport, 170
Isotope interactions, 228

J

Junction potentials, 181, 182
Junctional complexes, 15, 17

K

Kedem-Katchalsky membrane, 206
Kerkring folds, 2, 48
Kinetic parameters, 334
Kinetic studies, 412
Krebs cycle enzymes, 72
Kultschitzki cells, 35, 80

L

Lactase, 58, 59, 87, 98, 299
Lactate, 433
Lactate dehydrogenase isoenzymes, 71
Lactation and pregnancy, 458
Lacteals, 11
    structure, 39
Lactose, 433
Lamellar structure, 149
Lamellar-hexagonal transition, 147
Lamina propria, 11, 84, 182
Langmuir adsorption isotherm, 502
Large nonpolar side chains, 442
Lateral cell membrane, 128
Lateral membrane, 219
Lateral plasma membrane, 65
Lecithin, 140, 144
Leiberkühn crypts, 13
Length of human intestine, 252
Leucine, 364, 365, 372, 373, 381, 382, 383, 388, 398, 403, 407, 408, 409, 411, 413, 417, 418, 432, 434, 438, 450, 457, 462
D-leucine, 390
L-leucine, 381, 384, 388, 393, 411, 412, 415, 416, 421, 435, 437, 447, 522
Leucine aminopeptidase, 21
Leucylglycine, 367, 376, 377
L-leucylglycine, 366
Leucylglycylglycine, 376
Leucylleucine, 376
L-leucyl--L-tyrosine, 366
Leucyl-β-naphthylamidase, 65
Leucyl-β-Leucyl-β-naphthalamidase in development, 88
Leucyl-β-naphthylamidase in sprue, 100
Level flow, 231, 234
Linoleic acid, 141
Lineolinic acid, 141
Lineweaver-Burk, 406, 415
    analysis, 414
    plots, 402
Lipid-water, 146
Liposomes, 175
Lithium permeability, 185

SUBJECT INDEX

Localization of enzymes, 21, 298
London cannula, 256
Loop, circulation through, 258
Luminal cell loss, 7
Luminal digestion of ingested immunoglobulins, 494
Luminal fluxes, 244
Luminal perfusion, 262
Luminal transfer, 244
Lymph fistula, 257
Lymphatics, structure, 39
Lymphocytes, 84
Lymphoma, changes in, 106
Lyotropic series, 186
Lysine, 364, 365, 373, 388, 404, 407, 409, 435, 442, 450, 461
L-lysine, 383, 384, 391, 392, 398, 403, 407, 417, 420, 421, 423, 435, 447, 454, 455, 458
Lysosomes, 23, 27, 28, 66, 68, 82, 83, 129, 504, 506
  acid deoxyribonuclease, 89
  arylsulphatase, 89
  acid-β-galactosidase, 89
  acid hydrolases, 69
  acid phosphatase, 88
  enzymes, 88
  folate conjugase, 69
  functions, 69
  β-glucuronidase, 89
  E600-resistance esterase, 88
  pteroyl polyglutamate hydrolases, 69
  relation to fat absorption, 69
  relation to infections of intestine, 69
  significance of, 69
  Whippie's disease, 69

M

Macrophages, 84, 85
  and iron metabolism, 85
  cytochemistry, 84
Macula adherens, 15, 125
Malabsorption, 107, 451, 452
  congenital α-β-lipoproteinemia, 108
  glucose-galactose, 107
  Whipple's disease, 108

Maltase, 21, 59, 87
Manganese effect on electron micrograph, 133
Mannitol, 433
Mannose, 541
D-mannose, 437
Marker substances, 247, 248, 249, 253
Massive resection, 10
Maturation of cells, 7
Maximum efficiency, 233
Membrane ATPase, 423
Membrane degree of hydration, 189
Membrane digestion, 284, 285
  evolutionary aspects, 340
Membrane isolation, 137
Membrane lipids, 139, 172
  specificity, 139
Membrane negative changes, 184
Membranes in series, 212
Membrane potentials, 178, 181
Membrane proteins, 142
Membrane resistance, 170
  changed by sucrose, 182
Membrane, saturated fatty acids, 139
Membrane sites, 186
Membrane, unsaturated fatty acids, 139
Membrane viscosity, 162
Mesenteric blood analysis, 259
Mesenteric blood samples, 254, 259
Metabolic inhibitors, 392
Metabolism of gut, 268
Metachromasia, 83
Methionine, 364, 365, 368, 369, 373, 382, 383, 388, 389, 402, 404, 408, 409, 411, 413, 417, 432, 436, 438, 450, 455, 456, 462, 463, 464
D-methionine, 381, 389, 390, 398, 420, 427, 428, 437, 522
^{3}H-methionine, 463
L-methionine, 381, 384, 391, 392, 398, 401, 403, 405, 416, 423, 431, 437, 444, 454, 458, 460, 463, 522
Methionylglycine, 368
Methionylmethionine, 431

Method of calculating turnover, 8
Methods, disappearance from lumen, 250
Methods in human, 250
Methods, loop of intestine, 257
Methods of absorption, 239
Methylene group and permeability, 174
3-O-methyl-D-glucose, 394, 434, 437
α-methylglucoside, 435
α-methyl-D-glucoside, 461
Michaelis-Menten type kinetics, 394, 502
Microfilaments, 50
Microtubules, 50
Microvilli, 48
  size, number, 17
  structure, 17
Microvillous membrane, 303, 329
Microvillous zone, 86
  enzymes, 87
Migration, 504
  of cells, 7
Mitochondria, 23, 25, 69, 83, 126
  cytochrome oxidase, 70
  distribution of dehydrogenases, 70
  enzyme content, 70
  glucose-6-phosphate dehydrogenase, 70
  glycerophosphate dehydrogenase, 70
  β-hydroxybutyrate dehydrogenase, 70
  in development, 90
  isocitrate dehydrogenase, 70
  lactate dehydrogenase, 70
  malate dehydrogenase, 70
  6-phosphogluconate dehydrogenase, 70
  succinate dehydrogenase, 70
  tetrazolium reductase, 70
Mitotic activity 6
Mitotic counts, 6
Mobile carriers, 336
Mobile negatively charged carriers, 190
Mobile neutral carriers, 190

Mobilities of ions, 189, 224
Model membranes, 148, 149, 150, 153
  and dinitrophenol, 154
  antibiotics, 154
  electrical properties, 150, 155
  uncoupling agents, 154
IgG molecule, 485
Molecular configuration, 438
Molecular filtration, 523
Molecular radii, 171
Molecular volume and permeability, 176
Monoamine oxidase, 72
Monoglyceridelipase, 290, 291, 302
Mono-iodo-tyrosine, 385, 402, 406, 415, 433, 440, 461
L-mono-iodo-tyrosine, 390, 392, 446
Monolayer properties, 147
Mosaic structure of membranes, 132
Mucigen, 29
Mucopolysaccharide, 17, 21, 127
Mucosal scrapings, 268
Mucosal transfer, 244, 273
  calculation, 273
Mucus, 3
Muscularis mucosae, 13
Mycopolysaccharide, 11

N

Naphthylamidase, 64
Negative reflexion coefficient, 173
Negative staining, 135
Neutral pores, 190
Non-conjugate driving forces, 204
Non-electrolyte permeation, 160
Non-electrolyte reflexion coefficient, 171
Non-electrolyte selectivity, 173
Non-ionic diffusion, 180

SUBJECT INDEX

Non-specific phosphomonoesterase, 100
D-norleucine, 390
D-norvaline, 389
Nuclear magnetic resonance spectra, 143
Nucleus, 75
    deoxyribonucleic acid, 75

O

Oleic acid, 141
Onsager reciprocity, 203
Oral tolerance, 254
Ornithine, 373, 389, 404, 407, 435, 438, 442, 449
DL-ornithine, 384
L-ornithine, 383, 403, 437, 447
Osmium tetroxide, 20
Osmium effect on electron micrograph, 134
Osmotic flow, 165
Osmotic measurements, 165
Osmotic pressure, 215, 218
Osmotic water permeability, 166
OTAN, 50, 51
Ouabain, 398, 399, 423
'Overall' reflection coefficients, 222
α-oxoglutarate, 427, 433

P

Pancreozymin, 35, 84
Pancreatic and intestinal tryptic activity, 494
Paneth cells, 11, 14, 25, 32, 48, 78, 79
    acid phosphatase, 79
    autoradiograph studies, 31
    cytochemistry, 78
    destruction, 31
    electron microscopy, 78
    enzymes, 79
    function, 79
    granules, 79
    secretion, 79
    size, 31
    structure, 31, 78
    zinc, 79
Paneth function, 79
Parameter of interference, 501
Parameter of selection, 501
Parameters involving amounts, 272 et seq.
Parameters involving concentration, 274
Parameters to express absorption, 271 et seq.
Parathyroid, 456
Partial ionic conductance, 179
Partition coefficient, 162, 172, 209
Partition coefficients and permeability, 175
PAS reaction, 11, 20, 21, 50, 51, 52, 65, 66, 72, 79
Passive transport, 160, 180
Pathway for permeation of cations, 189
PEG, 249, 253
Pentose cycle enzymes, 72
Peptidases, 294, 300, 376
Peptides, 368, 369, 373, 376, 402
    absorption of, 363
Permeability, 207, 491
    and lipid solubility, 170
    and molecular volume, 170
    coefficient, 164, 172, 177, 179, 180
    measurements, 163
    of capillaries to chylomicrons, 39
    of cations, 184
    to ions, 183
    to non-electrolytes, 170
    to water, 168
Permeation via pores, 176
Permeations of alkali metal cations, 185

Peroral enterobiopsy, 44, 47
Peroxidase and ferritin as probes, 38
Persorption, 482
pH, 399
Phagocytic vesicles, 503
Phagocytosis, 328
Phagolysosomes, 505
Phagosomes, 506
Phase polymorphism, 147
Phases of regeneration of cells, 6
Phenomenological coefficients, 204
Phenomenological equations, 203
N-phenyl-glycine, 408, 441
Phenylalanine, 364, 365, 369, 381, 382, 383, 389, 398, 409, 413, 417, 418, 424 435, 450, 451
D-phenylalanine, 463
L-phenylalanine, 384, 391, 392, 402, 406, 415, 416, 424, 447, 463
Phlorizin, 543
Phosphatidic acid, 140
Phosphatidyl (N-dimethyl) ethanolamine, 140
Phosphatidyl glycerol, 140
Phosphatidyl (N-methyl) ethanolamine, 140
Phosphatidylcholine, 140
Phosphatidylethanolamine, 134, 140
Phosphatidylserine, 140
Phosphatidylthreonine, 140
Phosphoglycerate, 433
Phospholipids, 53
Physical adsorption, 312
Physical bases of cation selectivity, 186
Physostigmine, 464
S-piece, 518
Pig, 517
Pinocytic vesicles, 63, 66
Pinocytosis, 91, 148, 328, 489, 526, 527
Pinocytotic, 508
Pinocytotic vesicles, 524
Pituitary, 457
Plasma amino acid concentration, 371
Plasma amino acid levels, 449

Plasma cells, 84
Plasma membranes, 15, 125
Plasmocytes, 85
Polarity of absorption, 242
Polarity of enterocyte, 242
Polyethyleneglycol, 443, 463
Polyfunctional carrier, 436
Polyhydroxy alcohols, 176
Polyribosomes, 83
Polysaccharide synthesis, 25
Pore radius, 176
Pore radius calculations, 176
Pore size, 168
Pores, 210
Portal blood, 373
Portal blood samples, 257
Potential difference, 398, 417
Preclosure, 494
Pre-incubation, 397
Premature closure, 496, 519
Proline, 365, 368, 373, 382, 383, 404, 407, 408, 409, 413, 432, 436, 438, 450, 462
DL-proline, 384
L-proline, 392, 403, 405, 412, 421
Prolinuria, 408
Prolylglycine, 375
Propionate, 433
Protein, form of absorption, 362
Pteroyl polyglutamate hydrolase, 69
PVP, 495, 514, 516, 517, 519
Pyknotic cells, 7
Pyridoxal, 459
Pyridoxine, 459
Pyruvate, 433

R

Radiation enteritis, changes in, 106
Radioactive colloidal gold, 502
Radioactivity of liver, 254
Radioactivity of thyroid, 254
Radium permeability, 185
Rate of cell movement up villous, 7
Rate of loss of shedding, 8
Rate of production of cells, 6
Rectification of permeation, 213

Redox-electron transport theory, 149
Reflection coefficient, 166, 168, 207, 210, 218, 224, 225, 226, 227
Reflexion coefficient, 171, 172
Regeneration phases, 6
Relative permeability coefficient, 185, 188
Relative rates of transmission of immunoglobulins, 499
Renin, 510
Resection of gut, 445
E 600 resistant esterase, 82
Reticulum cells, 84
Reverse pinocytosis, 130, 309, 489
Rhamnose, 433
Ribonucleic acid, 129
Ribosomes, 25, 73, 83
Routes of transfer, 245
Ruminants, 510
Ruthenium red, 52

S

Sarcosine, 384, 407, 408, 454, 462
Saturable kinetics, 411
Scalar quantities, 205
Scanning electron microscopy, 14
S cells, 84
Scillaren A, 423
Scilliroside, 423
Secondary malabsorption, 109
Secretory system for IgA, 489
Selection, 496, 498, 501, 520
Selective intestinal transmission of antibodies, 528
Selectivity, 494, 499, 523
Selectivity isotherms, 188
Selectivity magnitudes, 187
Selectivity sequences, 186
L-selenomethionine, 447
Serine, 364, 365, 382, 383, 398, 409, 413, 417, 418, 450
D-serine, 389
Serosal transfer, 244, 273
Serotin, 82

Serum amino acid nitrogen, 372
Sg cells, 84
Shedding of cells, 7
Short-circuit current, 420
Shunt, extra-cellular, 194
Sialic acid, 139
SIgA, 518
Single file diffusion, 180, 228
 model, 152
Site of absorption, 442
Size of permeability of capillaries, 38
Small protein accelerator, 516
Sodium flux active and passive components, 183
Sodium-potassium-activated ATPase, 399
Sodium pump, 396
Solubilization, 306, 307
 of enzymes, 306
Solute-lipid intermolecular forces, 174
Solute-membrane interactions, 175
Solute permeability coefficient, 167
Solute-solute interactions, 227, 228
Solvent drag, 180, 208, 210, 219, 220, 221, 222
Sources of energy, 270
SPC cells, 108
m.n.r. spectroscopy, 145
Sprue,
 ATPase in, 106
 E cells in, 102
 changes in, 1—2
 changes in enterocytes, 101
 changes in histology of gut, 100
 dehydrogenases in, 102
 differences between tropical and non-tropical, 106
 endoplasmic reticulum in, 101
 goblet cells in, 102
 Golgi apparatus in, 101
 gut enzymes in, 100
 lysosomes in, 100
 mitochondria in, 102
 Paneth cells in, 102
 peptidases in, 100
 plasma cells in, 103

serotonin in, 102
Stages in absorption, 243
Standard chemical potential, 161
Standing gradient theory, 243
Standing osmotic gradient, 219
Static head, 231, 233
Stavernman coefficient, 166
Stearic acid, 141
Sterility of membrane digestion, 319
Streaming potentials, 169, 170, 171, 177, 180, 184, 224, 225, 226
Sucrase 21, 59, 71
Suicide bags, 27
Supranuclear region, 25
  of cell, 26
Supranuclear vesicles, 502, 503
Surface antigens as markers, 138
Surface area,
  effect of microvilli, 3
  glycocalyx, 3
  jejunum and ileum, 3
  of mucosa, 2
Surgical thyroidectomy, 456
'Sweeping away' effect, 165, 168, 180
Synthesis of enzymes, 308

T

Terminal bar, 15
Terminal web, 23, 51, 66, 128, 129, 504, 524
Terminology of absorption, 240
Ternary complex, 392, 393
Tetanus antitoxins, 490
Tetraglycine, 368
Tetrapeptidases, 305
Tetrazonium reaction, 51
Thermal agitation, 285
Thermotropic mesomorphism, 142
Thiamine pyrophosphatase, 82, 506
Thiry Vella loop, 255, 381, 402, 545
Thorium dioxide, 52
Three-layered lipoprotein complex, 287

Threonine, 364, 365, 373, 382, 383, 398, 403, 409, 410, 411, 413, 417, 450, 460
L-threonine, 384, 407
Thyroid, 455
Thyroid hormone, 456
Thyroxine, 455
L-thyroxine, 456
Tight junctions, 15, 16, 17, 128, 175, 193, 212, 219, 242, 482
Tight junctions as high conductance pathway, 190
Tolerance curves, 254
  chylomicrons, 253
  galactose, 253
  methionine, 253
  water, 253
Transamination, 424
Transcellular route, 244, 245
Transfer, relation to amount of gut, 274
Transition temperatures, 144
Transmission of passive immunity, 482
Transmural potential, 546
  difference, 397, 416
Transport number effect, 181
Trehalase, 59, 65, 87, 98, 299
Tri, 305
Tributyrinase, 290, 291
Triglycine, 368, 431
Triiodothyronine, 456
Trilaminar plasma membrane, 15
Trileucine, 376
Trimethionine, 369
Triparanol, 106
Tripeptidases, 305
Tripeptide hydrolase, 65
Triple-layered membrane, 126, 131, 132, 133, 134
Tritiated thymidine, 6, 7
Tropical sprue,
  changes in, 106
  enzyme changes, 106
Trypan blue-protein, 491
Trypsin inhibitor, 518
Tryptophan, 369, 381, 407, 450, 451, 453
D-tryptophan, 390

DL-tryptophan, 384
L-tryptophan, 383, 391, 439, 447, 455
Turnover time, 7
Tyrosine, 364, 365, 386, 402, 409, 435, 436, 450, 451, 456
D-tyrosine, 390
L-tyrosine, 384, 391, 392, 446, 447

U

Ultrastructure of mucosa, 14
Undifferentiated cells, 76
    changes in, 76
    maturation of, 76
Unidirectional flux measurement, 169
Unidirectional fluxes, 179
Unit membrane, 132
Unit membrane concept, 130
Unit pattern of membranes, 131
Unstirred layers, 152, 163, 164, 166, 167, 168, 172, 175, 176, 180, 181, 225, 545

V

Vacuoles, 526
Valerate, 433
Valine, 364, 365, 371, 373, 381, 382, 383, 388, 389, 398, 408, 409, 411, 413, 417, 438, 450
D-valine, 390, 437
L-valine, 384, 391, 393, 402, 407, 412, 415, 430, 454, 455, 461
Van der Waals forces, 174
Van't Hoff equation, 207
Van't Hoff flow, 166
Vascular perfusion, 260, 261

Vascular supply of villi, 13, 14
Vectorial forces, 205
Vesicles, 543
Vibration technique, 46
Villous shape, 3, 4
    diseases, 4, 5
    finger, 3
    infections, 4
    leaf, 3
    race differences, 4
Vitamin B_{12}, 22
Vitamin E, 461
'Vividiffusion', 371

W

Washing out effects, 223, 226
Water filled pore, 152
Water flow
    asymmetry of, 169, 170
    pathways for, 170
Water permeability, 152, 169
    of ileum, 226
    of jejunum, 226
Weaning, changes associated with, 87
Weighing abdomen, 253

X

Xylose absorption, 4
D-xylose, 437
L-xylose, 437

Z

Zinc, 79
Zona adherens, 15
Zona occludens, 15